RESOURCE ECONOMICS FOR FORESTERS

RESOURCE ECONOMICS FOR FORESTERS

G. Robinson Gregory
University of Michigan, Ann Arbor

JOHN WILEY & SONS
New York Chichester Brisbane
Toronto Singapore

Library of Congress Cataloging-in-Publication Data

Gregory, G. Robinson, 1915–
Resource economics for foresters.

Bibliography: p. 457
1. Forests and forestry—Economic aspects.
2. Forest management. 3. Forest products industry.
I. Title.

SD393.G74 1987 333.75 87-16069
ISBN 0-471-08521-9

Printed in the United States of America

10 9 8 7 6 5 4 3 2 1

PREFACE

When I began writing this book my intention was to update the original (1972) edition. However, half way through Chapter 2 I decided to combine the review of economic principles with the discussion of wood-using industries, which led to reviewing structural changes taking place in wood-based industries; this resulted in an evaluation of the changes taking place in forestry, which caused (eventually) a complete reorganization and redirection of the text.

Working on natural resource-based projects for government agencies of developing nations had made me realize that management of both private and public forests is—or should be—guided by identical economic principles, and that similar analytic methods are applicable. Results are seldom the same, but the differences stem from the use of divergent value systems—private values and public values are not the same. The project approach, however, is applicable to both sectors and to virtually every investment, large or small.

It was these similarities and differences that led to the final organization, one that emphasizes the underlying similarity in economic methodology but stresses differences in values and the identification and measurement problems created by these values. For example, benefit/cost analysis is not introduced in the early chapters simply as an alternative way of examining cost and revenue relations, and at a somewhat later point is compared directly to an after tax, discounted cash flow analysis.

Numerous changes and additions have been made. Probably the most obvious is the set of questions and problems following each chapter. Other changes may be less obvious at first sight, but are probably more significant: the presentation of derived demand has been clarified and expanded; the chapter on interest rates now includes the needed formulas and a more thorough treatment of how and why interest rates vary; major changes induced by the Tax Reform Bill of 1986 are incorporated in the chapter on tax (although capital gains are still discussed); willingness to pay and the several consumer's surpluses are examined and explained; linear programming is introduced, with an example, as a guide to multiple-use management; methods of treating non-market determined values are presented and discussed; far more attention is paid to foreign trade, tariffs, quotas, and impacts of foreign exchange fluctuations. An introduction to development theory has been included, and a (simplified) plantation project worked through with shadow prices. To a considerable extent, the emphasis has shifted from timber production per se to one of establishing an economic basis for managing the complex of natural resources called a forest.

Like all texts, the one draws on the work of hundreds who have brought economics and forestry to their present stage of development. There are, however, individuals who have contributed directly to this particular work. Dr. Robin S. Gregory (Decision Resources, Eugene, Oregon) gave the manuscript its first thorough critical review; Alice H. Ulrich (U.S. Forest, Washington, D.C.) went far beyond the call of duty to provide the latest available data; Dr. Henry Webster (Chief, Forest Management, Michigan Department of Natural Resources) provided much help through frequent discussion, particularly with problems of multiple use and non-industrial private forests; Dr. Bruce Fox (Professor of Forest Management and Economics, Northern Arizona University, Flagstaff) reviewed the entire manuscript and offered many useful suggestions; Sam Radcliffe (George Banzhaff and Co., Milwaukee, Wisconsin) examined several of the illustrative problems and suggested major improvements; Bill Siegel (U.S. Forest Service, Southern Forest Experiment Station) was tremendously helpful with timely information on the 1986 Tax Reform; finally Rosemary Mills, whose skill as editor should be appreciated by all who use this book. My thanks also to the School of Natural Resources of the University of Michigan for supplying secretarial help, and especially to Ms. Carole Shadley, who cheerfully typed and retyped the numerous drafts I needed and somehow found time to work them into her busy schedule. Most of all I must thank my wife, Ann, who always (although not always patiently) tolerated my being late for dinner, backing out of golf games, missing all the good movies, and spending far too much time in my office.

G. Robinson Gregory
Ann Arbor, Michigan
April 1987

Contents

RESOURCE ECONOMICS FOR FORESTERS

chapter one
INTRODUCTION: FORESTRY IN AN ECONOMIC ENVIRONMENT

The Changing Scene

HUNT, FISH AND TRAP   A PICNIC EVERY DAY!

BE A FORESTER!

Simply fill out and mail the coupon below for information explaining how you can become a member of this fascinating profession.

Advertisements such as this one are now part of forestry's folklore along with Robin Hood, Paul Bunyan and the Blue Ox, and 50¢ stumpage. Still, if you are a normal forestry student, you'll probably admit that a love of the outdoors and a desire to spend at least some of your life "in the woods" played an important part in your deciding to become a forester. So how come you are expected—probably required—to take a course in forest economics? What about the semester or two you spent looking at demand and supply curves for sugar and butter and gasoline and at how consumer expenditure patterns are changed as incomes increase? You still remember—in general, anyway—what GNP means. But what does economics have to do with growing trees?

This book answers these questions and, we hope, will do much more. Biology and ecology provide the necessary basis for understanding how trees grow and for developing ways to improve tree growth and quality. But to understand why, if, when, and where forests should be grown or cut, or to comprehend, and thereby make better decisions about, how and why forest resources are allocated and used, you must be able to bring economics into forestry.

Economics is the social science that deals with allocating scarce resources among competing means to satisfy human wants. It is, essentially, a **way of thinking**. To promote precision when treating these problems, economists have given certain words and phrases special meanings—words like demand, supply, and equilibrium, and phrases such as marginal utility and opportunity cost. Learning to use these terms should be no more difficult than learning the specialized terminology of botany and zoology. There is one way, however, in which economics differs, at least in degree, from most natural sciences. Economics is a way of thinking about how human societies allocate scarce resources, and these societies are highly developed organizations of complex, and different, individuals. Hence economists, though treating real problems, often find it necessary to abstract from reality. When analyzing a situation, few economists pretend to consider *all* the variables that might be important: in most cases it is impossible to consider even all the economic variables that are involved. The objective, however, is to produce a map of the realistic situation, not a photograph. Economic theory may be criticized as being overly abstract, but the thoughtful student will recognize that it is often precisely this abstraction that makes economics such a powerful analytical tool.

Forestry has changed greatly over the past years and the rate of change is increasing. When forests were recognized as a natural use of any land not needed to produce food, foresters could remain aloof from the everyday world and concentrate on growing trees and protecting game animals. A scant century ago, when Paul Bunyan was striding through the Lake states, increasing incomes meant bigger and better wooden houses and the lumberman needed to worry only about where he could find enough trees to supply his hungry sawmill. Later, when larger incomes and changing technologies brought houses of steel, glass, and brick instead of wood, the lumberman began to give serious thought to markets. As long as multiple use meant permitting other uses of forest land whenever and wherever it didn't interfere with timber production, the forester could continue along the path travelled by his European counterpart of the 1800s. But when camping in forests became financially feasible to most of the nation's families, when demands for municipal water took precedence over wood production, and when the per-acre income from hunting showed promise of rivaling that from timber sales, multiple use called for more than lip service. Management plans for major public forests (and some private ones) now recognize water, recreation, wildlife, and other non-timber products as significant outputs. But recognition may not be enough. In some countries (New Zealand, along with several Australian states, for example) man-made forests are now producing timber so efficiently that harvesting timber for commercial wood products will no longer be allowed on public natural forests. All are to be managed for recreational, aesthetic, and scientific purposes. Could public forest management in the United States undergo a similar change?

Great pressure for change is coming from outside our profession, and among the most important sources are those of an economic nature. Foresters must become familiar with both economic and political factors if we are to contribute our share to a more abundant life for the world's increasing population. Forest managers must understand and evaluate the significance of economic factors so that they may anticipate the inevitable changes of the coming years. This alone would be a large order, but it's not enough. If foresters are to continue as managers of one-third of the world's land, or only of that third of the United States occupied by forests today, they must develop competence in applying the basic principles of economics to the production of *all* the goods and services forests can yield—competence at least equal to that possessed by other production managers.

"Of course," you say. "Everyone expects the big timber corporations like Georgia-Pacific and Weyerhaeuser to estimate the financial implications of everything they do on their forests. But what about the thousands of small forest owners? What about the National Forests, the BLM lands, and the State Forests? They surely don't measure everything in dollars. The 'bottom line' of forestry isn't necessarily found on profit and loss statements!"

True. But "financial" and "economic" do not carry the same meaning. Economics is a way of thinking, and economic thinking applies to both public and small private forestry as well as to industrial forestry. Resources, whether public or private, and whether held by a large corporation or a small farmer, should be allocated efficiently. And it might be worth observing that the National Forest Management Act of 1976 requires the Secretary of Agriculture, when developing plans according to the Act, to "identify lands within the management area which are not suited for timber production, considering physical, economic and other pertinent factors to the extent feasible, as determined by the Secretary." It also directs the Secretary to "formulate and implement, as soon as practicable, a process for estimating long-term costs and benefits to supply the program evaluation requirements of this Act."[1]

There is no special brand of economics one must learn as a forester. The same principles of supply and demand, of production and distribution, that apply to manufacturing automobiles, Coca-Cola, cornflakes, or Scotch whiskey also apply to recreation, water, forage, wildlife, sawlogs, pulpwood, or fuelwood. Sometimes the application calls for a little ingenuity because of the peculiarities of the forest production process, and sometimes direct application becomes downright difficult, but the basic economic principles remain the same. And the principles of production and distribution, of supply and demand, apply just as surely in noncapitalistic countries as in those where the profit motive supplies the driving force, although application of the results may be quite different.

This does not imply there is no need for *forest* economics or for *forest* economists. Instead, it indicates that forest economics is an applied field and that its major problems stem from application difficulties rather than from theoretical ones. There are, of course, some problems of theory, usually the kind one expects to encounter whenever a general theory is extended to a specific instance. But forestry does have certain features that make economic applications difficult, or at least decidedly different. Other industries may share one or more of these characteristics, but only forestry has them all. A quick look at these distinguishing features may be enlightening.

DISTINCTIVE ASPECTS OF FORESTRY

Immobility of Standing Timber

Like most other natural resources, an established forest is fixed in space: it is essentially immobile. From some production aspects this fixity of location differs little from that of more usual commodities—automobile factories and forests, for example, are both difficult to move! But, unlike automobiles, timber is sold as it stands in the forest—as "stumpage"—and stumpage, like land itself, is immobile. This makes pricing the final product of the forester's effort decidedly different from the pricing of sugar or butter—or autos—for stumpage price is highly dependent on its specific location.

Length of Production Period

Timber production requires far longer time periods than most production processes. Under the very best conditions plantations may be ready for pulpwood cutting after 10 years, and in some tropical areas fuelwood is being produced on shorter rotations. At present in the United States, 25 years between planting and pulpwood harvesting is about average; sawlog and veneer log production periods are longer.

All production requires some time, but the periods involved in timber production are so long that the analytical techniques and decision-making processes must take time itself as a variable. Unfortunately, time is often ignored in economics—particularly in introductory courses—and for good reason. One of the most famous of classical economists, Alfred Marshall, wrote of time as "the center of the chief difficulty of almost every economic problem" (Marshall, 1947).

---

[1]Section 6(k) of the National Forest Management Act of 1976.

Yet the treatment of many major problems in forestry demands some familiarity with capital theory, the branch of economics dealing directly with time, and this will have to be dealt with at an early stage. The length of production period also has a critical impact on other factors, including tax effects.

Standing Timber's Dual Nature and Production Flexibility

When a forester marks timber for harvesting he or she is not only deciding the time is right to cut, but also that it is time to destroy that specific part of the timber-producing factory. The standing tree is at once the final product and the factory that has produced that product. While it takes years to increase the growth rate of a stand, it can be reduced or (under clear-cutting) eliminated very quickly.

Fortunately there are some offsetting flexibilities in product and in time. Once standing timber has achieved sufficient size, it can be sold in a variety of product markets: pulpwood, chips, veneer bolts and sawlogs, poles, piling, and so on, and because of its dual nature a forest will continue to produce a salable product over many years—there is great flexibility in the *time of harvesting*.

The dual nature of timber means that most long-term analyses of forest uses involve questions of dynamics. Although for analytic convenience we often treat different time periods as if they were independent, forest production is almost always time interdependent, and the interrelations between production periods can be crucial. The fact that the amount of timber harvested from a specific forest this year will probably affect the quantities available for harvest next year is easy to explain: certainly anything cut this year can't be cut the next time around! It is much harder to see, however, that the quantity *planned* for harvest in, say, the 2010–2015 period will almost surely affect the quantity to be removed during the current five years.

Full treatment of this problem is beyond the scope of this text, yet these interdependencies should be kept in mind, especially when policy problems are being considered.

Derived Demand

Some important forest products are producer goods—goods usually requiring further manufacture before consumption. Like most natural resources, forests produce basic raw materials for industrial use. Although there is nothing unique in this fact, it means that factors influencing the consumption of these products act indirectly: the amount of timber sold from a forest in Oregon (and, even more, its price) is heavily influenced by mortgage interest rates being charged for housing loans in Florida, Kansas, or Chicago, and the effect is transmitted through markets for plywood and lumber. In economic terms, the demand for certain forest products is derived from the demand for other goods. Derived demand, however, like time, is given little attention in introductory courses in economics, and it does pose some problems. An understanding of derived demand for basic products such as timber requires an understanding of markets for commodities like lumber, plywood, paper, and fuelwood. When the field is widened to forestry, however, the list must include many things for which organized markets do not exist—aesthetics, wildlife to view and to hunt, solitude, and a variety of "outdoor landscapes" ranging from simple riverside parks to wilderness areas that stretch unbroken over hundreds of square miles.

Externalities

An externality is created when something produced by one economic unit is used by or imposed on another unit without the permission or payment of the recipient.

Smoke from factory chimneys has been the usual example, but automobile exhaust emissions, the noise of jet airplanes, pesticide residuals, or blaring radios in campgrounds also serve as illustrations. Simply because forests occupy so much space, forestry necessarily involves many positive and negative externalities. Timber harvesting may increase siltation of streams, slash burning increases smoke pollution, heavy logging trucks often damage public roads, and pulp and paper manufacture is a major source of water pollution: all are examples of negative externalities associated with forest-based industries.

But there are also substantial entries on the positive side. Timber growing carries much of the cost of forest recreation, of wildlife and watershed protection, and of fire prevention and suppression. Logging roads provide the only access to many forest recreation sites. These are *positive* externalities of timber production. In many ways forests can be regarded as public utilities, and there have been suggestions that, like public utilities, forests should be subject to public regulation as well as being permitted to benefit from public investment.

Economic problems are often created by regulations designed to control harmful externalities of timber production; for example, restrictions on aerial application of certain pesticides. Other economic concerns arise from efforts to recover the costs of providing particular forest products, such as wildlife and watershed control, through special taxes or subsidies. Externalities will be treated more fully in subsequent chapters.[2]

Ownership of Commercial Forest Land

So far as timber production and sale are concerned, this nation is a "mixed economy": part of our timber is produced and owned by private industry operating under market forces and part is publicly owned and managed. Of the 482.5 million acres of commercial forest land in this country in 1980, some 99.4 million acres were federally owned, with 88.7 million acres, or 23%, administered as National Forests. States owned an additional 23.4 million acres, and the total area under public management of one type or another was 135.7 million acres. Private owners, on the other hand, controlled 346.7 million acres, including the 69 million acres owned by forest industries (U. S. For. Serv., 1982). Most public forests must, by law, be managed to produce a wide range of products, many of which are not market priced. Industrial holdings, along with many of the larger and some of the small private ownerships, are owned primarily for financial gain. Holders of farm woodlots and other small forest areas usually cite income as only one of many objectives of ownership. Different value systems and different economic objectives are likely to govern the manager's decisions on forests of different ownership status, differences that become significant as specific problems are examined.

PURPOSE OF ECONOMIC ANALYSIS

Is there any single objective toward which we might say our economic analysis will be pointed? In a very general way the answer is "Yes," for we will be striving to determine the optimum allocation of scarce forest resources, and of all those other resources that must be used in the utilization of forests, among the competing means by which society's wants can be met.

But optimum from whose standpoint? Suppose an owner—any owner—wants to know the best harvest pattern for his or her particular forest. To deter-

[2]It should be recognized that wide variation exists among economists about what should be called an externality. Baumol and Oates, for example, devote the major part of the first four chapters in *The Theory of Environmental Policy* to defining and classifying externalities (Baumol and Oates, 1975).

mine this requires that we first know the owner's objective. Even after learning that the primary interest is timber production (and it could as well be deer, fuelwood, assured privacy, or any of a variety of products), it would be necessary to spell out the objective more carefully. If the owner wants to maximize cash income from timber sales over a three-year period, then a particular harvest schedule could be established; to maximize income over a 20-year period would probably call for a quite different cutting schedule, while maximization of the property's present net worth could require a still different program. Clearly, the objective of a private owner may be very different from that of a public agency such as the U.S. Forest Service or a State Conservation Department, and these again would differ from the goals of a large paper corporation or the National Forest Service of Liberia, Mexico, or India.

The purpose of economic analysis is interpreted here to systematically examine the various options involved and demonstrate to the owner or the professional manager—the decision maker—the possible economic results of following alternative courses of action. The manager is expected to make the decision, quite frequently as much, or even more, on noneconomic than on economic grounds. Still, although a decision may be based on political reasoning, the economist has the responsibility to point out the economic results of the selected alternative. A decision to impose water or air pollution controls, for example, is usually made through the political process. An analysis of the economic consequences that would follow the imposition of such controls, however, should be available to the decision makers, whether they happen to be legislators or the voting public.

One may argue that it is not the forest economist who should decide whether a particular area will be devoted to recreation, sawlog production, fuelwood, or used as a wilderness. And this is correct! The task of the forest economist is not to *make* the decision, but to identify and analyze the costs and returns that are expected to *follow* a decision. The results of the analysis should then be made available to the decision maker in an understandable form. This could be a simple series of financial benefit/cost ratios. More often, however, the results will show not only the dollar costs and returns to individuals and to whom they accrue, but also the costs and returns to individuals and to society that cannot be valued in present-day markets, including the opportunities for future action that may be lost if particular alternatives are selected. In short, an economic analysis should present all those facts of an economic nature that will aid in making an intelligent choice between alternative courses of action.[3]

ECONOMIC MODELS

An economic analyst seldom pretends to consider all the variables that are important to the problem under consideration. In fact, it is generally impossible to consider even all the *economic* variables involved. Economic analysis aims at producing a realistic view of the situation, for only by abstracting from reality can the interactions of the key variables be worked out. This is probably best seen in the process of model building, a device we will use frequently.

Like models in other disciplines, an economic model usually begins from

[3]Having made the analysis, an economist will often have a strong opinion about the decision that "should" be made. Moreover, while the economist may not be the one who makes the decision, he or she may work hard to influence it. This is one reason why economists are so often involved in debates about policy matters. But they do so as citizens, rather than as economists, and the vote of even the best informed analyst is worth no more or less than that of any other person.

some observed or imagined event. The economist asks, "Under what set of assumptions could this event occur?" Those assumptions are tested by reasoning "If that is so, then this will follow." And one should also ask, "Under what conditions might this *not* follow?" As a very simple example, imagine a curious individual watching people as they buy fresh vegetables in a farmer's market on a series of Saturday mornings. As the season progresses it becomes quite clear that people tend to buy more of a product if the price is low than if the price is high. From this the observer concludes that the relationship between the price of a good and the quantity of the good sold might be "modeled" by Figure 1.1—or by an equation such as $P = a - bQ$, where P is the price of the good, Q is the quantity sold at that price, a is the intercept on the vertical axis, and b is the slope of the line DD. Conceivably, if enough data are collected, the observer might even attempt to quantify this relationship.

But someone, ideally the observer, is sure to ask, "Aren't you making some other assumptions? You should add that you also assume nothing else affects this relationship—and you know this isn't so."

At this point the observer may decide to stop worrying about price-quantity relationships, or might ask, "What other things might affect this relationship?" If the latter course is followed, the end result might possibly be something very similar to the familiar model of economic demand that will be discussed in Chapter 3.

We will be doing quite a bit of this model building in our study of forest economics. Quite frequently we will begin with assumptions that seem only distantly related to the real world. As we proceed, we may discover what happens as these assumptions are relaxed—as we allow the model to more closely approximate the world we live in. And, as the use of models becomes familiar, at least two points should become evident: (1) that theories may be very simple even though the phenomena they model are not; and (2) that a model (or theory) can be very general without being sterile.

PLAN OF APPROACH

The text is divided into five parts:

1. Wood-based industries: market organization, production, consumption, and prices (Chapters 2 through 5).
2. Timber production: time, taxes, production, and prices (Chapters 6 through 10).

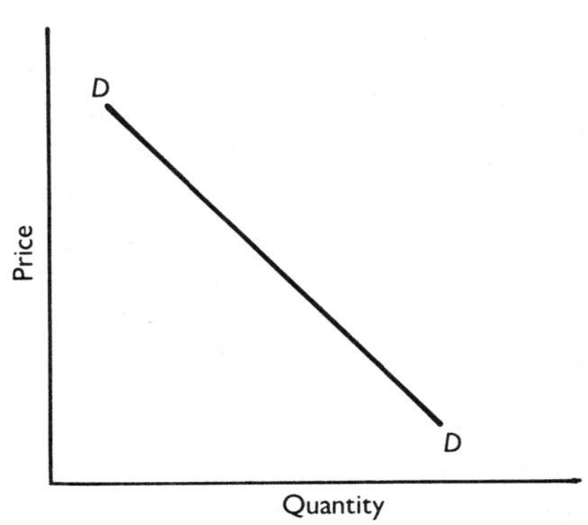

FIGURE 1.1. A market model.

3. Economic guides to management decisions (Chapters 11 through 14).
4. Nontimber products and multiple use (Chapters 15 and 16).
5. Foreign trade and economic development (Chapters 17 and 18).

As each section is covered, guides (*not* rules) for managerial decisions will be developed and explored through discussion intended to clarify the link between economic analysis of forestry problems and application of the analysis to achieve workable solutions. At times, opportunities for comments on forest policy may prove irresistible.

Each chapter will be followed by a set of review questions and problems. The problems are designed to require application of the materials and ideas that have been presented either in that chapter or in previous ones. By all means, WORK THE PROBLEMS. There is no better way of learning the principles and methods of economic analysis. Although all the problems can be worked satisfactorily by hand calculation, almost all are more quickly solved by computer. Special programs are becoming available for forestry problems, particularly those involving rotation determination, discounted cash-flow analysis, and benefit/cost analysis. These special programs can, of course, be used where appropriate, but all problems in this text can be satisfactorily handled by using one of the more usually available spread sheets, although memory capacity may cause difficulties if using a personal computer for problems involving long time periods.

Because graphs are a convenient and (usually) understandable shorthand for ideas involving functional relationships, they are extensively employed. A word of warning: **Don't try to memorize the graphs**. Learn the basic economics involved first, and the graphs will then seem easy. If you learn only the graphs, you will know only some simple geometry—and this is not intended as a geometry text.

Of the several reasons for beginning with timber products, the most important is that economic principles are thereby easier to illustrate. Dealing with nontimber products of the forest—particularly recreation, wildlife, water, and amenities—is much more difficult. To imply, for example, that demand curves for lumber, plywood, paper, fuelwood, campsites, and deer are equally simple to derive, or to understand, would be grossly incorrect, and quantitative illustrations for the last two are particularly difficult to find. Additionally, a knowledge of wood product markets is essential to an understanding of the role of forestry and the problems of forest management in most parts of the world. Because markets, at least in the usual sense, do not exist for many important nonwood forest products, the need for understanding how our market system functions and why it sometimes sends imperfect signals to producers is increased. Without this understanding one cannot hope to comprehend the difficulties inherent in the analysis of nonmarketed forest products.

The section on trade and economic development comes last because relations between different national economies can be appreciated much better if one understands how our own economy is organized and functions. International trade in forest products, especially that with Canada, has recently achieved substantial significance for forest operations in the United States.

The role of forests in economic development may seem of little importance to those expecting to confine their working careers to North America. But professional foresters, whether in public or private employment, are increasingly discovering that developments occurring half a world away can have an important impact on their own work. Forestry is no longer a European/American profession, in part because of the influential role forests can play in affecting the economic and social objectives of less developed nations.

AN OVERVIEW OF FOREST ECONOMICS

Distinctive aspects of forestry have been identified, economic models have been introduced, and the general plan of approach for this text has been presented. Now a quick look at forestry in this and other countries is needed, partly to gain a perspective, but more to know what the field of forest economics includes. Most readers already recognize that forests are important. But to what parts of forestry will economic analysis be applied? What kinds of problems will be explored?

Forest economics is the branch of forestry, and of economics, that deals with allocating scarce resources among competing means to satisfy human wants for forest products. We must therefore deal with two sets of resources: the products themselves and the resources used to produce those products. We'll start with the products.

Markets for Wood-Based Products

In the United States as in other developed countries, the major *marketed* forest product is wood of some type—the wood used for making a wide range of wood-based products. Because the organization of these markets has an important bearing on "who gets what for how much," quite a bit of time will be spent examining these markets—particularly markets for lumber, pulp and paper, and wood-based panels (plywood, flakeboard, particleboard, etc.). We'll learn more about the size of these industries, their competitive structure, how their products are priced, the outlook for their economic health, their stability, and their wood requirements. We'll also examine how technology is changing both wood and labor requirements and identify some basic changes that seem to be occurring in the nature of the forest industries. In the early 1950s, for example, the U.S. lumber industry provided one of the few good illustrations of a competitive market. Corporate mergers and increasingly large-scale production, however, have caused such substantial structural changes that it is questionable whether the industry can still be used to illustrate "pure competition." And to appraise the outlook for wood-based industries we will look not only at the industries per se, but also at the factors that affect consumer acceptance of their products.

Timber Production and Management

Production and management will require other concepts, for the time needed for timber production cannot be ignored. To take time directly into account calls for a good working knowledge of compound interest and discounting, as well as a basic understanding of the principles governing timber growth. Concepts such as internal rates of return and net present value are as essential as those of rotation length and allowable cut. Institutional factors, especially the various types of taxes, will be discussed, for these can play a decisive role in private timber management practices.

With an understanding of the markets for wood-based products and of timber production economics as background, the demand, supply, and pricing of standing timber can be explored. This will involve the direct application of derived demand and the concept of rent.

Economic Guides to Management Decisions

Because the effect of most forest management decisions becomes evident only after a considerable lapse of time, guides for evaluating the probable results of such decisions are extremely useful. Many of these are based on economic principles, and an understanding of why these economic guides are used and what they do and do not show is an important part of forest economics. As these guides are developed and illustrated, many of the differences between public and private forest management will be discussed. We will see that the basic analytic framework for management decisions is identical for both, although the results of the analysis may be very different. To understand the reasons for this seeming paradox necessitates a review of the objectives of public forest ownership and an introduction to "welfare economics," the branch of economics dealing with public production and with social values.

Nontimber Products

As already noted, wood is not the only product our forests provide. People want wilderness areas, national parks, campgrounds, hunting and fishing opportunities, wildlife preserves, nature trails, ski resorts, scenic highways, pure water supplies, and an almost interminable list of other goods and services that are provided by forests. Economic guides to both managerial and public policy decisions about how our forests might be handled for nontimber products and for multiple products will be developed, as will procedures for estimating the private and social costs and returns from management and policy decisions.

International Forestry

Finally, we will review the international forestry picture. This will include examining the extent and diversity of the world's forests, some of the effects of current management practice, and the economic opportunities open for forestry along with some of the problems facing anyone wishing to become directly involved in international work. Anyone choosing such a course quickly discovers he or she is dealing with a very large resource—one that covers one-third of the earth's land. Forestry is the principle land use of the world, and what happens on a third of the world's land surface must be of substantial significance to the way people everywhere live—and in some cases, *whether* people can continue to live!

In the "developed" part of the world the forested area is essentially stable, or even increasing slightly, but there are still questions regarding whether these areas will be able to continue supplying forest products in the quantities, and at the prices, that the industrialized nations apparently want. Can forests in the developing nations be used to satisfy these demands? Is it true that the tropical forests of the world are being cleared so rapidly that they really could be gone by the early to middle part of the twenty-first century? Some attention will be given to such questions, but the treatment of international forestry will concentrate on two issues: (1) where and how do imports and exports of forest products fit into the U.S. forestry picture and what are the prospects for increasing or decreasing our foreign trade in wood materials; and (2) when, and under what conditions, can forestry be used to promote economic development.

This material, however, is in the final chapters of the book, and there is much to cover first. We begin our study with wood-based products and questions of their demand, production, supply, price, and consumption.

PART 1

Price Theory and the Primary Wood-Based Industries

(Description, demand, production, supply, structure, and price)

The central problem of economics was defined in Chapter 1 as that of resource allocation — allocation of both natural resources and resources composed of produced goods and services. This first part deals directly with the concepts economics provides for studying these problems: the concepts of demand and supply, of production costs and revenues, and of price formation and profits. And it necessarily treats one of the major institutional innovations — the market system. The concepts are developed using primary wood-based products and their markets as examples.

Chapter 2 begins with a review of the market system and how (in theory) it functions. Then the major wood-based industries are described in some detail: how production has varied in recent years, the states in which the particular industry is important, its general development since 1950, its significance in terms of employment, and the extent to which foreign trade is involved.

Chapter 3 opens with a review of the assumptions underlying demand, and, after examining them, goes on to examine the demand concept. Using lumber as the example, factors believed important in affecting this demand are considered, along with some of the problems encountered in demand analysis.

Chapter 4 begins by presenting the fundamental production relationship, that between inputs and output. From this the various cost and revenue relations are developed, eventually leading to principles of profit maximization. Two variations of the general cost-revenue approach are presented — one making costs a function of inputs (net cash flow) and the second dealing with revenue/cost ratios that lead directly to benefit/cost analysis.

Chapter 5 goes directly into pricing wood-based products. The usual pricing models (pure competition, monopoly, oligopoly, etc.) are reviewed, then the major wood-based industries are examined in terms of their changing competitive structure. Price patterns of the principle products of these industries are presented in this latter context.

chapter two
WOOD-BASED INDUSTRIES AND MARKETS

Markets are the primary means through which our economy determines how much of each product to produce, how much of each resource should be used in its production, and who should get it. Any economic system must solve these three major economic problems. Like other western nations, the United States lets its market system play an important role in determining the results. In 1776, when Adam Smith explained how the "invisible hand" worked to bring about a remarkably fair and efficient result, England was a nation of farmers, small industrialists, and landowners. Things are no longer so simple: markets in industrialized nations are highly organized, with many steps involved in moving goods from producer to consumer. Governmental actions significantly influence production, both by setting the overall rules and by regulating many decisions made by producers. Moreover, the modern corporation has become far more powerful than Smith could have imagined, and can often influence or even control the quantity of a product that is offered for sale and thus influence the price consumers must pay.

THE MARKET SYSTEM

The U.S. market system has two major parts, or subsystems. One is for productive factors (labor, capital, and natural resources), the other for final products. The factor market signals what resources should be used and how, while the product market determines what to produce and who is to receive it. Both markets coordinate their activities by producing and responding to a set of signals—the market prices. It's worth reviewing this basic process before taking a closer look at how wood-based product markets are structured, how this structure is changing, and how these markets coordinate the needs and wants of producers and consumers. We'll also examine the major constraints that sometimes inhibit the market system from operating in a completely satisfactory manner.

Equilibrium Price

The success of both product and factor markets comes from the production of an **equilibrium price**—the price at which producers are willing and able to supply the market with precisely the quantity that buyers are willing and able to buy. This equilibrium price results from the interaction of many forces and many individuals. Each buyer is attempting to get the most satisfaction possible from his or her expenditure (the **demand side** of the market), while each seller is trying to do the same from his or her sale (the **supply side**)—and both are competing with other similar buyers and sellers. Their actions are summed up in the demand-supply diagram (Figure 2.1) familiar to almost everyone. The supply curve *(SS)* shows the varying quantities that sellers (producers) would offer at each indicated price, while the demand curve *(DD)* shows how the quantity purchased would change as the market price is altered. Significantly, both demand and supply curves show the quantities that would be offered or purchased as a *function of price*. The point of intersection of the two curves determines the **equilibrium price** (P') and **quantity** (Q'): no other quantity or price would simultaneously satisfy both buyers and sellers.

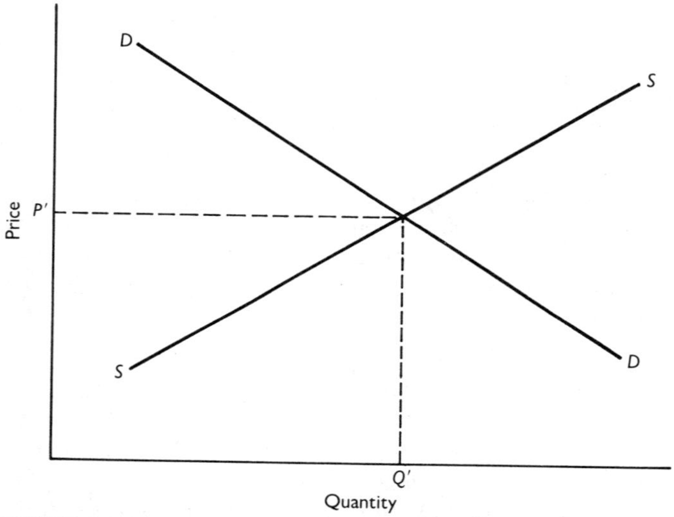

FIGURE 2.1. Market demand and supply curves and the determination of the equilibrium price and quantity.

The two market curves represent aggregates of individual demand and supply curves, and there is no reason to suppose that any particular seller or buyer will be perfectly satisfied with the equilibrium price. But by bringing together the offerings of the sellers and the wants and purchasing capabilities of the buyers, the market automatically produces this equilibrium price. Figure 2.2 depicts a mechanical analogy to the operation of this market system. When the market stock of some good begins to get low, price rises. The rise in price opens the valve controlling production which causes an increased flow of product into market stock. The same price rise also checks consumer buying, so the outflow from the pool is reduced. On the other hand, if stocks become high, prices fall, so production is checked and consumption encouraged.

Like the evening news, this type of market has some good and some bad features. First the good part: when prices are established as discussed above we can say

1. Consumers can purchase as much or as little as they want of any good or service offered *as long as they are willing and able to pay its cost of production.*

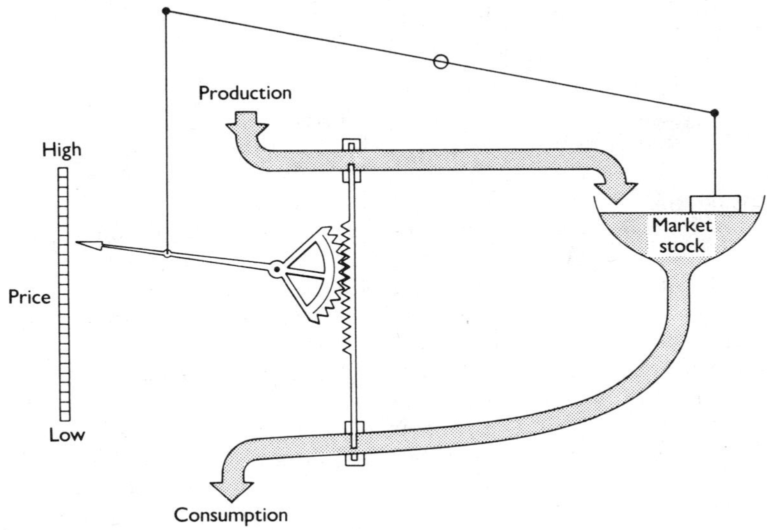

FIGURE 2.2. A mechanical analogy to the operation of a market system.

2. Producers will be happy to produce enough of their product to make the total quantity (the industry supply) just sufficient to assure the consumers of the above.
3. *Given the existing distribution of income,* there is no shift in the allocation of resources, or in the relative amounts of the goods being produced, that will permit any consumer or producer to be made better off without lowering the welfare of some other person.

Now the bad part. There are a few conditions that must be met if this happy outcome is to be assured. There must be pure competition in *all* production, complete knowledge on the part of consumers and producers, perfect mobility of all production factors, and acceptance of the given income distribution. Is this the type of market system we really have in the United States? No! Has such a market system ever existed, any place? Probably not. So why is it worth studying? Because by starting with this model, we can see how *departures* from the hypothetical system affect the results and, if these departures are undesirable, we may even design ways of correcting them. More important is the fact that once an understanding of the elements of this simplified system and how they interact is in hand, the "real-world" markets for forest products can be examined with a reasonable hope for understanding them, even though they are much more complex. Such an understanding requires knowledge of how the wood-based industries have developed and their significance to both forestry and our general economy. Of the many wood-based industries, our attention will be limited to what are actually broad aggregates—wood harvesting, lumber, wood-based panels, and pulp and paper. Changes in these major market areas are significant to forest managers everywhere. Data for relatively minor uses (i.e., cooperage, cross ties, fence posts, etc.) are available, but even substantial changes in these uses are not likely to significantly affect forest management decisions within the next decade or two.

We begin with the first step in the wood utilization process—the harvesting of the timber.

TIMBER PRODUCT MARKETS

Wood Harvesting (Logs)

This is the industry of Paul Bunyan and the rivermen of Maine and Michigan, the hard-working, hard-drinking lumberjacks of history. Today's timberman is (generally) a highly skilled worker, necessarily so because of the high degree of mechanization. Timber harvesting, however, is one of the more hazardous ways to make a living; accidents are frequent among those who work in the woods.

The Census of Manufactures calls this industry "logging camps and logging contractors," and includes all "establishments primarily engaged in cutting timber and producing rough, round, hewn, or riven primary forest or wood raw materials."[1] To avoid double counting (and also to simplify data collection), any woods operation carried out as part of a sawmill, pulpmill, or other manufacturing unit is counted as a part of that particular unit rather than as a wood-harvesting establishment. Most of the operations are small.

Employment

Of the 11,658 establishments (11,544 companies) so classified in 1982, only 657 had 20 or more employees. In the same year the industry had 80,900 employees (it peaked at 96,500 in 1980) with a payroll of just over $1.2 billion. Value added by manufacture totaled $2.5 billion, and shipments $8.3 billion.[2] Forty-two percent of the industry's 1982 employment was in Washington, Oregon, Georgia, and Alabama—a change from 1977, when Washington, Oregon, Georgia, and Maine had accounted for 48% of the industry's employment.

Probably because the industry has always had many small establishments, there is little solid information about its development. From the historical census data we know that the number of establishments has fluctuated over past years: from 12,800 in 1956 to 13,600 in 1963, up to 16,300 in 1967, down to 13,200 in 1972, and up again in 1977 to 15,500 then down to 11,658 in 1982. Employment, however, increased quite consistently from around 72,000 in the early 1960s to a peak of 96,500 in 1980, then dropped to 81,000 in 1982.

[1]The Census Bureau has developed the Standard Industrial Classification (SIC)—a four-digit system for classifying U.S. industries. Major Groups are identified by the first two SIC numbers. Forest products industries are in Major Industry Groups 24 (Lumber and Wood Products), 25 (Furniture and Fixtures), and 26 (Paper and Allied Products). Unless otherwise noted, the data and descriptive information of the forest industries in this chapter came from the 1982 Census of Manufactures (taken every five years), supplemented when necessary by the Annual Survey of Manufacturers.

[2]Value added by manufacture is obtained by subtracting all costs of materials, supplies, fuel, electric energy, sales, and miscellaneous receipts from the value of shipments (including resales) and other receipts. The result is adjusted for changes in inventory and work in process. Value added thus represents the amount available for wages, salaries, interest payments, profits (whether or not distributed), taxes, and depreciation and depletion charges. From an economic viewpoint, value added is important because it represents the contribution of a particular segment of manufacturing to GNP.

Foreign Trade

The wealth of U.S. forests and the fact that logs are expensive to ship long distances have made log importation by U.S. manufacturers relatively rare. Tropical hardwoods are an exception, but volumes have been dropping since the early 1950s and by 1980 were almost nonexistent (Figure 2.3). Much of the market for hardwood logs is apparently being filled by hardwood plywood (see page 26 and Figure 2.9). Softwood log imports are small, show no clear pattern, and since they have been less than 150 million board feet annually are not shown. Such imports are primarily cross-border transfers from Canada's British Columbia.

The log export picture is very different (Figures 2.3 and 2.4). Note that the volume scales of the two graphs are different: softwood log exports since 1960 have been over a *billion* board feet annually, while hardwood log exports are measured in *millions* of board feet. Virtually all softwood logs are of West Coast origin. Although Douglas fir has been the leading single species from a volume standpoint, substantial quantities of Port Orford cedar and western hemlock and firs are also shipped. In 1979, when log shipments reached a peak, over 80% went to Japan, with Canada taking another 10½%. In 1985 Japan took 49%, while Mainland China—a newcomer to the U.S. log export market—took 28%. Total volume of softwood log exports in 1979 was 3.77 billion board feet with a value of more than $1.6 billion—over twice as much as the value of all exported lumber. Softwood log exports reached 3.73 billion board feet in 1985 and were valued at $1.17 billion. Hardwood log exports, much smaller in volume, were valued at 138 million dollars in 1979 and at 91 million in 1986. At their peak in 1979 log exports accounted for 25% of the total value of forest product exports ($6.85 billion), but in 1985 accounted for only 18.8% of the $6.7 billion total (U.S. For. Serv., 1987).

FOREST PRODUCT MARKETS

Lumber (Sawnwood)

Since the first settlement on the Virginia coast, lumber (sawnwood) has played an important and often dramatic role in U.S. development. The industry has partici-

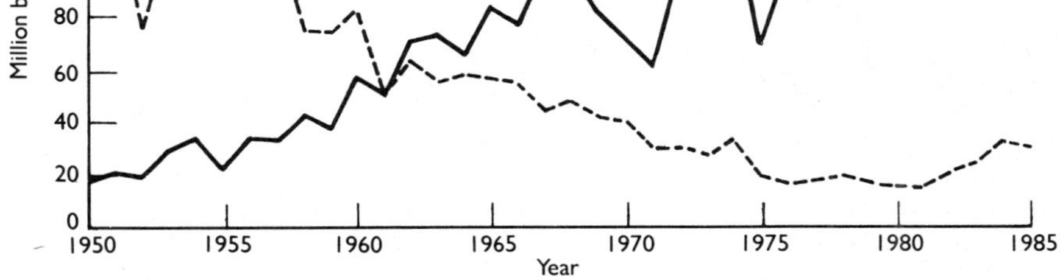

FIGURE 2.3. Hardwood log exports and imports, 1950–1985. *Source:* USDA Forest Service 1987. U.S. Timber Production, Trade, Consumption, and Price Statistics 1950–1985. USDA Misc. Pub. No. 1453, Tables 14 and 16.

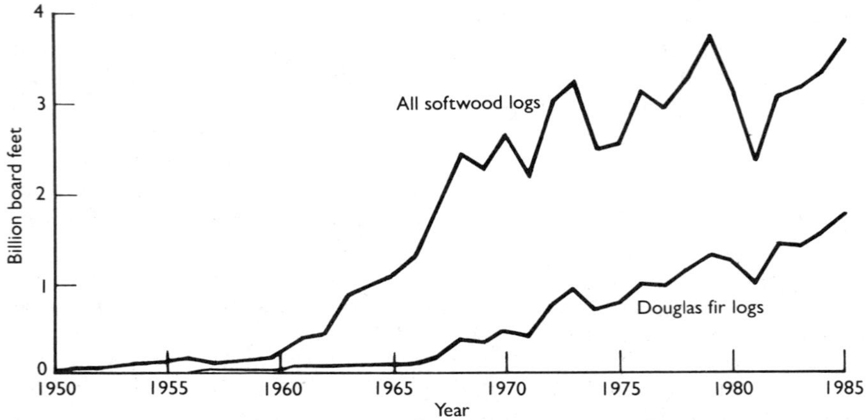

FIGURE 2.4. Softwood log exports and Douglas fir log exports, 1950–1985. *Source*: USDA Forest Service 1987. U.S. Timber Production, Trade, Consumption, and Price Statistics 1950–1985. USDA Misc. Pub. No. 1453, Table 16.

pated in all stages of U.S. industrial development: periods of pioneer daring and ruggedness, of dramatically rapid growth, sustained progress, precipitous decline and frequent failure, of comeback and reestablishment, and even, occasionally, of reasonable maturity.

In 1830 Maine led all other states in lumber production. As settlement moved westward, the raucous, hard-drinking loggers from Bangor found equally rugged companions in Michigan's Saginaw, AuSable, and Menominee during the last half of the 1800s, when the Lake states assumed production leadership. Michigan hit its lumber production peak in the 1880–1890 decade, and was followed by Wisconsin, then Minnesota. The lead passed to the Southern states during World War I, then went West to the Pacific Coast, where the production leadership is still maintained.

During this century the lumber industry has been by far the largest consumer of timber in this country. In 1950 lumber still accounted for nearly 70% of all timber used for industrial purposes. By 1969 lumber's share had dropped to 50%, in 1982 to only 43% but was over 45% in both 1984 and 1985 (according to preliminary 1985 data). Still, lumber remained the largest single industrial use of timber, although pulp and paper was very close (35%) and, if trends continue, may take over the leadership before 1990.

Production

An important part of the lumber industry's recent past, and possibly an indication of its future, can be read from the record of lumber production since 1950 (Figure 2.5). Note the difference between the peaks and valleys of hardwood and softwood lumber, the (somewhat uneven) plateau from 1950 until the recessions of 1974–1975 and (particularly) 1979–1982, the latter when housing suffered its worst drop since the depression of 1932. The ratio of softwoods to hardwoods has remained remarkedly stable since before 1950—very close to 80% of all U.S. lumber output is softwood.

Although production showed elements of stability from 1950 to 1979, significant changes occurred in the basic structure of the industry. By 1970 the portable mill, so important to lumber production before and immediately following World War II, had virtually disappeared. In 1958 there were 16,859 establishments classified as sawmills; by 1977 there were 9000, and in 1982 only 7403, with the number declining in every one of the census years since 1958.[3] Since production remained relatively constant during this time, the average sawmill size had increased.

During 1960, 1961, and 1962, Western states produced 60% of the U.S. lumber output, and from 1975 through 1979 Western production averaged nearly 57% of the U.S. total. Of the remainder, about 30% came from the South, while output from the North has remained almost constant at 12–13% of the total since 1950. Because stumpage supplies are declining in Western states while increasing in both North and South, it seems likely that the West's market share will gradually decline.

Employment

Between 1972 and 1982 employment in lumber production dropped from 180,000 to 158,000, an average annual decrease of 1.2%. The 1982 payroll came to $2.3 billion. During this same ten years the value added by manufacture (in 1967 dollars) decreased from $2.74 billion in 1972 to 2.56 billion in 1977, and to 1.25

[3]Includes SIC No 2421 (sawmills and planing mills, generally), No. 2426 (hardwood dimension and flooring), and No. 2429 (special products sawmills, not elsewhere classified). The U.S. Forest Service estimated there were 22,000+ in 1958. The difference between this figure and that of number of establishments stems largely from differences in definition; the Forest Service estimates included many of the small, often portable mills that were omitted by the Census.

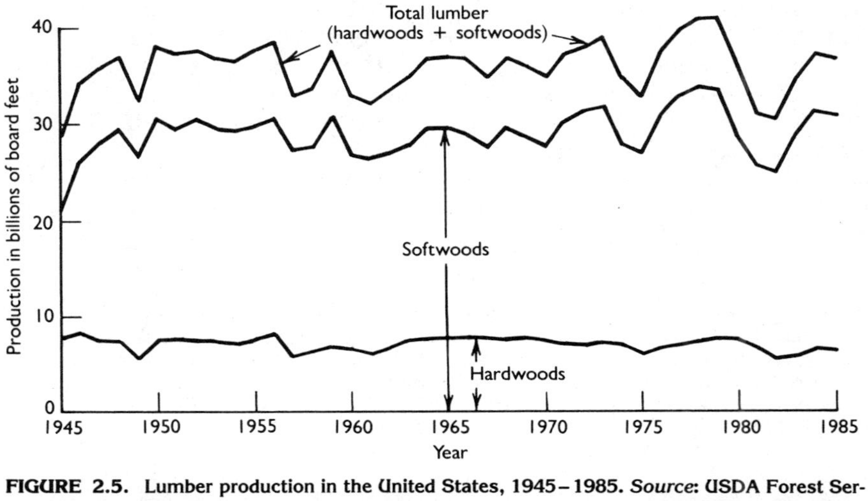

FIGURE 2.5. **Lumber production in the United States, 1945–1985.** ***Source*****: USDA Forest Service 1987. U.S. Timber Production, Trade, Consumption, and Price Statistics 1945–1985. USDA Misc. Pub. No. 1453, Tables 32.**

billion in 1982—an average annual decrease over the decade of fully 10%.[4] Value of shipments increased slightly (in real terms), from $6.08 billion in 1972 to $6.16 billion in 1977, then fell sharply to $3.72 billion in 1982.

Foreign Trade

Prior to World War II the United States was a net exporter of lumber. In 1941, however, imports began to exceed exports, and from that date forward U.S. lumber imports have grown. A new peak of lumber imports was reached in 1985, when 15 billion board feet (Figure 2.6) were imported to the United States. By far the largest part of this came from Canada's British Columbia. In 1985 the United States imported nearly 60% of Canada's *total* lumber production and U.S. imports from Canada had risen in each of the last four years. As a percent of total consumption imports have risen steadily—from 8% in 1950 to 11% in 1960, 15% in 1970, 25% in 1980, and to 30.4% in 1985. In part these imports stem from the Jones Act (requiring domestic companies shipping between U.S. ports to use U.S. ships), which in 1980 gave Canadian lumber producers shipping from Vancouver a cost advantage of $20 to $30 per thousand board feet over U.S. producers shipping from Northwest ports to the U.S. East Coast. Additionally, increases in U.S. overland rail costs have put Western lumber producers at a disadvantage compared to Canadian and U.S. Southern production. Perhaps most important has been the decline in the Canadian exchange rate; at the end of 1986 the Canadian dollar was worth 72¢ U.S.

As one might expect, the growing imports from Canada created strong efforts on the part of some Western U.S. lumber manufacturers to limit Canadian competition, although in 1983 seven of the top 13 lumber producers in North America were Canadian based (all in British Columbia) and at least four of the six U.S. corporations had operations in both Canada and the United States.

Plywood

No one knows who first decided to use a thin sheet of figured wood as decorative cover for other material, but an Egyptian mural done about 1500 B.C shows the steps in the veneering process used at that time, the tools required, and even a pot of glue heating on a fire (Perry, 1948). Successive sheets of veneer were combined to construct built-up wood panels of high strength; plywood, much as we know it today, was in use 3500 years ago. George Washington, William Penn, and Daniel Webster all used plywood-built desks, while Thomas Jefferson wrote the Declaration of Independence at a veneered desk.

Produced by hand, these early veneers and plywoods were extremely costly. Not until the late 1800s were mechanical means of cutting veneer introduced. In 1919 the Plywood Manufacturer's Association was organized and the name ply-

---

[4]Because the buying power of the dollar (or any other currency) has fluctuated substantially over the past fifty years, price time series usually require adjustment before meaningful analyses can be made. This is especially true for analyses involving post World War II data. The adjustment usually involves dividing each price by some index of purchasing power, with the index based on a selected base year. For example: if the price of a certain grade of newsprint in 1980 was $450 per ton, the value in constant (1967) dollars would be

$$\frac{450}{268.8} \times 100 = 167.41$$

since 268.8 is the producer price index for 1980 (1967 = 100). Such corrected prices are often referred to as deflated or real prices, in contrast to actual or current prices. Both consumer and producer price indices are published regularly by the U.S. Department of Labor, Bureau of Labor Statistics.

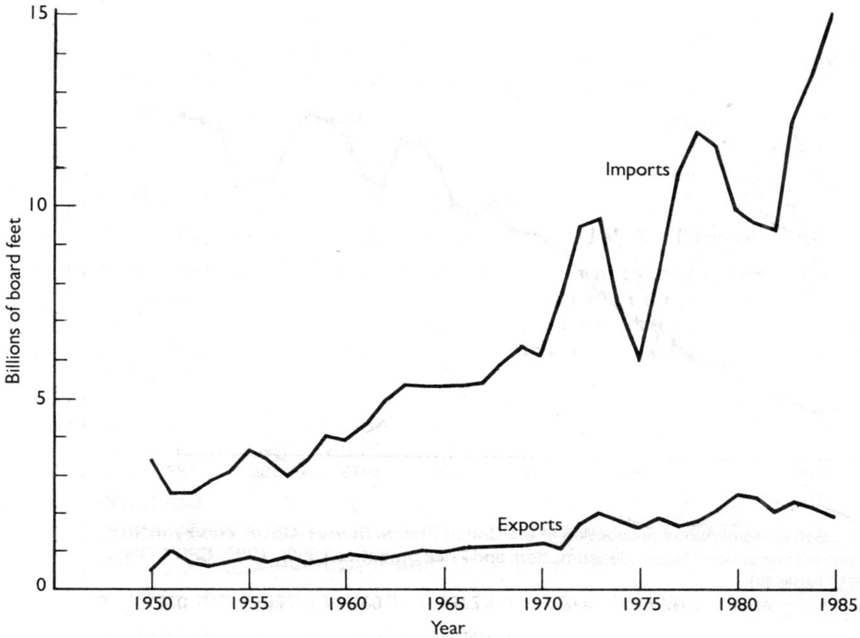

FIGURE 2.6. U.S. lumber imports and exports, 1950–1985. ***Source*: U.S. Timber Production, Trade, Consumption, and Price Statistics 1950–1984, Tables 33 and 34.**

wood replaced the older term veneered stock—a simple step, but one that may have been the key to the industry's later success, for it eliminated the unfavorable connotation the public had associated with the term veneer. From that time on the industry prospered, and plywood was increasingly accepted.

Two segments of the industry have long been recognized: hardwood plywood and softwood plywood. The two products are usually manufactured in different plants and there is little substitutability. Hardwood plywoods and veneers have a much longer history, but in recent years softwood plywood has become by far the largest segment.

Softwood Plywood

For many years softwood plywood has been the major structural panel board produced in the United States (in Europe, particleboard became more important than plywood shortly after World War II). Today's industry can trace its origin to the introduction of mechanized veneer-cutting techniques by the Douglas-fir industry of the Pacific Northwest, and to the discovery of moisture resistant and waterproof glues—the latter a product of the 1920s and 1930s. With these developments, only market demand was required to create an almost wholly new industry, and markets were beginning to boom just before World War II.

Production

Softwood plywood production had nearly tripled from 1938 to 1941 as the nation began to pull out of the Depression, but dropped back as wartime restrictions were imposed. With the end of the war in 1945 the industry began to grow, and for the

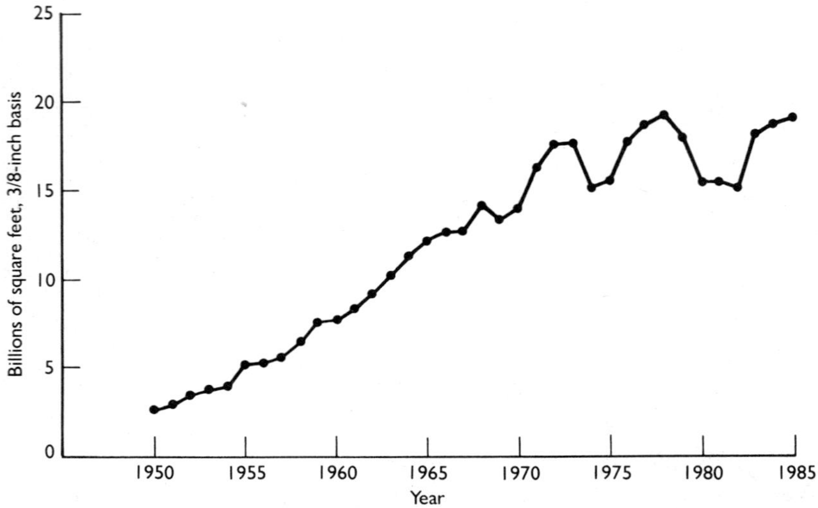

FIGURE 2.7. Softwood plywood production in the United States. *Source*: USDA Forest Service 1987. U.S. Timber Production, Trade, Consumption, and Price Statistics 1950–1985. USDA Misc. Pub. No. 1453, Table 38.

next two decades every year's production set new records (Figure 2.7). The importance of construction to softwood plywood markets is illustrated by the sharp drops in production in 1974 and during the 1979–1982 recession, both periods when new building was severely curtailed. The recovery, substantial in 1983, continued to a new production peak in 1985.

Prior to 1964 all U.S. softwood plywood came from Western states. Then production began in the South and grew rapidly. Sixty-one southern plants were in operation by 1977, and by 1982 southern production nearly equalled that from the West.

Employment

In 1977 four states (Oregon, Washington, Texas, and Louisiana) accounted for 67% of the industry's employment. By 1982 the leading four had changed to Oregon, Texas, Washington, and Arkansas, accounting for 64% of the industry's total employment of nearly 35,000 (down from 46,000 in 1977) and a payroll of $664 million. Total value of shipments in 1982 was $3.2 billion—down 29% from the 1979 high of 4.5 billion.

Trade

While exports hit a new high in 1981 of 686 million square feet, this was only 4.4% of production—but still more than 20 times import totals. Canada, although a major source of Douglas fir, exports little softwood plywood to the United States even though U.S. companies are important in Canada. Four out of the top five British Columbia softwood plywood producers, accounting for over 80% of the 1977 production from the top five B.C. mills, were either U.S. subsidiaries or affiliates (Ellefson and Chopp, 1978).

Other Structural Wood-based Panels

Particleboard and hardboard are now well established in wood panel markets. Their market success is due partly to their relatively low cost and to being a superior product for particular uses. Hardboard, in this country, is the older of the products, the original being Masonite which was protected by patent. This is a true wood product, made from exploded wood fibers pressed under high temperatures and pressures and using the natural adhesives of the wood itself. Production increaseu rapidly through the 1960s and 1970s, dropping at the onset of the slump in residential construction in 1979–1982 and again after its short recovery in 1983 (Figure 2.8).

Particleboard is made from small particles of wood that are dried, coated with an adhesive material, and then either pressed or extruded. Most particleboard is made from wood waste and can utilize a variety of species. Production has risen rapidly (Figure 2.8), to some extent by replacing softwood plywood but also by proving to be highly useful as a core-stock material for hardwood veneers, substituting for either lumber cores or veneer cores.

Trade in particleboard is relatively insignificant: while imports have exceeded exports since 1976, at their highest level (1980) imports were only 7½% of

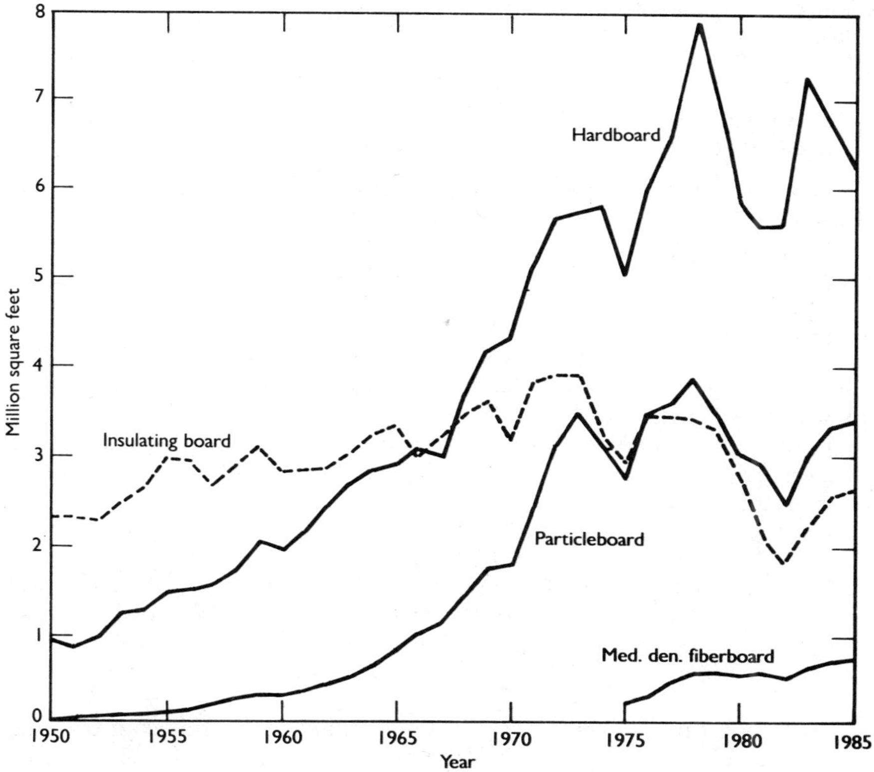

FIGURE 2.8. U.S. production of particleboard, medium density fiberboard, hardwood, and insulating board, 1950–1985. (Particleboard and fiberboard on ¾ in. basis; hardwood on ⅛ in. basis; insulating board on ½ in. basis.) *Source*: USDA Forest Service 1987. U.S. Timber Production, Trade, Consumption, and Price Statistics 1950–1985. USDA Misc. Pub. No. 1453, Tables 54, 55, and 56.

production and exports about half that level. Trade in hardboard, however, has been significant and reached nearly 19% of production levels in 1972 when over a billion square feet entered the United States. By 1980 imports had dropped to under 9% of production.

Insulation board (a soft, lightweight, wood fiber board often used for ceilings) (Fig. 2.8) has had a relatively steady history, increasing slowly until the early 1970s, falling slightly until 1979, then dropping precipitously until 1982. The decline was probably due to the use of styrofoam and similar panels offering superior insulating qualities, and to the expanded use of fiberglass for wall and ceiling insulation.

In 1980 over 75% of our structural panel material was still softwood plywood, but other panel products were being introduced: waferboard, chipboard, flakeboard, and, more recently, boards made from flakes or chips oriented to align the wood fibers, thus substantially increasing panel strength (often called oriented strand board or OSB). The industry was beginning to show signs of change—change of a type that moved Harry Merlo, Chairman and President of Louisiana-Pacific, to state that the new product, waferboard, was

> *the hottest, most exciting thing to hit our industry in many, many years. To make waferboard we take a lower cost resource—aspen, or southern pine pulpwood or lodgepole pine—and turn it into a high-quality, yet affordable product. It competes with plywood now for both builder and do-it-yourself applications. But, it isn't really fair competition because waferboard not only costs less, it does a better job as well. . . . I'm convinced that waferboard will eventually replace most of the plywood produced from public timber. (Forest Industries, 1984).*

It seems a bit premature, however, to count softwood plywood out of the panel picture during the 1980s. Technological changes are permitting use of smaller logs of lower quality that require less labor-intensive manufacturing and utilize a wider range of species—all of which serve to reduce costs.

Hardwood Plywood

In the United States the hardwood veneer and plywood industry is usually divided into two branches, one producing container and packaging plywood, the other making plywood for furniture, doors, decorative paneling, etc. Attention here will be focused entirely on the second type. Such plywood is identified according to the species used on the panel face, although other species—even softwoods—may be employed as an interior core.

The beauty and unique character of hardwood plywood vary greatly with the species but depend almost as much on the manner in which the face veneer is cut: rotary cutting, for example, yields one type of figure while sliced, sawn, and half-round veneers show quite different patterns. Most plywood veneer is rotary cut, for this yields by far the lowest cost material. Once cut the successive sheets of veneer can be arranged (matched) in various ways, yielding an almost infinite variety of designs.

The hardwood plywood industry was well established in this country long before World War I but gained new and large markets during World War II. New adhesives permitted far stronger, more durable panels to be made and techniques were developed for "molding" the material into various curved forms. PT boats and the Mosquito Bomber, both fashioned from molded hardwood plywood, provided more than ample publicity and new applications came rapidly in the postwar

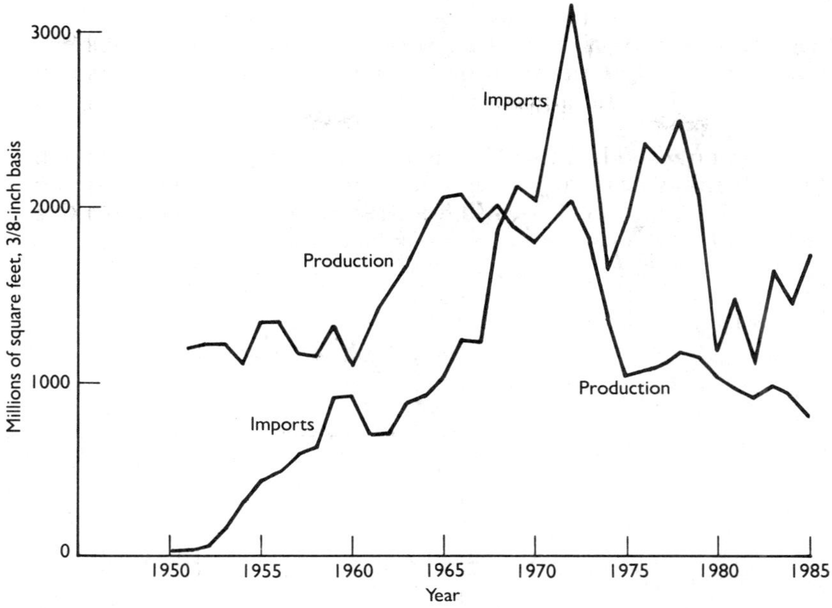

FIGURE 2.9. Hardwood plywood—U. S. production and imports. *Source*: **USDA Forest Service 1987. U. S. Timber Production, Trade, Consumption, and Price Statistics 1950–1985. USDA Misc. Pub. No. 1453, Table 38.**

period. For many of these uses, however, plywood was replaced by fiberglass during the mid-1970s, but hardwood plywood is still very popular. In construction it is used primarily for interior paneling, cabinet work, and doors.[5] In manufacturing, hardwood plywood finds its major use in furniture construction, store fixtures, and in radio and television cabinets.

Production

The production pattern of hardwood plywood is very different from that of softwood plywood. During the 1950s hardwood plywood production was essentially stable, then nearly doubled during the next five years. The following nine years were again almost constant, but then production dropped very sharply—back to a point even below that characterizing the 1950s (Fig. 2.9). By 1985 production had dropped to its lowest point since 1950—to only 824 million square feet.

Employment

In 1982 there were some 308 establishments (266 companies) producing hardwood plywood in this country—down slightly from the 1977 census figure—employing 18,000 people (down 19% from 1977). Nearly all these plants were small, but, typically, production was heavily concentrated in the larger mills. Total value of shipments in 1982 was $1,326 million—up 6% from 1977, but almost 50% higher than in 1972. The leading states were North Carolina, Indiana, Wis-

[5]"Door skins," a natural for hardwood plywood, became a major item with many plywood manufacturers immediately after World War II, when flush doors became standard for interior construction.

consin, and South Carolina, with the last having replaced Oregon from the 1977 census. Together these four states accounted for 45% of the total employment in the industry.[6]

Foreign Trade

Imports of hardwood plywood in 1950 were negligible, but by 1959 imported hardwood plywood amounted to 70% of the total U.S. production, and, in spite of the rapid growth of U.S. production during the 1960s, imports passed domestic production in 1969. In 1978 we imported more than twice as much as we produced, as we did again in 1985.

The Pulp and Paper Industry

The Chinese are usually given credit for discovering the art of paper making, probably about 2000 years ago. The industry was introduced from Europe to the American colonies around 1690 when the first papermill began operating near Germantown, Pennsylvania (Stevenson, 1940; Guthrie, 1950). By 1776 the industry was firmly established in Connecticut and Maine. Cylinder and fourdrinier machines were introduced in the early 1800s, making it possible to produce a continuous sheet of paper rather than making it sheet by sheet.

Early papermills were located on streams in or near cities, where access to fresh water, labor, and markets for both their end product and their primary raw material, rags, were assured. Once continuous sheets of paper could be manufactured, power also became important, again favoring location along streams.

The use of straw for making wrapping paper shifted production toward the Midwest. Wood was not used on a commercial scale for paper until after the soda process was introduced in Europe around 1855. The sulfite process was discovered in 1886, but nearly 25 years passed before it was adopted by U.S. manufacturers. Once wood became the usual raw material for making paper, the industry began to follow in the wake of lumber mills, a tendency that persisted as long as nonresinous woods were available.

Today's paper industry provides an outstanding example of how technological advances and mass-production techniques can change the pattern of a nation's living. Contrast the status of packaged goods in a modern supermarket with those of the old-time grocery store when the cracker barrel was a major attraction. As you thumb through the *New York Times*, think of the handbill-sized, four-sheet newspaper of 100 years ago. Pause before a display of paperbacks: low-cost paper may well be the most significant single development in news dissemination and education that the world has yet seen—at least until the advent of television and computers.

Pulp, Paper, and Board

The production operations for these three products are quite distinct and are often, but by no means always, carried out independently of one another. Both paper and board use some fiberous pulp material as their basic raw material and the manufacture of pulp is the primary step in producing both products. Paper and board manufacture can proceed only if pulp with the needed fiber characteristics is at hand.

---

[6]All data given here is from the 1982 Census of Manufactures, Industry SIC No. 2435—Hardwood veneer and Plywood.

The Pulping Process

Two resources are basic to pulp production: lots of water and a source of fiber. The first requirement virtually restricts pulp and paper plants to a location on, or very near, some body of water. Many materials have been and still are used as a fiber source. Bagasse (the plant residue from pressing sugar cane) is often used when available in the required quantities, and is important in Cuba, India, Mexico, and other semitropical or tropical countries. Where it grows well, bamboo is important—especially in India, southeast Asia, and China. Cotton waste, rags, and straw, once the major fiber source for Europe and North America, are still used for special papers. In recent years fiberglass and synthetic materials such as nylon and rayon have been added to the list. New wood, however, is usually the most economical fiber source if land is available for its production, and is the most widely used fiber source throughout North America and Europe. Waste paper, which averaged 30% of all fibrous materials used in manufacturing paper and board over the 1950–1954 period, accounted for only 19% during 1975–1979. Use of other materials has become almost negligible (Figure 2.10). Because of this dependence

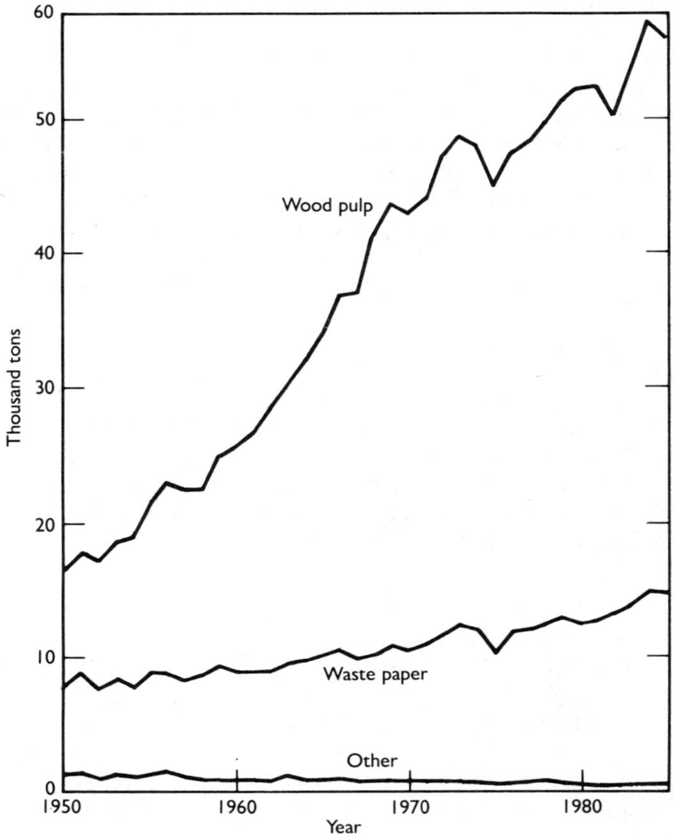

FIGURE 2.10. Fibrous materials consumed in the manufacture of paper and board, 1950–1985. *Source*: USDA Forest Service 1987. U.S. Timber Production, Trade, Consumption, and Price Statistics 1950–1985. USDA Misc. Pul No. 1453, Table 50.

on wood supplies, the pulp mills of North America are scattered throughout the continent's major forested regions.

To make pulp, the wood fibers must first be separated. Two major processes for this have been developed. **Mechanical** (or **groundwood**) **pulp** is made by grinding the wood—tearing the fibers apart by an abrasion process. Since the entire peeled pulpwood bolt is used, yields are high. This process, however, can only be used with certain species: spruce and fir are particularly favored in North America, and the paper produced is of relatively low strength and quality since the fibers are literally torn apart. Most newsprint is made from groundwood.

Chemical pulp, as its name implies, is made by mixing wood chips with chemicals, then "cooking" the mixture. This softens, or dissolves, the lignins binding the cellulose fibers together in the living tree. Several different chemical processes are used, each with particular advantages and disadvantages. Because softwood fibers are much longer than those from hardwoods, the former yield much stronger papers. Bamboo also has a long fiber, as does part of the bagasse. Hardwood pulps, however, produce papers that are more opaque and are much used for book papers. One of the most recent technological advances was the introduction of semichemical pulp: wood chips are first softened by chemical treatment, then ground. Most papers other than newsprint and those made from recycled papers are made from either chemical or semichemical pulps.

Since the chemical process separates the fibers by dissolving the materials (the lignins) that bind them together, making pulp chemically also results in large quantities of dissolved and suspended materials—the "black liquor." The important (and expensive) chemicals used by the process are recovered, but the lignins must be discarded somehow. Two or three decades ago they were frequently dumped in the river and the pulp industry held the unenviable distinction of contributing more to the pollution of U.S. waters than any other single industry. Between 1955 and 1980, however, the industry spent billions of dollars in an effort to control water and smoke pollution and both have now been dramatically reduced.

Two other technological developments triggered the growth of large Southern and Western pulp mills: (1) the move southward was preceded by the discovery of methods for pulping the highly resinous hard pines; (2) utilization of sawmill waste through the chipping of slabs and edgings further aided the Southern expansion and also made possible the use of Douglas fir mill residues by Western mills after World War II. Prior to that time the Western industry used primarily spruce and hemlock. The increased utilization of dense hardwoods has also been an important factor in expanding the total wood fiber supply.

Pulpwood Production

Figure 2.11 presents U.S. pulpwood production data and shows clearly the changing sources of wood fiber since 1950. Continued expansion of chip production is a matter of some doubt, (production in 1985 was only slightly higher than in 1979) for chip output is presently associated with lumber and plywood production. An increased use of dense hardwoods, however, seems almost certain, particularly as advancing technology permits increasing the proportion of hardwood pulp without serious reduction in paper strength. As indicated in Figure 2.11, prior to 1969 the United States was a net importer of pulpwood, but since then we have been a net exporter, primarily to Japan.

Pulp, Paper, and Board Manufacturing

Since World War II paper and board manufacture has been one of the major growth industries of the United States and the world. In sharp contrast to the

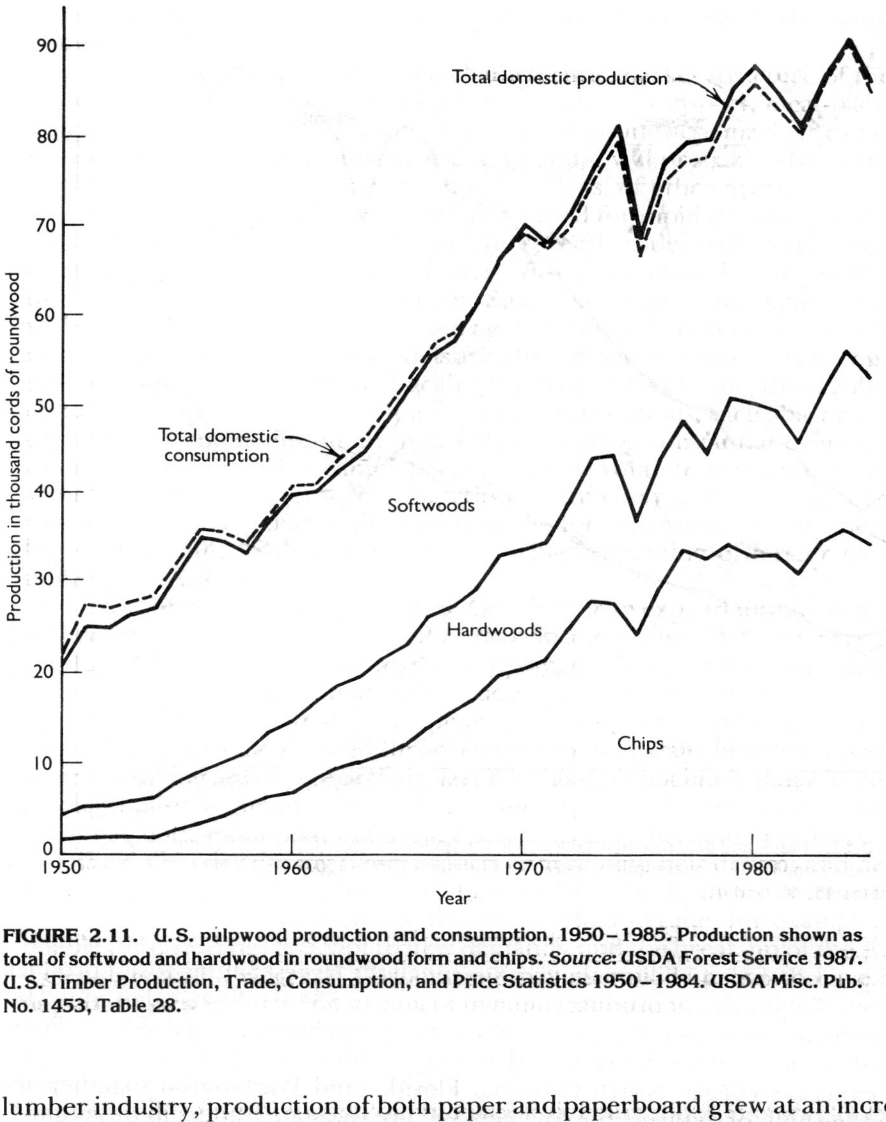

FIGURE 2.11. U.S. pulpwood production and consumption, 1950–1985. Production shown as total of softwood and hardwood in roundwood form and chips. ***Source:*** **USDA Forest Service 1987. U.S. Timber Production, Trade, Consumption, and Price Statistics 1950–1984. USDA Misc. Pub. No. 1453, Table 28.**

lumber industry, production of both paper and paperboard grew at an increasing rate from 1950 through 1973 (Figure 2.12)—from a total of 24.0 million tons to 59.3 million, an increase of nearly 150%. After the 1974–1975 recession the upward trend continued through 1984, when total output reached 70.25 million tons, then dropped slightly in 1985.

In 1982 the pulp, paper, and board industries of the United States as defined by the Census Bureau for their Survey of Manufacturers included 43 pulp mills, 299 paper mills, 222 paperboard mills, and 30 building paper and board mills—altogether some 594 establishments.[7] The 204,900 employees of these industries

[7]Because the Survey of Manufactures classified establishments on the basis of their "primary product," those producing pulp but converting it to paper or board would probably be classified as either a paper or a board mill, even though such integrated establishments include the largest producers of pulp in the world.

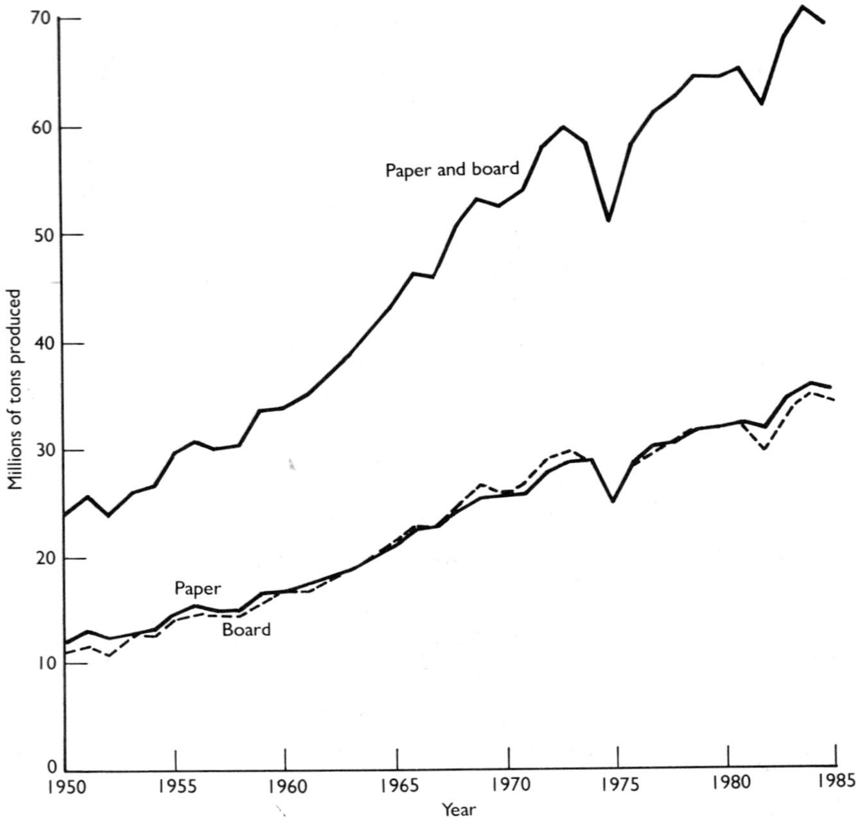

FIGURE 2.12. Paper and paperboard production in the United States, 1950–1985. *Source*: U.S. Timber Production, Trade, Consumption, and Price Statistics 1950–1984. USDA Misc. Pub. No. 1453, Tables 45, 46, and 47.

made a payroll of $5.47 billion that year—much the largest among forest product industries. Total value of product shipments came to $34.0 billion, value added by manufacture came to 14 billion, while new capital expenditures totaled 3.8 billion. Regional economic importance varied among the industry types more than most would expect. Georgia, North Carolina, Florida, and Washington together accounted for 55% of *pulpmill* employment and are the state leaders in this regard. This is a decided shift from 1977, when Washington, Florida, Georgia, and Maine led with 60% of the employment. In *papermill* employment, however, the leading states (Wisconsin, Maine, Alabama, and New York) accounted for 38% of papermill employment, while *paperboard* mill employment leaders were Georgia, Virginia, Alabama, and South Carolina, accounting for 31% of the 1982 employment in paperboard manufacture.

As with all aggregates, data describing the general U.S. paper and paperboard production pattern (Figure 2.12) tend to mask as well as to reveal. The tremendous diversity of paper products is a point of pride to paper producers:

> *Special papers are being fabricated for the manufacture of bags and beach slippers, cigarettes and cartridges, diapers and Dixie cups, filters and*

> *facial tissues, polishing cloths and patterns, roofing and roof trusses, soda straws and seat covers, towels and twine; wallpaper, sandpaper and building paper; meat wrappers, butter wrappers, candy wrappers; cartons, coathangers, and condensers. . . . all the multiplicity of products which aid transportation, sanitation, refrigeration, and. . . . "You name it, we got it!" (Ainsworth, 1958)*

Even though computer cards and printouts and the impact of the ever-busy copy machines are omitted, this classic recital of paper use gives the flavor of the diversity characterizing the paper market. To the manufacturer such diversity represents an opportunity for expanding markets. To an industry analyst, however, such diversity creates a problem: to what extent must these hundreds of paper types be treated separately? Can one meaningfully speak of the demand for paper? What is the "product" of a corporation that produces and sells 20 or more kinds of paper?

QUESTIONS AND PROBLEMS

1. What did Adam Smith have in mind when he said that humans, in pursuing their own gain, are "led by an invisible hand"?
2. Do you regard "the market" as an efficient mechanism for allocating resources in the United States today? If not, why not?
3. Define an equilibrium price. Why is the concept of an equilibrium price considered important?
4. Four major market groups were discussed in this chapter: wood harvesting, lumber, wood-based panels, and pulp, paper, and board. Discuss their relative significance to forestry, and their importance to national and regional interests in terms of employment and contribution to GNP.
5. How significant to the profession of forestry is foreign trade in forest products? How would you assess the future importance of wood imports from abroad? What about export?
6. If you were asked to discuss the probable returns to investors in forest product industries, how would you rank the industry groups, and what is your basis for so ranking them? What area of the United States would you select as the most probable site for expansion of the forest-based industries?
7. Are you optimistic or pessimistic about the future of forest-based industries? How about the future of forestry?

chapter three
FOREST PRODUCT DEMAND

During the decades immediately following World War II traditional uses of wood were often challenged. Plastic-sheathed geodesic domes, extruded forms that could be hardened as desired, pre-formed metal or plastic room units that could be grouped as one wished, and the "All Aluminum Home" were all proposed as replacements for the wood framed house. Cellophane and other plastics, along with fiberglass, made substantial inroads in packaging. Even the market for newsprint seemed (to some) doomed by the rapid takeover of television. Forest-based corporate executives as well as public officials found themselves wondering why they were growing trees—could it be that the future of forestry lay entirely with recreation, aesthetics, and watershed management?

The response of many foresters was to fall back on tradition. "People have always used wood! Why worry about future demand? There will always be *some* use for whatever wood we grow. It's nature's renewable resource."

Faith in the value of one's work is generally accepted as a necessity for healthy living, and for some the simple belief that "wood is good" may suffice. But there are better ways to establish a firm foundation for faith in future forest products than through public acclamation. Wood *is* good, but a forester should know what it is good for. Any thinking forester wants well-based estimates of how much and what kind of wood U. S. markets will be willing and able to absorb 5, 10, 20, or even 25 years from now, and the really thoughtful ones will also inquire about the potential of foreign markets and their possible competition. Before you say to yourself "Who can tell what will happen 20 years from now?," remember that the trees you plant this year will be used only after 20 to 25 years at the earliest. Are you—should you—be planting for pulp, sawlogs, plywood, or chipboard, or should you be improving the forest habitat for game animals and providing aesthetically pleasing landscapes?

Few economists claim to have definitive answers for such questions, but for anyone seeking even an understanding of the possibilities, some knowledge of forest product demand is necessary.

THREE BASIC ASSUMPTIONS

Before beginning an examination of forest product demand we should be clear about three assumptions underlying nearly all economic analysis: (1) that people are essentially rational creatures, (2) that we make decisions we believe will lead to the most satisfying outcomes for ourselves, and (3) that human wants may be regarded as being insatiable.

The first assumption (rationality) really needs little defense, for to assume the opposite would, by definition, effectively rule out any possibility of logically analyzing human actions. The assumption does not, however, mean that people make, or even try to make, *every* decision in a completely rational fashion. There are many important activities in life—religion and love are two examples—in which most participants would agree that careful, rational behavior would not only be inappropriate but also impossible. With many other activities, computational demands rule out strictly rational decisions. The assumption of economic rationality simply means that, when choosing between several available goods or services, an individual tries to select those that best serve the intended purpose at the least cost.

The second assumption can be put more directly by saying that humans are acquisitive and they are maximizers. Here a couple of qualifiers must be inserted: economics is concerned with a relatively narrow range of decisions—those that influence wealth. Altruism has not disappeared, but by and large business decisions are not made with an altruistic objective. And we know that few individuals literally *maximize* their financial returns, or minimize their dollar costs, because to do so would require spending so much additional time and effort that it wouldn't be worthwhile—it would cost more than it would be worth. Economists get around this difficulty by saying that people try to maximize their own welfare, and that any decision maker will select those outcomes most likely to yield the greatest "utility."[1]

The third assumption—that human wants are essentially insatiable—also requires qualification. It would be possible for anyone to get as much as he or she wants of almost any single thing if there were no financial constraints whatever. But few people ever become free of *all* material wants. If one has enough food, drink, and shelter there always seems to be other materialistic "things" to crave. It is in this sense that wants are regarded as being insatiable.

THE DEMAND CONCEPT

The term demand is a good example of how disciplines aid misunderstanding by giving words in regular use a special meaning. To most people the meaning of demand is obvious—it refers to the sale or prospective sale of products. If demand increases, sales go up; if demand drops, sales decrease. Demand may also be confused with **consumption**, but consumption is the **quantity consumed at a given price**, and may be either an historic or an anticipated amount. To an economist, however, demand is the functional relationship between the quantities of a commodity or service that buyers are willing and able to purchase at different prices during a specified time. It is the quantitative expression of expected buyer response to changing product prices, and can be described by a schedule, a graph, or an equation.

[1]Although so far no acceptable cardinal measure of utility has been created, the concept of utility has been of immense value to theoretical development in economics, especially in welfare economics. Herbert Simon's concepts of bounded rationality and satisficing, both important contributions to economic decision making, will be introduced in Chapter 11.

For example, suppose we had made a door-to-door survey of the market for fuelwood in Forestville (a small fictitious Northeastern city). Survey results indicate that if autumn and winter weather follow usual patterns, Forestville residents will buy firewood in the amounts shown in Table 3.1. This relation could be expressed by the schedule of Table 3.1, the curve of Figure 3.1, or the equation

$$P = 60 - 0.667Q$$

where Q is expressed in thousands of stacks.

TABLE 3.1
Schedule of Prices and Corresponding Quantities of Fuelwood That Could Be Sold in Forestville in 1988

Price per 4 × 8 ft Stack of 16″ Wood	Number of Stacks
20	60,000
25	52,500
30	45,000
35	37,500
40	30,000
45	22,500
50	15,000

One of the first things to be noted is the difference between a *movement along* the demand curve and a *shift* in the curve itself. If we want to know how raising the price from 40 to 50 dollars per stack would affect the quantity of firewood that could be sold in Forestville, we could check the schedule, move up the line on the graph, or plug the values into the equation—all would show sales dropping by 15,000 stacks. A shift in the curve, however, means the curve itself has moved, or changed shape, or both. If, for example, we had made the Forestville survey before OPEC quadrupled crude oil prices, we would have discovered a very different relationship: rising oil and gas prices dramatically changed fuelwood consumption in the Northeast.

Because factors do change, a time interval for any demand relationship is usually specified—the demand for firewood in 1988, the Canadian demand for lumber in 1973, or the anticipated demand for plywood in 1995. And when comparing demand at different times we usually speak in terms of *shifts* in demand.

With many forest products there are real difficulties in specifying the product itself. Take our fuelwood example: are we dealing with split wood or round? Seasoned or green? Hardwood or softwood? Actually, the curve would probably be different for each of these products, yet the more general concept of the demand for firewood also makes sense and has practical use.

Perhaps the most important assumption underlying demand is defined by the phrase *ceteris paribus*, usually translated as "other things being equal." The other things include the prices and quantities of everything else and the relative preferences or tastes of consumers. In our example it means that the demand curve describes the price-quantity relationship for firewood in Forestville during 1988 *as long as* the way people want their houses heated, regulations on the burning of wood, smoke abatement laws, the prices of fuel oil, natural gas, and coal, and the use of wood for other purposes all remain unchanged. Admittedly that's a lot of ifs,

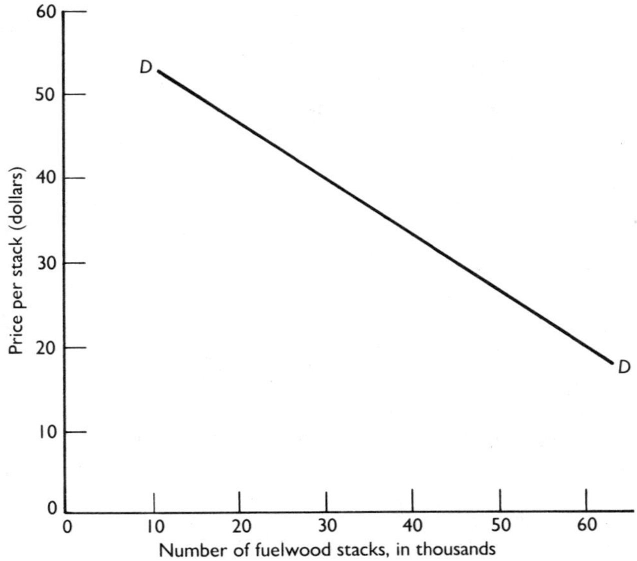

FIGURE 3.1. Curve of demand for firewood in Forestville, 1988.

but all we are really saying is that if any of these things change, our demand curve will probably shift.

Two other points about demand are important. We can interpret a demand relationship in two ways: (a) if price changes, the quantity taken in the market will vary according to the curve shown, or (b) if the quantity offered in the market is changed, the price that will just "clear the market" will vary according to the schedule described.

The second point is that the demand curve is also the curve describing the relationship of **average revenue** to the quantity sold. This is really a matter of economic semantics: since demand curves specify the quantities that can be sold in a given market at each possible price, then the average cash income *per unit* from the sale of any particular quantity (the average revenue) must be the price for that quantity shown by the demand curve.[2]

Elasticity

Now suppose we want to compare the demand for two different products. For example, how does the demand for walnut lumber compare with that for Southern pine lumber? If the demand curves were those of Figure 3.2 (they are not), one might conclude that the demands for the two products are very much alike—certainly the curves are similar. But note that the price and quantity scales for walnut are very different from those in the pine diagram. If the scales were actually the same, the two curves would seem quite dissimilar. The problem is compounded when one tries comparing the demand for, say, paper with that for plywood: paper quantities are measured in tons, while plywood is (usually) measured in thousands of square surface feet of ⅜ in. thickness equivalent.

[2]Total revenue (TR) is PQ, so average revenue is $(TR)/Q = (PQ)/Q = P$ under the conditions described here.

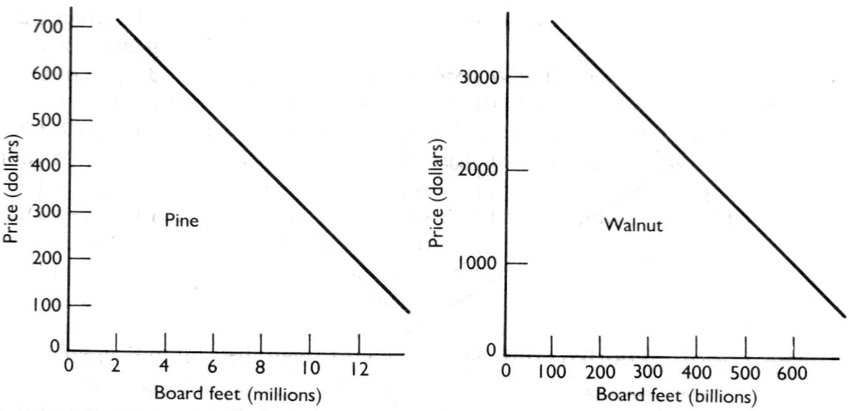

FIGURE 3.2. Two imaginary demand curves, the left for southern pine lumber, the right for walnut lumber. Although they look alike, they are very different in terms of their economic implications.

To overcome these and other difficulties, economists often use the concept of elasticity. Because many factors other than price can affect demand, the particular factor is usually identified: in the case mentioned above the needed measure is termed price elasticity, which is the relationship between the *percentage* changes in price and quantity. Price elasticity (E_P) can be precisely defined as:

$$E_P = \frac{\%\text{ change in Quantity}}{\%\text{ change in Price}}$$

Alternative formulations are easily derived:

$$\frac{\Delta Q/Q}{\Delta P/P} = \frac{\Delta Q}{Q} \cdot \frac{P}{\Delta P} = \frac{\Delta Q}{\Delta P} \cdot \frac{P}{Q}$$

where ΔQ and ΔP indicate, respectively, the change in quantity and the change in price.[3] Now, as ΔP and ΔQ approach zero, we can replace the ratio $\Delta Q/\Delta P$ with dQ/dP and are left with the formula for point elasticity:

$$E_P = \frac{dQ}{dP} \cdot \frac{P}{Q}$$

Normally price elasticity will be negative, since demand curves are usually negatively sloped. Frequently, however, the sign is ignored and elasticity is said to vary from zero to infinity. If E_P is less than one (so that the percentage change in quantity is less than the corresponding percentage change in price), then the curve is described as being *inelastic*; if it is zero, it is called *perfectly inelastic*. On the other hand, if elasticity is greater than one, demand is termed *elastic*, and if E_P tends toward an infinite value, the curve is said to be *perfectly* (or *infinitely*) *elastic*. **Unit elasticity** is just what the term implies—the special case when $E_P = -1$.

Because it is a ratio of percentages, elasticity is a pure number, completely

[3]The symbol Δ will signify a discrete change, while dQ/dP for example, will indicate the first derivative of Q with respect to P, or the limit of the ratio $\Delta Q/\Delta P$ as ΔP approaches zero.

unaffected by the scale used to measure either P or Q. This makes it possible to compare any two demand curves in a meaningful way: the elasticity of demand for recreation can be compared with that for timber even though the two products are measured in very different units. (We can't *add* apples and oranges, but we *can* compare their price elasticities!)

Arc elasticity is calculated from a discrete change in price and quantity (Figure 3.3.). Figure 3.3, for example, shows a hypothetical demand curve for lumber in Peru, with units converted to U.S. measurements. As price falls from \$210 per MBM to \$200 (MBM = thousand feet, board measure), the quantity sold in the market rises from 4500 MBM to 5000. The enlarged view permits a close examination of the geometry involved. The change in quantity is

$$5000 - 4500 = 500$$

so the percentage change in quantity could be 500/5000 or 10%. Similarly, the corresponding percentage price change would be −10/200, or −5%. The price elasticity (E_P) might therefore be calculated as

$$E_P = 10/-5 = -2.0$$

But, clearly, the percentage change might be based on either the beginning or ending values. If the beginning values had been used, we would have

$$\frac{\frac{500}{4500}}{\frac{-10}{210}} = \frac{.111}{-.048} = -2.31 = E_P$$

Which answer is correct?

To resolve this difficulty, one needs only to recognize that arc elasticity is a sort of "average" elasticity between any two selected points (p_1,q_1) and (p_2,q_2). It

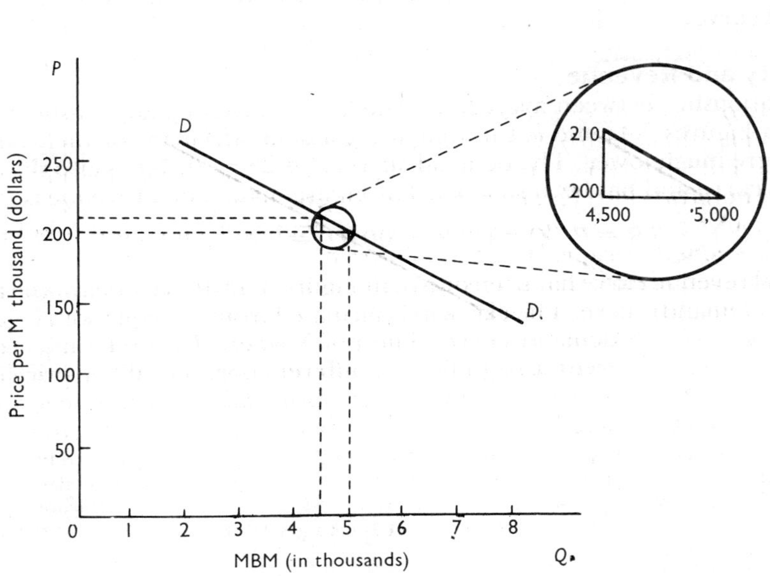

FIGURE 3.3. Arc elasticity.

should be calculated by basing the two percentages on the *average* values of p and q, on

$$\frac{p_1 - p_2}{2} \quad \text{and} \quad \frac{q_1 - q_2}{2}$$

In this example the calculations would be

$$\frac{\frac{500}{4750}}{\frac{-10}{205}} = \frac{10.53}{-4.88} = -2.16 = E_P$$

Point elasticity is a more precise measure, usually calculated directly from the equation of the demand curve. The ratio then becomes dq/dp, or the reciprocal of the slope.[4] To calculate elasticity of demand with respect to price at any point on the demand curve, one can simply multiply the reciprocal of the slope of the curve by the ratio p/q, using the values of p, q, and the slope at the desired point. In the example of Figure 3.3, the demand equation is $P = 300 - .02Q$ so the first derivative (or slope) is $-.02$. Point elasticity at a price of (say) $210 is

$$-\frac{1}{.02} \times \frac{210}{4500} = -2.33$$

and at $200

$$-\frac{1}{.02} \times \frac{200}{5000} = -2.0$$

Note that *slope* and *elasticity* are not the same thing: it is the reciprocal of the slope that is multiplied by p/q. For any demand curve such as that of Figure 3.3 (a straight line), slope is constant but elasticity is different at every point on the demand curve.

Elasticity and Revenue

The relationship between average revenue, total revenue, and elasticity is illustrated in Figures 3.4*a* and *b*, this time using a demand curve for fuelwood when prices were much lower. The demand curve, $P = 28 - .001Q$, is also the average revenue curve and has been so labeled in Figure 3.4*a*. Total revenue is $P \cdot Q$, so

$$TR = P \cdot Q = Q(28 - .001Q) = 28Q - .001Q^2$$

This total revenue curve has been drawn in Figure 3.4*b* directly below the average revenue (demand) curve. The elasticity, calculated from the equation, is given for three points on the demand curve. The point where $E_P = -1$ (unit elasticity) divides the demand curve into significantly different portions. When price is above this point any price reduction results in *increased total revenue*, as seen from the

[4]This assumes the demand equation is of the form $P = F(Q)$. For various reasons most demand analyses are aimed at quantifying equations in which quantity is made the dependent variable, or equations of the form $Q = F(P)$. Since the demand relationship is implicit, this in no way contradicts theory. In this latter form the derivative dq/dp is obtained directly. The simple equation used above could be written as $Q = 15{,}000 - 50P$, and $dq/dp = -50$. This -50 is the reciprocal of .02, and .02 is the slope of the equation $P = 300 - .02Q$ which was used above and for the demand curve of Figure 3.3.

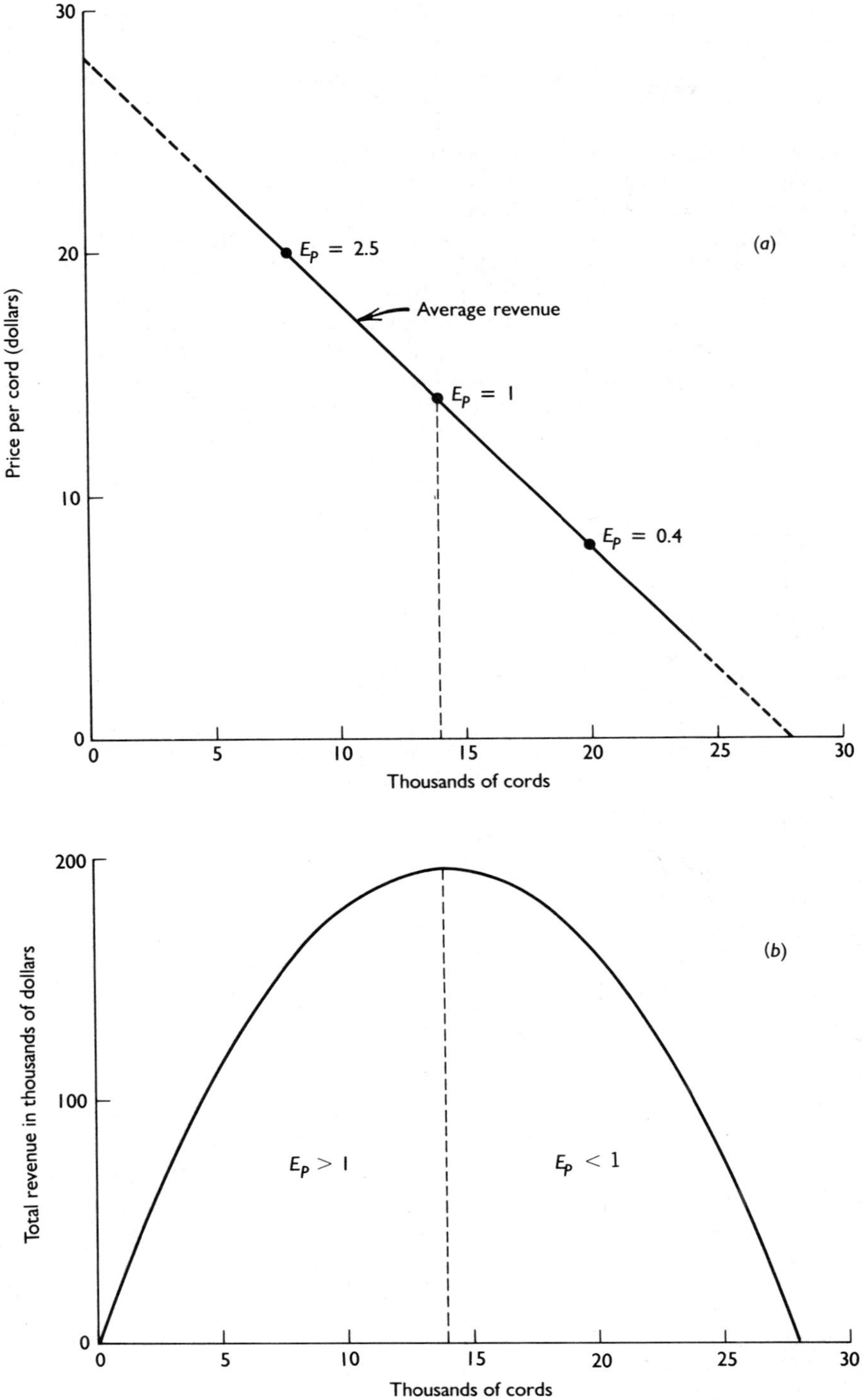

FIGURE 3.4. The relationship between average *(a)*, total revenue *(b)*, and elasticity.

curve of total revenue. Below this point of unit elasticity, any further reduction of price results in a *decrease in total revenue*, even though quantities sold continue to increase. Total revenue peaks where $E_P = -1$.

Figure 3.5 illustrates this important relationship in a somewhat different fashion. Let price and quantity be, respectively, p and q. Then $op \cdot oq$ (or $p \cdot q$) is the area of the rectangle *opaq*. But $p \cdot q$ also equals total revenue. Hence the area of a rectangle, such as *opaq*, inscribed under a demand (or average revenue) curve also represents the total revenue. Figure 3.5*a* illustrates a relatively inelastic demand curve. Here a price rise increases total revenue—the area of the rectangle *opaq* is substantially larger than the corresponding one formed at the lower price p_1.

The *b* part of the figure shows the opposite case—an elastic demand curve. Here an increase in price will obviously *decrease* the industry's total revenue.

All the curves in Figure 3.5*c* have a constant unit elasticity. They are rectangular hyperbolas and all rectangles formed by connecting a point on the curve to the reference points on the two axles will have equal areas—$p \cdot q$ is a constant. For any part of a curve over which $E_P = -1$, total revenue will not be affected by changing price.

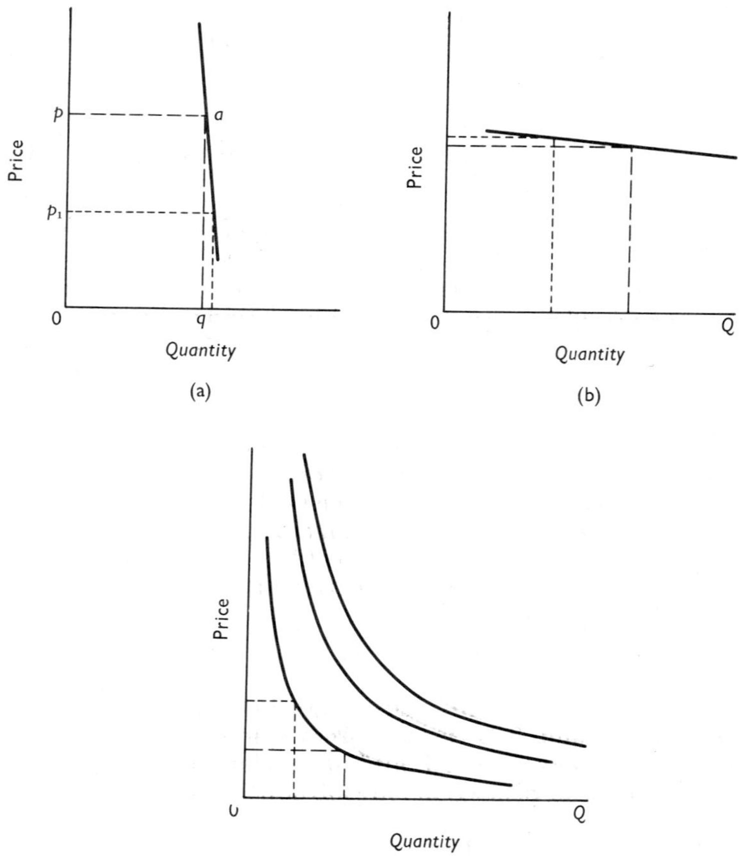

FIGURE 3.5. Relatively inelastic and elastic demand curves (*a* and *b*) and demand curves of unit elasticity (*c*).

Income Elasticity

So far our discussion of elasticity has been concerned only with quantity/price relationships—with price elasticity. We know, however, that many factors influence demand—all the things kept in place by the *ceteris paribus* restriction. These factors either shift the demand curve or change its shape, and often do both.

Among the many factors known to affect demand for forest products, one of the most important is income because the ability of people to buy influences the economic demand for any product that must be purchased by consumers. Generally, the income-consumption relationship will be curvilinear and resemble that of Figure 3.6—increasing at a declining rate and eventually becoming virtually horizontal—since an individual's want for almost anything begins to be satisfied at some point of consumption. Although most forest products are of this type, there may be a few where increasing incomes lead to increasing rates of consumption. Some of these, such as solid walnut or rosewood paneling, are often used for what Veblen called conspicuous consumption, similar to mink coats or diamonds. And there are other possible reasons: if increased incomes permit people to substitute one product for another, the original item may show a decline in consumption as incomes increase. If, for example, wealthy people build their homes of glass and brick instead of wood, their income-wood consumption curve could decline. A far more important case is that of low-income groups in developing nations: as incomes rise, kerosene or butane is substituted for fuelwood. (During the 1970s, of course, we saw how this curve works in reverse.)

APPLICATION

By now you should have the general ideas of economic demand and elasticity reasonably in mind. The fuelwood market of Forestville is unlikely to be a major concern of forest managers but the U.S. demand for fuelwood certainly will be. We now need to see how the concepts of demand and elasticity can be used to aid

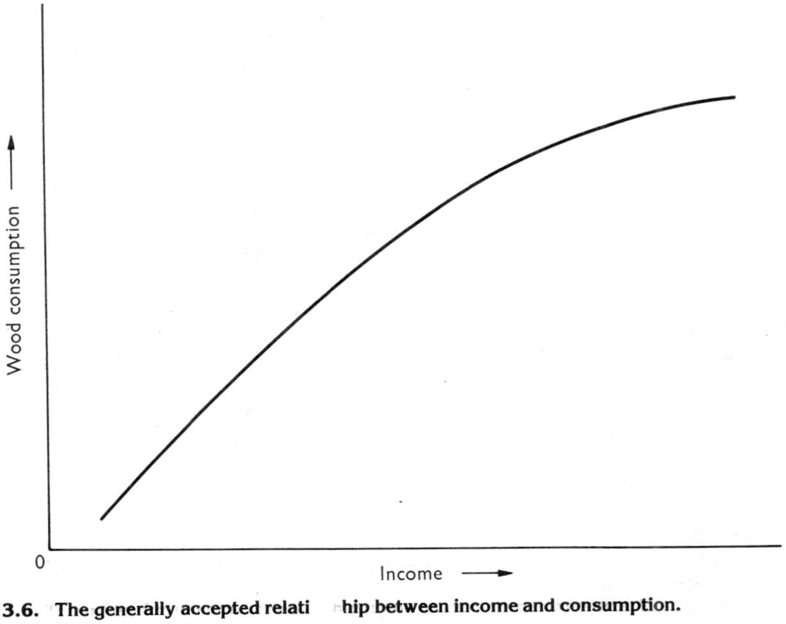

FIGURE 3.6. The generally accepted relati hip between income and consumption.

decision making. While any of the major forest industries could illustrate a variety of applications, the lumber industry seems a logical choice.

Two approaches could be taken to a broad study of U.S. lumber demand. As with any product such as bread or automobiles, one might regard it directly. Alternatively, one might recognize that lumber is a producer good and reason that its demand should be treated as "derived" from the demand for the final product into which lumber is manufactured. Because a broad quantitative and historical viewpoint is believed the first requisite, the initial possibility will be explored. Later, in Chapter 9, the derived demand alternative will also be examined.

Annual lumber production, imports, and exports since 1950 were given in the previous chapter (Figures 2.6 and 2.7). Lumber prices have followed a consistent trend for a remarkably long period. In Figure 3.7 deflated lumber prices from 1800 through 1985 have been plotted on a logarithmic scale.[5] For 180 years lumber prices have increased at an almost constant 1.8% annual rate. Such increases are one sign of relative scarcity in a free market, and it is worth remembering that timber is the only natural resource used as a raw material that has had a consistent price increase over time. When these same deflated prices are plotted arithmetically over the 1950–1985 period, however, they seem much less consistent (Figure 3.8). After a nearly 15-year mildly downward trend (1950–1965),

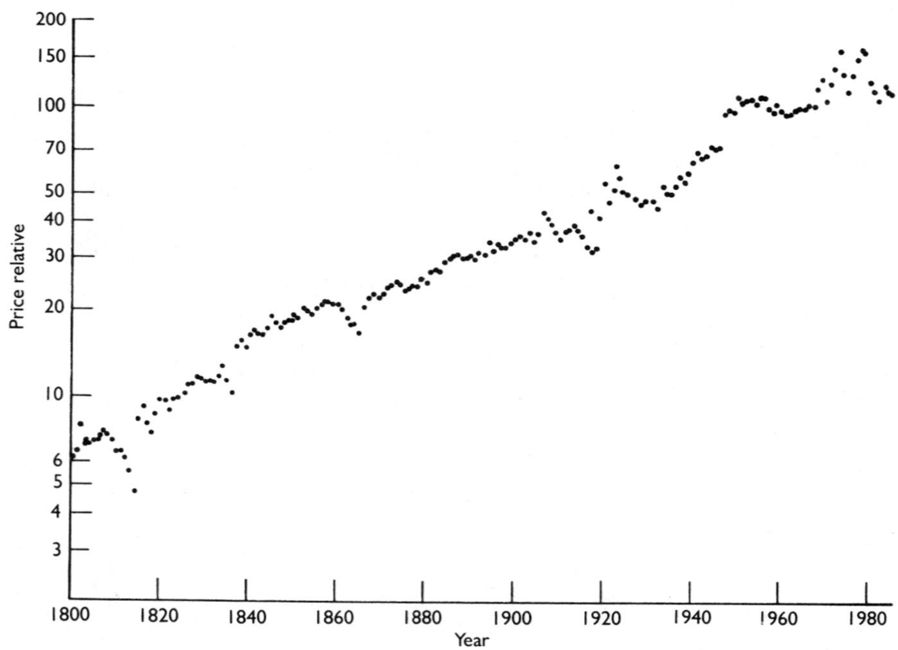

FIGURE 3.7. **Trend of U.S. lumber prices (deflated), 1800–1985. Relative producer price index for lumber, 1967 = 100.** ***Source:*** **U.S. Timber Production, Trade, Consumption, and Price Statistics 1950–1985. USDA Misc. Pub. No. 1453, Table 37.**

[5]For Figure 3.7 the deflator was the all commodity producer price index as calculated by the Department of Labor, Bureau of Labor Statistics. The data were plotted on semilog paper (logarithmetic scale for one axis, the other arithmetic) because any straight line on such a graph indicates the variable plotted logarithmically is increasing (decreasing) at a constant rate.

FIGURE 3.8. U.S. lumber prices (deflated), from 1950 through 1985. *Source*: U.S. Timber Production, Trade, Consumption, and Price Statistics 1950–1985. USDA Misc. Pub. No. 1453, Table 37.

prices turned upward and became very erratic, with rapid swings of unprecedented magnitude.

For demand analysis it would be preferable to have market sales data, but, unfortunately, information on lumber sales at the consumer level has not been systematically collected. To approximate sales, imports are added to production, then exports are subtracted.[6] The result is appropriately called apparent consumption. The diverging curves of production and apparent consumption shown in Figure 3.9 should help establish the difference between the two, in this case the result of the negative lumber trade balance.

When the deflated price is plotted over apparent consumption (Figure 3.10), the result obviously bears little resemblance to the neat price/quantity relationship of textbook demand! But no such relationship should be expected, for the specifications of the demand model have not been met. The market area has been reasonably constant, but in 1950 the U.S. population was 152.3 million, while in 1985 it was 239.3 million. One might certainly expect that with a 50% increase in the number of people to house, lumber consumption would change.

To get around this obvious problem we might use per-capita lumber consumption (Figure 3.11). While far from uniform, this figure indicates that lumber use by the average U.S. consumer declined from 1950 through 1958—a continuation of a long-term trend that began before 1950 as the result of a movement toward smaller houses and changing construction techniques. The sharp recession of 1973 dropped per-capita consumption to its lowest ever level in 1975. It bounced back to a 22-year high in 1978, then went into a sharp decline and sank to

[6]If data were available, changes in mill stocks should also be entered—subtracted if stocks have increased, added if they have decreased.

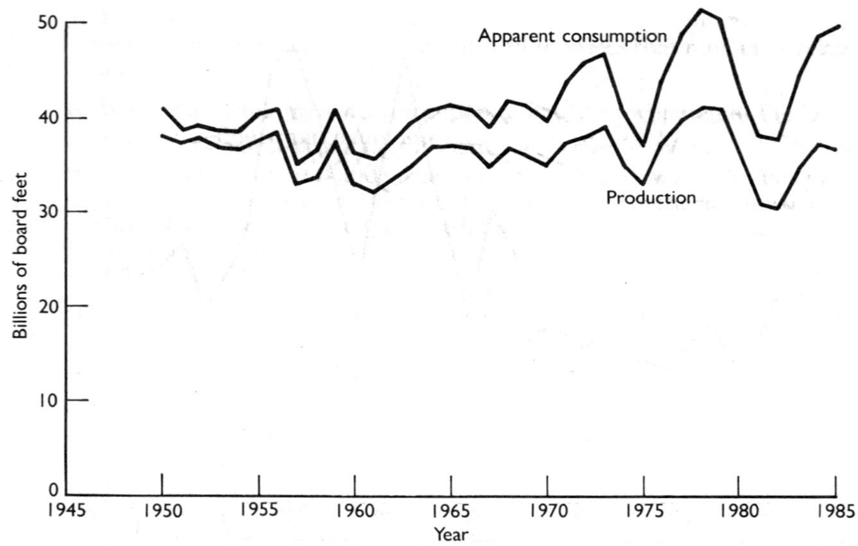

FIGURE 3.9. Production and apparent consumption of lumber in the United States, 1950–1985. *Source:* U.S. Timber Production, Trade, Consumption, and Price Statistics 1950–1984. USDA Misc. Pub. No. 1453, Table 31.

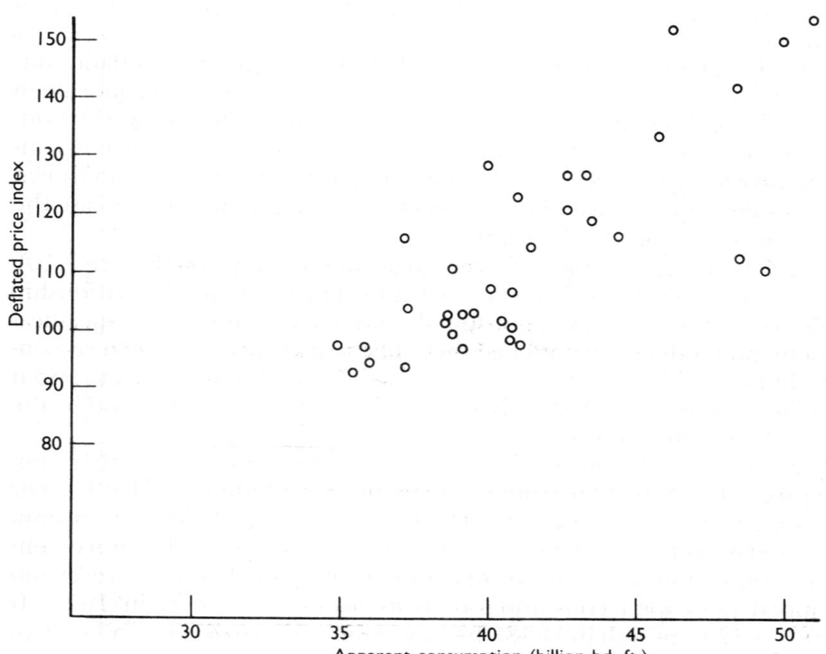

FIGURE 3.10. Relative producer price index for lumber plotted over apparent lumber consumption, 1950–1985. *Source:* U.S. Timber Production, Trade, Consumption, and Price Statistics 1950–1985. USDA Misc. Pub. No. 1453, Table 31 and 37.

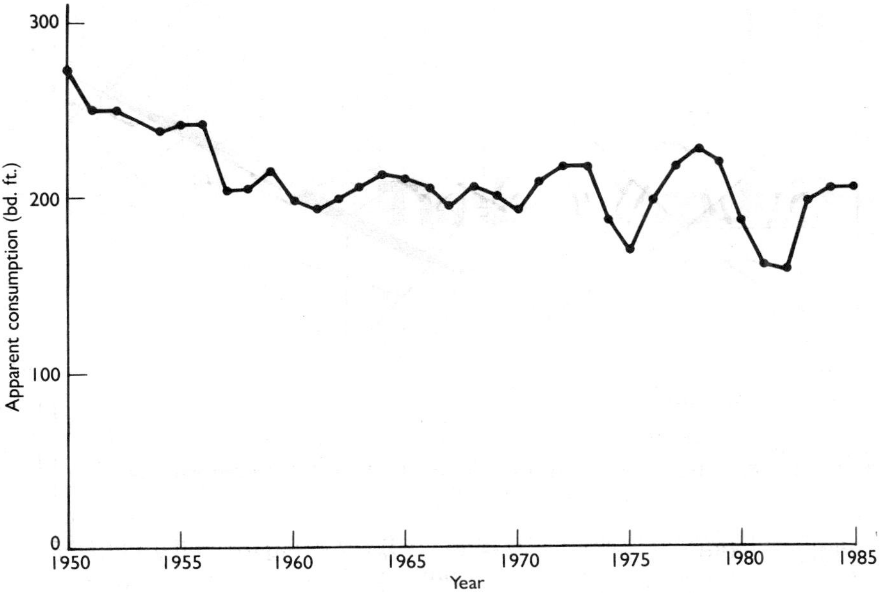

FIGURE 3.11. Trend of apparent lumber consumption per capita, 1950–1985. *Source*: U.S. Timber Production, Trade, Consumption, and Price Statistics 1950–1984. USDA Misc. Pub. No. 1453, Table 38.

an all-time low in just 4 years, but quickly recovered—to the same level as in 1956 and 1957—in 1984 and 1985.

The changing population, however, is not the only deficiency in our model. Perhaps we should look more carefully at some of the other items included in that *ceteris paribus* assumption made for the demand model. Are there some other variables that might affect this price/quantity relationship?

Other Variables

We began with the simple two-dimensional price/quantity relationship of *a* in Figure 3.12. We might well use per-capita consumption data to counter the effect of population increases, but not all factors can be handled so easily. The influence of any additional factor on the demand curve means adding a new dimension to the model. Income, for example, might be included as shown in Figure 3.12*b*; as incomes increase the per-capita demand curve shifts upward.

The problem is that our data consists of a series of points, each one representing the (average) intersection of two curves—demand and supply—that have been subjected to constantly changing forces. So far we have been looking at factors that might produce shifts in the demand curve, yet, clearly, a shift in supply could produce an equally pronounced effect on the observed price/quantity intersection. Each observation of price/quantity is a point of intersection of demand and supply functions (Figure 3.13). The analyst of such an array of points is faced

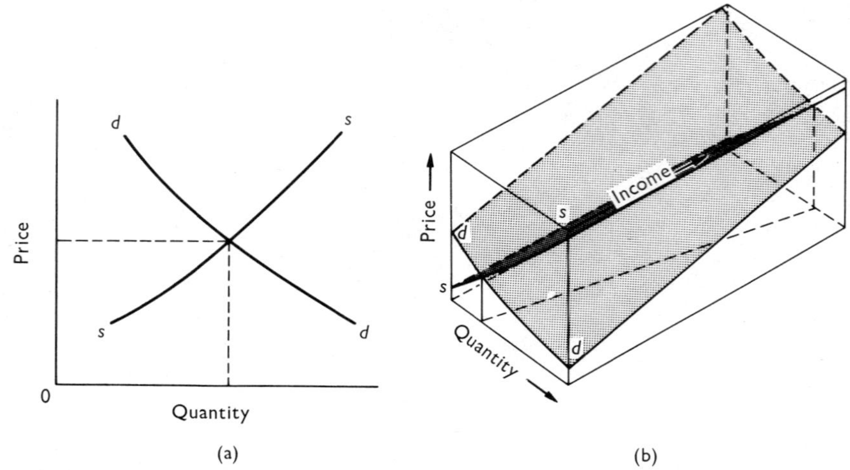

FIGURE 3.12. The effect of income: An increase in income tends to shift the demand curve upward.

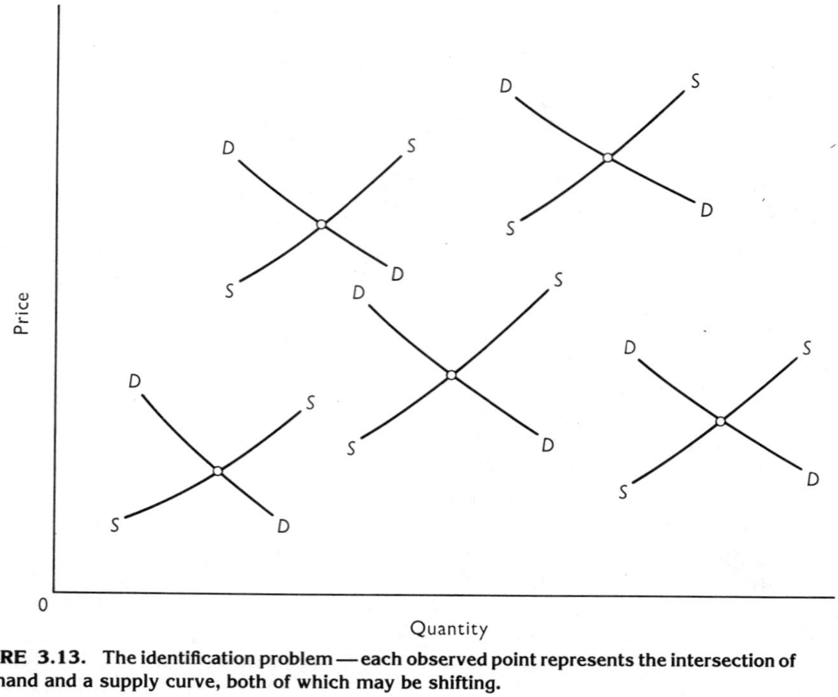

FIGURE 3.13. The identification problem—each observed point represents the intersection of a demand and a supply curve, both of which may be shifting.

with the task of identifying whether demand or supply factors, or both, are responsible for the shifting point locations.

The Identification Problem

This is the essence of the problem of identification that has troubled econometricians since research in quantitative economics began. There are, however, some occasions when the problem is not so confusing. Suppose, for example, one were reasonably certain that supply conditions had been unchanged over the same time span. If this were the case, then the changing points of intersection caused by shifts in the *demand* curve (provided, of course, that it did shift) would trace out the location of the *supply* curve, as in Figure 3.14.[7] And, of course, the reverse would also be true: during periods of constant demand conditions a shifting supply curve, by its intersections with the demand curve, would trace out the stationary demand curve.

But what if both shift? Then things get difficult! To solve this problem requires more sophisticated analytical techniques and additional information; from the data given there is no means of "identifying" whether these prices were produced by shifting demand curves or shifting supply curves—or both. Such means exist, but any detailed development would take us considerably beyond the scope

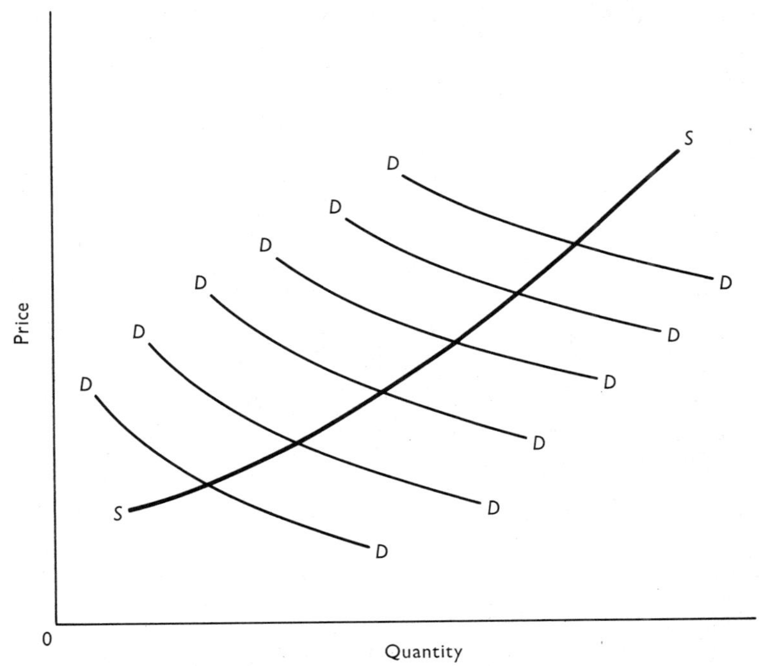

FIGURE 3.14. If supply is stable, while demand shifts, the observed points indicate a supply curve.

[7]Look again at Figure 3.10. Does this resemble data describing a demand curve or a supply curve? We'll take another look in the next chapter.

of this book. Most approaches call for the simultaneous solution of systems of demand and supply equations (Adams and Haynes, 1980; Lange, 1981; Luppold, 1982).

Does this mean that the concepts of demand, elasticity, supply, and equilibrium price are of little use to anyone except the research-oriented forest economist? No! But it does mean that we must think carefully about the factors that can shift the demand—and supply—curves. We'll look at supply shortly, but we can go considerably further with demand. So, besides population and income, what factors affect the demand for lumber?

First, as stated before, we must keep in mind that the demand for lumber, as with most forest products, in reality is a derived demand—derived from the demand for the products into which lumber is manufactured or fashioned. Hence, factors affecting consumption of these products will also influence consumption of lumber. There are, literally, hundreds of these, but we can identify the major ones.

In 1976, over half of the lumber consumed was for housing of some kind: 39% for constructing new units, and 13% for maintaining and repairing existing housing. Fully 16% was used for shipping—for boxes and other containers, for dunnage, pallets, and so on. Nonresidential construction and manufacturing took about 10% each, while all other uses combined took the remaining 11% (USDA, For. Serv., 1983).

Clearly, anyone involved in producing or marketing lumber or the trees from which it is obtained will be affected by the actions of the construction industry, particularly residential housing. That new housing is an industry of substantial fluctuation is well demonstrated by Figure 3.15. Partly because it is a major form of

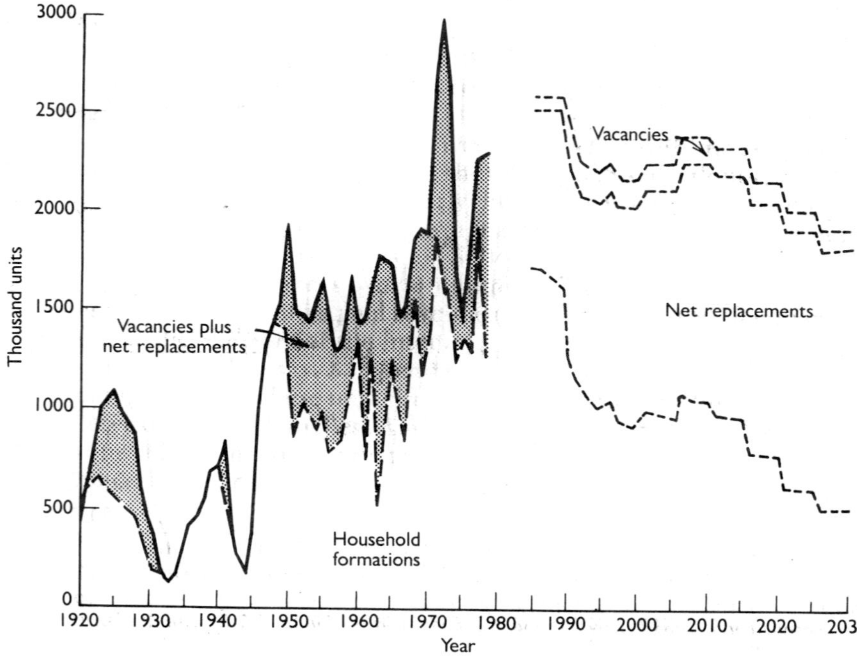

FIGURE 3.15. Demand for new housing by source of demand, 1920–1979, with projections to 2030. *Source*: USDA Forest Service, Forest Resource Report No. 23, Fig. 3.3.

investment in the United States, partly because it is definitely cyclical in nature, and partly because it plays such a significant role in employment, housing has been the subject of much study. Zivnuska's early work (Zivnuska, 1952) demonstrated the tie of lumber to the construction and housing cycle and the consequent cyclical nature of the lumber market. A long series of Forest Service studies of wood consumption have continually emphasized this interdependence.

As a result, any shock (either positive or negative) to the housing or construction industry is almost sure to be transmitted to lumber markets. The most recent example was the unprecedented increase in interest rates between 1979 and 1984. Loans for house buyers became prohibitively expensive as mortgage rates climbed, from 4½ to 5% in the 1960s to 9% in 1975 and to 12% in 1979. By 1981 fixed rate mortgages were rarely available at less than 17½%. At the same time prices for new houses climbed as inflation became a major national problem. The result was the most precipitous drop in residential construction—and in lumber production—since the Depression.

The repercussions in forestry were drastic. Few doubted that the major direct cause was high interest rates.[8] This belief was strengthened when interest rates dropped (briefly) in 1982, permitting mortgage contracts to be written for 12½% or less; both housing and the lumber industry spurted ahead, outpacing recovery of general manufacturing. But when rates climbed again, lumber prices resumed their slide and so did the lumber industry. Interest rates, and especially mortgage rates, can be major factors in determining the demand for housing and, consequently, the demand for lumber.

But interest rates are only one of many factors affecting the relationship between housing demand and use of lumber. Other important ones are:[9]

Type of house: Single family, multifamily, or mobile home. In the 1950s over 85% of new houses were single family types—which use more wood than others. By 1969 only 45% were single family, but by 1977 this type again accounted for two-thirds of all new dwellings.

Size of house: Partly a function of income, the average house size (sq. ft. of floor space) has been increasing.

Construction technique: Architectural design and construction method influence amount of lumber used. Plywood, for example, has almost completely replaced wood as a sheathing material; carpeting has replaced much of the hardwood flooring, and so on.

Number of households: Changes in lifestyle, in the age structure of the population, and in income all strongly affect the number of separate households seeking dwelling places. These same factors also affect the size of the house desired (it has been increasing), whether the unit will be single family, multifamily, mobile homes, and so on.

Vacancies: The number of units available for rent or purchase, or the existing stock of housing. Included in this are second dwellings—vacation homes, second apartments, and so on.

---

[8]In 1981–1983 North America, along with the rest of the industrial world, was experiencing a severe recession—a difficulty not entirely divorced from either high interest rates or inflation.

[9]Most of this information is from the U.S. Forest Service publication: *An Analysis of the Timber Situation in the United States 1950–2030.* (Forest Resource Report No. 23, 1982).

Replacement: Houses eventually must be replaced, due to loss through fire, flood, windstorm, to shifts in population, delapidation, etc. Between 1970 and 1975, replacement averaged slightly over 500,000 units annually.

Demand Estimation

By analyzing these and other factors, the Forest Service has projected the demand for new single family units to average about 1.7 million per year during the 1980s, then gradually decline to about 1.2 million. Multifamily units are anticipated to increase to about 570,000 per year before declining in the late 1980s and 1990s. Mobile home demand is projected to remain stable.

As pointed out previously, housing is only one of the major uses of lumber, and Forest Service researchers examine each use in much the same fashion as that described here for lumber. Combining these estimates of future use yields an estimate of the nation's prospective need for lumber, and this permits deriving estimates of the quantities of sawtimber needed. Such analyses play a highly significant role in planning the management of our National Forests as well as providing a basis for forest policy. And through using essentially the same data, our major lumber corporations also plan their operations—their investments in machinery and equipment and in forest land and silvicultural practices. Only if the long-term outlook for lumber demand is strong will investment opportunities be searched for and taken.

Estimating changing demands is particularly important for many developing nations. International agencies often use very general estimates of income elasticity and per-capita consumption to forecast the need of particular areas or nations for timber products. An example provides the best explanation.

Suppose an agency such as AID was considering financial assistance to permit establishing a stronger National Forest system in a friendly and needy nation. One of the questions that should be raised concerns that nation's future forest product needs, against which its production potential might be weighed and the desirability of the assistance logically determined.[10] The data available show the country has a present population of 5 million growing at 3% per year, a per-capita income of about \$400 that is expected to increase at 3% annually, and a current annual consumption of 300,000 cubic meters of sawnwood (0.06m^3, or about 25.4 bd. ft. per capita). How much sawnwood will be needed in 15 years—the estimated time it will take to get forest management in place?

In 15 years this nation will have nearly 7.9 million people. If incomes do rise at 3%, per-capita income will be at about $\$400(1.558) = \$623$. What will this mean for sawnwood consumption? If income elasticity of sawnwood consumption is around 0.8—a figure often used in such situations—then per-capita consumption would increase by $623/400 \cdot 0.8 = 25\%$, or to $0.06 \cdot 1.25 = 0.075\text{m}^3$, and the new requirements for the country would be $.075\ (7{,}700{,}000) = 577{,}500\text{m}^3$. This means an *increase* of $277.5/300 = 92.5\%$—practically a doubling of the present sawnwood consumption in the 15-year period.

This estimated need can now be compared with the potential forest production to provide some basis for determining the amount of help this nation could

[10]International assistance for forestry projects is not always logically determined, although some UN agencies make this type of analysis. One must also recognize that governments occasionally provide financial assistance with objectives quite different from those stated in official news releases.

use—providing, of course, the other needed political and institutional factors are in place.

Although an understanding of demand analysis per se can help the forest manager on occasion, the real role of demand analysis can only be played when demand is teamed with its partner—supply. Demand is only one blade of the pair of economic shears. We need now to examine the other blade, then fit them together as a single analytical tool.

QUESTIONS AND PROBLEMS

1. A survey of contractors and other homebuilders indicates the following quantities of Douglas fir plywood ($\frac{3}{4}$ in., AC exterior) could be sold over the coming year at the prices listed below.

Prices per 4 × 8 ft Sheet	Number of Sheets
$32	4,000
28	6,000
24	10,000
20	12,000
16	14,000
12	16,000

a. Plot the demand curve indicated by the data.

b. Calculate the (arc) price elasticity if prices change from: (i) $22 to $18 (ii) $32 to $28

Determine the equation for the demand curve, then

c. Calculate the price elasticity at the points $P = \$20$, $P = \$30$, and $P = \$12$.

d. If price is set at $20 per sheet, what would the combined gross revenue from plywood sales be for the four local lumberyards that constitute this market?

e. What is the total revenue function (the equation)?

f. At what point does elasticity of demand with respect to price equal −1?

2. Using the above data, suppose the number of families in the market increases by 50%, and that tastes in housing, family size and age, income, and so on all remain unchanged.

a. Plot the new demand curve.

b. What is the elasticity of the new curve at $P = \$20$?

c. Do you suppose the price schedule would remain as given? If you expect it would change, why?

3. Interested in expanding their overseas operations, planning officers of a certain paper corporation have calculated the income elasticity of paper consumption in Cocala, a small developing country, to be about 2. Population of Cocala is currently estimated to be 4 million, and, at its present growth rate, is expected to be 5.6 million in 10 years. Recent oil discoveries have stimulated Cocala's economy and officials now estimate that per-capita income, currently $250, will increase by $125 over the coming decade.

At present, all paper is being imported and there are no import restrictions of any kind on paper. Current per-capita paper consumption is estimated (from import data) to be 10 kg (less than 4% of U.S. per-capita consumption).

Assuming there are no limitations on supply or labor, and that the stated conditions remain unchanged over the coming decade:

a. Calculate the expected annual paper consumption in Cocala 10 years from now.

b. Do you believe the paper corporation should invest in building a mill in Cocala?

c. How large a mill would you recommend?

d. What other factors would you want the company to consider before deciding the build?

(**NOTE:** Save your answers to this problem. We will have a similar one later on after you have become familiar with other economic considerations.)

chapter four

PRODUCTION AND SUPPLY: PROFIT AND RENT

Chapter 2 began with a review of how an equilibrium price in a free market results from the interaction of buyers with sellers, both attempting to maximize their own utility (or satisfaction, or return) through a market exchange. The essence of economic demand—the subject of Chapter 3—is found in the *buyer's* response to a change in the market price. The analogous concept for sellers is supply, for a supply curve shows the *seller's* response to market price changes. Before a product can be sold, however, it must be produced, and it is with production of the good—lumber, plywood, recreation, or whatever—that supply begins. The chain of development runs from production, to production costs, to cost changes (marginal costs), to supply. Then, given the market price, the firm's supply curve can be used to determine the profit-maximizing output level.

But the supply of a firm is not the same as the supply of an industry, so this must also be determined. And to develop the industry supply when some or all the firms can change their productive capacity takes one into long- and short-run differences—which eventually lead to economies of scale and to external economies and diseconomies, now generally called externalities. Finally, the returns to land and labor will be defined and discussed, after which we can proceed with Chapter 5 to see how industrial structure might affect industry prices. We begin with the production process, for production relationships determine economic supply.

PRODUCTION

Production will be defined as the process of combining natural resources, labor, and equipment (capital) to produce specific products for sale or use. A study of timber production, therefore, requires an examination of how forest land is made to produce sawlogs, pulpwood, or other wood material by using labor and capital available to the producing unit. Studies of wood-based industries require a similar approach. Initially, we will assume that the firm's objective is the relatively simple one of maximizing net revenue (profit), the positive difference between total revenue (income) and total costs (expenditures). To further simplify the problem we will begin with a firm that produces a single product.

Production Functions

Production functions describe a relationship between inputs and outputs. **Inputs** are anything used in the production process that has a cost to the firm; **outputs** are anything the firm produces for sale or internal use.[1] Specification of a production function requires that the output (product) be identified and measured, and that we indicate both the process and the materials (inputs) that it requires. One can easily see why production functions are often referred to as engineering functions. In many instances, they may be viewed as a recipe for producing the desired product.

One of the least complicated types of production functions describes how the quantity of a product changes as the amount of just one factor is changed. Suppose, for example, we plan to produce one-year-old loblolly pine seedlings and want to know how the number of plantable seedlings will be affected by the amount of fertilizer used. To determine the optimum application level we might experiment by applying measured amounts of fertilizer to a series of test areas on our nursery site and recording the resulting yields. Everything else — soil type, watering schedule, seed variety, and so on — would be held constant. These are our **fixed** factors, while the fertilizer (always of the same kind) is the **variable** factor. A graph of the results would probably resemble Figure 4.1, a typical biological production function or yield curve — usually called a total product *(TP)* curve by economists.

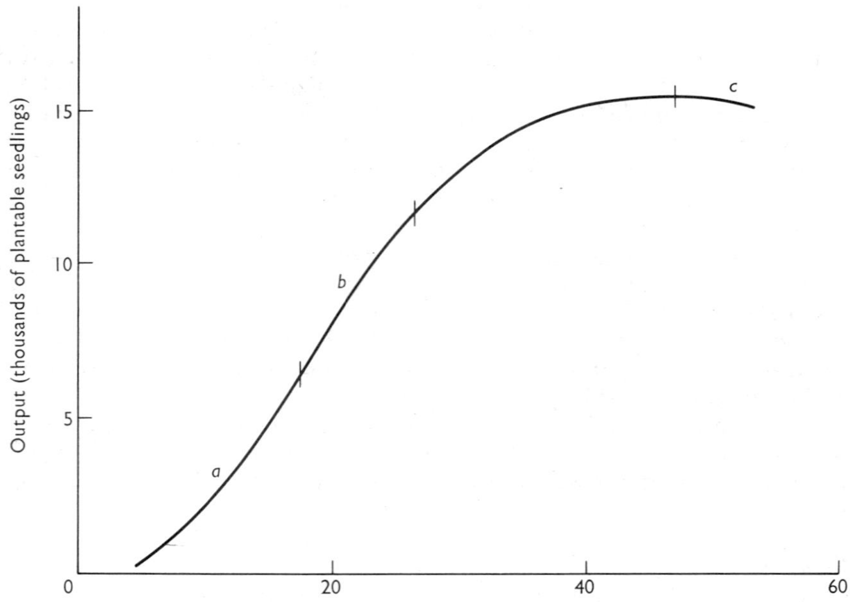

FIGURE 4.1. Production function: the relationship between input (fertilizer) and output (seedlings).

[1]Many production processes also produce outputs not intended for sale or internal use. These are usually referred to as externalities and can have either positive or negative values: air and water pollutants are negative externalities; flood control, watershed protection, and wildlife are often positive externalities of forest production. More will be said about these in later chapters.

This graph simply shows, quantitatively, how increasing amounts of fertilizer affect seedling production under the conditions established for our experiment.[2] Apparently, the test areas we used lacked some essential nutrient, for without fertilizer no plantable seedlings were produced. However, this need not be true for all cases: production functions don't *have* to start at the origin.

Average and Marginal Product

From these same data two other curves, or functions, can be derived, called the **average product** *(AP)* and the **marginal product** *(MP)* **curves**. The relationships between these two curves and the total product curve are shown in Figure 4.2. Note the change in the vertical scale used for the average and marginal curves—a change made to accentuate the curve forms and their points of maxima or intersection.

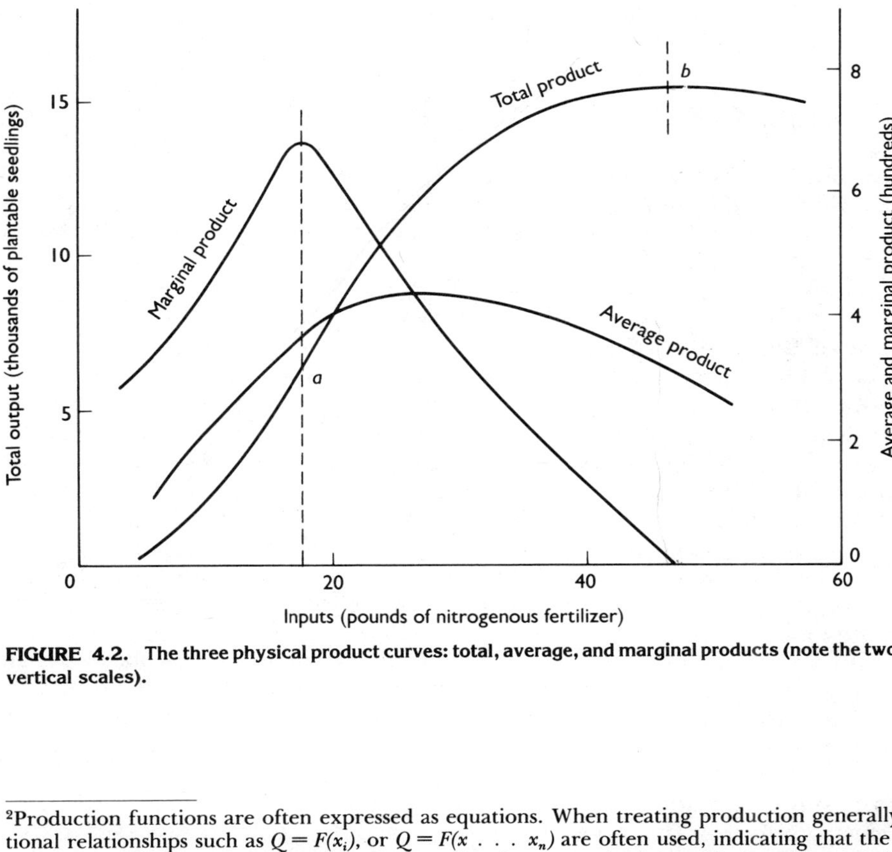

FIGURE 4.2. The three physical product curves: total, average, and marginal products (note the two vertical scales).

---

[2]Production functions are often expressed as equations. When treating production generally, functional relationships such as $Q = F(x_i)$, or $Q = F(x \ldots x_n)$ are often used, indicating that the output quantity (Q) is a function of n inputs. The equation form may also be specified as (for example)

$$Q = F(C + b_1x + b_2x^2 + b_3x^3)$$

a simple polynomial of the third degree that often takes a curve form resembling that of Figure 4.1. Logarithmic curve forms are much used in research, in part, because they are easily fitted by ordinary regression techniques. Finally, a specified equation can be used, as will be done with the total and average cost curves.

Average product curves describe the average output per unit of input for all tested quantities of the variable factor. It can be established arithmetically by dividing the total product by the input required to produce that amount.

$$AP = \frac{\text{Total product}}{\text{Required input}}$$

The amount of the variable factor needed to produce the maximum average product may be located graphically by drawing a line from the origin tangent to the total product curve. Beyond this maximum point (a) additional quantities of fertilizer will reduce the average output per unit of input. The average product will reach zero only when the total product is zero—a result that is entirely possible but scarcely desirable in a forest nursery.

Marginal product curves show the change in output that accompanies the change in input quantities. Mathematically, the marginal product is the ratio of these two changes and is the slope of the physical product curve.

$$MP = \frac{\text{Change in output}}{\text{Change in input}}$$

The marginal product curve rises only as long as the yield *per unit of input* is increasing—that is, up to the point of inflection (b) of the total product curve. Beyond this point the marginal curve drops (as the rate of increase in the total product decreases), crosses the average product curve at its maximum, becomes zero when total product is a maximum, and is negative thereafter (when total product is decreasing).

The Law of Diminishing Returns

The total product curve of Figure 4.1 also provides a graphic illustration of a major constraint nature places on all production, a constraint so universal it is called the **Law of Diminishing Returns**. The important part of the curve is the decreasing phase—part *b* in the figure. The increasing phase (part *a*) may be very small or even nonexistent. As long as the conditions are met—as long as one continues to increase the quantity of some one factor or group of factors while holding the quantity of some other factor(s) constant—there is no exception to this "law" in the physical world: there will *always* be a decreasing phase.[3] As stated decades ago, if this were not so, it would be possible to produce enough food to feed the world from a single flower pot.

Production functions for forest products tend to be more complicated than the one just described; more variables are involved, often several products may be produced at once, and sometimes it is impossible to control the quantities of some variables, or even to measure those quantities with any precision. For example, even though great care is used in establishing a plantation, rainfall, sunshine, insect pests, and similar variables don't always behave as expected.

Most production functions should be regarded as boundary curves in that they describe the maximum output (the upper boundary of output) the specified input levels will yield. Of major significance to forestry applications is the fact that frequently time is not explicitly considered: as in the present seedling example,

[3]The Law of Variable Proportions is simply a restatement of the above in terms of changing *proportions* of variable to fixed factors.

production is treated as if it were instantaneous. Time will be incorporated later, but for now we will continue with the seedling example.

Short Run and Long Run

Engineering production functions are peculiarly short-run phenomena. Although sometimes the short run is defined as a period of time during which no changes can be made in plant capacity, short run and long run are best treated as purely intellectual concepts, completely divorced from clock or calendar time. A very general definition, particularly useful to forestry, relates the terms to the presence or absence of fixed factors: a short run is always characterized by the presence of some fixed productive factor; a long run is devoid of any fixed factor. This difference reflects a peculiarity of forest economics, for in this field a "clock period" of as much as a century can be a short run for analytical purposes. The production cycle for sawtimber from some species, for example, can be 100 years or even more—scarcely a short run by usual standards. But if one of the productive factors is fixed, such as the area of the forest concerned, we are dealing with a short run problem.

It should now be clear why the production function is a short-run phenomenon; this function describes the quantitative response (output) generated by varying the amount of certain inputs (the variable factors) that are applied to a fixed factor or a group of fixed factors, and, although the function may specify only the variable factors, a fixed factor is always implied.[4] Note, however, that the output level at which the manager should produce has not been identified.

COSTS AND REVENUES AS A FUNCTION OF OUTPUT

Costs

Suppose our fertilizer costs 40¢ per pound. Simply multiplying the weights indicated along the horizontal axis of the graph (Figure 4.1) by .40 generates a new scale, this time of variable costs. And when using this new scale, the total product curve could be taken as the **total variable cost curve**: output becomes a function of the cost of the single input factor, fertilizer.

Total Variable Cost

For a variety of reasons, economists like to make cost a function of output, even though this cost arises directly from the input. To express costs in this fashion requires that we rearrange the axes of the diagram. As a first step we could rotate the whole diagram counterclockwise 90°: starting with *a* in Figure 4.3, turn the diagram to the position shown in *b*. Now observe its mirror image—or look at the diagram from the back side of the page (through the paper). What one sees is a total variable cost curve like that found in virtually all price theory texts. This curve is reproduced in diagram *c*.

[4]Basically, this is the reason for the usual assumption that production functions are "homogeneous and of the first order." A doubling of the right side of the equation (the inputs) results in a doubling of the left (output) side, which, of course, is only true if the fixed factor is also doubled. The same idea is sometimes expressed by assuming constant returns to scale. Thus a function such as that shown in Figure 3.1 is applicable to a 1 square meter plot, or a 50-acre nursery. If the inputs are scaled to the fixed factor, outputs will be proportionately scaled.

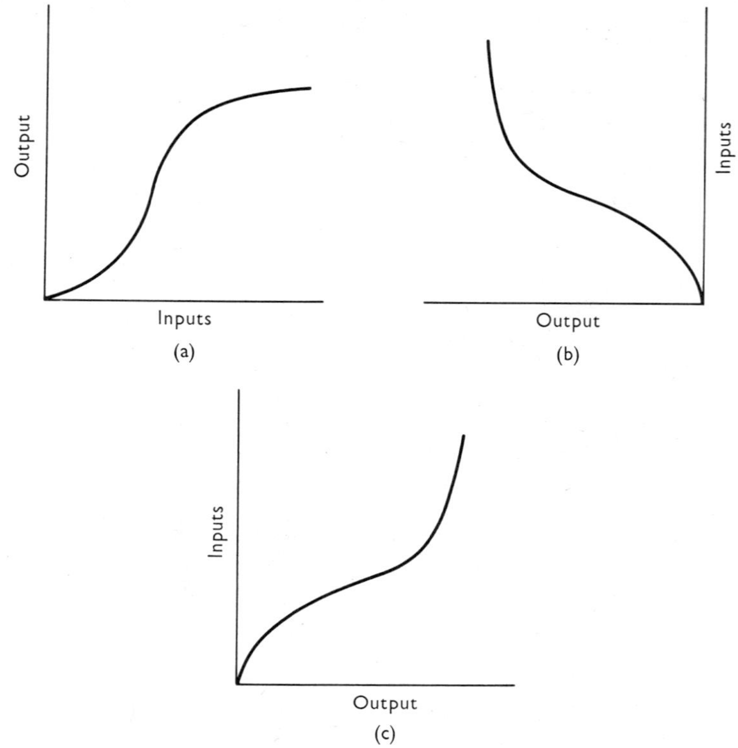

FIGURE 4.3. The relationship between the production function and the variable cost curve: *(a)* depicts a production function; *(b)* shows this rotated 90 degrees; and *(c)* is the mirror image of *(b)*, which, when scales are appropriately changed, is the total variable cost curve.

It is difficult to overemphasize the relationship between production functions and cost curves. Any change in the production function—a new technology, a different species or adoption of a genetic strain that grows faster, a different form of fertilizer, or even a new hand tool that makes work easier—may significantly shift the cost curve of a production process.

Cost curves can also be expressed mathematically, and it is often desirable to do so since this permits broadening the cost concepts to include as many variable factors as used in the production process.[5] After converting the quantity of each variable factor to its cost, the costs of all inputs are summed at each output level, thus giving the total variable cost of the product at each level of production.[6] As an

[5]Students often ask "Why wasn't the equation for the production function given so the process could be followed from beginning to end as a set of equations?" The difficulty is mathematical: for equations other than straight lines the process of reversing variables is complex, and for many equations beyond a simple quadratic there is, literally, no solution!

[6]It is important to recall here that production functions are "boundary" curves. Our cost curve, derived from the production function, is thus also a boundary curve: every point on the curve is the minimum cost possible for that particular output when using the techniques described by the particular production function.

example we will take a set of costs representative of a small pallet operation in the Lake states, beginning with the total variable cost equation (in 1000s) of

$$TVC = 9.75Q - .225Q^2 + .0025Q^3$$

which is shown in Figure 4.4.

Fixed Costs

Bringing the group of fixed factors into the cost picture poses no problem. By definition, these fixed factors have no effect on the quantity produced. Consequently, fixed costs are represented graphically by a horizontal line at the appropriate level—in our pallet example, $100,000. The equation for total fixed costs therefore becomes

$$TFC = 100$$

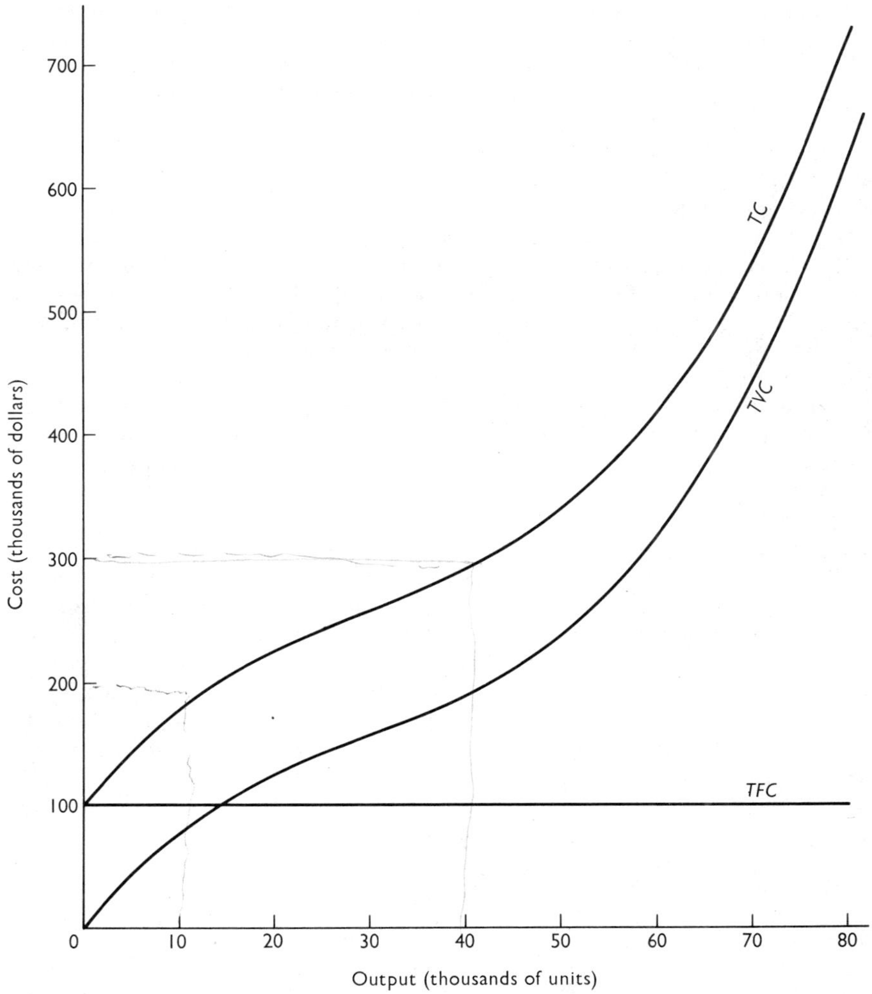

FIGURE 4.4. A set of total cost curves.

Now the two cost curves can be added together to form the Total Cost Curve (TC), also shown in Figure 4.4. Clearly, all that has to be done graphically is to raise the entire cost curve by \$100,000. Deriving the equation for the total cost is equally simple, since we need only add the two equations.

$$TC = 100 + 9.75Q - .225Q^2 + .0025Q^3$$

To determine the optimum level of production (the level that will maximize net revenue) we need one further set of information—the revenue side of the picture.

Revenues

Total revenue can also be represented as a function of output. The relationship here is direct, for revenue is generated through the product sale. We will assume that our producer's output will not be large enough to affect the price received in the market. Total revenue will plot as a straight line from the origin, since each pallet produced sells for the same price. Algebraically, $TR = P \cdot Q$, where P is the pallet price and Q the quantity sold. At a price of \$8 per pallet, the equation for total revenue becomes $TR = 8Q$ and is shown in Figure 4.5. Note that the slope of the total revenue curve—the tangent of the angle in Figure 4.5—will be equal to the price of the product.[7]

The cost and revenue parts of our problem can now be combined by superimposing the revenue and cost curves (Figure 4.6). The break-even point, often used by engineers, is at b in the figure, and the reason for the name is immediately obvious.[8]

To find the profit-maximizing production level we need to locate the greatest vertical distance between the total revenue and total cost curves, making sure *net revenues* are positive (negative net revenue is shaded in the diagram). A bit of geometry helps here: the point of maximum vertical distance between any two curves is located where the curves are parallel at a given value on the horizontal axis. In the example, since TR is a straight line, this point can easily be found by drawing a line parallel to TR and tangent to TC. The point of tangency on TC identifies the output that maximizes net revenue—an output of approximately 56 (thousand) pallets.

Net Revenue and Profit

We will use the terms net revenue and profit interchangeably, although the second word is subject to misinterpretation. One should recall that fixed costs include whatever return on invested capital is believed necessary to keep that capital employed in this particular production effort. Hence the net revenue (profit) of Figure 4.6 is a return over and above that necessary to pay the operator's wage or salary, to yield the required rate on all pertinent investments, and to make the necessary payment for anything owned by the entrepreneur and being used to produce the product. It is, in fact, true economic profit.

[7]Since $TR = P \cdot Q$, the tangent of ϕ will be

$$\frac{d\,(TR)}{d\,(Q)} = \frac{d\,(Q)}{d\,(Q)} \cdot P = P$$

whenever P is constant.

[8]Note that if the cost function were linear, as it would be if the production happened to be of a kind performed with machines yielding constant costs per unit of output (automatic stamping machines, for example), then the straight line of total cost would intersect total revenue at only one place.

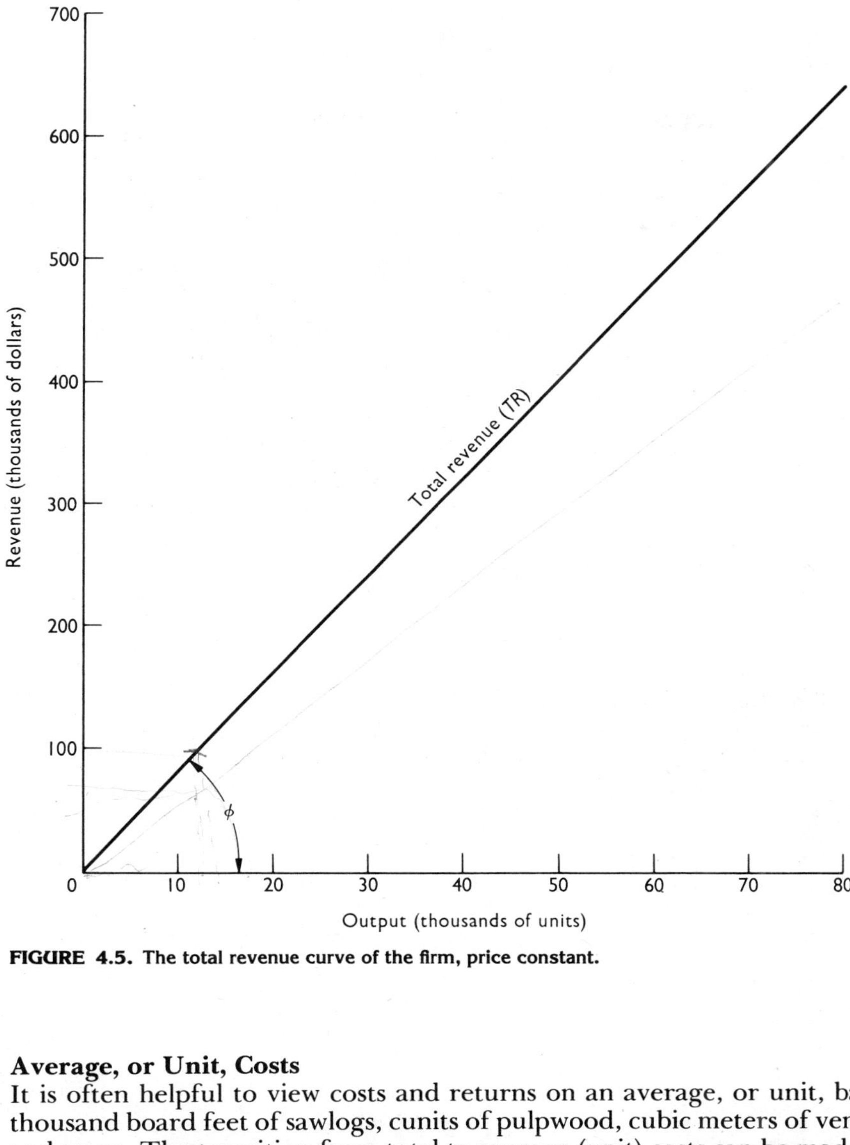

FIGURE 4.5. The total revenue curve of the firm, price constant.

Average, or Unit, Costs

It is often helpful to view costs and returns on an average, or unit, basis—per thousand board feet of sawlogs, cunits of pulpwood, cubic meters of veneer bolts, and so on. The transition from total to average (unit) costs can be made directly.

Since we have three total cost curves, three average cost curves can be derived: one for fixed costs *(AFC)*, one for variable costs *(AVC)*, and one for total costs *(ATC)*. The relationship between them, and their derivation from total costs, can be seen in Figure 4.7.

Average fixed cost can be defined by

$$AFC = \frac{TFC}{Q}$$

and, viewed geometrically, is simply the tangent of the angle ϕ (Figure 4.7a) where $Q = q$. If one visualizes the behavior of this tangent as q is shifted outward from the

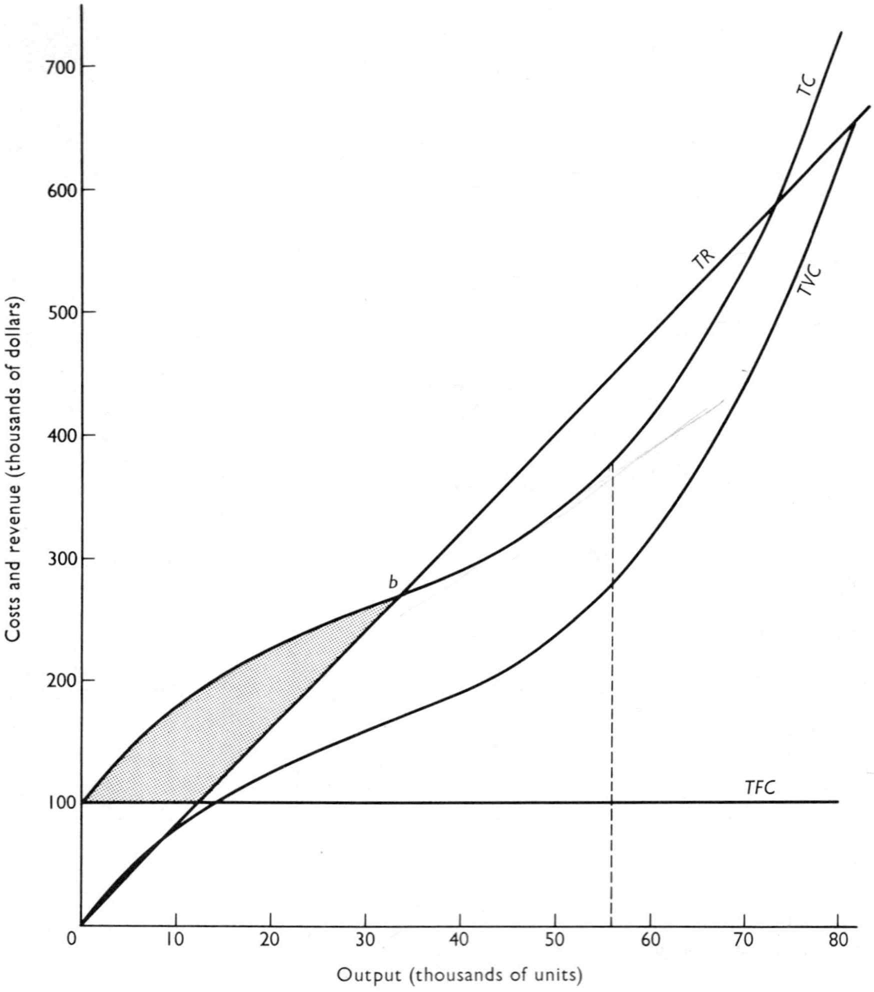

FIGURE 4.6. **Total cost, total variable cost, total fixed cost, and total revenue (for the equations of the curves, see text).**

origin, it becomes clear that it will drop from a high value to a low one—from infinity to a figure approaching zero. The average fixed cost, therefore, is a constantly declining curve, as shown in Figure 4.7*b*.

Similarly, the average variable cost and the average total cost curves can be defined as

$$AVC = \frac{TVC}{Q} \quad \text{and} \quad ATC = \frac{TC}{Q}$$

respectively, and these two ratios also represent tangents of angles. In these cases, however, the angle is one formed by the horizontal axis and a line connecting the origin to a point on the *TC* or *TVC* curves—line *Ob* in Figure 4.7*a*. The behavior of these tangents as the point moves along the particular curve is helpful in understanding the relationships between the two curves. At first, as the point moves out (as Q increases), the angle, and hence its tangent, declines (angle B is an example). Eventually, a point is reached on the two curves, TVC and TC, where, as Q increases, the angle and its tangent *increase*. Since the two ratios TVC/Q and TV/Q are at once the tangents and the values of AVC and ATC for the particular levels of Q, the two average cost curves also at first decline, and then eventually increase, giving rise to the familiar U-shaped average cost curves of Figure 4.7*b*.

The mathematical derivation is equally simple. The equations of the three total cost curves of Figure 4.7 can be readily transformed to the equations of the three average cost curves by dividing each equation by Q. Hence

$$AFC = \frac{TFC}{Q} = \frac{100}{Q}$$

$$AVC = \frac{TVC}{Q} = \frac{9.75Q - .225Q^2 + .0025Q^3}{Q} = 9.75 - .225Q + .0025Q^2$$

$$ATC = \frac{TC}{Q} = \frac{100 + 9.75Q - .225Q^2 + .0025Q^3}{Q}$$

$$= \frac{100}{Q} + 9.75 - .225Q + .0025Q^2$$

The divisions accomplish more precisely the steps taken in the graphic derivation of the average cost curves of Figures 4.7*a* and *b*.

Similarly, total revenue curves can be converted to average revenue curves and added to the unit cost and revenue picture. In this case, starting with the

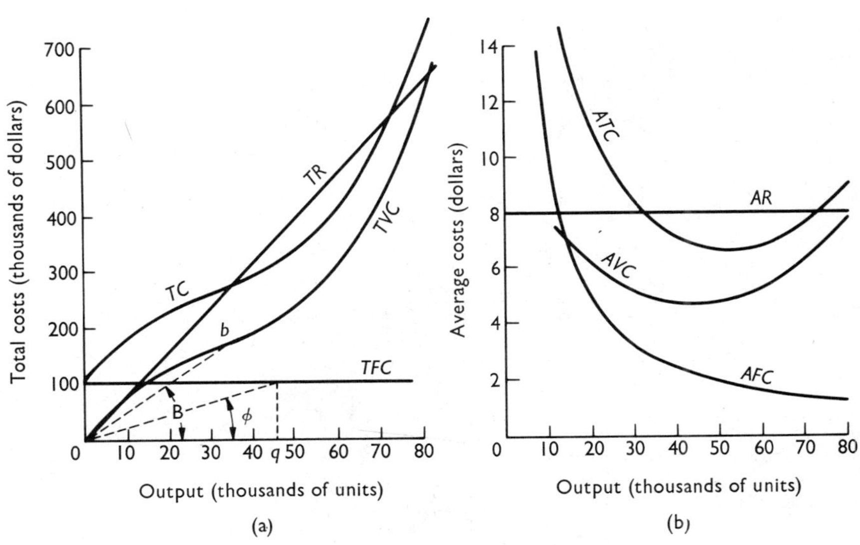

FIGURE 4.7. Derivation of average cost curves from total cost curves (for equations of curves, see text).

known relationship $TR = PQ$, the average revenue is derived by dividing by Q. In our example

$$ATR = \frac{TR}{Q} = \frac{PQ}{Q} = P = \$8 \qquad \text{(again, } Q \text{ is in thousands)}$$

Marginal Cost

Marginal cost is defined (at least for a start) as a ratio of changes:

$$MC = \frac{\Delta TC}{\Delta Q}$$

This ratio is seen in Figure 4.8 to be the tangent of the angle ϕ in the lower left corner of the small triangle, and it is also the tangent of the angle ϕ (formed by extending the hypotenuse of the small triangle until it intersects a prolongation of the base of the graph). Note that this ratio is also the *average slope* of the total cost curve as costs rise from c_1 and c_2, and output increases from q_1 to q_2.

As ΔQ is allowed to approach zero, the hypotenuse of the triangle approaches

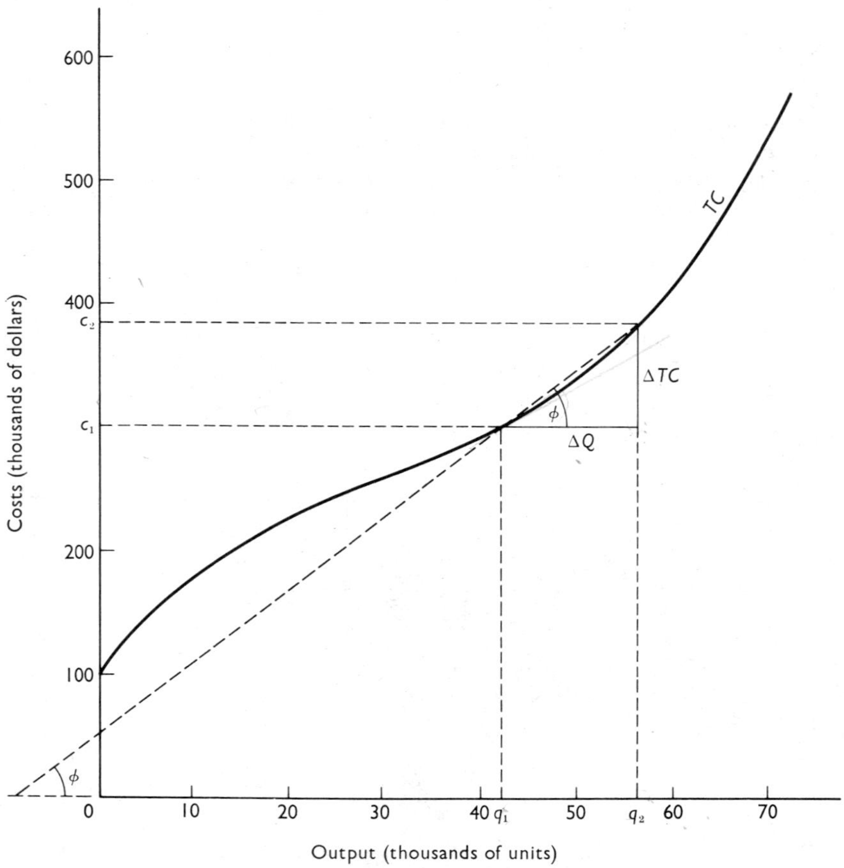

FIGURE 4.8. The derivation of the marginal cost curve from the total variable cost.

the true slope of the total cost curve at that particular point. As we move from a discrete to a point concept we find that

$$MC = \frac{\Delta TC}{\Delta Q} = \frac{d\,(TC)}{d\,(Q)}$$

Marginal cost is equal to the *change* in total cost that accompanies a *change* in output. It also equals the *slope* of the total cost curve at any particular point, which is given by the first derivative of the total cost function. Looking at marginal cost this way permits us to clear up several points.

1. When marginal cost is calculated as the discrete change in either total or variable cost (as in most classroom problems and many real-life situations), it should be plotted at the midpoint of the two quantities involved. For example, if costs were to change from (say) \$100 to \$110 as output changed from 45 to 50 units, marginal cost per unit could be calculated as

$$\frac{110 - 100}{50 - 45} = \frac{10}{5} = \$2$$

Since this would represent the average slope, or change, between the outputs of 45 and 50 units, it would be plotted over 47.5. Note that this is an approximation. Only in special cases would this average represent the actual slope of the total cost curve at its midpoint. It is, however, considerably more accurate than plotting over either the beginning or ending output level.

2. The marginal cost curve can also be derived directly from either the total cost or total variable cost equations. Since $TC = TVC + TFC$, and TFC is a constant, both equations will have identical first derivatives. Returning to the example used earlier,

$$MC = \frac{d\,(TC)}{d\,(Q)} = \frac{d\,(100 + 9.75Q - .225Q^2 + .0025Q^3)}{dQ}$$
$$= 9.75 - .45Q + .0075Q^2$$

and to find values for MC one needs only to insert the desired values for Q and solve. The resulting MC curve, together with all the average cost curves, is shown in Figure 4.9.

3. The MC curve always cuts the two average cost curves at their lowest points and the geometry shows why. Average total cost may be represented as the tangent of the angle formed by connecting the origin to some point on the total cost curve, while the average total cost curve was derived by moving this point along the curve as output changed, and then plotting this tangential relationship. Marginal cost, on the other hand, is the actual slope of the total cost curve—the tangent of the angle formed by the intersection of the horizontal axis with a line just touching, or tangent to, the total cost curve at the particular output. As the point connected to the origin begins to move along the total cost curve, the angle formed—and hence its tangent (which is, numerically, ΔTC)—is becoming smaller and smaller. Eventually, however, it starts to increase. This point of transition from a drop to an increase marks the low point of the average total cost curve, and here the line from the origin is *just tangent* to the total cost curve. Hence the slope of this line is also the slope of the total cost curve, and $ATC = MC$ at this point.

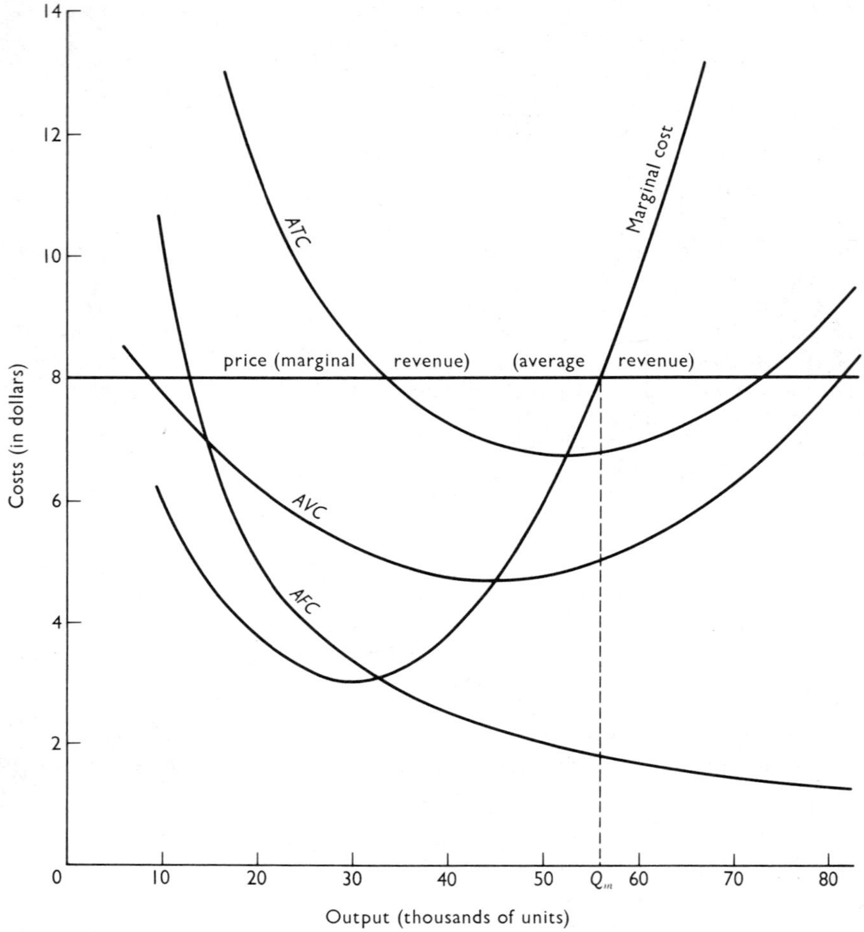

FIGURE 4.9. Marginal cost curve, together with the average cost curves and average revenue (or price).

Marginal Revenue

Marginal revenue may be defined as *the change in revenue that accompanies a change in output.* This ratio of changes may be expressed mathematically as

$$MR = \frac{\Delta TR}{\Delta Q} = \frac{d\,(TR)}{d\,(Q)}$$

which makes marginal revenue the *slope* of the total revenue curve (its first derivative with respect to output) at any selected output level. We have assumed our market price is constant, unaffected by how much is produced by our firm. Since $TR = P \cdot Q$, one can also see that

$$MR = \frac{d\,(TR)}{d\,(Q)} = \frac{d\,(P \cdot Q)}{d\,(Q)} = P \cdot \frac{d\,(Q)}{d\,(Q)} = P$$

and also because we have assumed it is constant, price will be equal to the *average* revenue. A three-way equality thus exists for our firm when operating at its profit-maximizing point: average revenue will equal marginal revenue which will also equal market price. This equality is one of the results of a purely competitive market, as we will shortly see. Marginal revenue in other types of markets will be treated in a subsequent chapter.

Marginal Revenue Product

The amount of revenue resulting from changing the amount of a single factor while holding all other factors constant is called the *marginal revenue product* of that factor. For example, suppose you could hire two additional workers each for $15,000 per year, including all fringe benefits, expenses, and so on. They will cut and bundle slabs that would otherwise be wasted at your mill. You figure the two will produce not less than 15,000 firewood bundles annually. If you can sell these at the mill for two dollars or more per bundle you should put the two to work: the marginal revenue product of this production factor—the two additional workers—would be 15,000 × $2 = $30,000 and this is exactly equal to their marginal cost.

The marginal revenue product is often used as a simple decision guide when it can be easily isolated and can also guide factor selection when several different factors can be used to change output—a problem discussed in detail in Chapter 6.

MAXIMIZING PROFIT

With both marginal revenue and marginal cost plotted on the set of average cost curves (Figure 4.9), the profit-maximizing output level can be identified by the intersection of the rising marginal cost curve with price. The profit-maximizing output of Q_m is exactly the same as the output determined from the set of total cost curves, for in Figure 4.9 the *slopes* of the total cost and revenue curves were plotted and the point of equality determined by their intersection, while in Figure 4.6 the output corresponding to equal slopes of the total cost and revenue curves was determined by constructing a parallel tangent.

Mathematical Maximization

It is often helpful (and at this time also serves as a summary) to view profit maximization as a mathematical problem. Given the equations for total cost and total revenue as functions of output,

$$TC = F(Q) \qquad \text{and} \qquad TR = F(Q)$$

we can define total net revenue as a function of output also, since

$$TNR = TR - TC$$

The output level maximizing *TNR* (profit) can now be determined by taking the first derivative of the net revenue equation with respect to *Q*, setting it equal to zero, and solving for the value of *Q*. (Remember that this procedure establishes *either* a maximum or a minimum.) Hence,

$$\frac{d(TNR)}{d(Q)} = \frac{d(TR)}{d(Q)} - \frac{d(TC)}{d(Q)} = 0$$

and

$$\frac{d\,(TR)}{d\,(Q)} = \frac{d\,(TC)}{d\,(Q)}$$

These derivatives, $d\,(TR)/d\,(Q)$ and $d\,(TC)/d\,(Q)$, are, respectively, the slopes of the total revenue and total cost curves and are the marginal cost and marginal revenue functions. To find maximum (or minimum) net revenue, locate the point at which the slope of the total cost curve equals the slope of the total revenue curve. This was done geometrically in Figure 4.6, while the equation system was solved graphically in Figure 4.9. Both solutions established the maximizing output as approximately 56,000 pallets.

The corresponding mathematical operation consists of setting the marginal cost equation equal to that of marginal revenue. In the example used here, since marginal revenue is equal to price, the equality becomes

$$8 = 9.75 - .45Q + .0075Q^2 \qquad \text{or} \qquad .0075Q^2 - .45Q + 1.75 = 0$$

Now, using the formula $Q = (-b \pm \sqrt{b^2 - 4ac})/2a$, where a, b, and c are the coefficients for Q^2, Q, and any constant, respectively, the two roots of this quadratic equation are found. In this case, both are positive, and the two corresponding solutions are

$$Q = 55{,}820 \qquad \text{and} \qquad Q = 4{,}180$$

The first solution, $Q = 55{,}820$, corresponds to that found graphically in Figures 4.6 and 4.9. The second solution, $Q = 4{,}180$, represents the output that would *minimize* net revenue (would maximize loss). This same conclusion could be reached from the graphs: in Figure 4.6 the equality of the two slopes at the smaller output is readily seen, but was avoided by choosing only positive values of net revenue. In Figure 4.9 the left arm of the (parabolic) marginal cost curve must be extended until it intersects the price line to find the output minimizing net revenue. Because two solutions do exist, we must add a condition to the statement regarding profit maximization: profit is maximized when marginal cost equals marginal revenue—*provided the marginal cost curve is rising* (or while the second derivative of total cost is positive).

Profit equals total revenue minus total cost, and can be expressed as

$$NR = 8Q - (100 + 9.75Q - .225Q^2 + .0025Q^3)$$

$$= -100 - 1.75Q + .225Q^2 - .0025Q^3$$

Now, substituting the maximizing output level of 55,820 pallets, we find that the maximum net revenue (profit) we can receive under the prices and production methods specified is \$68,560. If the second root of the equation (4.18) is used, the profit turns out to be −\$103,566—a substantial loss, due in large part to the \$100,000 fixed cost.

ALTERNATIVE APPROACHES TO COST AND REVENUE RELATIONSHIPS

Throughout this chapter we have followed the conventional approach of making costs and revenues functions of output. To deal with forestry problems, however, one should be familiar with two other approaches, the first making both costs and revenues functions of *inputs*, the second directly relating revenues to costs.

Costs and Revenue as Functions of Inputs

The starting point, as before, is the production function, but instead of using this relationship as a basis for making cost a function of output, the total physical product curve is converted directly to **revenue**. As long as the amount produced has no effect on price, the conversion can be accomplished by rescaling the vertical (output) axis in value terms—a process that requires nothing more than multiplying output by the price per unit (assumed here to be \$8). The curve remains unchanged, and total revenue thus becomes a function of *inputs* (Figure 4.10).

Using the same (vertical) dollar scale, the cost of the inputs can also be shown. Costs and revenue are *both* plotted as functions of input levels, and the problem of determining the optimum level of inputs and output directly tackled. Since total fixed cost *(TFC)* is, by definition, unchanged as either input or output changes, fixed cost becomes a horizontal line in a graphic representation such as Figure 4.10, its level fixed by the amount (here \$100,000). But there are several important differences between this approach and the previous one of making costs and revenue functions of output.

1. If (a) either a single variable factor is used, or the variable factors are always used in the same proportions, and (b) their prices remain unchanged, then variable cost *(TVC)* becomes a straight line with a positive slope equal to the unit cost of the factor (or group of factors).

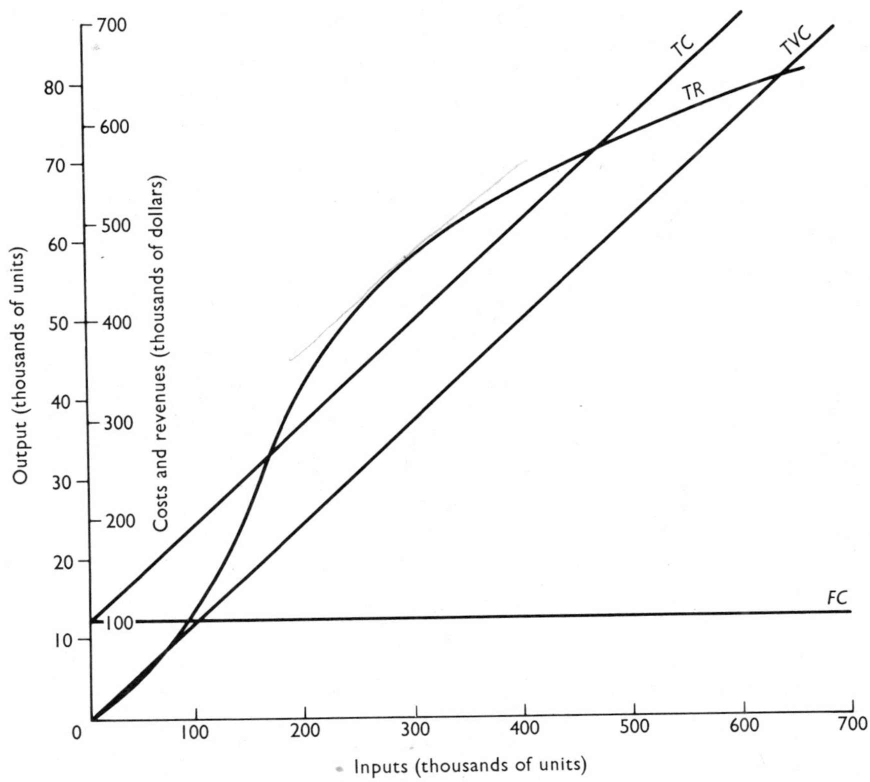

FIGURE 4.10. Total revenue and costs as functions of inputs.

2. Total Cost *(TC)* will also be a straight line, paralleling the total variable cost but shifted directly upward by the amount of the fixed cost.
3. Total revenue is no longer linear: it has exactly the shape of the production function.
4. As previously, net revenue (or profit) is represented by the area between the total revenue and total cost curves. The point of maximum vertical distance between these two curves identifies the profit-maximizing input level and is located by the input level at which the slopes of the two curves (*TR* and *TC*) are identical. A line parallel to the total *cost* curve, and just tangent to the total *revenue* curve (Fig. 4.10), will identify this point.
5. A complete set of average cost and average revenue curves can also be constructed in terms of input units, as can corresponding marginal cost and marginal revenue curves (Figure 4.11). Average variable cost per unit of input *(AVC)* and the marginal cost *(MC)* (here the change in cost as the *input* level is changed) will, in this formulation, be identical horizontal lines, their level determined by the cost (or price) per unit of input. Average fixed cost will be a constantly declining curve, as will average total cost, since it is the sum of the average variable and average fixed costs. Note that the average total cost will always lie *above* the marginal cost.

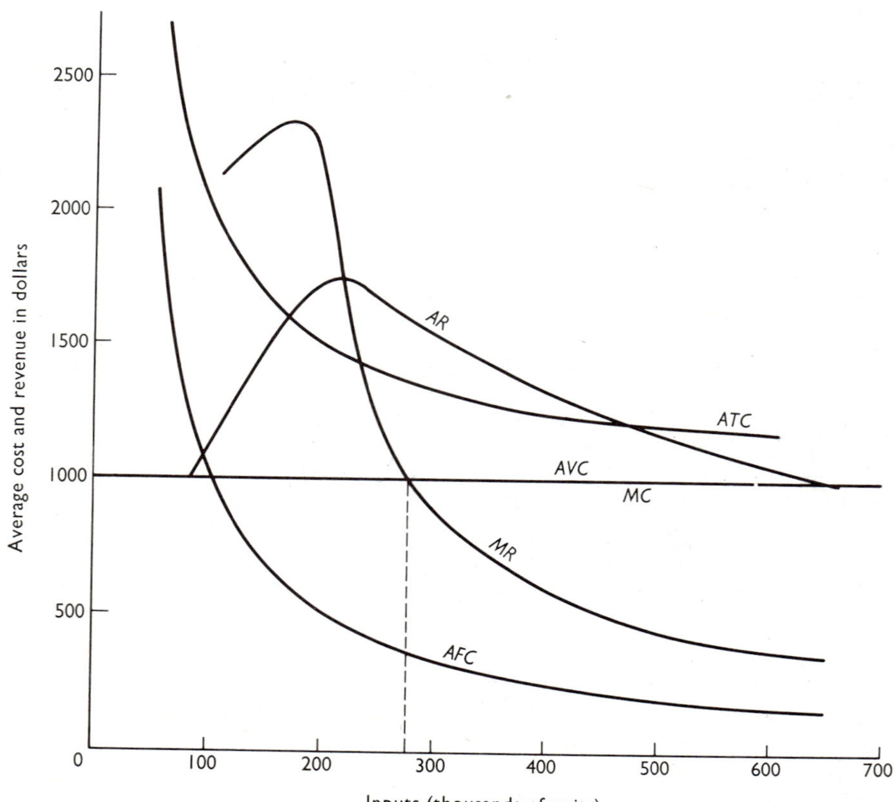

FIGURE 4.11. Average costs and revenues, and marginal costs and revenues, per thousand units of inputs.

6. In terms of input units, average revenue (total revenue/input units) and marginal revenue (the change in total revenue accompanying a change in *inputs*) now become curvilinear, their shape depending directly on that of the original production function. As long as prices are assumed constant, these curves (providing the scale is maintained) will be identical to the curves of **average and marginal physical product**, and are referred to as the curves of **average value product** and **marginal value product**.
7. The input level maximizing profit is again indicated by the intersection of the marginal cost and marginal revenue curves (Figure 4.11) and is identical to that obtained earlier.
8. When using this approach with a production function that has a parabolic shape and with constant product prices, the average revenue and marginal revenue curves are both straight lines with negative slopes. Moreover, the marginal revenue curve will always have exactly twice the slope of the average revenue curve. This approach is frequently used in forestry and wildlife problems, and unless one recognizes that these curves are based on taking *input*—not *output*—as the independent variable, more than a little confusion can result.

Cash Flow or Benefit/Cost Analysis

The third approach is in many ways the most interesting and potentially the most valuable to foresters since it provides the basis for both discounted cash flow and benefit/cost analyses. In the conventional framework (the first approach discussed), costs and revenues were made a function of output; in the second they were regarded, essentially, as functions of inputs. The third says "Since we are primarily interested in the relation between costs and revenues, why not look at this directly—why not plot revenue on the vertical axis and costs on the horizontal one?"

Figure 4.12 illustrates the use of this approach. Each production level incurs some costs and some revenues, although generally production will be zero until costs reach some minimum level. Both axes use identical scales and are in value units—in this case, dollars. The data are the same as that used in the two previous approaches—the Lake states pallet operation.

Since the horizontal and vertical axes are similarly scaled, a ray from the origin at a 45° angle contains all points of equal costs and revenues. The 45° line divides the space between the axes into two equal parts: revenues exceed costs for any point lying above the line, while costs exceed revenues for all points below. The **slope** of the curve at any point (Figure 4.12) is the change in revenue divided by the change in costs—which is the marginal revenue divided by the marginal cost. And when marginal revenue *equals* marginal cost, this ratio will be unity—which is the tangent of a 45° angle. Since net revenue is maximized when marginal revenue equals marginal cost, we can locate the point of profit maximization by determining the point on the curve having a 45° slope—where the curve is parallel to the 45° line we drew from the origin.

Now, if we regard all revenues as benefits, we have the basic framework of the usual financial benefit/cost analysis. The optimum point of operation for any "project" is that which maximizes net benefits, or $B - C$, and this is the point we have just located. Note that it is *not* the point maximizing the B/C ratio: that point can be located by drawing a line from the origin that is just tangent to the B/C curve, and will always indicate a somewhat smaller output and expenditure level than the one maximizing net benefits.

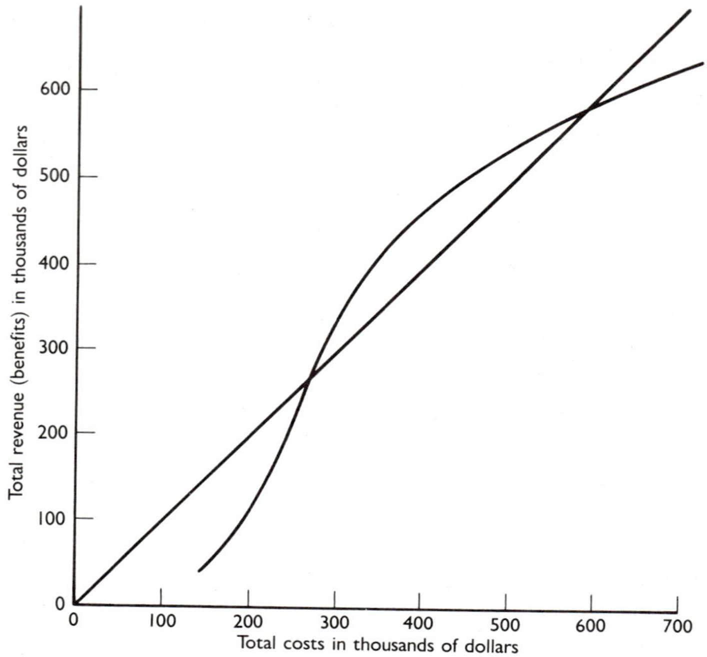

FIGURE 4.12. Revenues plotted over costs: the benefit/cost relationship (data basis same as that used in previous examples).

We will see in a later chapter that applied cost/benefit analysis is not always quite as simple as outlined here. Both costs and benefits, particularly those affecting society as a whole, are often difficult to measure and sometimes even difficult to identify. The basic analytical framework, however, is nothing more than a straightforward marginal analysis.

SUPPLY CURVES

Chapter 3 stressed the point that, when used in an economic sense, demand does not refer to any given quantity but rather to a specific functional relationship between price and quantity, showing the varying amounts of a given commodity or service that buyers are both willing and able to purchase at a given time and at different prices. The word supply has a parallel connotation in economics. Also a functional relationship between price and quantity, supply refers to the varying amounts of a commodity or service that producers are willing and able to supply at different prices to a market over a given time. The short-run supply curve for an individual firm shows its output schedule under a variety of prices if no significant changes in its fixed factors (its capital equipment) are made.

Again paralleling the demand concept, this functional relationship of supply can be represented by a schedule, by a graph, or by an equation. Table 4.1 is the supply schedule for the pallet producer of Chapter 4. The same data can be presented in graphic form (Figure 4.13) or by an equation, in this case $P = 9.75 - .45Q + .0075q^2$, where $Q \geq 45$ (outputs in 1000s, and output of minimum AVC = 45,000 units).

TABLE 4.1
Supply Schedule of the Firm (rounded to nearest 100 units of output)

Price (dollars)	Quantity (units)
4.70	45,000
5.00	46,300
5.50	48,300
6.00	50,000
6.50	51,600
7.00	53,100
7.50	54,500
8.00	55,800
8.50	57,000
9.00	58,300
9.50	59,400

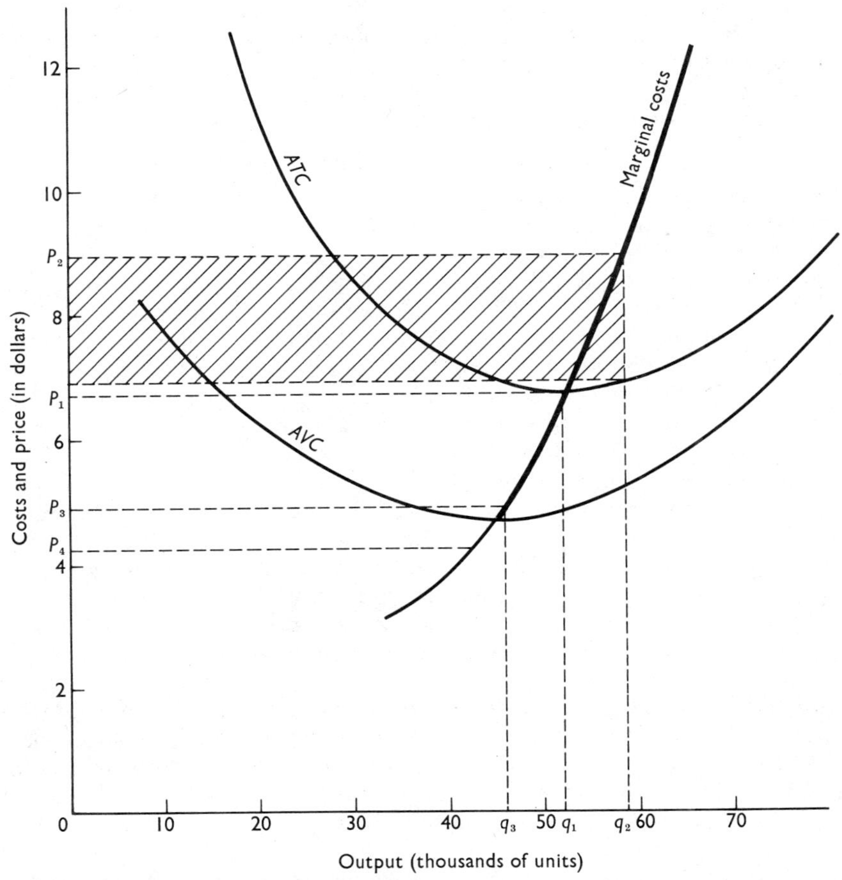

FIGURE 4.13. The short-run supply curve of the firm.

The Firm's Short-run Supply Curve

In the short-run, purely competitive case, the supply curve of the firm is that part of its rising marginal cost curve that lies above the minimum average variable cost, identified in Figure 4.13 by the heavier portion of the *MC* curve. The cost curves are again those of the pallet manufacturer. If price is P_1, then all costs are being covered: the producer is making an adequate return on his or her investment, is covering all salaries and all fixed costs, and is also meeting all other expenses. No *profit* is being made, but the producer will be content to remain in business. He or she is maximizing net revenue; it happens to be zero at this particular output level, but at any lower output level net revenue would be negative. If price rises to P_2, output will be increased to q_2. Here, not only will all expenses be covered but a profit will also be made, the amount of which is indicated by the shaded rectangle in the figure. Note that average total costs have risen, but because of the higher price the increased costs are more than covered.

If price should drop below P_1, the producer would be far less satisfied. At a price of P_3, for example, production would continue, but at the reduced level of q_3. By operating at this output level all variable costs will still be covered and there will be some return on fixed costs—not as much as the producer would like, but some return is better than none at all!

If price should drop to a level below the average variable cost—say to a level of P_4—production would stop entirely, thereby eliminating the necessity of meeting variable costs. True, there is no return on investment; in fact, the producer earns no production revenue whatever. But now there are no production *expenses* that must be met by cash outlays. Fixed costs continue, but less is lost by stopping production than by continuing to operate.

Industry Short-run Supply Curves

The industry short-run supply curve is simply the horizontal sum of the short-run supply curves of the firms making up the industry. Suppose, for example, the entire industry were composed of the four firms, *A*, *B*, *C*, and *D*, whose marginal cost curves are given in Figure 4.14. (Four firms would hardly meet the specifications for a purely competitive industry, but we will assume that each of these four producers *acts as if* he or she were in a purely competitive industry.) Now assume market price is at the level P_1. To maximize net revenue, firm *A* will produce the quantity q_a, and firms *B*, *C*, and *D* will produce q_b, q_c, and q_d, respectively. Since the output of all four firms will constitute the total market offering, the supply on the market will be $q_a + q_b + q_c + q_d = Q_1$.

If price should increase to P_2, each of the four firms would expand their output, moving up on their own marginal cost curve until $MC = P$. The new quantity offered on the market by the industry would be Q_2, again the sum of the individual firm's outputs. In the same fashion the response of producers to any price can be determined. The short-run supply curve represents the horizontal aggregation—the "summing across"—of the short-run marginal cost curves of all the firms in the industry.

In 1982 the pallet and skid industry of the United States included 1649 companies—enough so one might reasonably assume this would approximate a competitive industry. However, one would scarcely expect all these to have identical cost curves. If we were to randomly choose five of these firms and examine their total cost, marginal cost, and price situations, we might find something resembling Figure 4.15*a* and *b*.

The industry supply curve *(b)*, formed by aggregating the marginal cost

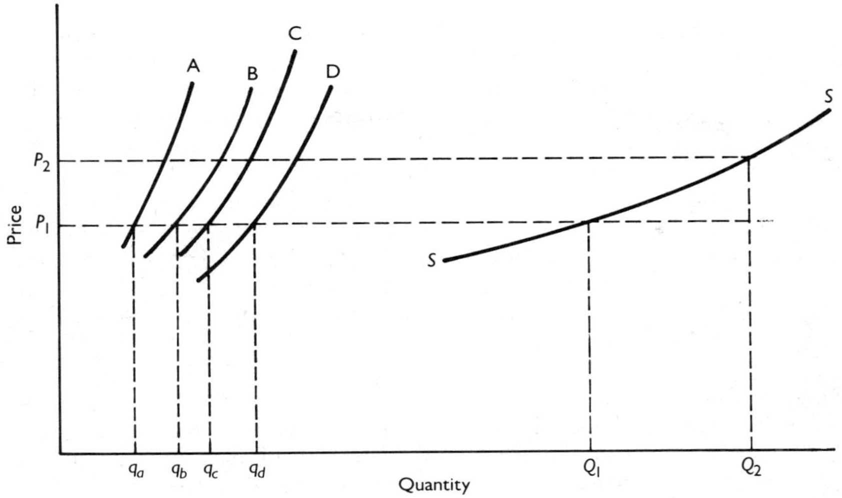

FIGURE 4.14. **The industry supply curve *(SS)* is formed by aggregating the marginal cost curves of all the firms in the industry.**

curves of the 1649 companies, intersects the demand curve D_1 at P_1 (the equilibrium price). Firm 4 of Figure 14.5*a* is exactly at equilibrium—just covering its total cost, but no more. This is a truly "marginal firm," making just enough to warrant its continued operation. The other four firms, however, are showing a true economic profit. If demand should shift back to D_2, price would drop to P_2, at which price level only Firms 1 and 2 would be profitable. Firm 4 would continue to produce only if price was above its average variable cost, but it would not be covering all its fixed costs. If price remained at this level, Firm 4 might struggle along for some time, thus preventing the supply curve from shifting somewhat

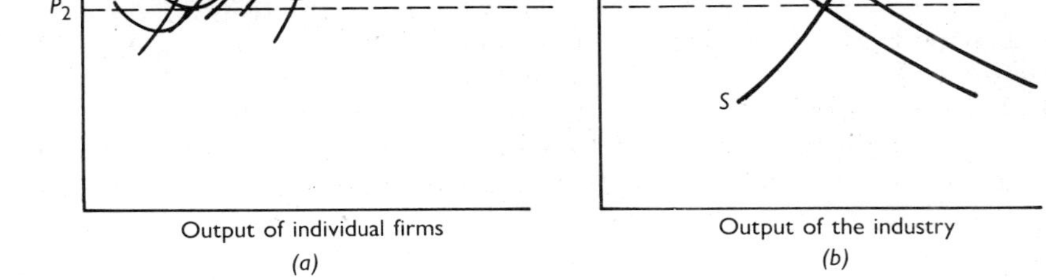

FIGURE 4.15. **Five firms of the industry whose supply and demand curves are shown in *(b)*. Only the average *total* cost and marginal cost curves of the firms are shown. Even when the industry is in equilibrium, some firms may be making a profit (or quasi rent).**

leftward. Eventually, however, Firm 4 would drop out, with its place taken either by new entrants or by the expanded output of one or more of the remaining firms.

The important point here is to realize that even though the *industry* is at equilibrium, some firms may be making a true profit—in this case sometimes called a quasi rent.[9] The marginal firm, the one just covering all its costs (no profit), can be visualized as the one producing the last "marginal unit" of product. It is the output of this marginal firm which, when added to the output of all other firms in the industry, pushes the total industry output to the equilibrium level. The other firms, with lower costs, *are* making a profit. This is the only kind of profit possible for a firm in a competitive industry—a profit stemming from the firm's ability to produce at a lower cost. And because of this profit potential, new firms that are at least equally efficient will enter the market, eventually forcing prices downward until a new equilibrium is established.

Long-run Supply Curves

In the long run the firm is able to adjust its *scale* of operation to whatever seems most desirable—there are no fixed factors. Reductions in unit costs brought about by a larger scale of operation are called economies of scale and are usually divided into two types: **technological** (bigger machinery, faster equipment, more efficient management, better organization, etc.) and **pecuniary** (reductions in cost due to bulk purchasing, etc.). Such economies of scale give rise to the so-called "envelope" curve, or the **long-run average cost curve** *(LRAC).*

This envelope curve is strictly a planning concept, not a chronological description of the development of any particular firm. For example, suppose one were planning a new firm and plans were drawn for plants of five different scales of production. If these plants were built, they would have operating characteristics that would result in the five short-run average cost curves of Figure 4.16. The marginal cost curve (short-run supply curve) of each proposed plant is also shown, but remember we are still in the process of deciding *which* plant to build.

Suppose Plant No. 1 were selected. As long as price remained at or above the level p_1, profitable operation could continue satisfactorily. But this is a competitive industry, and the owner of this plant would be hard pressed to compete with the owner of Plant No. 2, for Plant No. 2 can operate profitably as long as price remains at or above p_2. Similar reasoning leads one not only to reject Plant No. 1, but also No. 2 and No. 3. Only when the scale represented by Plant No. 4 is reached can there be any hope of success, for in a purely competitive industry it is certain that plants as large or larger than No. 4 will be built. Because each prospective producer would go through a similar analysis, we can be sure that all firms in the industry will be at least as large as Plant No. 4, where all the advantages of scale are present.

If a curve is drawn so that it is tangent to all the possible short-run average cost curves, this "envelope" (*LRAC* in Figure 4.16) will indicate the trend of *long-run average costs* as scale is increased. If a *long-run equilibrium* is to be reached, it will necessarily come only when all firms in the industry are large enough to have achieved all the economies of scale—when all firms are at the "bottom" of the long-run average cost curve. In this happy state of affairs all firms in the industry will have identical long-run average costs, short-run average costs, marginal costs,

[9]Quasi rents are returns determined by the price level, not by costs. They will be met again toward the end of this chapter.

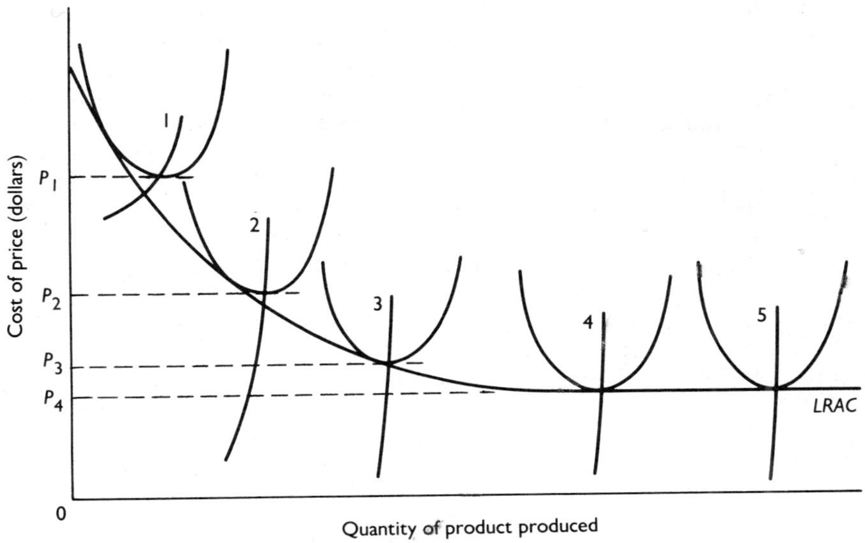

FIGURE 4.16. When economics of scale exist, the long-run average cost curve (LRAC) is an envelope curve.

and product price. What is more, all four of these will be equal. This is the familiar, but ephemeral, condition of long-run equilibrium.

But what happens if, before all the economies of scale can be attained, one firm becomes so large that it can influence product price? In this case it simply means that the technological facts of production rule out the purely competitive model. It also illustrates the idea that competition tends to create the forces that eventually result in its own demise: the drive toward lower costs imposed by a purely competitive industrial structure may force the firms to get so large that they can influence price. Once this latter position has been reached, the purely competitive model is no longer applicable.

Constant Costs and Increasing Costs

If the *LRAC* curve flattens out at a scale such that the industry remains purely competitive, the industry long-run supply curve will be perfectly elastic—a horizontal line. In such an industry output is expanded or contracted by firms entering or leaving the industry, as has occurred in sawmilling in the United States, and still continues in harvesting, in some segments of building, and others.

If diseconomies of scale exist, and if they outweigh the possible economies, then costs could increase with increasing scale. But increasing costs can usually be avoided by duplicating the plant of most efficient size. For example, one doesn't *have* to build a bigger sawmill to double output; instead, two sawmills could be built. Foresters should hold the possibilities of these diseconomies in mind, however, for forest land cannot be moved about and it is often difficult to assemble exactly the pattern of ownership desired.

External Economies and Diseconomies: Externalities

In contrast to the *internal* economies discussed so far—economies arising from the action of the firm itself that to a large extent give rise to the declining average costs observed as firms grow larger—are the so-called *external* economies and diseconomies. These economies depend on the general development of the industry. In the Pacific Northwest, the South, or the Northern Lake states, for example, the supply of woods labor is large and the tradition of woods work strong. Hence, if one crewman leaves his job, another can be picked up with relative ease—an economy certainly not attributable to any single firm but rather to the size of the industry itself, and hence an "external" economy. The network of woods roads, many of which are used in common, provides another example. Perhaps most important is the fact that any industry providing substantial employment can attract public attention and political support. The forest industry, for example, has been given valuable aids such as federal and state forest experiment stations, forest product laboratories, forest fire protection systems, and so on.

The opposite side of this coin must also be examined. The general public is often aware of the **external diseconomies** that plague forest industries. Such diseconomies are often divorced from the question either of firm or industry scale of operation: some writers refer to these as **social costs** (Kapp, 1950) while the more modern term is externalities. As used today, externalities are either costs or benefits that do not enter into the accounting system on which the producer bases production decisions. Logging roads and landings, for example, have been located along stream bottoms because they are easier to build and considerably less expensive. But this practice results in the loss of high-quality forest sites and the silting up of reservoirs—a negative externality of the logging industry. Water pollution by pulp mills, damage to roads by logging trucks, the destruction of waterfowl nesting sites, the silting over of salmon spawning beds, the destruction of scenic values—all are examples of negative externalities associated with timber production.

All forest industries show both positive and negative externalities. The timber industry presently shoulders substantial costs of outdoor recreation—increased fire prevention and protection charges, highway maintenance, hunter patrols, and so on. It also bears increased costs of planting due to damage by deer and bear, as well as other costs associated with wildlife maintenance programs.

Public resource policy is often directed toward either preventing externalities harmful to the public or providing some means of recovering the costs from those who create the problem. Such actions are referred to as "internalizing the externalities." Taxes could be levied according to the amount of effluent discharged into the river, for example, or subsidies paid for "proper" forest management; various other means of "internalizing" the costs of benefits may also be employed.[10]

Externalities are far too important to be passed over in a paragraph or two. They are introduced at this point only to show how and where they fit into the general scheme of economic theory.

PROFIT AND RENT

Ricardo, Marx, Pigou, Pareto, and other famous early economists were primarily concerned with distribution issues—with who gets how much of whatever our economic system produces. Small wonder that in earlier days the term political economy was used to describe the field now usually referred to as economics.

---

[10]Many attempts to solve these kinds of problems through regulation have been made, especially when public policy issues were involved—and they usually are. Unfortunately such regulations often incur substantial costs—costs that may, and sometimes do, exceed the benefit sought.

Although this text is no place for a detailed review of distribution theory, three of its principal topics are of great importance to forestry: profit, rent, and interest. All three terms identify the return to a specific kind of production factor. Profit and rent will be discussed immediately; the discussion of interest is the subject of Chapter 7.

Categories of Productive Factors

When speaking in general terms, economists usually divide productive factors into at least three, and sometimes four, broad categories: land, labor, and capital are almost universally recognized, while management is frequently cited as a fourth. The return to land (or natural resources) has traditionally been called **rent**; the return to labor is called **wages**, while the return to capital is usually termed **interest**. Some early economists dubbed this latter return "profit," and some defined interest as part of profit—after all, don't both of them go to capitalists? Management was added to the classic trio in the early 1900s on the basis that without management, land, labor, and capital are often combined ineffectively. Few agreed, however, that **profit**, the fourth type of return, was the return to management. This reluctance stems in part from the knowledge that management, ownership, entrepreneurship, and risk bearing are all closely interrelated but distinct functions. Perhaps even more important is the growing awareness that management in the modern corporate sense is better regarded as a highly professional form of labor.

The source and magnitude of rent, interest, profit, and wages have been a major topic for economists, but the ethical problems of distribution are not to be settled by economics. A capitalistic system distributes rents, interest, and profits on the basis of private ownership of resources and capital, and includes these payments as part of personal income along with any wage payments that might be due the individual. A socialistic system makes personal income primarily dependent on wages. The choice between the two systems must be based on a very complex system of value judgments: there is no *a priori* "justice" that can be accorded either.

Profit[11]

Profit is the difference between total cost and total revenue, and is synonymous with net revenue. It should not be confused with accounting profit, which is defined as sales receipts less the sum of payments for hired factors plus depreciation. This latter residual, the accounting profit, generally goes to owners of an enterprise and may include interest, wages, and rent, for owners often supply much of the capital, furnish at least part of the management, and may also own resources used by the firm.

As long as this residual consists entirely of wages, rent, and interest, whether they are identified or only imputed to the owners, there is no particular problem of distribution. But after the owners have received payment for services supplied there may still be something left over. *This* remainder is true profit in the economic sense.

The purely competitive equilibrium position of the firm is a no profit position. There will be an accounting profit, since all costs are covered and total costs

[11]Probably the best overall review and analysis of profit is that of Schumpeter. Beginning with the ancient Greeks, he traces the evolution of economic analysis through Keynes. His discussion of profit as developed by Ricardo, Mills, Marx, Marshall, and others is most revealing (Schumpeter, 1954).

include payment to *all* productive factors, including those to owners for wages, interest, and rent. There is no economic profit to a firm when in equilibrium.

But equilibrium conditions are seldom achieved and, if achieved, are not long lived: demands shift, factor costs change, technologies alter, new products appear, competitive products are invented, and profits arise whenever market price exceeds the firm's average total cost. Profit can also result from monopoly, oligopoly, and similar nonperfect market structures. With such profits there is a real distribution problem: who should get it and why.

Marx maintained that profits could only arise from the exploitation of labor. Other economists have argued that profit should be regarded as a reward for innovation or as a payment for particular astuteness in assessing the true risk of some specific venture. The opportunity to earn such rewards has been regarded as a mainspring of capitalistic economies.

There is still much ethical concern over the distribution of profits that arise from market imperfections. Yet monopoly profits accruing from the discovery of new products or new processes are also a reward for innovation, and the possibility of establishing such a position is surely a major incentive for technological progress in any capitalistic system.

Wages

This term will be defined here as any payment made for human services. Wages thus include many kinds of payments—to singers and other artists, doctors and school teachers and bank presidents and ditch diggers and garbage collectors—and even those made to professional foresters. Also included are such nonmonetary benefits as health insurance, vacation time, and so on. The term is thus somewhat more broadly defined than in some other works.

Rent

"Rent is that portion of the produce of the earth which is paid to the landlord for the use of the original and indestructible powers of the soils." So wrote David Ricardo (Ricardo, 1817). Since that time it has become quite obvious that these powers are not entirely indestructible, and the concept of rent has been extended to all factors of production. Rent plays a highly significant role in forestry and foresters should be familiar with the concept. Like many economic concepts, it is most easily explained by an example.

Figure 4.17*a* gives the basic short-run average total cost curves for a firm operating in a purely competitive industry. Price is at the equilibrium level P_1, established by the intersection of the industry demand and supply curves, shown in Figure 4.17*b*.

Now introduce a couple of basic changes: (1) suppose the product involved requires a very special type of land and that all this land is already in use by the industry to produce Q_1; (2) suppose the land in use by the particular firm illustrated in Figure 4.17*a* is being *rented* from a landlord.

If for any reason the demand curve shifts out to D_2D_2, price will rise; our firm will increase its output (as will all other firms in the industry); and eventually industry price and output will reach P_2 and Q_2. At this price the firm will be producing q_2 and will not only be covering all fixed and variable costs (including the rent) but will also be making a very handsome profit of $q_2\ (p_2 - c)$.

Under normal conditions one would expect new firms to enter the industry or old firms to enlarge their scale of production (or both) and the former equilibrium to be restored. But this normal reaction to shifting demand cannot take place, for there is no more land of the required type. Now, observing the large profit

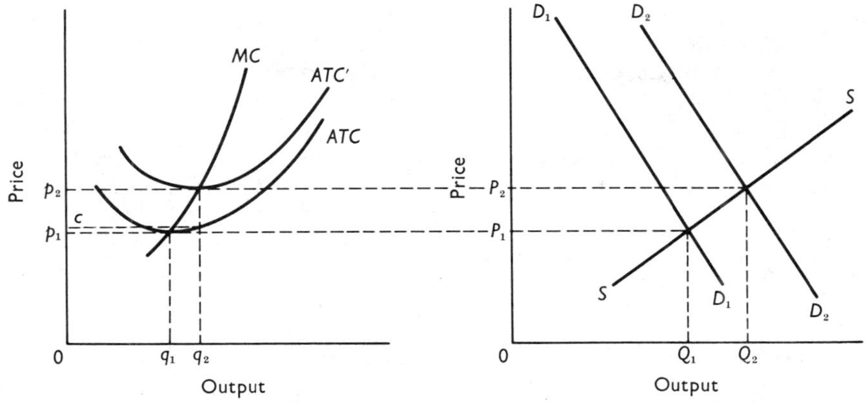

FIGURE 4.17. Ricardian increasing rent—rent arising from an increase for product demand when one productive factor is absolutely fixed.

being enjoyed by the firm, the landlord may well decide to increase the rental charge. The firm will have to meet the increase or go out of business since there is no more of the needed type of land. The only remaining question concerns the amount of the rent increase, and this can be settled quickly; the landlord can take the entire profit of the firm as rent for the scarce factor of production he owns. After including this new, increased rent in his production costs, the average cost curve of the producer will be moved up along the marginal cost curve to the new position indicated by *ATC′*—differing from the former *ATC* by precisely the increased rental payment and exactly equal to the (new) market price.

Quasi Rent

Earlier in this chapter it was shown how, when an equilibrium price had been established in a competitive industry, the more efficient firms in that industry could be earning a profit—sometimes called a **quasi rent**. The basic distinction between a land rent and a quasi (or economic) rent is that land rent is a payment required to bring the land into production, while quasi rent is a payment to any factor *above* that required to bring it into, or keep it in, production.

For example, a couple in western Washington planted two acres in front of their residence with fir trees. To preserve their view of the mountains they planned to sell them as Christmas trees after eight years. The trees were purchased from the Soil Conservation Nursery for ten cents each and planting costs came to no more than 15 cents per tree. With 700 trees per acre, total costs came to \$350, and they calculated that if they sold 1200 trees at one dollar each they would be well repaid. However, two weeks before they planned to sell the trees they received a call from the nearby city: someone had seen their trees and wanted to buy all 1200 at six dollars each—\$7200 cash. The sale was consummated, with the owners gaining not only their expected return, but some \$6000 over and above the anticipated amount—\$6000 of quasi rent.

Land rent (and remember that here land means any natural resource) is a payment necessary to make that factor available for production, and is part of its price. Quasi rent is determined *by* the price: it is not part of the production cost.

Summary

This chapter started with a review of the basis of all production—the physical relationship between inputs and outputs called a *production function*. From this relationship the firm's *variable cost* function was derived and, after developing the concept of *fixed costs*, the two were combined to produce the *total cost* function. Next, *marginal cost*—the change in cost accompanying a change in output—was developed.

Revenues were introduced by combining product price with output. Maximization of profit (net revenue) was then shown to result from adjusting the firm's output so marginal cost and marginal revenue were equal.

Two variations on the general cost-revenue picture were then presented, the first deriving the firm's revenue directly from the production function and making costs a function of *inputs*, the second dealing entirely with the revenue/cost ratio or financial *benefit/cost* ratio.

Economic supply, both short and long run, and the relation of supply to production costs were discussed along with the significance of economies of scale, the likely presence of externalities, and the effect of both on costs and supply.

Finally, the concepts of profit and rent were reviewed and discussed, with economic profit distinguished from accounting profit, and rent identified as the payment for the use of a natural resource as a productive factor. As such, rent is a cost influencing supply. Quasi rent, being a payment over and above that necessary to bring (or keep) a productive factor in its present use, does not enter into the supply function.

QUESTIONS AND PROBLEMS

1. Worried about the possibility of a water shortage, a Western municipality has employed a consulting firm of forest ecologists to investigate possibilities of increasing water yields from their watershed. The recently received report indicates that water yields could be increased substantially by cutting brush from the area each year, thereby reducing evapotranspiration. "Our research," the report concludes, "indicates that water yields can be increased over the present quantities in accord with the following table."

Acres To Be Treated Annually	Increased Water Yield (millions of gallons)
560	8
640	12
694	16
730	20
761	24
800	28
852	32
927	36
1043	40
1241	44
1600	48

The city fathers show you the table and ask, "Should we go whole hog and treat the entire watershed, or just treat part of it?" You glance at the data and ask, why they are interested in a little problem like this—and are promptly told that the research report is from a pilot study: these data are from a carefully drawn sample. The area involved actually runs well over 100,000 acres. The *problem*, however, is just as illustrated here. You decide it's worth your time, and become their consultant in forest economics.

After a couple of days work you discover that:

a. The city pays 8¢ ($0.08) per 1000 gallons of water—or $80 per million gallons.

b. The brush can be chipped and sold to cover part of the cost, making the annual net cost of treatment just $2 per acre.

c. Fixed charges (scaled down from the watershed manager's books) are calculated to be $800/year.

Assume that the city will manage its watershed to maximize annual net revenues under the given price structure.

To present your analysis to the town fathers you are to prepare the following:

1. A table, giving variable cost, fixed cost, total cost, average fixed, variable and total cost, and marginal revenue—all with reference to output.
2. Plot the physical product curve.
3. Plot all total cost and revenue curves. Plot fixed, variable, and total cost, along with total revenue, with cost on the vertical axis, output in millions of gallons on the horizontal.
4. Plot the average cost and the marginal curves, showing average, fixed, variable, and total cost curves. On the same graph, show the marginal cost and revenue curves. As above, plot dollars on the vertical axis, output in gallons on the horizontal.

Now, using these tables and graphs, determine the following:

a. The production level in millions of gallons that will maximize annual net revenue. Identify this point on all appropriate graphs.

b. How many acres would you treat (of the 1600) to obtain this output? Identify this point on all appropriate graphs.

c. How much additional net revenue would be generated if your recommendations are followed?

d. Suppose fixed costs were to drop to $500. How will this change your answers to the above?

e. If the price received by the city for water were to drop to $60 per million gallons, would you change your recommendations? If so, in what way?

f. Assume the marginal cost of treating an additional acre is two dollars. What is the marginal revenue product of the treatment if, after treating 800 acres, an additional 52 acres are treated?

g. Calculate the sum of the marginal costs in your first table. Multiply this sum by 4, and add the initial variable cost (the cost of treating 560 acres). The result is, or should be, the final variable cost—the variable cost of treating 1600 acres. Explain why.

h. Finally, suppose that just as you were completing your report the Mayor looks over your graphs and says (in effect), "Anyone can look at that first chart—the one you've called a 'production function'—and see that you'll produce the most water if you treat the whole works. That's what I suggested without any analysis!" What would be the result of following his "experienced" view?

2. Using the same data given for Problem 1, prepare an elementary financial benefit/cost analysis of the project by:

   a. Constructing the benefit/cost curve.

   b. Identifying the optimum area to be cleared, the costs, the (financial) benefits or revenue, and the net benefits.

3. Assume that the following data apply to the fuelwood market in a nearby city:

   a. There are only three fuelwood producers: Tom, Dick, and Harriet.

   b. Including delivery charges, their total production costs are as shown in the following schedule:

Number of Face Cords Produced	Total Cost (in dollars)		
	Tom	Dick	Harriet
250	$ 7,000	$ 7,000	
350	10,000	8,000	
450	14,000	10,000	
550	19,000	13,000	$10,000
650	25,000	17,000	11,000
750	32,000*	22,000	13,000
850		28,000*	16,000
950			20,000
1,050			25,000
1,150			31,000

*Capacity.

   c. Total fixed costs for the three producers are:

   Tom—$5000
   Dick—$7000
   Harriet—$10000

   d. A market survey reveals the schedule of prospective fuelwood purchases at various prices to be as follows.

Number of Face Cords	Price per Face Cord
1200	$100
1400	90
1600	80
1800	70
2000	60
2200	50
2400	40
2600	30
2800	20

   e. The fuelwood market in this city is purely competitive.

f. Each producer seeks to maximize profit.

After plotting the supply curve for each producer, the industry supply curve, and the market demand curve, answer the following questions:

a. What will be the price of fuelwood in the city?

b. How much fuelwood will be purchased?

c. How much will each supplier produce?

Present the answers to questions on the graph, and show all calculations.

4. In the previous problems a graphic approach was used to illustrate how an individual firm chooses the output (and in the monopoly situation, the price) that will maximize its net revenue or profit. The following two problems show how easily output and price decisions can be made by using mathematical rather than diagramatic techniques.

Case 1—Pure Competition

Given the following equations for a firm:

TVC $= 25 - 2Q + .2Q_2$
Fixed cost $= 200$
Product price = \$4 per unit
Q = number of units to be produced

a. Write the equation for the total cost *(TC)* of this firm.

b. Write the equation for the marginal cost *(MC)*.

c. At what output would you recommend the firm operate?

Case 2—The Monopoly Situation

TVC $= 25 - 2Q + .2Q_2$
Fixed cost $= 2000$
Demand curve = price = average revenue $= 42 - 2Q$

a. Write the equation for total revenue.

b. At what output would you recommend the firm operate?

c. What will be the market price?

QUESTIONS

1. What are externalities and of what significance are they to forestry?
2. When someone says, "Pollution is easy to cure! Just internalize the externalities" what does that mean? To what extent would this individual be correct?
3. Distinguish between rent and quasi rent.

chapter five
MARKET STRUCTURE AND PRICING OF FOREST PRODUCTS

Prices play an important role in any market-oriented economy. On the one hand, consumers decide what and how much to buy on a basis of their incomes and product prices. On the other hand, these prices reflect to a considerable degree the relative cost to society of producing and marketing the items made available. This balancing of societal costs and consumer wants makes a free market price system an efficient means of allocating resources. But this is only part of the picture.

Lack of production know-how, imperfect knowledge on the part of consumers, or important externalities may result in some price signals being poor guides to either personal or social decisions. More important, certain firms may so influence a product's price that it no longer indicates comparative production costs.

Competitive markets were assumed throughout the preceding discussion of demand, production, and supply. Because some wood product markets are essentially competitive (and because it offers the best starting point), this chapter begins with a review of the market structure called pure competition. Monopoly, oligopoly, and monopolistic competition will next be reviewed, after which the changing structure of present day wood product markets will be examined. As part of this discussion we will also look at the way prices in these industries have behaved since 1950.

PRICES UNDER PURE COMPETITION

Economic theorists usually attach many qualifications to pure (or perfect) competition, but for nearly all work in forest economics three major conditions will be sufficient to identify a purely competitive situation:

1. *Large number of producers and buyers.* The number of firms and the number of buyers within the market area concerned must be large enough so that no one firm or group of firms, and no one buyer or group of buyers, can significantly influence the product price by changing the quantity produced or purchased.
2. *Product is undifferentiated.* The product of one firm must not be recognized by buyers as being significantly different from that produced by other firms.

A Southern pine 2 × 4, for example, produced by any one of several hundred sawmills, is equally desirable to the average buyer.

3. *No barriers to entry or exit.* There can be no real or artificial barriers to firms wishing to enter or leave the market. In effect, this means no controlling patents, no secret processes that significantly affect costs, and a relatively low required capitalization.

Short-run Prices

In the short run the individual producer in a purely competitive market will maximize net revenue (profit) by adjusting production until the cost of changing output by one unit just equals the price received for that unit—that is, until marginal cost equals price. Since every producer acts in this fashion, the industry supply curve can be visualized as the horizontal aggregation of all the individual firm's marginal cost curves. On the demand side, consumers adjust their purchases until the satisfaction gained from purchasing one more unit is just equalled by the price they must pay for that unit. Market price will be determined by the intersection of the supply curve with the market demand curve.

In Figure 5.1 the short run supply curve (or marginal cost curve) of one firm in the industry is shown on the left, while on the right are the industry demand and supply curves (D_1D_1 and S_1S_1). Under the equilibrium conditions illustrated, the industry is producing the output Q_1, price is at P_1, and the particular firm shown is producing at q_1.

Now suppose the demand curve shifts outward to D_2D_2. This will result in a price rise, causing the firm to expand its output to q_2. The similarly increased output of all firms will raise the industry's output to the new level of Q_2, where price will stabilize at P_2. If the demand shift is temporary, prices will drop back when the downward shifting demand returns to its original condition. If the demand shift is to the left, prices and profits fall and the less efficient firms have to struggle for survival until demand increases.

Long-run Prices

Once we leave the short-run conditions, adjustments can be made in plant capacity, new plants can be built, new technologies employed, and so on. Since at P_2 (Figure

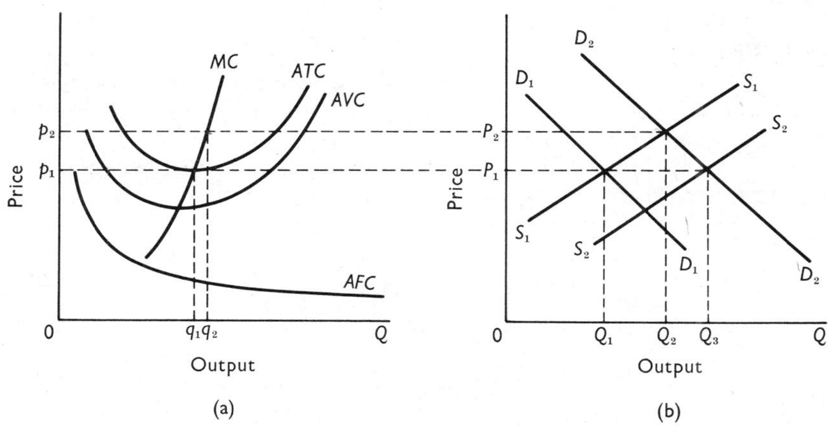

FIGURE 5.1. Short-run adjustment of the firm to demand change in an industry operating under pure competition.

5.1) profits are high, this kind of adjustment will take place and the increased capacity of the industry will shift the supply curve outward. If there are no restrictions on supplies of productive factors, the supply curve will eventually reach S_2S_2, with the new equilibrium output of Q_3 and with price restored to its former level of P_1. If the demand curve should shift left, conditions would be reversed: prices and profits would fall, and although the less efficient firms would struggle for awhile and prevent the needed reduction in industry output, eventually they will drop out of the market. With the now reduced capacity of the fewer remaining firms, the supply curve shifts left until price is restored to its original level of P_1.

From the above it would seem that no matter what happens to demand, prices will eventually return to the same equilibrium level where Price = MC = Average Cost. This is the situation of an industry of constant costs, and it comes about only when there are no limits on the supply of any productive factor (so costs can remain constant). If there *are* some limits, if something—skilled labor, land, raw material supplies, and so on—is fixed, then costs and prices will rise with increasing demand unless offset by decreasing costs of some factor(s). The steady rise in U.S. softwood lumber prices since 1820 (see Figure 3.7) provides an excellent illustration. The many cost reductions in harvesting, manufacturing, and transportation have not been enough to offset the facts that softwood sawtimber is limited in supply and is fixed in location.

PRICES UNDER NONPURE COMPETITION

Pure competition describes the limit of market competition at the competitive end—there are so many producers or consumers that no one can influence the product price. **Monopoly** describes the limit at the other end of the competition spectrum; monopoly exists when a single firm is the only producer of a good. (A situation with only one buyer is termed **monopsony**.) As with pure competition, examples of pure monopoly are rare—in the United States prior to World War II one could cite the aluminum manufacturer, ALCOA, but this monopoly ended under pressure of wartime needs. The fact that monopolies are rare, however, does not make analysis of monopoly behavior useless, for many oligopolistic firms, particularly groups of such firms, act as if they were monopolies. Monopoly power is an important factor in the U.S. economy—important economically, politically, and socially, and an understanding of how monopolies behave is a key to understanding market behavior in the United States.

Monopoly Pricing

As an example, we can take the Forestville fuelwood market data used previously (Figure 3.1 and Table 3.1). Now assume that a single producer has established a monopoly. Because an industry demand curve is also its average revenue curve, total revenue can be derived directly by multiplying the average revenue (or price) by the quantity marketed. Hence, since average revenue is given by the equation

$$AR = P = 60 - .667Q$$

and total revenue equals $P(Q)$, the equation for total revenue will be

$$TR = P\,(Q) = 60Q - .667Q^2$$

and marginal revenue, the first derivative of total revenue, will be

$$MR = \frac{d\,(TR)}{d\,(Q)} = 60 - 1.33Q$$

Note that the price intercept of *MR* (60) remains unchanged from that of average revenue while the slope coefficient is exactly doubled. Hence, if the demand curve is represented by a straight line as in the example used here, the marginal revenue curve will also be a straight line, will intersect the price axis at the same point as the demand curve, and will have exactly twice the slope of the demand curve. Under these (somewhat unrealistic) conditions, therefore, the marginal revenue curve will always bisect a horizontal line drawn from the price axis to the demand curve (Figure 5.2).

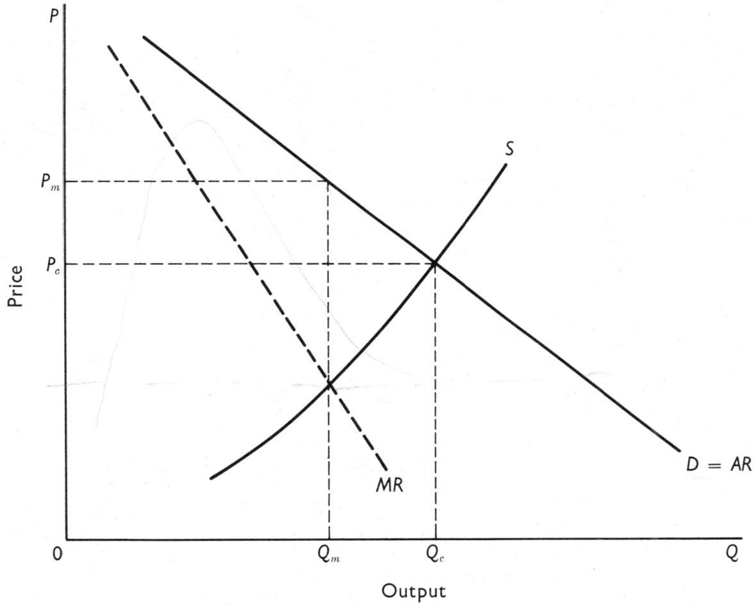

FIGURE 5.2. Price and output decisions under monopoly.

The supply curve of our Forestville monopolist will still be the same as previously calculated—the *production* aspects are unchanged. But the monopolist's supply curve will intersect the *marginal revenue* curve at an output level of Q_m, a decidely reduced output! And this output can be sold at the price P_m. These price and output levels can be contrasted in Figure 5.2 with the competitive price and output levels of Q_c and P_c. Clearly, the tendency of monopoly will always be to restrict production and to increase prices. How much production will be restricted, and how much prices will be increased, will depend on the shape and the slope of both the demand and supply curves involved, as a little experimentation on scrap paper will readily confirm.

Oligopoly Pricing

True monopolists, having no competition, are free to adjust their production and prices as they see fit, while the producer in a competitive market must accept the market price as given. An oligopolistic structure poses a more difficult pricing problem. An oligopoly consists of a few firms, any one of which produces enough of the particular product to influence product price. When selling in this sort of market, oligopolists must consider not only their own price decision but also two further questions: what will other firms do when they learn of their competitor's price change, and how should the original price changer react if and when other firms change their prices?

Suppose we have an industry of four firms, each producing essentially the same product by essentially the same technique. Figure 5.3 shows one of the firm's cost curves *(a)* with the industry demand curve on the right *(b)*. If the four could "cooperate" completely, they might operate as if they were a single firm with four producing units. Profit for the group would be maximized by producing a total output of Q_1, price would be P_1, and the unit on the left would produce q_1.

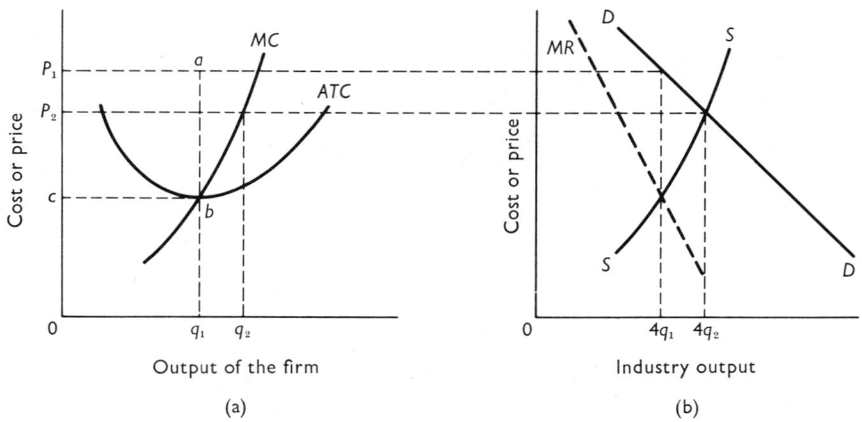

FIGURE 5.3. Price and output decisions under oligopoly.

But these are not units in a single firm, and the industry demand curve is not the demand curve of any single unit. If our particular producer prices a product at p_1, any or all of the other firms will probably set their prices just a little lower, thus capturing our producer's customers. If price is raised and the others do nothing, customers are lost. It's as if our producer faced an average revenue curve (demand) like that of Figure 5.4, one with a pronounced "kink" at the more or less established price level. (Remember: this is the demand, or average revenue curve for an *individual firm* in an oligopolistic market—not the *industry* demand curve.) If our producer raises prices above those of the other producers, sales will quickly drop. On the other hand, if prices are cut to increase the firm's market share, then competitors will probably follow suit. The firm's average revenue curve is more inelastic to a price drop than to a price rise. **Given this kink in average revenue, a producer's marginal revenue curve is discontinuous.** Costs can change substantially, yet the marginal cost curve of the producer will indicate no change in output—which helps explain why prices in an oligopolistic market tend to be "sticky."

Business executives do their best to avoid uncertainty and usually find some way to cope with uncertain situations. In Europe and many other parts of the world where such actions are perfectly legal, the usual strategy in oligopolistic markets has been to form a cartel—an organization specifically designed to foster price and production agreements that will eliminate wasteful competition. OPEC, the Organization of Petroleum Exporting Countries, is probably the best known of these, and from 1972 through 1983 this world cartel successfully set world oil prices and established national production levels for its members.

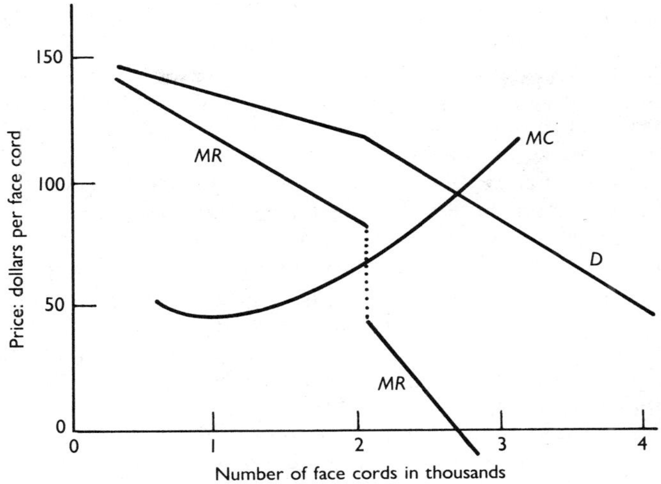

FIGURE 5.4. A kinked average revenue (demand) curve and the resulting marginal revenue function.

Cartels are illegal in the United States.[1] But if producers accept the price set by the largest or most aggressive firm—accepting that firm as the price leader—they can gain many of the advantages of a formal cartel while staying within the law. Such arrangements, however, lack any possibility of enforcement. As a consequence, in times of economic stress price wars may occur, with wave upon wave of price cutting, sometimes followed by a substantial reorganization of the market structure. Domestic airlines provided a good illustration of this during the winter of 1982–1983 and 1984–1985.

In some cases it appears that a different but effective strategy has been adopted. For example, assume an oligopolistic industry. Further assume Congress or a state legislature, recognizing that the public interest needs safeguards against possible collusion under oligopolistic conditions, establishes a regulatory commission empowered to regulate prices of the industry's products. The commission can prevent excessive price increases and may also eliminate much of the price uncertainty that previously existed. However, some commissions have tended to become more interested in protecting the industry than in protecting the public. The well-established custom of having ex-officials of utility companies appointed to public utility commissions is not above suspicion. The rapid change in price structure of the airlines and trucking industries following their deregulation in the late 1970s and early 1980s illustrates the possibility that regulators may serve the industry as well as, or even to the neglect of, the public's interest.

What does the supply curve for an oligopolistic firm or industry look like? The answer: it looks like nothing at all! There can be no real supply curve for either firms or industries under oligopolistic conditions. The quantity to be supplied by an oligopolistic firm depends partly on the firm's costs, but also very much on the firm's anticipation of their competitor's pricing behavior. And since any one firm's

[1]For a most interesting discussion of U.S. antitrust policy and its effect on wood-based industries, see Jay O'Laughlin's *Antitrust in the Era of Diversication: The Wood-Based Industry as a Case Example* (1981). Also see O'Laughlin and Ellefson, 1979.

production response to industry price changes will not be known in advance, the price outcome of the industry is equally uncertain.

There is one interesting exception to this "no supply curve" situation. Suppose a single price leader can supply around one-third of the total market demand. (The proportion is unimportant as long as it is high enough to maintain price leadership.) If the management of this corporation can estimate the approximate marginal cost curves of the other firms, the price which will maximize the leader's profit can be determined. Figure 5.5 illustrates the possibility.

D_i is the industry demand curve and S_o the supply curve of all the other firms—the horizontal sum of the marginal cost curves of all firms in the industry *excepting* that of the price leader. D_L, the demand (or average revenue) curve for the price leader, is determined by the difference between the quantity demanded of the industry (D_i) and the amount that the other firms can supply (S_o). Thus, at a price of P_o (where S_o intersects D_i), the other firms will supply the entire market demand and there would be no demand for the price leader's output. On the other hand, at a price of p_1 the other firms would be willing and able to supply only the quantity a_2, leaving the quantity $a_3 - a_2$ available as market demand for the price leader. So the price leader's demand curve, D_L, will begin at P_o and pass through the point $p_1 a_1$ $(0 - a_1 = a_3 - a_2)$.

With a demand curve of D_L, the price leader's marginal revenue curve will be MR_L. The price leader's marginal cost curve, MC_L, intersects this marginal revenue curve at b, identifying the price leader's profit maximizing output of Q_L which will sell at the price P_L. At this price the combined output of the other firms will be Q_o and the total industry output will be Q_i—which is the sum of Q_o and Q_L.

Is this ever done in real life? That, like most of the pricing problems in oligopolistic markets, is uncertain, but price leadership is not uncommon, and in

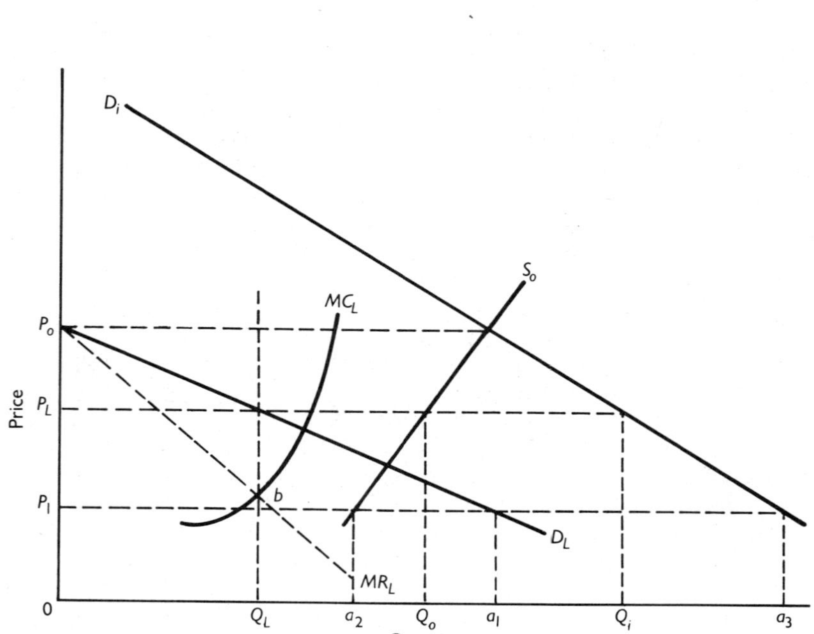

FIGURE 5.5. Determining the profit maximizing price for the price leader in an oligopolistic market.

such situations the general production and cost relationships of the firms are well known.

Oligopoly has been called the characteristic market structure of the United States, with the automobile, steel, oil, trucking, air transport, meat-packing, and other industries providing ready examples. Firms in these markets can influence the price of their own products, and in so doing will probably also influence the sales, and hence the prices and production plans, of other firms.

Oligopolistic markets are admirably suited for rapid growth, and the following scenario is not unusual. Sooner or later several producers become large enough to attain all significant scale economies, with cost curves that are essentially flat over a substantial production range. Since such firms can expand production without cost increases, they compete aggressively for market shares: advertising expenditures grow, multiple models are offered, services increases—competition is fierce in every way but price. Lacking the economies of scale, the smaller, higher cost firms drop out, leaving only a few large corporations, each able and willing to expand, and competing in an industry now having substantial overcapacity. As these large companies accumulate capital a new route to expansion is opened—acquisition of, or merger with, competitors. Ultimately, as the number of competitors is still further reduced, the opportunities for expansion, even by these means, become difficult; further mergers may bring restraining orders or legal action under anti-trust laws.

Now there are two possibilities. One firm may expand into foreign markets and become a multinational. Another may decide to enter new and very different production areas—to move into the conglomerate status. Either way they become larger, usually gaining in both economic and political power.

One should note that the comments on growth of oligopolies were directed toward the advantages gained through economies of scale: bigger firms may grow larger by taking advantage of their size. But this does not imply that a large firm, or an oligopolistic structure per se, fosters innovation. There is some reason to suspect that both monopolies and oligopolies, as they become very large, tend to become dull in developing new products or accepting new designs and new ideas. Growth in this latter sense may in fact be hindered by the industry structure. The U.S. automobile industry during the 1975–1985 decade, for example, seemed unable (or unwilling) to adapt to new demands, and the break-up of the American Telephone and Telegraph Corporation may eventually prove to have been an important step in fostering advances in communications.

Monopolistic Competition Pricing

In a very real sense, pricing under monopolistic competition is simply an extension of the oligopoly case. With many producers, each with some small measure of control over price because of some real or fancied difference in their particular product, there is far less drive toward mergers simply because there are too many firms. For similar reasons, collusive attempts are patently useless. Hence, each producer strives to intensify the degree of product differentiation in the buyer's eyes. Brand names and trademarks are a common means of product differentiation. The Weyerhaeuser and Georgia-Pacific brands of Douglas-fir lumber provide an easily observed example from the timber industry, while writing-paper watermarks illustrate a similar approach by the paper industry.

One way of expressing this is to say that each producer has a demand curve that is *slightly* sloped; it is not perfectly elastic (as it would be under pure competition) nor is it steeply sloped (as it would tend to be under monopoly). As product differentiation increases, the product demand becomes more inelastic to price changes.

Need for Market Control

This last point permits bringing up a final concern about monopoly pricing under all degrees of market imperfection: monopoly, in the sense of controlling production, is not particularly significant unless it is accompanied by the power to control price. This latter power is dependent on the *demand* situation as well as that of *supply*.

Suppose, for example, that by some stroke of fate all Southern pine lumber production were to come under the control of a single producer. To be sure, price would be affected, but increases would be limited substantially by the fact that Douglas fir, ponderosa pine, white pine, and other species could be—and *would* be—substituted for Southern pine if prices were raised very far out of line. Monopoly *position* and monopoly *power* are not the same thing!

TRENDS IN WOOD MARKET STRUCTURE AND PRICES[2]

With an understanding of the usual market models as a basis, we are ready to look at the structure and, so far as possible, price behavior, in the markets for major forest products. Three market structure aspects seem particularly important: the number of producers, the extent to which production is concentrated, and the availability of substitute products. The basic information on both numbers and the degree of concentration in an industry, and on assessing the trends in these, is that published by the Department of Commerce from the Census of Manufactures, taken every five years (U.S. Dept. Commerce, Census Bureau, 1977, 1982).

The ratio most often used in market structure studies is the percent of the total industry's *value of shipments* accounted for by the 4, 8, 20, and 50 largest *companies* of the industry.[3] Economists usually regard an industry with a four-firm concentration ratio of 75% or more as being extremely concentrated, from 50 through 74% as highly concentrated, from 25 to 49% as moderately concentrated, and 24% or less as being of relatively low concentration.[4] We will use these ratios in two ways: first to assess the relative degree of concentration in each of the major timber markets, and, second, to investigate the direction of any trend in concentration that might be evident.

Wood Harvesting

Harvesting is more a service than a specific product, but is an important member of the wood-based industry group. The service it provides is highly dependent on the

---

[2]The work of Paul Ellefson, Jay O'Laughlin, Edward Bilek, and Michael Chopp of the University of Minnesota has been an important source of both data and ideas about market structure for wood-based products. See, for example, O'Laughlin and Ellefson (1982a and 1982b), O'Laughlin (1981), and Bilek (1981, 1983). Others are listed in the Bibliography.

[3]Similar percentage ratios are available but based on the percentage of (a) total industry value added by manufacture, and (b) employment accounted for, by the 4, 8, 20, and 50 largest companies in the industry.

[4]The several limitations of these concentration ratios were well summarized in a letter from Senator Estes Kefauver to J. O. Eastland, then Chairman of the Senate Committee on Judiciary, and reprinted in "Concentration Ratios in Manufacturing," Dept. of Commerce. Of major significance is (a) the fact that the Department relies entirely on statements of company affiliation made by the company itself, (b) the ratios apply to the nation as a whole, so may seriously understate concentration in particular regions or states, and (c) imports are excluded, hence the ratios tend to overstate in certain cases (U.S. Dept. Comm. Concentration Ratios, 1979, 1986).

nature and location of the timber being harvested. Douglas fir sawlogs, for example, pose vastly different harvesting problems and require far different materials and methods than does the harvesting of pulpwood—or sawlogs—in the South.

Concentration

The number of companies in the wood harvesting industry (SIC No. 2411—Logging Camps and Logging Contractors) has fluctuated substantially since 1950, with no clear trend in evidence (Table 5.1).[5] And with over 15,000 companies in the industry one would normally expect concentration to be very low, particularly since there are no great barriers to entry or exit. There are, however, substantial differences in the nature and quality of the service rendered. The data (Table 5.1) indicates a rather surprising trend. Although the total number of companies increased by 20% between 1954 and 1982, the four-company concentration ratio increased from 8 (a very low concentration) to 30 (moderately concentrated). The four largest of the 11,541 companies shipped 30% of the industry total, while the 50 largest shipped somewhat over half of the total. Most of the increase in concentration between 1972 and 1982 stemmed from the growth of the four largest companies. Although the number of companies dropped by 15% between 1977 and 1982 the concentration ratios increased only slightly. The data do not permit identification of the four companies; they are probably logging divisions of major forest product corporations.

Prices and Price Trends

Because harvesting costs and log prices are so closely tied to stand location, to species and to log quality, any national or regional averages would carry little or no meaning with respect to market structure. Probably for these reasons no data of

TABLE 5.1
Share of Value of Shipments Accounted for by the 4, 8, 20, and 50 Largest Logging and Logging Contractor Companies, 1982 and Earlier Years

		Percent of Value of Industry Shipments Accounted for by			
Year	Number of Companies	4 Largest	8 Largest	20 Largest	50 Largest
1954	12,789	8	13	20	(NA)
1958	12,730	13	19	25	32
1963	13,522	11	19	25	31
1967	16,265	14	22	30	35
1972	13,152	18	25	35	41
1977	15,344	29	36	45	50
1982	11,541	30	37	47	52

SOURCE: U.S. Dept. of Commerce, Census Bureau, 1986. "Concentration Ratios in Manufacturing." MC82-S-7 Subject Series (SIC No. 2411).

[5] A company is defined as a business organization consisting of one *establishment* or more under common ownership. The SIC Number (Standard Industrial Classification Number) is a number from the four-digit system developed by the Census Bureau of the Department of Commerce. SIC Numbers are used in virtually all publications of the Department of Commerce, including those concerned with imports and exports, employment, production, shipments, value added, and so on.

regional or species price (or cost) trends are readily available for the harvesting industry, nor would comparisons over time or between regions be useful as indications of structural changes.

Lumber

The most inclusive category of lumber producers for which data are available is the one titled "Sawmills and Planing Mill, General" (SIC No. 2421). In 1977 this major segment of the lumber industry included 6966 companies operating over 7500 separate establishments. By 1982 the number of companies had declined to 5810, (one out of every 6 lumber companies had gone out of business) and the number of establishments to slightly over 6300 (Table 5.2). Most of these establishments are relatively small; only 25% of the total had 20 or more employees.[6]

Concentration

Concentration in the lumber industry is relatively low, with the four largest corporations accounting for a combined total of only 17% of industry output. The data indicate a definite trend toward increasing concentration: all four of the concentration ratios increased from 1963 through 1977, but not between 1977 and 1982. One should note, however, that 1982 was a bad year for the lumber industry.

Part of the increased concentration has undoubtedly been due to mergers and/or acquisition. Nearly 350 acquisitions of other companies had been made by lumber concerns over the 1960–1979 period, and 29 of 1970's "Top 100" producers had been merged into other companies by 1980 (Bilek, 1981, 1982). Of the several reasons for mergers and acquisitions, the desire to increase timber supply seemed most prominent (O'Laughlin and Ellefson, 1982b). Trends are clearly toward an increased concentration, but lumber manufacturing still remains an industry of low concentration when compared to other U.S. industries (Major, 1985).

Markets are primarily concerned with product *sale*, and, for a bulky product such as lumber, regional or local concentration could be high even though on a national basis concentration is low. Unfortunately, good data as to regional con-

TABLE 5.2
Share of Value of Shipments Accounted for by the 4, 8, 20, and 50 Largest Sawmills and Planing Mills, 1982 and Earlier Years

		Percent of Value of Industry Shipments Accounted for by			
Year	Number of Companies	4 Largest	8 Largest	20 Largest	50 Largest
1963	11,931	11	14	20	29
1967	10,016	11	15	22	31
172	7,664	18	23	33	45
1977	6,966	17	23	36	49
1982	5,810	17	23	34	46

SOURCE: U.S. Dept. of Commerce, Census Bureau, 1986. "Concentration Ratios in Manufacturing." MC82-S-7 Subject Series (Industry 2421).

[6]The number of establishments came from the preliminary reports issued in 1984 (MC82-I-24A-2[P]); the number of companies is from the 1986 "Concentration Ratios."

centration of lumber markets are not available, but one would expect to find high concentration ratios in the Pacific Northwest and California and perhaps also in the South. Probably more important is the concentration of market power in certain species that have special characteristics: redwood and Western red cedar may be two examples, though no studies of this possibility have been found.

One of the outstanding characteristics of the lumber industry is the great range in the size of lumber companies. In 1983, Weyerhaeuser, the largest company, produced a little less than the combined output of the next two largest firms. Furthermore, Weyerhaeuser not only continued to be the largest individual producer (a position it has held as long as statistics permitting the ranking of firms have been available), but also increased its own size and its lead over its nearest competitor (Forest Industries, 1984). From 1974 through 1980 the four top U.S. firms (Weyerhaeuser, Louisiana-Pacific, Georgia-Pacific, and International Paper) kept their rankings, but below this position there was much change. Only 13 of the top 25 lumber corporations of 1966 maintained a place in that group through 1980 (Bilek, 1981).

Prices and Price Trends

The trend of lumber prices over the past 150 years (Figure 3.7) has been very consistently upward. The price/quantity point for each year represents the intersection of a demand with a supply curve; the problem, discussed in Chapter 3, is how to "identify" where a change in consumption was caused by changing demand, by changing supply, or by changes in both. Econometric studies, utilizing the ability of modern computers to process large quantities of data, have led to an improved understanding of the many variables involved.[7] Even very simple models, however, can be quite revealing if coupled with a little understanding of economics.

In Figure 5.6 the deflated price index of lumber has been plotted over U.S. lumber production for each year beginning with 1932 and the points connected chronologically—from 1932 to 1933, to 1934, and so on. The resulting pattern indicates quite clearly that from 1932 through 1938 the lumber industry provided a classic example of a purely competitive industry with essentially constant production costs. Changes in output came about primarily by the exit or entry of firms. The small price increases from 1932 through 1938 are probably insignificant; deflated lumber price in 1938 was almost identical to that of 15 years earlier.

From 1938 through 1943, however, production increases brought rising prices, halted by the imposition of price controls throughout World War II. Relaxation of these controls in late 1946 resulted in what was virtually a completely new price structure for the industry: prices rose by nearly 40% by 1948, and they have never returned to anything approaching the prewar levels.

For the next 20 years, 1948 through 1967, production varied from 32 to 38.2 billion board feet, with shifting demand apparently tracing out a slowly rising supply function that was decidedly less elastic than that of any previous period. And from 1967 through 1982 the supply curve continued to shift upward and to become increasingly inelastic. The impact of increased energy costs from 1971 through 1973 led to sharply increased real costs for lumber with little change in production. Perhaps the most traumatic period since 1950 began in 1978, the year lumber prices hit their all time high. For the next four years both price and production declined, with 1982 output being the lowest in 37 years! It should be

[7]See, for example, Adams and Haynes (1980) and Rockel and Buongiorno (1982).

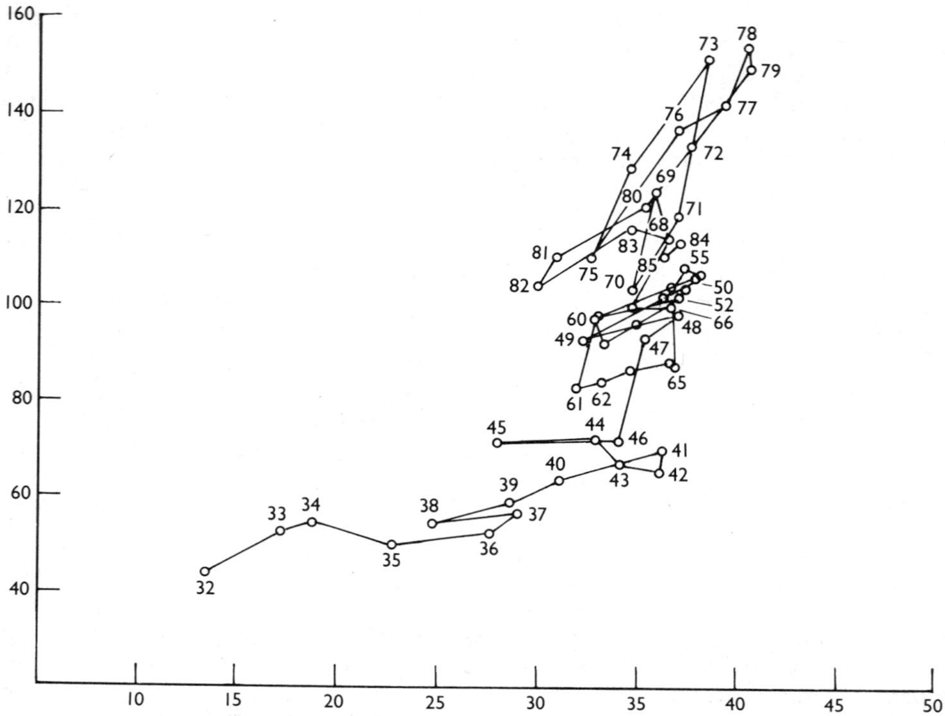

FIGURE 5.6. Deflated price index of U.S. lumber prices plotted over U.S. lumber production, 1932–1985. *Source*: For 1932–1950 data: USDA Forest Service 1982. An analysis of the timber situation in the United States, 1952–2030. Forest Resource Report No. 23, Table 1.15, p. 292. For 1950–1985 data: USDA Forest Service 1987. U.S. Timber Production Trade, Consumption, and Price Statistics 1950–1985. Misc. Pub. 1453, Tables 32 and 37.

noted that from 1967 through 1979 imports of softwood lumber increased from 4.8 billion board feet to 11.5 billion—an increase of 140%—and to 15 billion in 1985!

Finally, the deflated lumber price index plotted over consumption for the period 1950–1985 (Figure 5.7) further supports the idea that a shifting demand curve has "identified" not only an upward shifting supply curve for lumber since the end of World War II, but also that supply has become decidedly less elastic. The effect of the increasing imports from Canada is obvious when one compares Figure 5.7 with Figure 5.6 (U.S. consumption with U.S. production). Imports did not prevent the increase of inelasticity, but they apparently were a mitigating factor. From Figure 5.6 it appears that even the long-term supply curve of U.S. producers is decidedly inelastic.

There seems little or no evidence, however, that lumber industry concentration, per se, has had any appreciable effect on lumber prices. The rising supply curve is, rather, an indication that when lumber production is increased, costs rise—probably because of increased bidding for the (now) limited stumpage, as well as increased transport costs, utilization of lower quality (and probably smaller) logs, and so on.

It should also be noted that the upward shift of lumber supply prices following World War II coincides closely with the very large increase in sawlog stumpage price. This rise in stumpage price is often cited as a contributing factor to the increase in lumber price, and there can be little doubt that this is true. One can easily argue, however, that the causal direction is really the other way around—that high lumber prices will inevitably lead to high stumpage prices, because stumpage prices are *derived from* the price of lumber. This will be discussed more fully in Chapters 9 and 10.

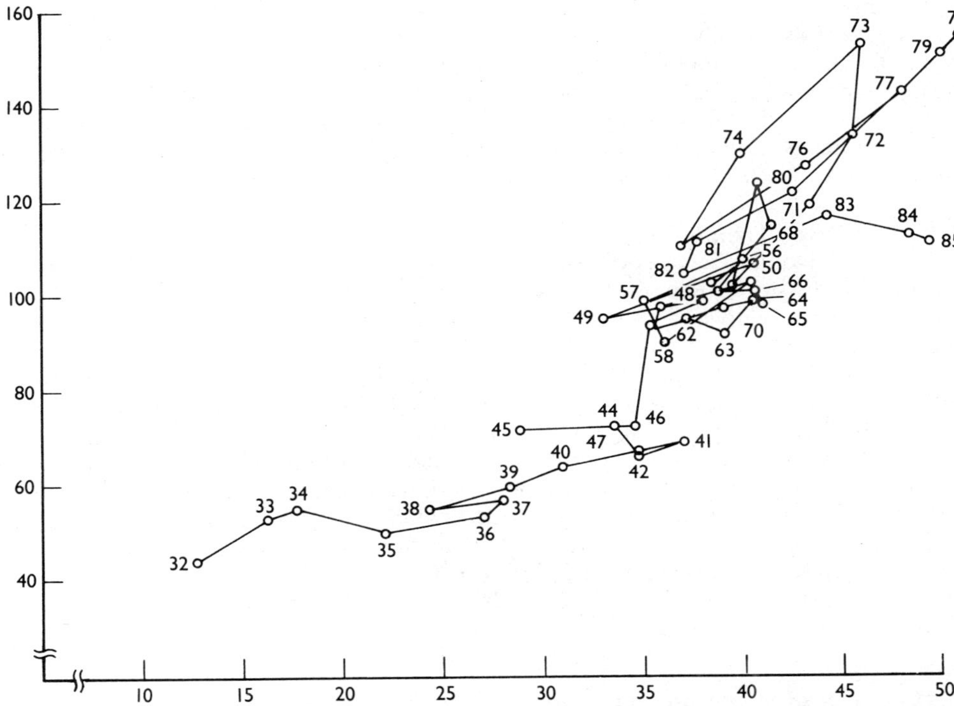

FIGURE 5.7. Deflated price index of U.S. lumber prices plotted over lumber consumption, 1932–1985. *Source*: For 1932–1950 data: USDA Forest Service 1982. An analysis of the timber situation in the United States, 1952–2030. Forest Resource Report No. 23, Table 1.15, p. 292. For 1950–1985 data: USDA Forest Service 1987. U.S. Timber Production Trade, Consumption, and Price Statistics 1950–1985. Misc. Pub. 1453, Tables 32 and 37.

Wood-based Panels

Softwood Plywood

Prior to World War II the industry consisted of fewer than 35 plants, nearly all in Washington and Oregon. Without far better information than currently available, one can only observe that 35 mills would scarcely meet the requirements of a purely competitive industry, especially since several firms operated more than one plant. Capital requirements probably served as a moderate barrier to entry, but certainly not an insurmountable one. The product was, essentially, undifferentiated, and there were no monopsonistic concentrations of buyers. This was apparently an industry in which economies of scale were such that the existing market could be supplied by a small number of mills and the structure was probably oligopolistic. Trade associations may have played an important role but we have no way of knowing.

The rapid expansion following World War II would indicate little restriction on entry; in four years (from 1946 to 1950) the number of mills doubled, and in less than another decade it had quadrupled. The situation in 1972 (the first year for which comparable data is available) is indicated in Table 5.3.

By 1972, there were 121 companies were operating 232 plants, by 1977 the number of producing companies had grown to 129, and by 1982 to 135. Nearly all of the indicated growth was in the South.

From these data it would seem that concentration is increasing slightly: all four ratios show an increase and, according to the four-firm concentration ratio, the industry would be termed moderately concentrated. The leading firm in 1977 was Georgia-Pacific, followed by Champion International, Boise Cascade, Willamette Industries, and in fifth place, Weyerhaeuser (Forest Industries, 1981). Comparable data for 1982 were not available.

Price and Price Trends

Softwood plywood price/production relationships from 1950 through 1967 (Figure 5.8) follow the pattern one would expect for an industry expanding its production to catch up with existing demand. Here was a product offering substantial labor-saving cost advantages in addition to technological superiority and it soon replaced lumber as the primary structural material for roofing, sheathing, and subflooring in most residential construction. As new mills came on line (the number of mills quadrupled between the end of World War II and 1967), the supply curve shifted outward and prices dropped. Then conditions changed: from 1970 increased output came only as prices increased, while each significant price reduction brought cuts in supply. If the industry had remained dependent on high quality Douglas fir peeler logs, one might argue the implied rising supply curve (we have not even tried to derive the supply curve) from 1970 on was due to restrictions on raw material supply. But in 1962 the first southern mill began marketing plywood made from Southern pine, and Southern production has more than filled

TABLE 5.3
Share of Value of Shipments Accounted for by the 4, 8, 20, and 50 Largest Hardwood and Softwood Veneer and Plywood Companies, 1982 and Earlier Years

		Percent of Value of Industry Shipments Accounted for by			
Year	Number of Companies	4 Largest	8 Largest	20 Largest	50 Largest
Softwood mills					
1972	121	37	50	67	87
1977	129	38	53	72	91
1982	135	41	56	74	92
Hardwood mills					
1954	326	34	43	61	77
1977	274	27	40	60	78
1982	264	25	39	57	74

SOURCE: U.S. Dept. of Commerce, Census Bureau, 1986. "Concentration Ratios in Manufacturing." MC82-S-7 Subject Series (Industries 2435 and 2436).

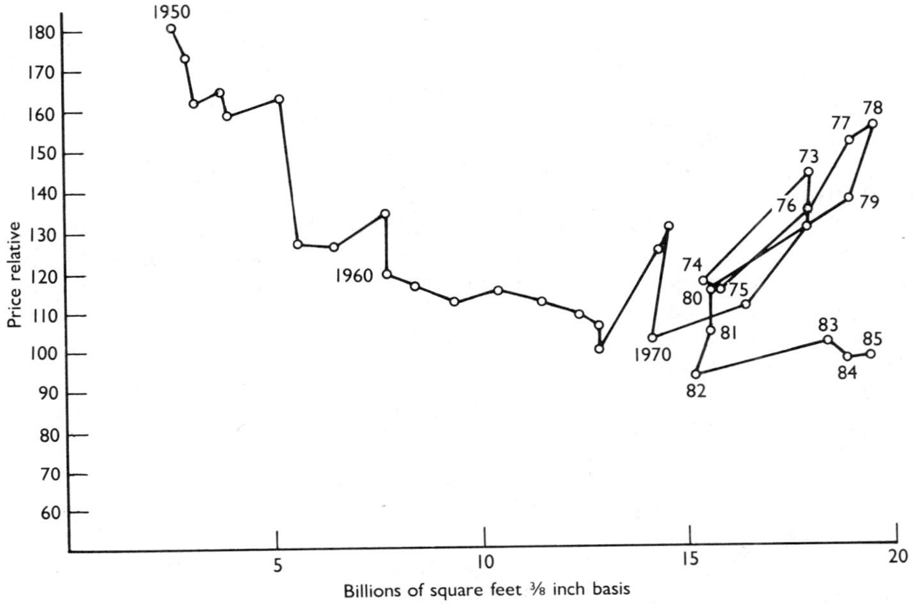

FIGURE 5.8. Softwood plywood: price relatives and production in billions of square feet, 1950–1985. Points connected chronologically. *Source*: For 1932–1950 data: USDA Forest Service 1982. An analysis of the timber situation in the United States, 1952–2030. Forest Resource Report No. 23, Table 1.15, p. 292. For 1950–1985 data: USDA Forest Service 1987. U. S. Timber Production Trade, Consumption, and Price Statistics 1950–1985. Misc. Pub. 1453, Tables 38 and 44.

the volume gap that would have otherwise developed. Plywood price/production behavior since 1970 can hardly be attributed to timber supply shortages. Instead the patterns suggest an industry that since 1978 has struggled with overcapacity problems. In 1982 prices (deflated) were at an all-time low. Output increased 21% in 1983, while prices went up 8.5%, to a level just 2.4% above that of 16 years earlier.

Existing data do not permit any conclusions to be drawn, but two observations seem pertinent. First, this is an industry of moderate production concentration: over a third (41%) of total softwood plywood is produced by the four leading companies and 56% by the eight largest. Second, regional concentration may be more significant to market structure, particularly with products for which transportation can constitute a significant percentage of market price. Using data from trade journals, Bilek gave regional four-firm concentration ratios as being 36% for the Pacific Coast, 83% for the Rocky Mountains, 52% for the North, and 53% for the South (Bilek, u.d.)[8] The nation's leading plywood producer, Georgia-Pacific, controlled 30% of the southern production, and Louisiana-Pacific an additional 7%.[9]

The above figures would make three of the four regions "highly concentrated" in terms of production—sufficiently concentrated to expect softwood

[8]As the source of this information Bilek cites *Forest Industries*, 1978, p. 30.

[9]In 1972 Louisiana-Pacific was separated from Georgia-Pacific as a result of a Federal Trade Commission antitrust suit.

plywood today to be a reasonable example of an oligopolistic industry. Although such industries are unlikely to have a definable supply curve, they can be expected to develop a price strategy.

Hardwood Plywood

Concentration

Hardwood plywood operations in the United States present a much different picture from that of their distant softwood relatives. Hardwood plywood and veneer are manufactured in smaller plants. Both the number of companies and their total output have been declining, and so have the concentration ratios (Table 5.3). The four-firm ratio dropped from 34 in 1972 to 27 in 1977, to 25 in 1982, and the decline came primarily in the larger mills. The industry has never been highly concentrated and is now one of relatively low concentration.

Prices and Price Trends

The price/production ratio pattern (Figure 5.9) describes an industry declining under the pressure of severe competition. During the 1950–1960 decade both production and price remained essentially constant. Then, during the next five years (1960–1965), production nearly doubled while imports from Asia were

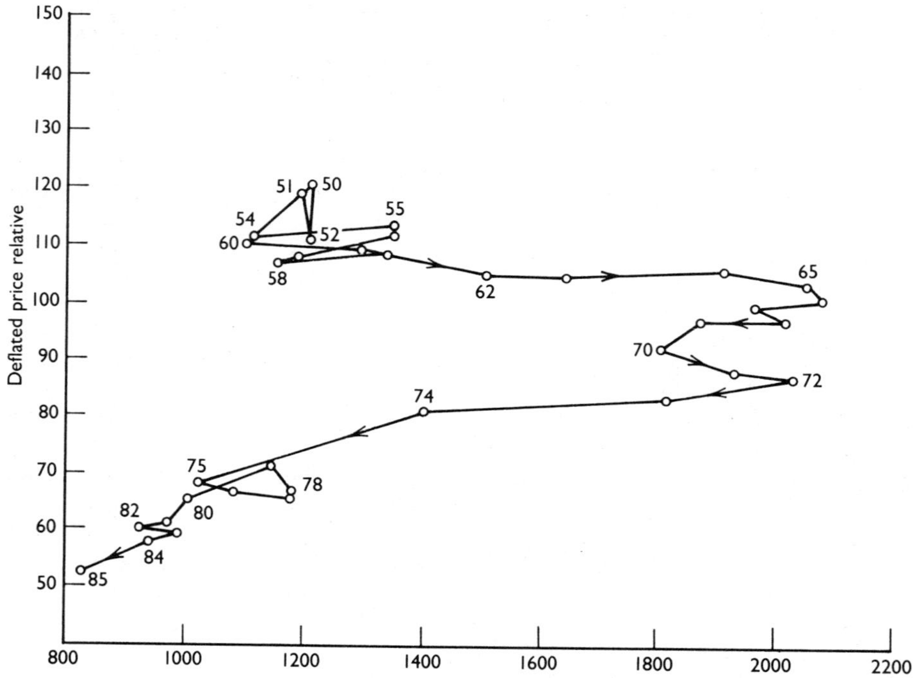

FIGURE 5.9. Price/production relationships for hardwood plywood, 1950–1985. *Source*: For 1932–1950 data: USDA Forest Service 1982. An analysis of the timber situation in the United States, 1952–2030. Forest Resource Report No. 23, Table 1.15, p. 292. For 1950–1985 data: USDA Forest Service 1987. U. S. Timber Production Trade, Consumption, and Price Statistics 1950–1985. Misc. Pub. 1453, Tables 38 and 44.

growing rapidly (refer to Figure 2.10). As imports approached, then surpassed, U.S. production, prices began to drop. Since 1972 U.S. producers have cut output by fully 50%, but prices have continued to decline and by 1980 were down 50% from their 1950 levels. By 1985 U.S. production was only 40% of that in 1966—nearly 20 years earlier—and over half the hardwood plywood sold in the United States came from Asian producers—primarily from Taiwan and Korea in 1980, with the Phillipines a distant third, followed by Japan and last Indonesia. Four years later, Indonesia was producing more than half of all U.S. hardwood plywood imports, while Korea had dropped to fifth place.

Particleboard

For the first decade or so after World War II many believed particleboard was destined to become the wonder product of the century. Production techniques had been thoroughly developed and tested in Europe, where the market expanded rapidly. Particleboard could be made from almost any species, could utilize sawmill waste, and its competitor (Douglas fir plywood) was expected to encounter a shortage of good quality logs. But by 1980 much of the excitement about particleboard had gone. Production in the United States had increased every year until 1973, when nearly 3500 million square feet was produced, but this was still only about a fifth the production of softwood plywood. No concentration ratios were published until 1972, when 39 companies were producing. The four-firm concentration ratio for that year (Table 5.4) was 48 and remained at that level through 1977, then dropped to 43, when only 31 companies were producing. The industry had been approaching the highly concentrated category, but the ratio was decreasing between 1977 and 1982. Ratios for the 8 and 20 firm categories remained virtually unchanged.

The price/production pattern (Figure 5.10) for the early years bears a distinct similarity to that of the 1950–1960 softwood plywood picture—also a rapidly expanding industry. As production increased, prices gradually dropped, but with decided unevenness. Prices then leveled off as production continued to rise through 1973, but as prices dropped further, production was curtailed. Then, as prices recovered, so did production—only to drop back again as prices again weakened. While the evidence is certainly far from overwhelming, the price pattern appears to take on the general outline one might anticipate from an oligopolistic industry that is quite concentrated, but producing a product having close substitutes.

TABLE 5.4
Share of Value of Shipments Accounted for by the 4, 8, 20, and 50 Largest Particleboard Mills, 1982 and Earlier Years

		Percent of Value of Industry Shipments Accounted for by			
Year	Number of Companies	4 Largest	8 Largest	20 Largest	50 Largest
1972	39	48	69	92	100
1977	36	48	70	95	100
1982	31	43	69	94	100

SOURCE: U.S. Dept. of Commerce, Census Bureau, 1986. "Concentration Ratios in Manufacturing." MC82-S-7 Subject Series (Industry No. 2492).

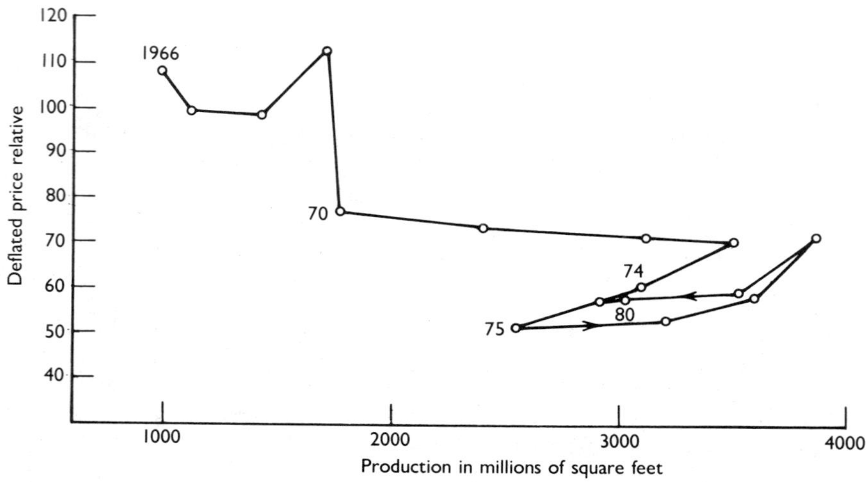

FIGURE 5.10. Price/production relationships for particleboard, 1960–1981. *Source:* USDA Forest Service 1987. U.S. Timber Production Trade, Consumption, and Price Statistics 1950–1982. (Corresponding data not available beyond 1981.)

Pulp, Paper, and Paperboard

Pulp Mills

Data of pulp mills tend to be somewhat misleading, since much of the pulp is produced by integrated companies and is not marketed. Paper and paperboard, however, are usually produced in separate establishments since these two products require different raw materials and manufacturing technologies.

Concentration

Between 1958 and 1972 the number of pulp companies was quite stable (Table 5.5). By 1977 it had dropped substantially (by 40%) but increased slightly in 1982. Surprisingly, over this same period the four-firm concentration ratio decreased from 59% to 45% in 1982. It seems likely that both mill closings and company acquisitions or mergers were responsible for the change, with the latter probably involving smaller companies. While there is no obvious trend toward greater concentration, the pulp industry is already close to being in the highly concentrated category.

Prices and Price Trends

When looking at historical price patterns in the pulp industry it is important to realize that pulp is traded on a *world* market; U.S. mills compete with those in Canada, Japan, Sweden, Finland, and the United Kingdom. For this reason the high concentration of production does not necessarily mean these companies have much power over product price. Moreover, because prices of pulp vary by species mix and pulping process, and because much of the pulp produced by integrated mills is unpriced, the price relationships, even if plotted, would be of little value.

The paper and paperboard industries, although producing what are really quite different products, show remarkably similar production and price patterns over the 1950–1985 period (Figures 5.11 and 5.12). Combined with pulp manu-

TABLE 5.5
Share of Value of Shipments Accounted for by the 4, 8, 20, and 50 Largest Companies in the Paper and Allied Products Industries, 1982 and Earlier Years

		Percent of Value of Industry Shipments Accounted for by			
Year	Number of Companies	4 Largest	8 Largest	20 Largest	50 Largest
Pulp mills					
1958	49	46	68	94	100
1963	34	48	72	97	100
1967	45	45	70	97	100
1972	46	59	83	99	100
1977	26	48	76	99	100
1982	29	45	70	99	100
Paper mills, except building paper					
1963	186	26	42	63	85
1967	203	26	43	65	86
1972	194	24	40	66	88
1977	171	23	42	70	92
1982	135	22	40	71	93
Paperboard mills					
1963	146	27	43	67	90
1967	148	27	42	67	92
1972	135	29	44	69	93
1977	120	27	42	70	95
1982	106	26	45	75	96

SOURCE: U.S. Dept. of Commerce, Bureau of Census, 1986. "Concentration Ratios in Manufacturing." MC82-S-7 Subject Series (Industries 2611, 2621, and 2631).

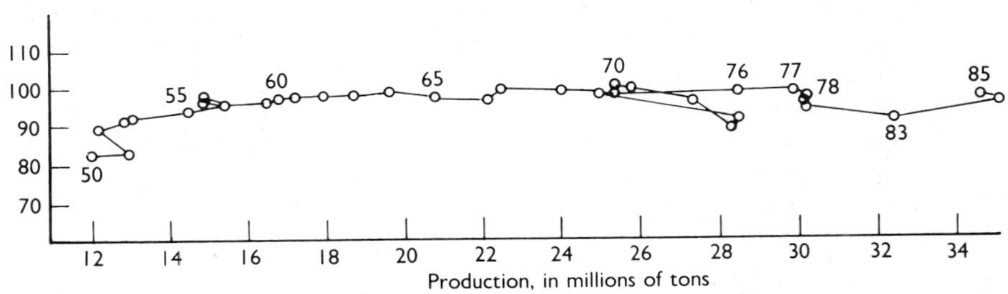

FIGURE 5.11. Price/production relationships for paper, 1950–1985. Price index, 1967 = 100, deflated by all commodities price index. Points connected chronologically. *Source*: USDA Forest Service 1987. U.S. Timber Production, Trade, Consumption, and Price Statistics 1950–1985. USDA Misc. Pub. No. 1453, Tables 46 and 53. (1985 data preliminary.)

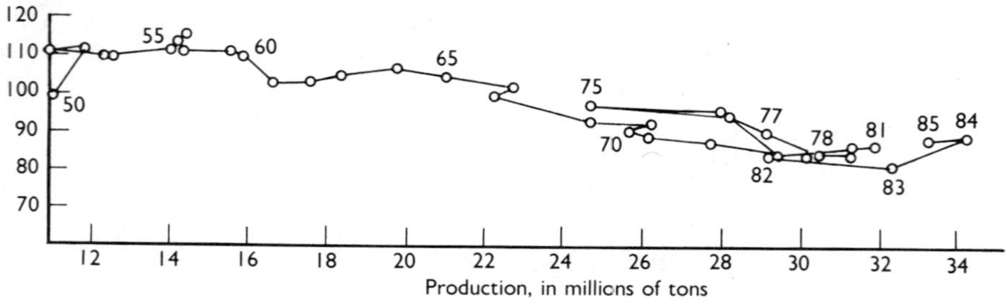

FIGURE 5.12. **Price/production relationships for paperboard, 1950–1985. Price index 1967 = 100, deflated by all commodities price index. Points connect chronologically. *Source*: USDA Forest Service 1987. U.S. Timber Production, Trade, Consumption, and Price Statistics 1950–1985. USDA Misc. Pub. No. 1453, Tables 46 and 53. (1985 data preliminary.)**

ture, they constitute one of the very large industries of the United States and of the
facture, they constitute one of the very large industries of the United States and of the world. Because there are unusually large economies of scale in manufacturing paper products, capital requirements are correspondingly large and constitute a genuine entrance barrier to the industry. (After all, not everyone has a half-billion dollars to invest.) Still, although the number of both paper and paperboard companies is slowly decreasing (Table 5.5) the two four-firm concentration ratios are very similar and very close to or below the "low concentration" limit.

Price/production patterns are also similar (Figures 5.11 and 5.12). The paper industry increased output by over 10 million tons, nearly doubling its production in the 1952–1968 period while increasing real prices only 10%. Then, during the next 13 years, it increased output by another 9 million tons while prices slowly dropped back to about their 1953 level. The paperboard industry managed slightly larger production increases in these same periods, during the first of which prices fell by 10%, and by 15% in the second. If it were not for the knowledge that very large corporations are involved, one would either judge these to be patterns of purely competitive industries, or, alternatively, industries in which price competition was being held to a minimum.

Actually, even though the industries include large companies, competition is by no means absent. Two things, however, should be kept in mind. First, as with pulp, this market is international in scope: companies in Finland, Sweden, Canada, Britain, and the United States all compete for essentially the same world markets. Second: there are, literally, hundreds of different types and grades of paper and paperboard. Many of the companies specialize their production and in this way reduce the number of competitors.

In 1981 the largest paper-making corporation in the United States and Canada was International Paper, which had held this position since 1969, when the listing began. Their production was 60% larger than that of the next largest company, St. Regis, which was followed by Champion International, Weyerhaeuser, and Boise Cascade. (Bilek, u.p.).[10] St. Regis and Champion International have since merged (Major, 1985) and Crown Zellerbach was purchased by James River Corporation in early 1986, but International Paper was still the largest in mid-1986.

Summary

Market structure and product pricing were introduced by reviewing the general economic models of pure competition, monopoly, monopolistic competition, and ologopoly. Particular attention was given to ologopoly, the characteristic structure of industry in the United States. The major wood-using industries were then examined through use of data from the Census of Manufactures and the concentration ratios calculated by the Census Bureau. Evidence of structural change was developed, then historic price/production patterns were examined for each of these industries over the 1950–1985 period.

The lumber industry shows clear evidence of increasing concentration but may still be characterized as one of low concentration. Lumber price/production patterns indicate rising supply curves and increasing inelasticities. Softwood plywood, an industry of moderate concentration, also shows a trend toward increasing levels of concentration, and its price/production pattern is that of an industry with a market that: (a) initially expanded rapidly while prices declined, (b) consolidated as supply caught up with demand, (c) entered a period (in 1970) when output could be increased only under rising prices (increasing inelasticity of demand), and (d) finally began to show declining output as prices dropped. Hardwood plywood, in contrast, has never been highly concentrated and is an industry with decreasing numbers of plants and decreasing concentration ratios. Hardwood plywood prices remained essentially constant for the 1950–1965 period while output doubled. During the next eight years while output remained essentially constant, prices dropped sharply, and from then on the industry declined in both production and price levels—an almost classic case of an import takeover of domestic markets.

The pulp and paper industries were found to be highly concentrated, but because of their great size and the fact that they sell in an international market, the price/production patterns closely resemble those one would expect from a highly competitive industry.

QUESTIONS AND PROBLEMS

1. What are the three major conditions sufficient to identify a purely competitive market?
2. Under purely competitive conditions, how is the market price determined?
3. How does a firm adjust to price changes in the short run? In the long run?
4. How does a firm in a monopoly market determine the selling price for its product? What limits exist?
5. Oligopoly is usually said to be the characteristic business structure in the United States. Describe this market form, showing how and why it operates as it does.
6. How would you characterize the four major wood-based product markets of the United States in terms of market structure? Can you give reasons for your answers?

[10]According to Bilek, the "top 50" firms in the paper industry is published annually in the June 30 issue of *The Paper Trade Journal*. It is worth noting that Weyerhaeuser was No. 1 in lumber, No. 5 in softwood plywood, No. 1 in particleboard, No. 6 in hardboard (Masonite is still the leader here), and in 1983 became second in paper production.

7. What is a "kinked" demand curve, and under what circumstances might it exist? Show how a kinked demand curve might lead to prices that remain at relatively constant levels.

8. What is a cartel? Why do you suppose they are forbidden in the United States but allowed to flourish in many European nations?

9. Some say that mergers and takeovers are moving the wood-product industries toward monopoly. Moreover, in the past few years several of the major wood-product corporations have been accused of price fixing or attempts to fix prices in their markets. Do you agree or disagree with this assessment and, even if it is true, will it make any difference to forestry?

PART 2

Timber Production

Along with a quick review of production, supply, and demand theory, Part 1 provided necessary information on the primary wood-using industries — their past production and price patterns, trade, and apparent trends in market structure. Part 2 begins the application of economics to the production of timber — the raw material on which wood-using industries depend.

Chapter 6 has two parts: the first applies the tools of marginal analysis to basic timber harvesting decisions, the second part expands the analytic framework to multiple inputs and finally to multiple products. This expansion results in a considerably more complicated model, but one capable of dealing with most forest production problems.

One of the major complicating factors of timber production is time. Forestry requires some method of comparing costs and returns that are incurred or received at different times, perhaps even in different generations. This is accomplished by discounting future values — using compound interest. Chapter 7 deals with interest, its computation, how interest rates are determined, and the effects of risk and uncertainty.

Chapter 8 takes up another complicating factor — taxes. The property tax and its effect on timber production decisions are first examined, along with the alternative most often mentioned for forestry, the yield tax. Both personal and corporate income taxes are then discussed, with special attention to capital gains provisions.

The demand for standing timber is the subject of Chapter 9, with that demand being derived from the demand for sawlogs. Factors affecting both demand and supply are treated, with supply curves for both public and private timber developed along with a general stumpage price model. The influence of market structure, taxes, and other land uses on stumpage price is examined in Chapter 10.

chapter six

MARGINAL ANALYSIS IN FORESTRY

Industrial foresters have a marked tendency to close discussions of economic problems with their favorite phrase: "It's the bottom line that counts!" Profit is of great importance to any enterprise, yet marginal analysis may be the more significant concept. It is of particular importance to foresters, especially those concerned with producing primary forest products, for there are many situations where marginal analysis is critical to sound decision making.

In many respects, the marginal principle is almost universal in its applicability. A classic example was given by Wicksteed more than 50 years ago (Wicksteed, 1933). He described the actions of an individual who, though dreading any sort of high dive, would readily leap into a river from a five-foot bank to save a drowning stranger. The same man, however, would be conscious of the conflict of two forces—the costs and revenues—"as he took off from eight feet." At a height of 10 feet he might screw up his courage and "take a header"; he would walk away "with some measure of self-contempt" if the bank were 12 feet high. But at a height of 20 feet he would feel no pangs of self-reproach; the "costs" would have become too high!

The same principle applies to many of the decisions we make daily. Should you study another hour before going to bed? Only if the expected gain is at least equal to the "cost" of the lost sleep (but don't stop reading here!). Should you take another year of collegiate work and gain a Master's degree? Only if the benefits to be derived are more than the costs.[1] Should you . . . the list is almost endless.

The range of application to forestry and forest production problems is also great. An operator watches a crew piling lumber in the yard. Will the addition of another crew member speed the work enough to warrant the hiring? If the required additional costs are equalled or exceeded by the marginal revenue product, the new crew member should be added to the payroll. A sawmill owner looks critically at a piece of equipment in his mill. Will a new edger increase either quality

[1]Here both benefits (revenues) and costs are more difficult to quantify, but the principle remains the same. And in this case one must also take into consideration the fact that most of the benefits are expected in some future period.

or quantity enough to cover the added costs incurred by its purchase and installation? (Will *MC* be exceeded by *MR*?) These cases also illustrate that in practice the changes tend to be incremental, for both productive factors and product may be, practically speaking, indivisible. One adds a person to a crew, not part of a person; one replaces an edger, not part of the machine (although edgers of different speeds may be purchased).

THE MARGINAL LOG OR TREE

One of the most discussed economic problems of timber production, and one of the most simple in theory (though not in practice), is that of determining which logs or trees should go to the mill and which should be left in the woods. There are many variations on this same theme.

Marginal Sawlog Size

Consider first the case of an operator who has purchased cutting rights to a farm woodlot for a lump sum.[2] He receives \$100 per MBF Doyle scale for all logs he delivers to the sawmill.[3] At the moment we meet him he is examining the four logs of a tree that has already been felled and bucked. The three lower logs are unquestionably worth taking, but he studies the top log, 14 inches in diameter. He has already paid for the tree, for its felling, and for the bucking and limbing. How should he decide whether to take or leave the log?

This is a clear marginal case. If taken to the mill the log will increase his total revenue by \$100 × 100/1000 or \$10. The only costs to be considered are the *additional* ones involved from this point forward. If the cost of skidding, loading, and delivering the log to the mill is less than \$10, the log should be taken, but if these costs amount to more than \$10, it should be left in the woods. If costs are exactly \$10, then it makes no difference to the operator if the log is taken or left, because the added cost is just covered by the added revenue (although it could make a difference to others).

Digression on Sunk Costs

But what of the costs already incurred in felling and bucking? Hasn't the operator already invested money in this log? Is he simply to leave the log to rot and lose whatever he has already invested? Providing there are no restrictions or special slash disposal requirements, the answer is a very definite "Yes, leave it!" For at this point all past costs are irrelevant. They are "sunk" or "historic" costs and should have no effect on the decision at hand.

A second example: suppose you were to buy a new programmable, printing calculator for \$400. One month after your purchase you learn that a new computer is available. For \$100 you can now buy something that will outperform your \$400 machine in every way. Should you buy the new computer or not? If you decide you can't afford to throw out the calculator because there is so much invested in it, then you may be making a serious mistake. So far as the question of purchasing the new

---

[2]This involves a purchase of timber that entitles the buyer, by payment of a given sum, to cut any (or all) timber he wishes from the area not specifically reserved by stated terms in the contract. Usually contrasted with a purchase on scale, wherein payment is made according to the measured (scaled) volume removed.

[3]The Doyle rule for 16-foot logs is: Board Foot Content = $(D - 4)^2$ where D is diameter inside bark at top of log, in inches. For the above example, board foot volume is $(14 - 4)^2 = 100$. Other lengths are proportional. This rule is widely used in the Southern United States. It grossly underestimates small log volumes so is to the advantage of the buyer.

machine is concerned, whether the old one cost $400, or $4000, or $4 is completely unimportant. If the new machine will be worth more to you than its cost, buy it! If the additional (marginal) cost is covered by the additional (marginal) revenue created—by the marginal revenue product—the cost is not only warranted, it should be incurred. Past costs should never be permitted to influence present decisions except that, of course, one always hopes to learn from mistakes.[4]

This is why our operator could and should properly ignore all costs already incurred.[5] At any point in a logging operation, or in any undertaking, the question should always be: "Will the additional costs incurred because of the particular decision being made be matched by the resulting revenue?" But one must bear in mind that this question may be answered differently by people in different circumstances.

Marginal Pulpwood Size

Of somewhat greater complexity is the case of a pulpwood contractor with cutting rights on a 200-acre tract of loblolly pine. Wanting to convert the old field stand to pasture, the owner has sold the timber for a lump sum ($40,000), with no specifications about cutting diameter limits, trees to be left, and so on. The contractor must now decide how to cut the timber so as to maximize net revenue.

Since it is seldom a problem when cutting pulpwood, the quality aspect can be ignored. The major question becomes one of size: should the contractor specify some minimum diameter to the cutting crew? Very probably he or she will have developed a rule of thumb that will identify this diameter sufficiently closely for the present purposes, but how might the minimum diameter be determined with reasonable precision?

Again, as you should certainly suspect, the answer can be found by taking a marginal approach. In this instance the problem may easily be related to output, for output will increase directly as cutting diameter is decreased. If typical logging costs in this type of stand had been found to vary as shown in Table 6.1, then the marginal costs may be calculated by completing the three right-hand columns, as in the table.

The same data may be portrayed graphically with the relationship between diameter cutting limits and output shown by means of a double scale (Figure 6.1). In this case the marginal cost is clearly the change in cost that accompanies a change in output, even though the decision itself is made in terms of minimum cutting diameters. At a delivered price of $30 per cord the operator will maximize net revenue by taking all trees over a minimum diameter of 7 inches, where marginal cost is about equal to marginal revenue (price). (Recall also that marginal costs are plotted at the *midpoint* of the interval.)

If price *rises* to, say, $35 per cord then clearly the cutting instructions should be changed. Under these new circumstances the timber should be taken down to a diameter of around 5 inches. And if, instead of rising, price should *fall* to $25 per cord the minimum cutting diameter should be increased.

---

[4]The value of the asset to someone else may enter into the decision, but only because it could reduce the marginal cost of the new item. (If you can unload your calculator on an unsuspecting, or unthinking, friend, for, say, $100, you would have the means to buy the new machine. But your trade might cost you the friendship of your friend, which you may value at more than $100.) The original cost, however, is unimportant.

[5]There are many cases in which sunk costs can be used to offset gain when making tax calculations, so they should not be forgotten!

TABLE 6.1
Total Costs of Logging 200-acre Tract to Different Minimum Cutting Diameters, Exclusive of Stumpage Payments

Minimum Cutting Diameter (inches)	Total Volume Delivered to Mill (cords)	Total Cost (dollars)	Change in Cost (dollars)	Change in Volume (cords)	Marginal Cost per Cord (dollars)
20	3,300	82,956			
18	4,200	92,316	9,360	900	10.40
16	5,000	101,916	9,600	800	12.00
14	5,700	111,856	9,940	700	14.20
12	6,320	122,520	10,664	620	17.20
10	6,920	135,000	12,480	600	20.80
8	7,500	150,000	15,000	580	25.86
6	8,000	165,000	15,000	500	30.00
4	8,400	179,400	14,400	400	36.00

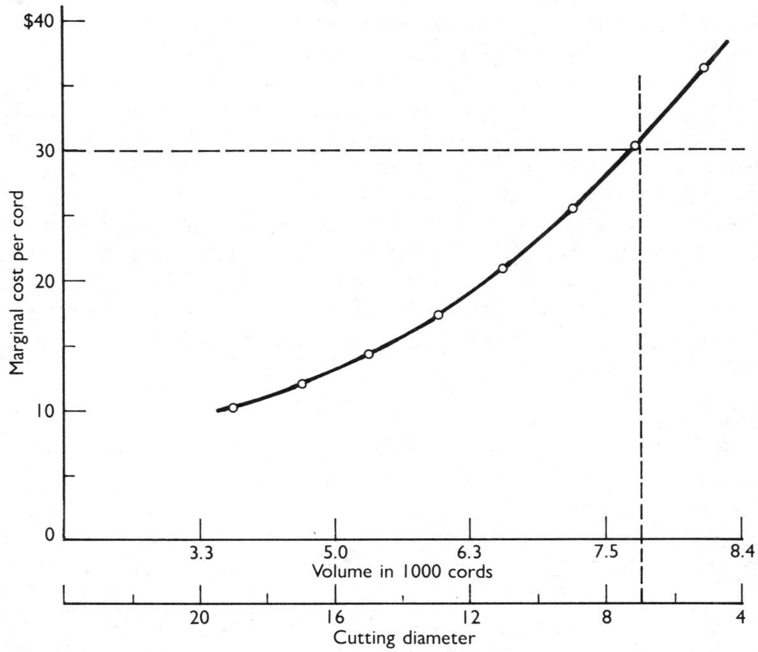

FIGURE 6.1. **Marginal cost as a function of cutting diameter and volume (from data in Table 6.1).**

What about all the other costs — salaries of assistants, depreciation that must be charged to trucks, power saws, and other equipment, as well as all the overhead costs that are involved in any contracting business? These costs do not change under the impact of a change in cutting diameter and, hence, should have no effect on the decision. They can and should be ignored for the purpose of determining the marginal log or tree. In fact, if overhead costs are apportioned to the trees by any means whatever, there is great risk that their inclusion will lead to a wrong decision.

Before leaving this case it may be worth noting that the landowner would have quite a different viewpoint. Hoping to convert the area to a cattle ranch, the landowner would realize that everything left on the ground will have to be removed at his or her own expense. The smallest possible lower diameter specification would thus be preferred. On the other hand, if the landowner intended to stay in the timber business, a cutting limit well above that of the contractor might be preferred.

Effect of Scale Used

Now change the picture a little. Instead of a pulpwood contractor, visualize the problem faced by an operator who has paid cash for unrestricted cutting rights to a tract of pine which she intends to harvest, selling the logs to a local sawmill at $50 per thousand board feet, log scale. She pays all her laborers by the day while her assistant, her two foremen, and her bookkeeper are salaried.

The operator must answer the same question that faced the pulpwood producer. To what minimum diameter should the stand be cut in order to maximize net revenues? The same basic approach should be taken (and approximately the same data must be in hand) on the costs and returns from trees of different diameters.[6] Because of the peculiarities of timber measurement, however, a few additional complications arise. The basic information is presented in Figure 6.2, where the required cost data have been plotted over thousands of board feet using both the traditional Doyle scale and the International $\frac{1}{4}$-in. rule. Clearly, the way in which the timber is measured will influence the minimum tree diameter to which the stand should be cut. Or, viewed strictly in price terms, one could say that a price quoted in board feet, Doyle scale, is a different price from the same dollar figure quoted in terms of the International rule.

In this illustration it was specified that the contractor paid her workers at a *daily* rate. If instead, wages had been paid by log scale, the relationship between cost and log diameter would again change. And, like revenues, the costs would depend not only on tree diameter per se, but also on the method used in measuring volume. Similarly, if payment were to be on a piecework basis (so much per tree or per log) the analysis would change.

Effect of Purchase Method

The **method of purchase** by the contractor can also have a significant effect. If, for example, the contractor in the previous example had bought the timber according to scaled volume instead of for a lump sum, then she would tend to cut to a higher diameter limit. This is easily seen from the total cost and revenue picture (Figure 6.3). A double horizontal scale has been used, as in the example of Figure 6.1, to

---

[6]There is a great deal of carryover from one stand to another. Once this analysis has been made, the application to other stands can be made with little new effort required. Costs must be adjusted to the new conditions, of course, but this seldom requires any time-consuming new study.

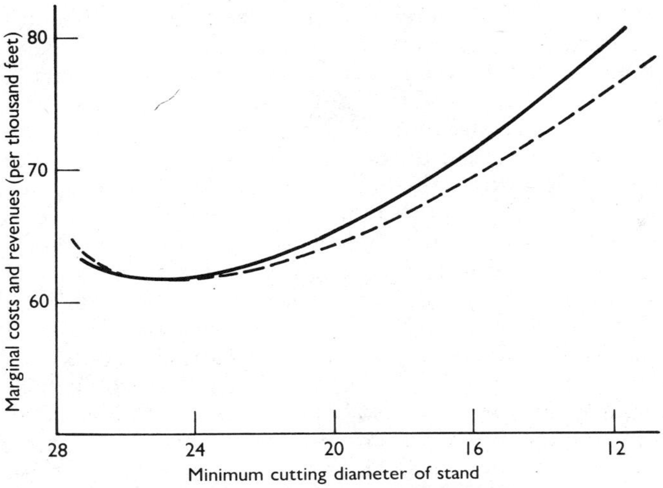

FIGURE 6.2. Marginal costs, per thousand board feet, of delivering logs to sawmill if stand is cut to various diameters. Solid line indicates Doyle scale; dotted line, International ¼-in. scale.

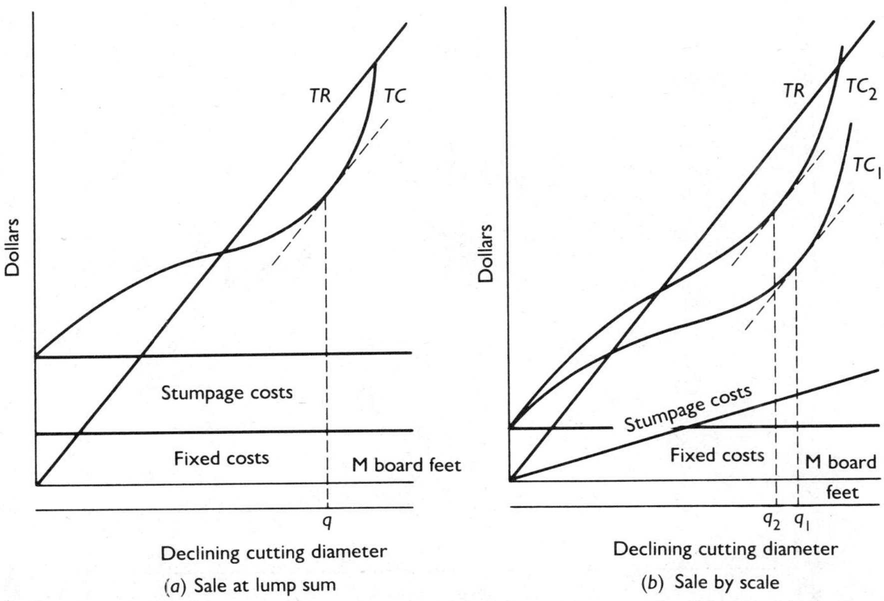

FIGURE 6.3. The method of sale may also influence the optimum cutting diameter (see text).

show both volume produced from the 500 acres and the corresponding diameter to which the stand would necessarily be cut. Both the effect of stumpage price and the manner of payment can be seen from these graphs. With a lump-sum purchase, as in Figure 6.3*a*, stumpage cost is fixed and, as with any fixed cost, its level will not influence the operating decision; once the lump sum has been paid it is also an *historic* cost, so it clearly can have no effect on the cutting diameter. But in the present instance, Figure 6.3*b*, each additional 1000 board feet of timber cut must be paid for at the agreed-on price. In Figure 6.3*b*, where the output scale is regularly graduated, the cost of stumpage will appear as a straight line passing through the origin, with its slope equal to the price per thousand board feet of stumpage. Adding this cost to the variable cost curve will clearly "attenuate" the total cost curve, pushing it upward from TC_1 to TC_2. Since the slope of total cost will be increased, optimum output will be shifted to the left, toward lower outputs and, hence, toward higher diameters. In the present example production drops from q_1 to q_2. From the operator's standpoint there has also been a gain in flexibility. The choice of diameter is no longer so critical a decision, for the cost curve has been stretched out and become more "flat," increasing the zone of profitable operation.

The same result will be attained if one regards the stumpage price as a reduction in total revenue. The stumpage charge simply "swings" the total revenue curve downward, from TR_1 to TR_2 in Figure 6.4, again indicating a reduced output and consequently a higher diameter limit for cutting. Even though the timber might be purchased under any one of many log rules, the effect will *always* be such as to make payment by log scale tend to reduce output and increase optimum cutting diameter (and also the operator's cutting flexibility), as compared with a lump-sum purchase without specification of diameter limits.

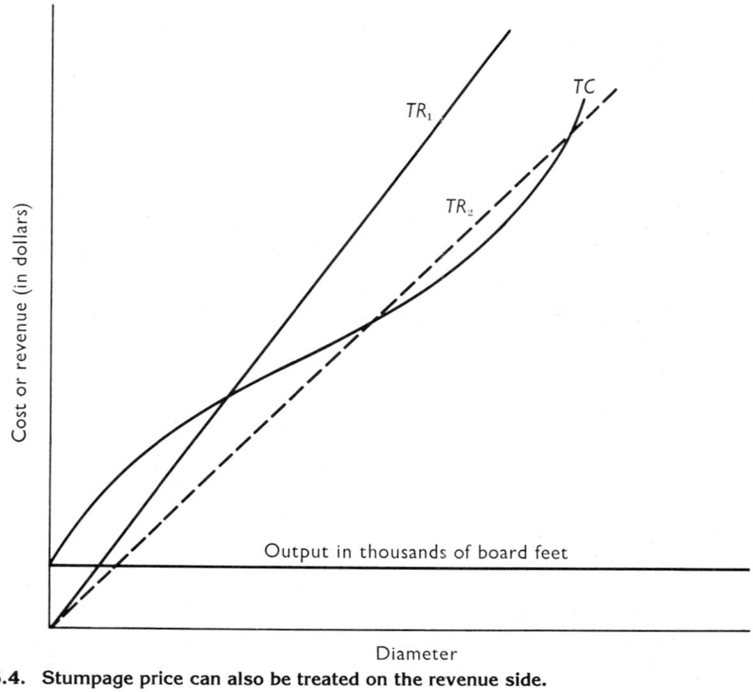

FIGURE 6.4. Stumpage price can also be treated on the revenue side.

This does *not* mean that the operator's net revenue from the tract will necessarily be reduced by reason of sale by log measurement. It may, in fact, be increased. Net revenue is the difference between total cost and total revenue. In the comparison being made here it depends on the level of the lump-sum payment on the one hand and the level of stumpage price on the other. In the case of a lump-sum purchase, the *level* of the purchase price has no impact on the optimum diameter limit for cutting, but it does influence profit. In the case of sale by scaled measure, the higher the stumpage price the higher the diameter limit to which the stand should be cut, provided that the relationship between log diameter and costs is the usual one in which costs per 1000 board feet rise as log diameter decreases (as shown in preceding figures).

The variations on this same general theme of minimum cutting diameters are almost endless. Perhaps the most instructive case to think through is that of an integrated operation, where the corporation owning and operating the sawmill also has its own logging crews, its own trucks, and so on. In that case the cost data and analysis must extend all the way through to the final sale of lumber, provided there are factors that cause costs to vary as diameter changes. Sawing costs, of course, also change with log diameter—and so does overrun,[7] but few of the other costs not associated with harvesting will vary with diameter. Marginal *revenue*, on the other hand, may vary quite sharply, for lumber quality and seasoning degrade may both act to decrease the value of lumber cut from smaller diameters, while typically the overrun will increase the value of the smaller diameters by raising the yield.

Effect of Quality

Variation in tree or log quality must be given some attention simply because it is so very important in attaining maximum net revenues from a lumber operation. Neither tree nor log grade have a very significant effect on cost, but the effect on revenue is pronounced. So long as the logging operator is paid on a flat per M basis (or on any other volume basis, such as cubic feet or cubic meters), the matter of log or tree grade will not enter into calculations of profitability or influence in any way the minimum cutting diameter. If a minimum diameter or log quality is specified by the mill operator, the contractor will modify the cutting accordingly. But within this constraint, quality will have no bearing on *that* cutting. However, log quality will greatly influence the profitability of a sawmilling operation whenever the product is sold by grade. Many mills could improve their profit position substantially by paying different prices for logs of different size and quality, as long as logging contractors would accept the innovation and no undue transaction costs arose. And the logging contractors would also be helped. If these marginal revenue curves were to be applied to cutting limits in field operations, decidedly different tree or log diameters should be specified for the different log or tree grades, as is illustrated in Figure 6.5. Note that in this illustration only the revenue curves are shifted.

The increasing proportion of plantation-grown timber is having an impact on both sawlog and pulpwood quality and could eventually influence the method of purchase. "Fast-grown, short-rotation conifers lack the wood fiber needed for products requiring structural strength" (Senft, Bendtsen, and Galligan, 1985).

---

[7]Overrun is the amount by which the actual amount of lumber cut exceeds the log scale. With the Doyle scale, for example, it is not unusual for a mill to get as much as 200% overrun on 8-inch logs, and to cut slightly *less* than scale on logs over 25 inches in diameter. Doyle scale is seldom used on really large logs.

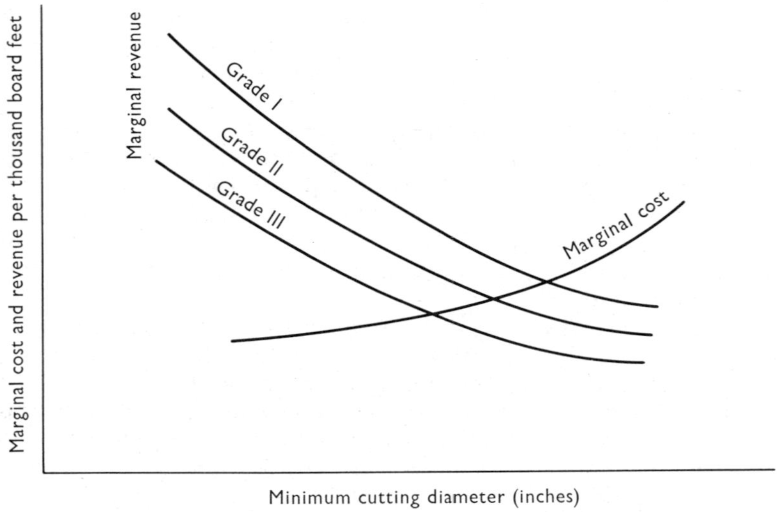

FIGURE 6.5. Effect of tree grade on minimum cutting diameter as long as revenues are made effective for logging operator.

Although the weak wood apparently results from the higher porportion of juvenile wood rather than from rapid growth per se (Kellison, 1985), the problem remains. Wood density can be estimated by weighing logs as they are delivered, and purchase by weight has been used by some pulp mills. As plantation-grown sawlogs become more common, weight may become an important quality indicator to sawmill operators as well as to pulpwood buyers.

Opportunity Cost

The mill owner of earlier days faced a different kind of situation. In the Lake states of the 1880s, for example, tremendous stands of timber were waiting for anyone willing and able to take them. There was no shortage of timber or of loggers. There was, however, a premium on getting the timber out quickly, for if one stayed very long in one place a rival logger would be cutting over the next section.[8] Every 1000 feet of lumber cut from 20-inch timber would yield an average profit of, say, $5, while 1000 feet cut from 40-inch trees might yield $20 in profit. The "opportunity cost" of cutting 20-inch timber under these conditions would be $15 per thousand.

This principle of opportunity cost has wide applicability. Consider the following examples. A hydroelectric development creates a sizable lake bordering a privately owned forest. Highly productive timberland suddenly becomes desirable for an outdoor recreation development. The owners are offered $300 per acre for the forest property, but decide to maintain ownership for timber production, reasoning that since they paid $20 per acre for the tract the present yield of $1.20/acre means they are earning 6% annually on their investment. From an

[8]The situation is repeated today in many of the developing countries. True, most of these nations have strict legislation regarding timber removal, but these laws are seldom more stringent than those existing in the Lake states during the 1870s and 1880s. And, unfortunately, far too often they are no better enforced.

economic standpoint, however, the conditions have changed: $20 per acre is no longer the relevant cost, for each acre of this forest property now has an opportunity cost of $300. The return of $1.20 per acre is not yielding 6%, but instead a paltry 0.4%. This does not necessarily mean that the property should immediately be sold or even converted to recreational use. For some people a return of 0.4% from timber production may be preferable to sale for recreational purposes. Such individuals are receiving a return (satisfaction) in addition to monetary income. This return, although nonmonetary, is a very important part of the revenue from the property.

The minister of agriculture in a South American country recommends to the planning commission that the overgrazed brushland on the rolling high plain be gradually converted to improved pasture. Land prices are presently so low that the return from beef cattle could be as much as 10%. But the chief of the forest service states flatly that *Pinus radiata* will yield at least 25 cubic meters per hectare per year at $2.40 per cubic meter—an annual return of $60 per hectare. To solve this controversy, the planning commission utilizes the opportunity cost concept, calculating land values and also product values under alternative uses. Using the area for producing beef cattle may force importation of newsprint and thereby impose a serious drain on the country's foreign exchange balance, an opportunity cost of beef production not foreseen by the Ministerio de Agricultura. On the other hand, conversion of the brushland to timber may mean a reduction in protein intake for the nation's people, an opportunity cost of timber production completely ignored by the Jefe de Servicio Forestal.

The Marginal Property

The marginal log and marginal tree are concepts quite readily accepted by most operating managers, even though an analytical approach to determining just which logs or trees are marginal may represent something quite outside their experience. The extension of this reasoning to determine marginal properties is less easily accomplished, for it involves more than one dimension—precisely the reason for redefining marginal cost as the change in cost under the impact of some particular decision. There are many such dimensions, and we can consider only a few here.

Consider, for example, a sawmill operator looking for timber to supply his mill. How far from the mill can he afford to go? What represents the "marginal tract" to him? Clearly, distance is only one of the several factors involved: a 50-acre woodlot of scrubby second growth within a mile of his mill could well be submarginal, while a tract of high-quality old growth 150 miles distant but near the expressway that skirts the mill yard could be highly profitable. To the sawmill operator, then, the marginal property is one that returns just enough after conversion to lumber to cover all the expenses of acquisition, logging, transport, milling, seasoning, sale, and any opportunity costs. This can be stated very directly: the operator must decide whether the additional costs involved in turning the trees on the property into salable lumber will at least be covered by the additional revenue created by these actions. If so, and if existing opportunity costs have been included, the timber should be acquired.[9]

With a timber *grower*, however, a number of additional factors will be in-

[9]This does not mean that the operator will necessarily receive his anticipated profit level. Uncertainties remain even though opportunity costs are correctly assessed.

volved, for everything affecting timber growth or its cost must be considered. Two of the more important of these factors are site and transportation costs.

Perhaps because forestry has long been regarded an extensive land use there has been a tendency to disregard basic productivity when acquiring land for growing timber. Yet timber yields may be more affected by site than agricultural yields, for the timing of the timber yield can be changed—a point emphasized in later chapters. A property consisting entirely of site five land, for example, may well be submarginal for timber growing at a price of $50 per acre, while land of site one quality could be a good buy at anything under $150 per acre.

Transportation can be equally important. Other things being equal (which is seldom the case), land values decrease as their distance from the manufacturing point increases. For acquisition purposes, or for tentative evaluation of a region's potential for forest development, it is often helpful to map areas of timberland according to cost contours based on transport costs, thus establishing the timbershed of the mill and at the same time dividing it into cost zones. Figure 6.6 gives an example of this kind of map and illustrates also the effect of existing transportation routes.

If the mill is located so that land might be purchased in different states (or even counties, in some cases), then any difference in tax levels must be taken into consideration. Similarly, such items as insurance, workmen's compensation, and minimum wage legislation are other factors that tend to vary with political boundaries.

EXPANDING THE ANALYSIS

These examples show that problems in producing forest products, especially those of a primary nature, can often be solved by a straightforward application of marginal analysis. Before becoming involved with the time aspects of production, however, the analytic framework must be expanded. Forestry frequently involves balancing several inputs, and often the simultaneous production of several products as well. Multiple inputs and outputs are easily fitted into the marginal analysis.

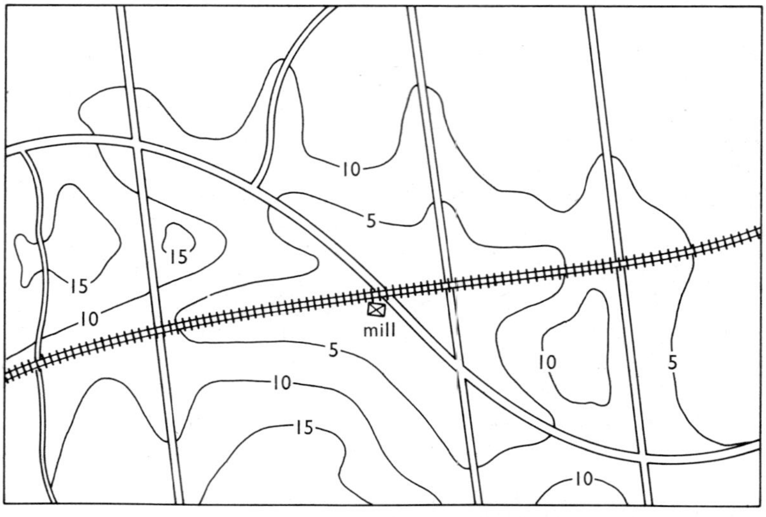

FIGURE 6.6. The timbershed of a mill. Note the influence of roads, railroads, and so on. Cost contours in dollars of transport costs.

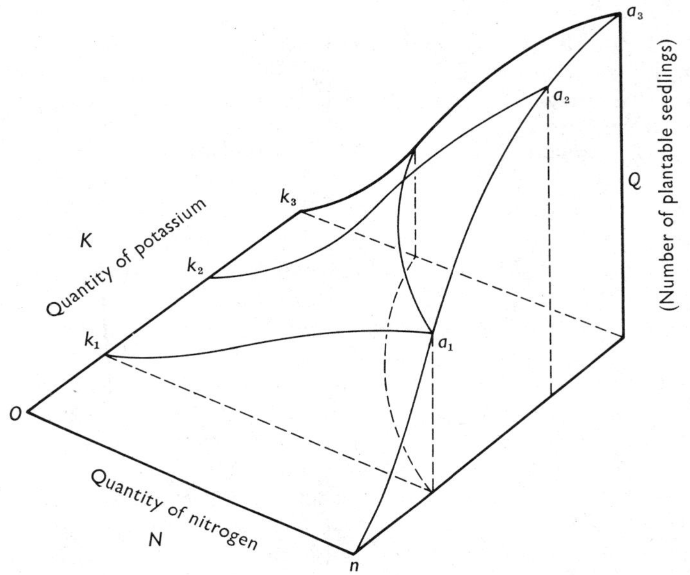

FIGURE 6.7. A production surface, generated by two input factors.

Production with Several Inputs

The problem of determining the optimum combination of several productive factors is most easily understood by first examining the special case of two inputs, which has the advantage of lending itself to graphic treatment. The two-factor example can be illustrated in two steps:

1. Given two productive factors, what is the optimum combination to use when producing any quantity of the product?
2. What combination of inputs will assure maximization of net revenue?

A Production Surface

The effect on output of varying the combination of two inputs is shown in Figure 6.7, again using a forest nursery example. The two sides of the base indicate increasing inputs of two factors—nitrogenous and potassium fertilizers—while the vertical axis represents output measured in terms of plantable seedlings.

These two inputs are added to a fixed factor (a given amount of land) and all other inputs—seedlings, water, other fertilizers, herbicides, and so on—are held constant. If no potassium is used, additional nitrogen will have no effect—in fact, without some potassium, no seedlings will be produced. And the same goes for nitrogen.

If the level of potassium is held constant at, say, k_1, the output response as the input of nitrogen is increased will be shown by the curve k_1a_1. Similar total physical product curves are identified on the diagram for two other levels of potassium input, k_2 and k_3. The same general pattern of output follows changes in potassium levels while nitrogen inputs are held constant, although the figure shows only the curve $na_1a_2a_3$.

Now, with a "production surface" such as this, any vertical cross section parallel to either the N or K axis will show the characteristic production function of a single input (see Fig. 4.1). This curve indicates the response of a single input with all other inputs held constant. If, however, the surface is cut vertically but with the "slice" passing through the origin, the resulting curve of output will show the effect of changing *both* N and K in some precise proportion: the ratio N/K is a constant on any line drawn through the origin. This is an important special case and will be discussed later.

Isoquants

The portrayal of a production surface on a two-dimensional diagram creates no problem for foresters, who are necessarily familiar with topographic mapping. If lines of equal quantities of product are traced on the production surface, they will form "contour" lines, as is true for any two-dimensional representation of surface topography. Labeled according to the quantities represented, these lines connecting points of equal output quantities are called isoquants, much as lines connecting points of equal temperature are called isotherms. As with contour lines, the number of these isoquants drawn on a given "map" is limited only by the detail required. And, like contour lines, these isoquants can never cross. Figure 6.8 shows only four, indicating outputs of 25, 50, 100, and 200 plantable seedlings. Their significance, however, is not the volumes per se, but that each isoquant identifies all the combinations of N and K that can be used with the given set of fixed factors to produce the indicated quantities of seedlings.

Isocosts

Suppose someone were to ask "If I have only a certain amount of money for buying nitrogen and potassium, how much of each should I buy?" Now the problem has an economic slant, for the questioner is asking "How should I allocate my scarce resource?" Assuming that the producer wants to produce as many seedlings as

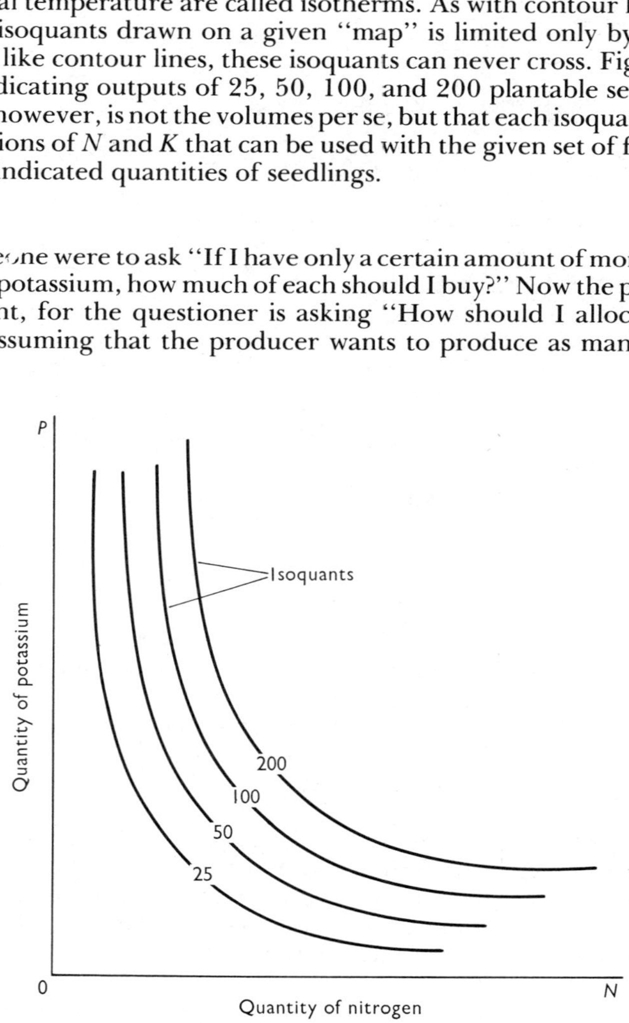

FIGURE 6.8. Two-dimensional representation of a production surface similar to that of Figure 6.7. Outputs of 25, 50, 100, and 200 seedlings are shown.

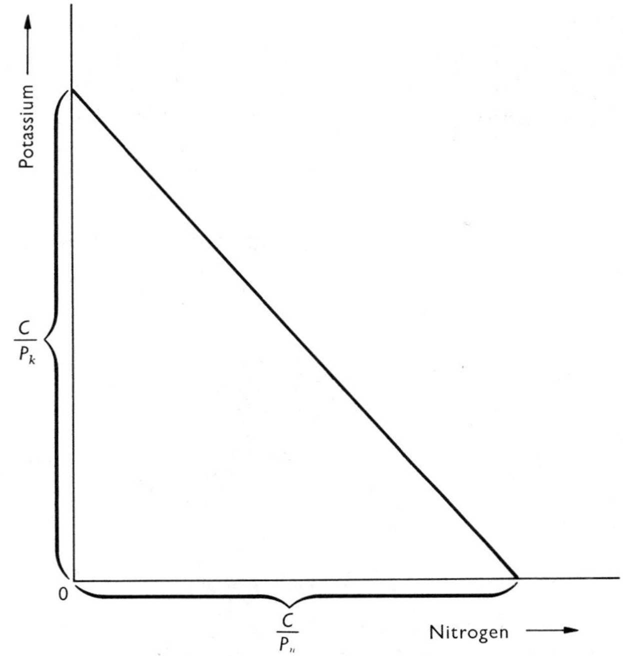

FIGURE 6.9. The quantities of the two inputs that can be acquired for a given expenditure.

possible for his expenditure, the answer can be determined easily. Let C represent the amount of money—any amount—he has for buying N and K, let P_n and P_k be the price of nitrogen and potassium fertilizers, respectively, and let C/P_n and C/P_k represent the quantities of nitrogen and potassium fertilizers that can be purchased for C dollars (Figure 6.9). By connecting the two quantities with a straight line, all possible combinations of the two inputs factors that can be purchased for C dollars will be identified. This line is a type of isocost (a line of equal costs) and is only one of a whole family of such curves, since a different line will be generated for every different value selected for C. But no matter how many of these we draw, as long as the two factor prices remain unchanged all the isocosts will have exactly the same slope, and will be parallel.[10]

Least Cost Combinations

If we combine the two figures, Figures 6.8 and 6.9, by superimposing one on the other (Figure 6.10), the first part of our problem can be answered. The isocosts in the figure show all combinations of K and N that can be purchased for a given level

[10]The slope of the isocosts of Figure 6.9 will be

$$\frac{\frac{C}{P_k}}{\frac{C}{P_n}} = \frac{\not{C}}{P_k} \cdot \frac{P_n}{\not{C}} = \frac{P_n}{P_k}$$

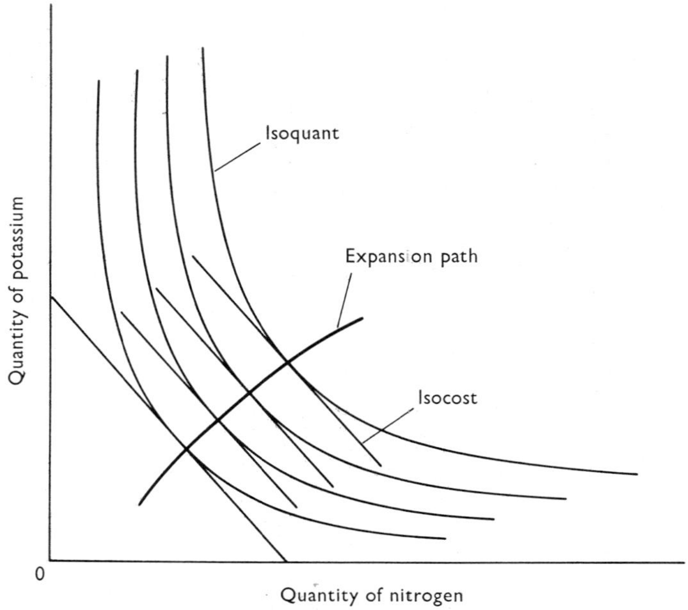

FIGURE 6.10. Optimum combinations of the two inputs that, under the given conditions, can be used to produce any feasible level of output are indicated by the expansion path.

of expenditure, while the isoquants show how different combinations of *K* and *N* can be used to produce various quantities of seedlings. The points of tangency of isocosts and isoquants identify the combinations of *K* and *N* that will permit producing those quantities at lowest possible cost—they identify **least cost combinations**. In the figure, a curve connecting these points of tangency has been drawn—a curve called the **expansion path**, or the *scale line*—which shows the least cost combinations of input factors for the whole range of operating scale. The manager will always minimize production costs by using the combination of inputs shown in this curve, so the variable production costs will always be those indicated by the isocost curves at their points of intersection with the expansion path. Outputs are also shown, but must be read from the isoquants. The relation of cost to output can be read directly from Figure 6.10 and, if desired, replotted in conventional form as in Figure 6.11. In each case, the output indicated by the intersection of the expansion path with an isoquant has been plotted at the cost level indicated by the corresponding (and also intersecting) isocost curve.

Maximizing Net Revenue

The maximization problem is now exactly the same as the one already solved in the single input example. Fixed costs must be added along with the total revenue curve, and the profit maximizing output will be found where marginal cost and marginal revenue are equal—where the slopes of the total cost and total revenue curves are identical (Figure 6.12). Before leaving this problem, however, recall the assumptions and the significance of the production surface. Any change in technology, and any error or uncertainty introduced by lack of knowledge of how seedlings might respond to the particular inputs, will be reflected by a shift of the isoquants, and this in turn will dictate a new combination of the productive factors.

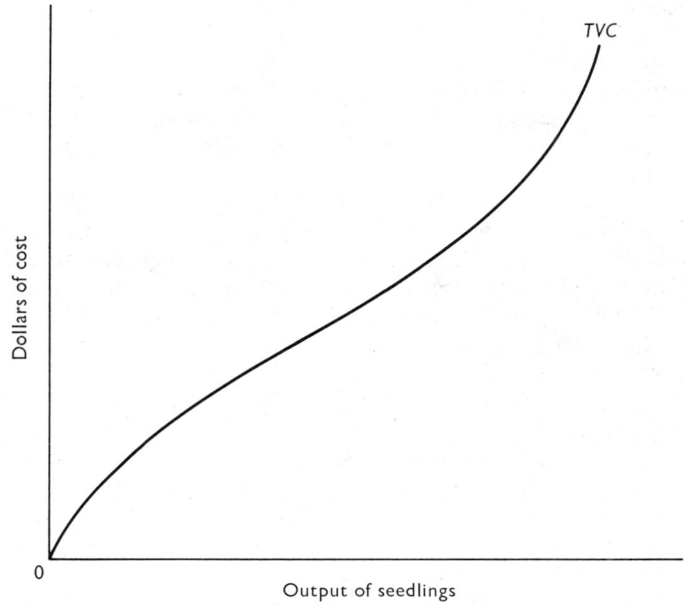

FIGURE 6.11. The total variable cost curve derived from Figure 6.10.

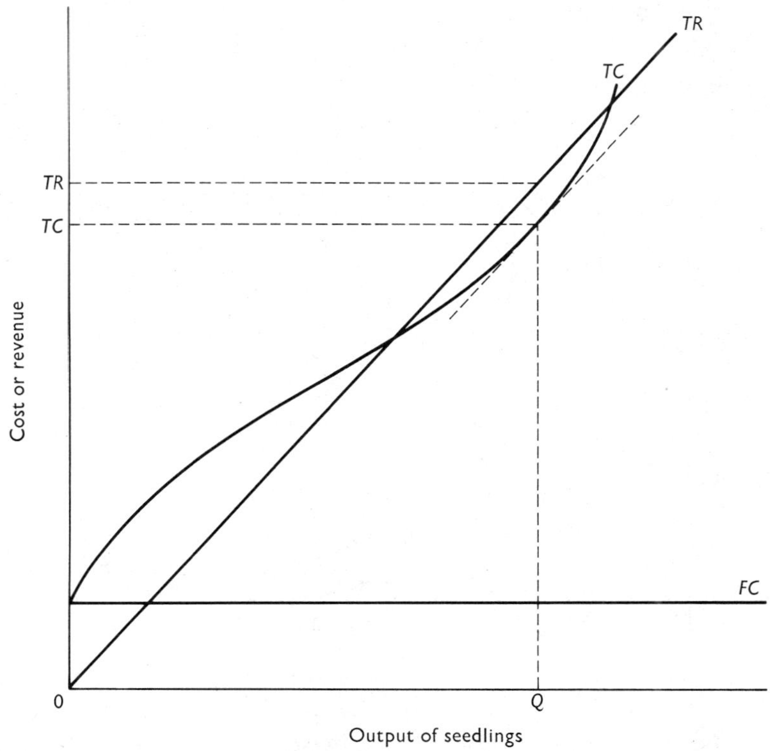

FIGURE 6.12. The solution to the maximization problem.

A General Guide

Suppose we have the isoquant pattern of Figure 6.13. If we take any two points of an isoquant—say *a* and *b*—then ΔK is the *change* in the amount of potassium that must be "balanced" by an increase in the amount of nitrogen (ΔN) in order to hold output constant, and $\Delta K/\Delta N$ represents the ratio of these changes. At any point on the isoquant, $\Delta K/\Delta N$ is the **marginal rate of substitution** of potassium for nitrogen (while holding output constant).

Previously, it was seen that the slope of the isocost curve (P_n/P_k) was dependent on the relative prices of the two factors. At the point of tangency between the isocost and isoquant curves $\Delta K/\Delta N = P_n/P_k$, or **the marginal rate of substitution of the factors is equal to the inverse of the ratio of their prices** (costs). Hence, if the price (cost) of nitrogen (P_n) rises relative to that of potassium (P_k), one should use relatively *less* nitrogen and *more* potassium. In simple terms, substitute the cheaper material for the more expensive.

Intuitively, one can follow this rule with any number of inputs. Substitute one input for any other input (or any other group of inputs) until their marginal rate of substitution is equal to the inverse of their price ratio. (Substitute cheap inputs for expensive ones as long as output can be kept constant!). The *n* input case can be developed easily by using differential equations for isoquants.

Degree of Factor Substitutability

A little reflection will reveal that the isoquant pattern depends directly on the physical relationships involved. The degree to which one factor may be substituted for another varies tremendously—from perfect substitutability to no substitut-

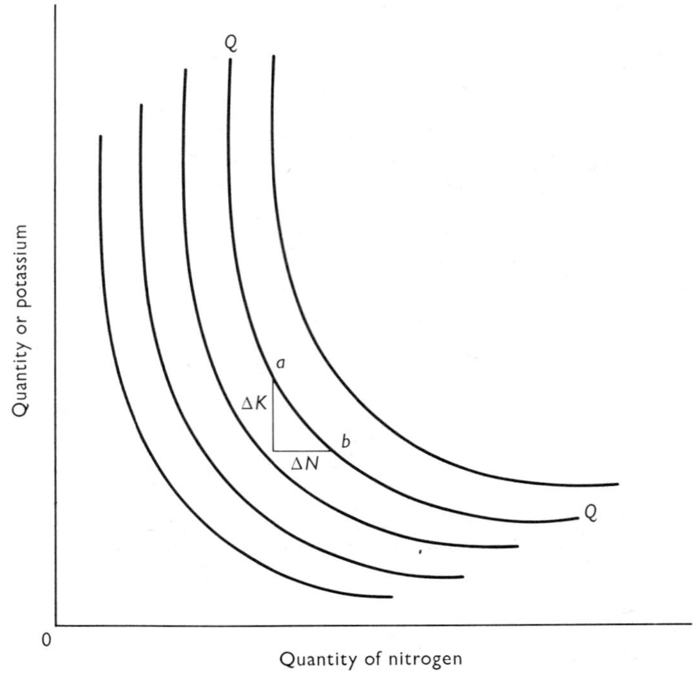

FIGURE 6.13. The slope of an isoquant (here $\Delta K/\Delta N$) is equal to the marginal rate of substitution of potassium for nitrogen.

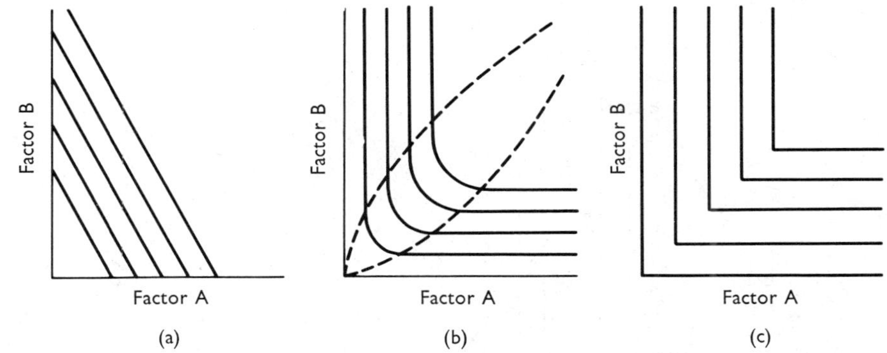

FIGURE 6.14. Isoquant patterns showing different degrees of substitutability; *(a)* depicts perfect substitutability; *(b)* shows intermediate substitutability (the two factors are substitutes over a limited range); *(c)* shows no substitutability.

ability whatever. Perfect substitutability of productive factors yields an isoquant pattern similar to that of *a* in Figure 6.14 while *c* gives the opposite case—that of no substitutability whatever.

If the situation is that of *a*, three alternatives exist, with choice dependent on the relative prices of the two factors. One would use either all *K* and no *N*, or all *N* and no *K*, or (if the isocost curve had exactly the same slope as the isoquant) the choice of input combinations would be immaterial. In the situation shown by *C* (no substitutability), regardless of the relative prices the optimum combination is that indicated by the "corner" of the isoquant; there is no other possible point of tangency.

The more usual case is the intermediate one of Figure 6.14*b*. Here the two factors can be substituted for each other, but over the limited range indicated by the dotted lines. Outside this range there is no substitutability. No matter what the relative prices of the two factors are, some of each will be required, but there is an opportunity to substitute one for the other if relative prices change.

Constant Proportions

The case of constant proportions is of special interest. Given any isoquant map or diagram relating output to two productive services, any straight line through the origin connects input combinations of **constant proportions**. In many cases productive services can only be used in specific proportions. Tractors and trucks require a driver; power saws require an operator; and so on. When factor proportions are absolutely fixed there can be no problem of finding the optimum combination. Only one combination makes sense, so it can be treated as a single factor.

There are, however, situations where several combinations are possible. Figure 6.15 provides a simple illustration of production under conditions of constant proportions where there are two possible ratios. For this example, suppose either one-person or two-person crews can be used for felling, limbing, and bucking timber, with each *crew* equipped with a power saw and an axe. If (1) these are the only combinations of workers and equipment possible, (2) returns to scale are constant (each crew performs equally well and output per crew is the same regardless of the number of crews), and (3) production is as shown by the isoquant pattern of Figure 6.15, then along either of the two "rays" from the origin the proportions of labor to equipment are constant, and indicate the two possible combinations.

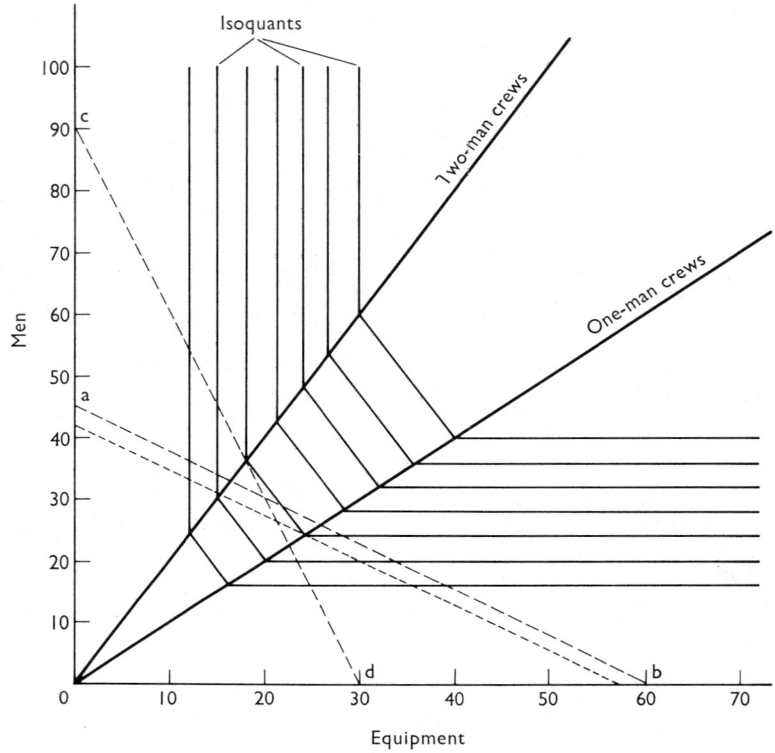

FIGURE 6.15. Constant proportions, where two combinations are possible.

Note that, as drawn, production increases at a constant rate as one moves out along either ray.[11]

Costs of the two inputs can now be introduced. With equipment costs of $12 per day for each "package" (one power saw plus one axe), and with *wages* at two dollars per hour ($16 per day), the dashed line *ab* becomes an isocost, showing all combinations of workers and equipment that can be purchased for $720 daily. Any desired number of isocosts can now be constructed, all paralleling line *ab*, and any number of isoquants can also be drawn if helpful. But the point of tangency between isocost and isoquant can only be at the lower corner of the isoquant (as shown by the dotted line parallel to *ab*) indicating that under the stated conditions, one-person crews would be best for producing *any* output.

Now suppose wages were only $4 per day while equipment costs remained at $12. The isocost pattern would shift to that shown by *cd* and the upper corner of the isoquant would become the only possible tangency point. At these relative costs, the two-person crew would be best.

In this particular illustration a wage rate of $6 per day (with the same equip-

[11]The assumption of constant returns to scale is not necessary at this point. *Any* pattern of isoquants could be substituted so far as the *progression* of output is concerned. The assumption is, however, necessary if the problem is to be treated by usual linear programming.

ment costs) makes the choice between one- and two-person crews immaterial—the isocost curves parallel the isoquants along the significant part of the production pattern. This is another way of saying that if wages are less than $6 per day, the two-person crew should be chosen, while at any higher wage a one-person crew will be less costly (assuming equipment costs and productivity remain unchanged).

MULTIPLE PRODUCTS

Even though sawtimber is the only objective of management, a forest produces logs of different grades from which lumber of a variety of grades and sizes is manufactured. And in today's economy it is rare to find a forest managed only for sawtimber; at the very least, pulpwood will be sold at some stage of the production process, and most often other products as well. Clearly, if economics is to be applied to forestry, the framework must cover simultaneous production of several products.

Joint Production

The essential features of the needed expansion follow directly from the two-input case. Consider, for example the diagram of Figure 6.16. With the quantity of pulpwood measured along one axis and that of sawlogs along the other, all possible combinations of the two can be shown. In a manner directly analogous to that used in the two input problem, point *A* indicates a combination of 400 cords of pulpwood and 200,000 board feet of sawlogs, while point *B* indicates 200 cords of pulpwood and 800,000 board feet of sawlogs.

If the two products had been produced by entirely separate processes, no special problems would arise—each product would have its own production function, its own set of cost curves, and so on. The really interesting cases, however, are those in which the same production facility—the same "factory"—is used to produce two or more products. This is *joint production*, so called because at least one

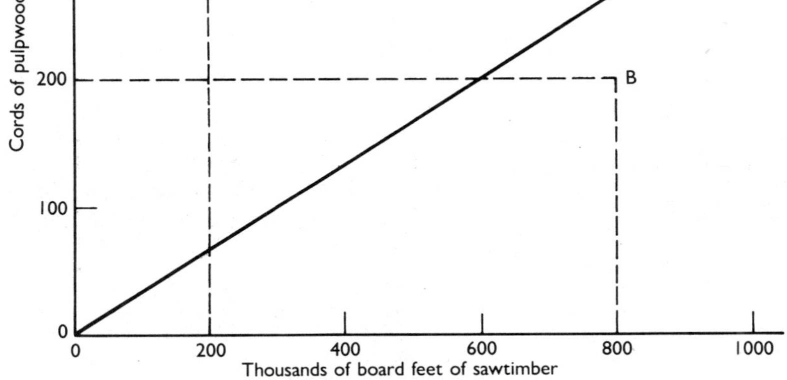

FIGURE 6.16. Product combination.

of the factors, often the entire group of fixed factors, is jointly used. One of the two types of joint production is that in which products are produced in technically fixed proportions; in the other, the product proportions are technically variable.

Technically Fixed Proportions

With production of this kind, all technically possible product combinations hold a constant ratio to each other. When plotted on a diagram such as Figure 6.16, the product combinations always fall on a straight line passing through the origin. The line OC, for example, indicates pulpwood/sawlog combinations that are always in a 1 cord to 3000 board foot proportion. Forestry offers few examples of truly fixed proportions, although some forest areas are handled as if the proportions were unchangeable. Naval stores (from which turpentine and other chemicals are extracted) and sawtimber, or pulpwood and sawtimber, might occasionally be so considered, although these proportions can be drastically altered.

Wood products offer many examples of fixed proportions: most chairs require four legs and two arms for each seat; some papers require specific proportions of long and short fibers. In such cases the problem can always be reduced to the single product case. If, for example, a sawmill produces one ton of pulpwood chips with every 1400 board feet of sawnwood, one can treat this as a sort of "combination product." The total cost and revenue picture might then appear as in Figure 6.17, where single cost and revenue curves are used while the horizontal axis shows quantities of both lumber and chips at all relevant outputs. The profit-maximizing solution is obtained as before, and a full set of total cost curves could be drawn. But **average costs cannot be derived**. In the above example one cannot calculate an average cost for either chips or lumber. It doesn't exist! Try to derive it. This is the major point to remember about joint production with technically fixed proportions, and it is a point that holds for *all* joint production.

In purely competitive situations (where prices of both products are given to the producer) no particular treatment of the firm's demand is necessary. But if one is making a market analysis of an *industry*, especially if industry prices are being studied, the prices of the two jointly produced products may behave quite differently. Even though the demand curves may be completely independent, the supply curves are not, and price is a function of both demand and supply. For example, the derived demand for sawlogs in a particular area—derived from the demand for lumber—might be D_s of Figure 6.18*a*, and the demand for chipable waste material from these same logs D_c of Figure 6.18*b*. The industry demand for sawlogs *and* chipable waste would then be regarded as the vertical addition of the two demand curves, the curve labeled D_{sc} in Figure 6.18*b*. If the supply curve for the combination product were that shown as S_{sc}, the price of logs would eventually settle at P_{sc}.[12] Actually, in almost all cases *some* costs can be separated from the total and identified as applying only to the output of one or the other product. Product sales and transport costs are typical examples of such "separable" costs. These can be brought into the supply schedules (see footnote 12) but serve at this time primarily to emphasize that, while average costs of production cannot be calculated for jointly produced products, it is often possible to calculate *marginal* costs for both

[12]This represents a decided oversimplification. It is introduced here primarily to indicate the existence of problems on the demand as well as on the production side of joint operations. Full treatment of these problems, and especially of price determination, often involves the theory of discriminatory pricing and necessitates recognition of the coexistence of pure and nonpure competition in the product markets. Although it's highly interesting, such treatment is beyond the scope of this text.

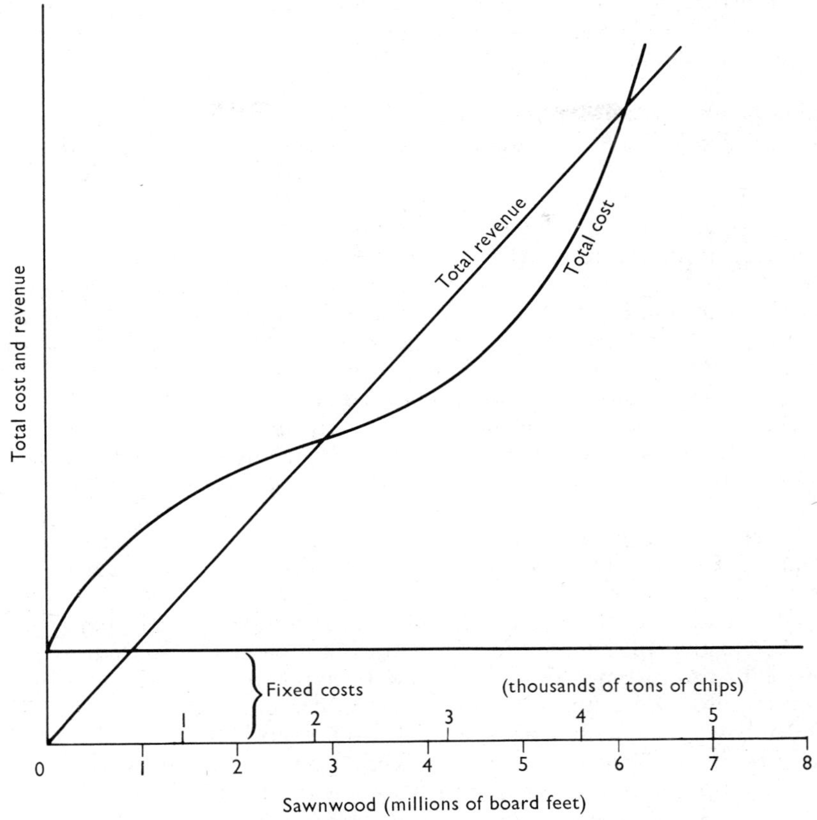

FIGURE 6.17. Joint production with fixed proportions. Essentially a "combination" product, this requires no change analytically.

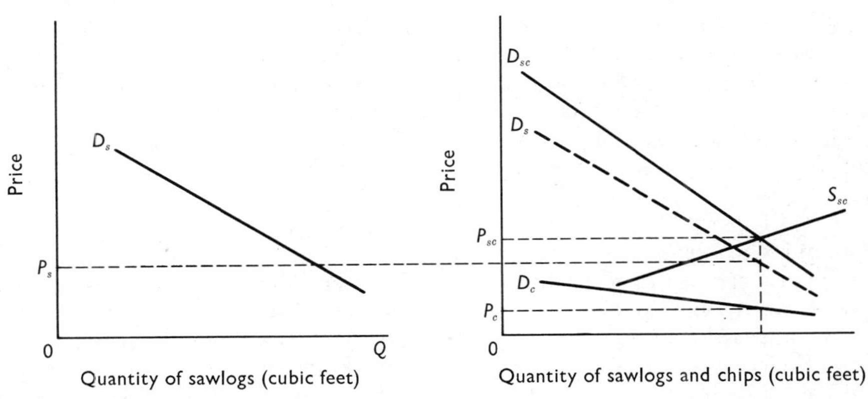

FIGURE 6.18. The demand curves for sawlogs and chippable waste would be added vertically to form the total demand *(D_{sc})*.

items. These marginal costs include all the additional costs incurred by the production of the particular product, and it is the *marginal* costs that are relevant to the production decisions.

Technically Variable Proportions

The second type of joint production—and the more interesting to forestry—is joint production with technically *variable* proportions. Such products have the particular characteristic that the proportion in which the products can be produced may be varied. This kind of production may be illustrated by Figure 6.16 if one assumes it is technically possible to produce either *A*, *B*, or any combination on the line *OC*. More often, one finds the range of product proportions limited.

Forestry abounds with examples of such joint production. A given acre of forest can be made to yield sawtimber and pulpwood in varying proportions. The same holds true for a particular tree; nearly every log can be cut to yield varying proportions of high- and low-grade lumber. In practice, lumber and chips provide a prime example of variable proportions. A different type of joint production is that involved in any mixed forest, where widely varying species combinations can be produced from the same acre or tract; in fact, a primary problem of management is that of determining the optimum species combination. Management of tropical forests provides the extreme case here, with so many possible product combinations for most tropical forests the silvicultural techniques have yet to be established. And, as we will see in Chapter 15, even the much maligned term "multiple use" can be analyzed as a problem in joint production where the principal difficulty is that of determining the optimum product combination.

Since maximization of net revenue has been taken as the primary objective of management, both costs and revenues must be brought into the picture, which so far deals only with the technical aspects of product combinations. Costs and revenues may be introduced, at least, in a fashion again analogous to that used in determining input combinations, but with a few twists of its own.

Production Possibility Curves

The starting point can be that of Figure 6.16, without the line *OC*, a "map" of various combinations that can be produced with the given set of productive factors. *Costs* are introduced by connecting all those product combinations that might be produced for some specified amount. Take, as an example, the situation described by Figure 6.19. Here the output of sawlogs in board feet is shown on the vertical axis and pulpwood output in cords on the horizontal axis. The curve labeled $16 connects all combinations of the two products that can be produced (from each five acres) for a total cost of $16, and similar curves are identified for other total cost outlays. Such curves are called production possibility curves, or isocosts, here relating products instead of inputs. There would be an isocost for any total cost one wished to identify (given complete divisability of inputs), and hence a "family" of isocosts.

Isorevenue Curves

Since the objective of production is taken to be maximization of net revenues, it is still necessary to bring revenues into the picture. Revenues stem from product sales, so competitive conditions in the markets for the two products determine the shape of the isorevenue curves. There are many conditions that can and should be recognized, but for the moment we will deal only with the one most frequently found in forest enterprises. Typically, no one forest producer influences product price by his or her production decisions. Under such (purely competitive) condi-

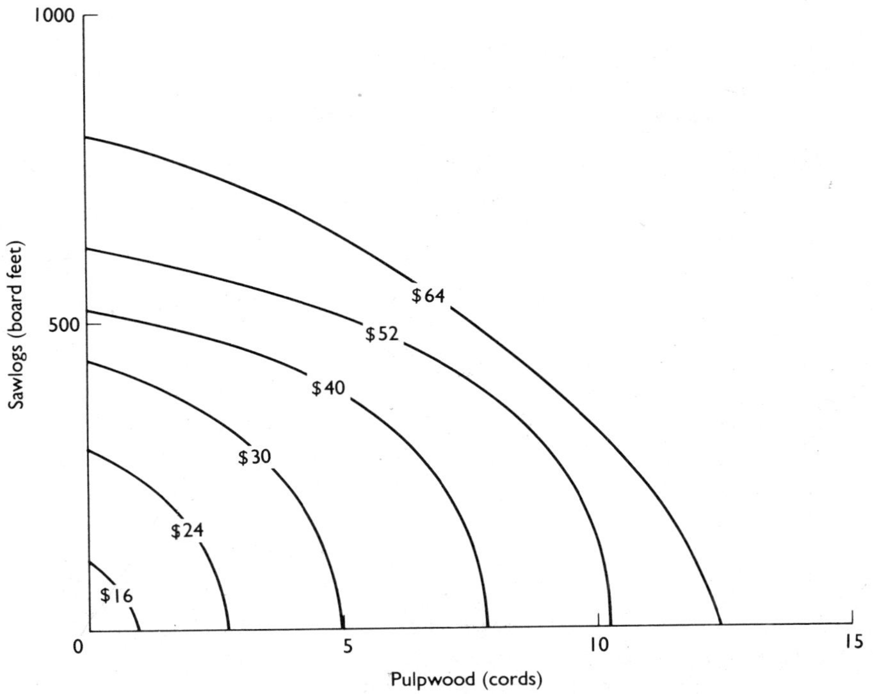

FIGURE 6.19. Production possibility curves, or isocost curves, for jointly producing sawlogs and pulpwood (five-acre basis).

tions the family of isorevenues will be parallel straight lines. If, for example, the two products are priced in a 1 : 12 ratio (as they would be if the price of pulpwood were $5 per cord and sawtimber $60 per thousand board feet), then the isorevenues would be as shown in Figure 6.20.

The Expansion Path

Combining these isorevenues with the set of isocosts previously developed in Figure 6.19 yields the diagram of Figure 6.21. And now the tangency points will identify the points where revenues are at a maximum for any particular cost outlay. Hence, any rational producer would always operate in such a manner as to produce one of the product combinations identified by a line connecting all these points of tangency; that is, the producer would always operate at some point on the expansion path.

Maximizing Net Revenue

The final step is identical to the single product case, differing only because two products are now involved in varying combinations. The complete solution can be shown as in Figure 6.22—a combination of the previous figure (Figure 6.21) with the resulting total cost/total revenue curves. The profit-maximizing combination can be read from the point on the expansion path that falls directly below the point where the slopes of the total cost and total revenue curves are identical. In the particular case illustrated, this occurs with an output of 340 board feet of sawlogs and 8 cords of pulpwood.

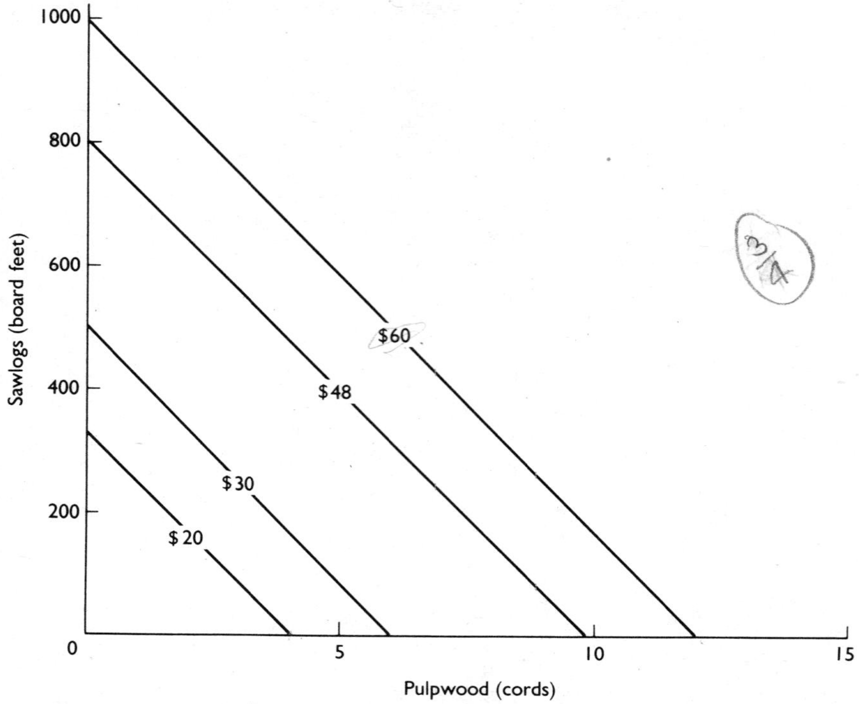

FIGURE 6.20. Corresponding set of isorevenue curves for sawlog and pulpwood, jointly produced.

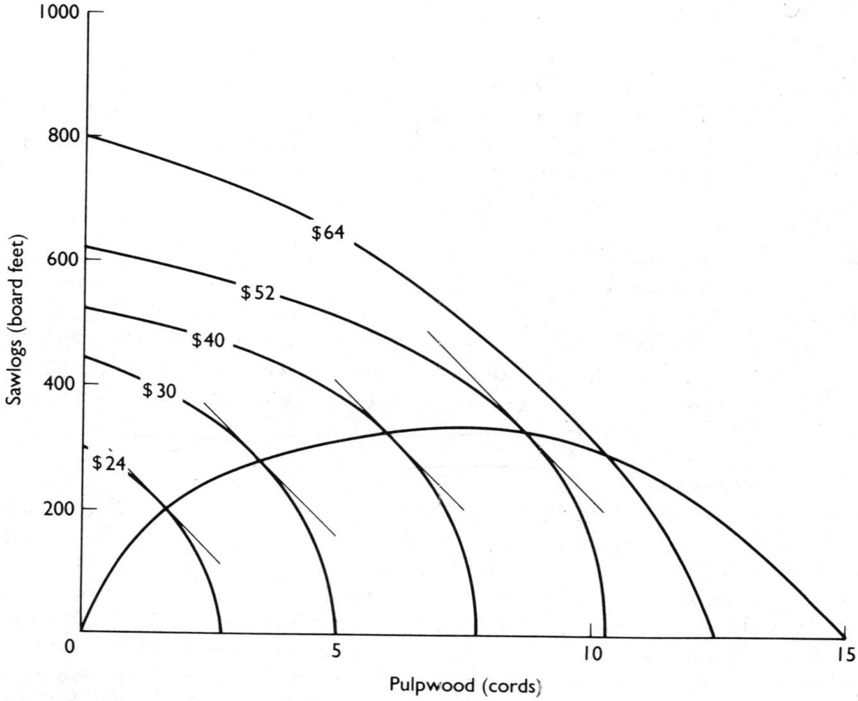

FIGURE 6.21. Isocost and isorevenue curves and the expansion path.

In such cases no set of average costs can be shown for either product. An average cost of pulpwood production *or* sawlog production simply does not exist. Conceivably, one could construct a pseudoaverage cost diagram by employing a device similar to that of Figure 6.22, but the idea of an average cost has been completely destroyed by the joint nature of the production process: average cost equals total cost divided by output, and one cannot divide by output when the makeup of that output is constantly changing. In joint production, average costs are meaningless.

But marginal costs do exist and can be determined for either product. Because the marginal cost equation for either product will contain the output of the jointly produced product as a variable, the two marginal cost curves must be solved simultaneously.[13] Joint production, therefore, always leads to making output decisions on *both* (or all) products simultaneously, and this is true whether or not it is recognized by the manager or decision maker. There is probably no more basic point to be made in forest economics than this, because so many products *are* produced jointly by a modern forest in today's economy. Decisions affecting one product affect the output of all other jointly produced products by definition.

Product Relationships

Although the principal problem of joint production has been "solved" (at least conceptually), there is still much to be gained from a consideration of some complications that might occur. For example, the shape of the isocosts reveals something of the nature of the relationship between the two products from a production standpoint. Consider, for example, *a* and *b* in Figure 6.23. With the relationships indicated in *a*, one might produce either 200 units of *X* and a quantity of *Y* ranging from 0 to 200 or, alternatively, and with the same total cost, an output of 200 units of *Y* but with the quantity of *X* ranging from 0 to 200. Isocosts such as these are characteristic of joint production with technically *fixed* proportions: the corners of the isocosts follow a straight line from the origin as shown in the earlier diagram, Figure 6.16. Clearly, no matter what the price relationship between the two products might be, the isorevenues could only be "tangent" at the corners of the isocosts, that is, on the straight line passing through the origin.

Isocosts such as those pictured in *b*, while quite possible, are of genuine interest only when they appear after some other pattern has been experienced or expected. Whenever isocosts are convex to the origin, as in *b*, then the two products should be produced separately, not jointly. At either end of the isocosts a small increase in yield of one product is possible only through decreasing the output of the other product by a large amount. Again, one needs only imagine an isorevenue superimposed on these convex isocosts to realize immediately that the only possible rational combinations are along the two axes. This necessarily means a "combina-

---

[13]Given the total cost as a function of the output of two products, *X* and *Y*,

$$TC = f_c(X,Y)$$

then the marginal costs of *X* and *Y* become, respectively,

$$MC_x = \frac{\delta(TC)}{\delta X} \quad \text{and} \quad MC_y = \frac{\delta(TC)}{\delta Y}$$

Note that here the marginal costs are defined as *partial* derivatives. Hence $\delta(TC)/\delta X$ is the change in total cost that accompanies a change in the output of *X*, *when the output of Y is held constant*. In any marginal cost equation involving joint production of two or more products the output of all products produced jointly will appear as variables.

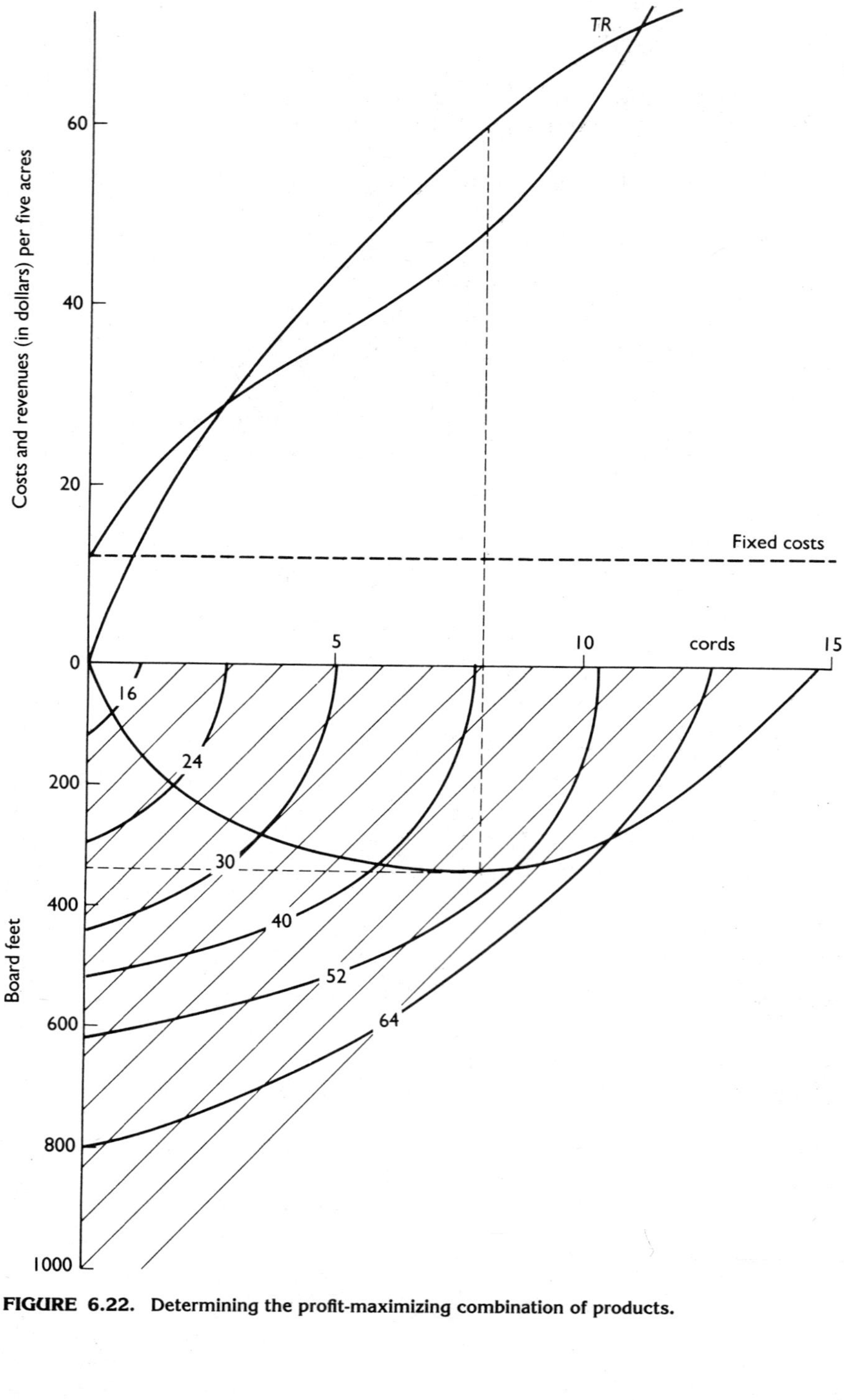

FIGURE 6.22. Determining the profit-maximizing combination of products.

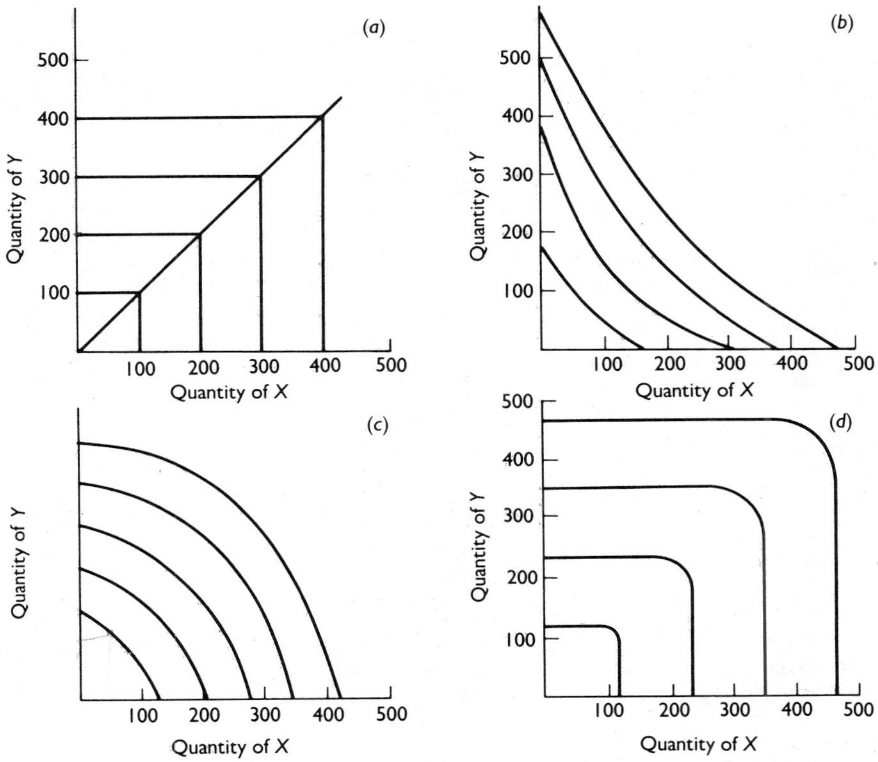

FIGURE 6.23. Four possible shapes of isocost relationships (production possibility curves).

tion" in which the output of one or the other is zero. The slope of the isorevenue will determine which of the products is to be eliminated from the enterprise.

Isocosts that are concave to the origin (bowed outward) as those in diagrams *c* and *d*, indicate a process that *should* be joint: the two products can be produced *jointly* at a lower cost than would be possible if separate facilities were to be established. In contrast to the situation in *b*, at either end of the isocosts in *c* a small decrease in the yield of one product results in a large increase in the output of the other. The introduction of isorevenues in this case results in precisely the type of solution already outlined in Figure 6.21.

In real life most products produced jointly are not completely variable in proportions. The isocosts (if one were to plot them) would be curved, but only for certain combinations—as in diagram *d*. When isorevenues are superimposed on such isocosts one sees immediately that production will always be joint, and that the optimum product combination can only vary over a limited range.

Form of Isorevenue Curves

The shape of the isorevenues is controlled by the competitive market situation for the products. As long as the markets are purely competitive the isorevenues will be straight lines and will be parallel to each other. Their slope will always be the inverse of the product prices. Take, for example, the isorevenue of Figure 6.24.

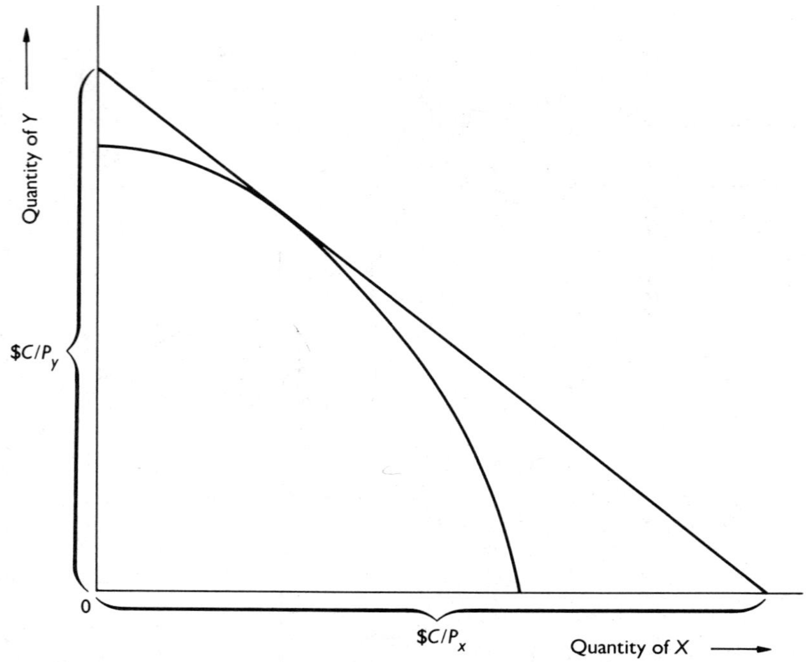

FIGURE 6.24. Determining the slope of an isorevenue curve. (The slope will always be inversely proportional to the prices of the two products.)

To construct such an isorevenue the easiest approach is probably that of taking any value, $\$C$, and determining the quantity of each of the two products that can be sold for this amount. These amounts are then measured out on the respective axes and the two points connected with a straight line. The appropriate quantity of X is thus $\$C/P_x$ and of Y is $\$C/P_y$.

The slope of the isorevenue is, therefore,

$$\frac{\dfrac{C}{P_y}}{\dfrac{C}{P_x}} = \frac{C}{P_y} \times \frac{P_x}{C} = \frac{P_x}{P_y}$$

and is the inverse of the product prices. This being so, any change in the price of either product that results in a change in the price ratio will influence the optimum combination of the products. If the price of Y rises relative to that of X, the optimum combination will consist of a higher proportion of Y than before the price change, and vice versa.

If the markets for the two products are *not* both purely competitive, then the isorevenues will not necessarily be parallel, or even straight lines, but a few generalizations can be made. If increased production of product X, for example, led to a lower price for X while the price of Y remained constant (or if the situation were reversed), then the isorevenues would become concave toward the origin. But if such increased production lowered the price, then production would no longer be purely competitive since the producer has at least some control over price. Under these circumstances the producer would cease sale of the product at the quantity producing the maximum revenue and the excess quantity would be destroyed (wasted). In earlier years the mines of Boliden in Sweden, producing gold and copper as principal products, also yielded more arsenic than the entire world's consumption (Carlson, 1937). The disposal of arsenic actually became a major production problem. While no such clear-cut examples can be cited from the forestry field per se, much the same thing could be said of the production of vanillin from the black liquor of sulphate pulp, of garden mulch from redwood bark, or of fuelwood as a by-product from sawmill waste in many mill towns.

In much of the world, including the United States, the forest owner most able to influence prices of sawtimber and pulpwood is the State, Provincial, or National Forest Service. This is because they usually control by far the biggest block of stumpage in the nation—in many nations, the entire forest area. Hence, the production policies favored by public forest managers often have a decided influence on supplies and prices of jointly produced products, even though there is no deliberate effort in such directions. In certain African nations, for example, where early colonial foresters directed their efforts toward classical plantation management with long rotations aimed at producing high-quality sawlogs, there is presently inability to supply needed pulpwood. In the United States, where public forest management has favored sawlog over pulpwood rotations for many years, it is doubtful that there has been any major effect of government management practices on pulpwood prices. As long as pulpwood demand is met at prices no higher than production costs, additional pulpwood production (that might have resulted from a policy of shorter rotations) would serve as an indirect subsidy to the pulp industry. Quite probably the policy of long rotations has aided in keeping sawlog prices from showing an even more rapid increase than that exhibited in the 1950s and 1960s. In specific locations, however, there may well have been quite different results. Long rotations can also *restrict* sawlog supplies over the short term. Public forests of the West Coast supplied many such examples during the 1960s and 1970s.

Quality is frequently of prime importance when choosing between one or another managerial step and, provided it is measured, fits into this multiproduct framework with practically no special treatment. A particular product that has two quality levels, for example, can also be called two separate products and treated accordingly.

Linear Programming

For multiproduct or multiquality situations a *linear programming* approach will often be preferable. Real-life problems of this general type are characteristically subject to *constraints* of one type or another. Workers, or machines, or storage space, or shipping space are limited; some given amount of one product must be produced under any circumstances; marketing agreements restrict production to certain levels, and so on. Mathematical programming is a powerful tool for many problems, especially those that can be treated under assumptions of *linearity* (constant proportions, constant returns to scale, etc.), and those subject to a series of constraints.

A brief introduction to the technique will be given as part of the discussion of multiple use, although no attempt will be made here to provide a full treatment of this important tool.[14]

[14]Anyone interested in gaining a good understanding of linear programming would do well to read Baumol (1965), Leuschner (1984), or Clutter et al. (1983).

Summary

The possible applications of marginal analysis to forestry and forest product decisions are virtually inexhaustible; the objective here has been only to illustrate the variety of problems amenable to the approach rather than to present details of a specific application. Every problem calls for a careful identification of costs and returns that stem from the particular decision being made. Sometimes, as will be seen when nontimber products are discussed, this very problem of identification and measurement makes decisions difficult.

Viewed strictly from a forestry standpoint, there is one major addition necessary, for production has been treated as if it were instantaneous. Most forest production involves substantial time periods. Even with firms producing wood products, questions of how to take time into account frequently arise. Since a timeless framework is of limited applicability, we now turn to this problem.

QUESTIONS AND PROBLEMS

1. Assume that you are a logging contractor in Louisiana, dealing with a single mill that places no limit on the quantity of logs it will take from you. The mill pays you $230 per thousand board feet, Doyle scale, for pine logs delivered to the mill.

   A landowner offers you 500 acres of good pine, to be purchased on a lump sum basis. You cruise it, but before all the data has been analyzed the prospective seller telephones you, saying, "I've got an offer of $250,000 for that timber. What will you bid?" You ask for a few minutes and pull the material together. You see that the volume in trees 10 inches and up averages about 11.5 thousand board feet per acre. After a few rough calculations you call, up the bid, and eventually buy the cutting rights to the area for $300,000, with the provision that you complete the job in 6 months.

   Your assistant finishes working up the data and you find that volume per acre, by diameter classes, is as follows.

DBH Class	Volume
20 inches and over	4.0 M board feet
19	5.9
18	6.8
17	7.5
16	8.3
15	9.0
14	9.4
13	10.0
12	10.5
11	11.0
10	11.5

Note that the above is a cumulative table. The 11.5 M board feet in trees 10 inches and over, for example, includes the 12-inch class and all classes above, while the 9 M board feet in trees 15 inches and over includes the 15-inch diameter class and all classes above.

You will use tree length logging, and, on the basis of data collected from similar operations, you estimate volume per tree and cost per tree of felling, limbing, skidding, and bucking (all costs incurred before loading), by diameter classes, to be as follows.

DBH Class	Volume per Tree (MBF)	Cost per Tree for Felling, Limbing, Skidding, Bucking
20 inches and up	600	27.00
19	340	22.41
18	285	20.91
17	238	19.80
16	195	18.54
15	160	18.18
14	125	17.82
13	105	17.64
12	90	17.50
11	60	17.37
10	40	17.28

You contract with a trucking outfit to load and haul your logs at $30 per board foot. Road construction is estimated at $10,000. Your assistant, who will supervise the job, gets $16,000 per year, and you expect to take most of the 6 months to get everything cleaned up.

Since you have purchased cutting rights with no diameter specifications, you can take as much or as little of this timber as you like.

QUESTIONS

Part I

a. What diameter limit should you give your cutters (assuming that you are trying to maximize profit)? That is, you will direct them to cut everything from some minimum diameter and up. (You may find it convenient to solve this graphically, but you can stick with figures as long as you show how you arrived at your decision.)

b. How much, if anything, will your venture contribute to your overhead, depletion charges, and *your salary and / or profit*?

Note: Your choice of diameter limit will, in actual fact, affect the time required to complete the job and perhaps the road construction cost also. For the sake of simplicity, however, ignore these effects and assume that it will take 6 months to complete the job, and that the roads will cost $10,000 regardless of the diameter limit.

Part II

Six months later you drive up in your new Porshe to watch your men finish the operation. The landowner comes out to chat and says, "Hey, I've got the other half of this tract to sell. It's just like this one you're finishing, but I was talking to the County Forester the other day and he says I should sell this one on a log scale basis and get $60 per M board foot. How about it?"

Having suspected something like this might happen, you are well prepared and close the deal within 5 minutes at his figure, providing your cruise shows the timber is essentially the same as the previous purchase. The cruise checks out OK; in fact, you conclude there is no need to change either the stand table or the logging cost data from the previous operation.

c. What change will you make in your cutting instructions, if any, and how much "profit" will you make?

In all your work, be sure to show how you arrived at your decisions.

2. Define opportunity cost, and give an example other than those used in this text.

3. What is a sunk cost? Why, if a cost has already been made, should it not influence your decision with regard to a future event?

4. What is meant by the term joint production? Give an example from forestry of joint production with technically fixed proportions and one with variable proportions.

5. Given the following set of relationships (isoquant curves) between two different inputs, *A* and *B*, used to produce a particular product, and given that the cost of *A* is \$3 and that of *B* is \$4, in what proportion should the two factors be used if you plan to produce 15 units of the product? If input costs were reversed (if *A* costs \$4 and *B* costs \$3), what would be the least cost combination to produce the 15 units?

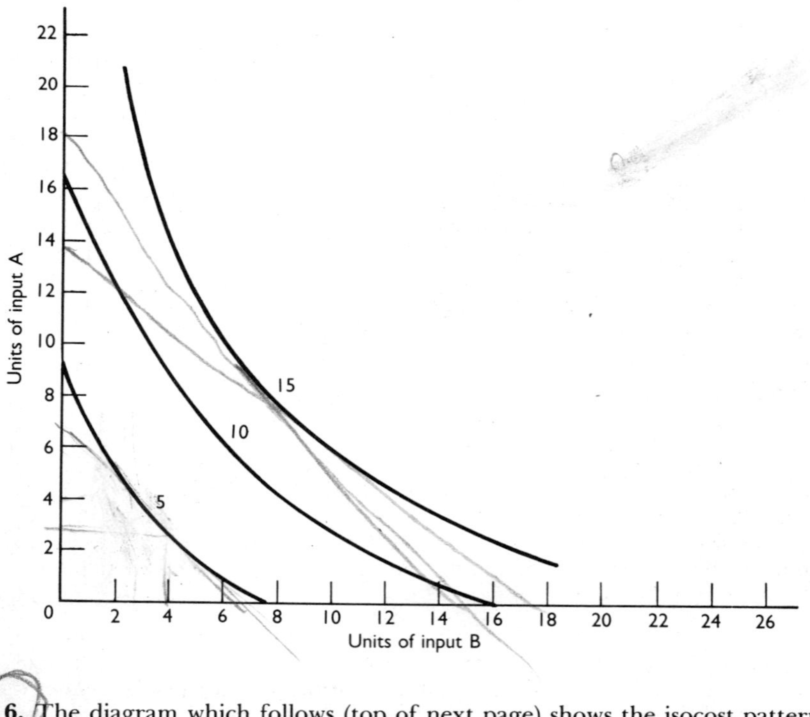

6. The diagram which follows (top of next page) shows the isocost pattern of a firm producing two products jointly—products A and B. From this diagram determine answers to the following questions: (*Note*: You may want to trace the diagram on another sheet of paper first.)

a. If product A sells for \$4 per unit and B for \$3 per unit, draw the expansion path on the diagram.

b. If you were the acting manager and are directed to spend \$40,000 next month on production, what combination of the two products would you produce in order to maximize your net revenue (or minimize your losses)?

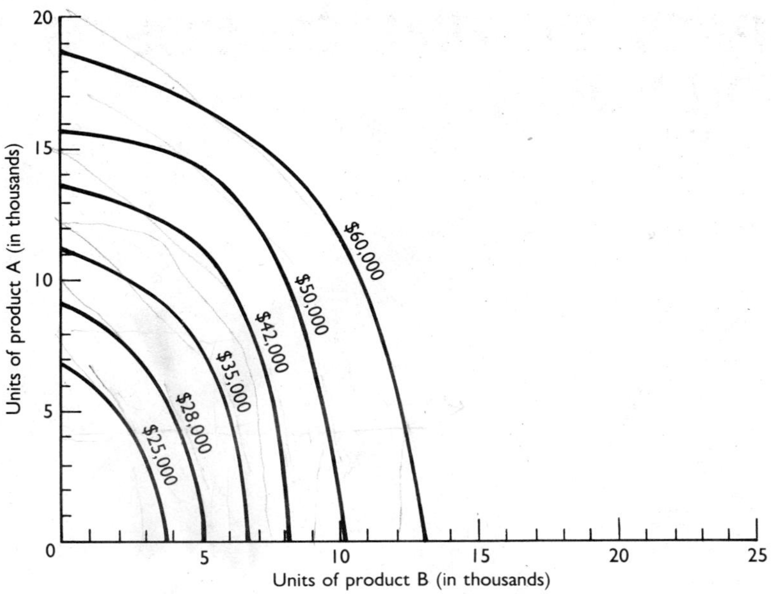

c. Assuming these costs are *total* costs, what would be the net revenue if all goes according to the production plans you decided on in question b above?

7. As a pulpwood contractor, you find that in a stand you have purchased the volume per acre, by diameter classes, is as follows.

12 inches and over	6 cords
10	10
8	14
6	17
4	20

This is a cumulative table. The 14 cords in the 8 inch and over diameter class, for example, includes *all* trees over 8 inches.
A trucker will load and haul your pulpwood for \$6 per cord. Your cutters are paid by the piece—30¢ for each bolt stacked by the road. From past work you have compiled a table of the number of bolts per cord for each diameter class.

DBH Class	Average Number of Bolts
14 inches and over	26
12 to 14	30
10 to 12	40
8 to 10	54
6 to 8	70
4 to 6	85

You can take any, or all, of the timber.
a. To maximize profit, what diameter limit should you give your cutters?
b. How much, if at all, will you profit?

chapter seven

THE IMPACT OF TIME ON COSTS AND REVENUES

Chapters 3 and 4 were concerned primarily with quantitative aspects of production—the *quantity* of product that leads to maximum net revenue and *how much* of each necessary input to be used. The *when* aspect—especially the timing of costs and revenues—was given no specific attention. Yet the nature of forest production demands that some rational means be accepted for comparing costs and revenues paid or received in different time periods, whether these are measured in dollars or in units of social or psychological satisfaction. Interest has long been used for this purpose. Because most financial analyses of forestry involve compound interest calculations, mastery of this tool and the underlying concepts is necessary.

INTEREST

What Is It?

Suppose you had an eccentric aunt who, after entertaining you at her home, informs you that she intends to make you a gift, but that you will have to decide between several alternatives. First, she asks, "Would you prefer to have $1000 tonight, in cash, or $1000 one year from now?" After checking your wallet, you (probably) reply, "I think I'd better take the cash tonight!" Then your aunt says, "OK—but would you prefer $1000 now, or $1200 next year?" Now the choice is more difficult, and as you hesitate she asks, "How about $1000 tonight or $1800 next year?" The point at which you would delay the gift is not important just now; it would be very rare, however, if there were not some value—$2000, $5000, $10,000—that would persuade you to defer the gift. The point is that money has a "time value."

It is often said that interest is the price of money. This is valid as long as one keeps in mind that the "price" varies not only with the amount, but also with the time over which it is retained and with the perceived probability of loss (risk). If you were willing to take $1100 a year from now instead of the $1000 today, then the time value of $1000 to you for that year was no more than $100, or 10%. But you would probably not have been willing to wait two years, or certainly not five, to gain the extra \$100—especially if your aunt is getting along in age. Interest is always a *rate* concept, a payment of so much per unit of time. The conventional time unit is one year, and when one quotes an interest rate of, say, 12% it is normally understood that the payment (price) will be 12/100 of the borrowed amount for each year the money is held.[1] Put simply, interest is the return on one dollar loaned for one year (and note, it is a dollar *loaned*, not *saved*).

Interest may also be defined as the rate of growth of capital. If, for example, I obtain needed funds for my corporation through selling 8% corporate bonds of \$1000 with a maturity date 10 years ahead, I am promising to pay the bond holder \$80 annually until the bond is redeemed (the \$1000 is paid back to the bond holder). The *nominal* rate of interest is 80/1000 or 8%. But if you purchased this same bond from a broker for \$900 you would be receiving 80/900, or 8.9%. In the rather unlikely case that you purchased the bond for \$900 just one year before it matured, you would receive not only the \$80, but also the promised \$1000—a total gain of \$180. This means you have earned a yield of 180/900, or 20%, on a bond with a nominal 8% yield. The 20%, however, is the real rate of growth of your capital, for you would now have a total gain of \$180, any dollar of which is as available for spending as that received as wages or salary, profit, rent, or any other income. The interest (\$80) and the \$100 difference between purchase price and sale (a "capital gain") both represent income, although the two are treated differently under current income tax legislation.

Compounding and Discounting

To students of biology, and especially to forestry students, the basic principles underlying compounding and its counterpart, discounting, should be almost self-evident. In discussing the volume growth of a tree, or the growth of a deer herd, percentages are commonly used and create no particular problem. Asked, for example, to predict the deer population five years hence for an area having a present herd of 1000 deer, and told that the herd is increasing 10% annually with no harvest, most students would probably set up a table much like Table 7.1. If a tree with a volume of 1000 board feet were assumed to growth 10% annually, this table would be equally valid for predicting its volume over a five-year period. When applying the growth rate to a tree, no forester would worry about removing the 10% increase *each year*: the increase is simply added to the previously existing volume and provides an additional base for growth over the following period.

TABLE 7.1
Growth of Deer Herd over Five-year Period at Net Increase of 10% per Year

Year	Population	Calculation
Present	1000	$1000 + (1000 \times .10)$ or $1000\,(1.10)$
After one year	1100	$1000 + (1100 \times .10)$ or $1000\,(1.10)^2$
After two years	1210	$1210 + (1210 \times .10)$ or $1000\,(1.10)^3$
After three years	1331	$1331 + (1331 \times .10)$ or $1000\,(1.10)^4$
After four years	1464	$1464 + (1464 \times .10)$ or $1000\,(1.10)^5$
After five years	1610	

[1] The price is not necessarily paid each year since the interest charge can, and often is, allowed to accumulate. When this is done the charges also accumulate on the unpaid amounts, giving rise to compound interest.

The same reasoning applies to the growth of any capital fund. If \$1000 were invested at 10% and the proceeds allowed to accumulate as part of the investment, then the preceding table would serve to predict perfectly the capital value of the investment at the end of each year. Naturally, if the growth of the capital were paid out each year (simple interest) it could not be "compounded," nor would the deer herd increase as predicted if 100 deer were harvested each year. But as long as the increase is not removed it serves to increase the base for growth over the following period.

A calculation of this type yields a quite typical form of growth curve, as shown in Figure 7.1. Two interest rates (6% and 10%) are graphed in the figure, and two initial starting values (\$1000 and \$500). Both these factors—the rate and the amount at the beginning of the period—influence the position and shape of the resulting curve. (Incidentally, there is a simple and convenient rule of thumb for estimating growth of any sum under compound interest: divide the number 72 by the selected rate and the result is the number of periods required for any sum to double at that particular rate. For example, at 6% it takes 72/6, or 12 years, for an invested sum to double.)

Discounting is simply the reverse of compounding. As an example: a rate of 10% must be earned by \$1000 invested today if this amount is to grow to \$1610 at the end of a five-year period (Table 7.1). At 10%, therefore, the *present value* of \$1610 to be received five years in the future is \$1000. Why? Because if \$1000 were to be invested today at 10% it would "grow" to \$1610 at the end of five years. Since discounting is nothing more than "compounding in reverse," curves appropriate to one concept can be used equally well for the other. In Figure 7.1, for example, the present value of a future sum is found by moving "backward" along the curve—downward to the left.

Computation

No attempt will be made here to discuss in full the mechanics of compound interest calculations. Those seeking a more detailed exposition of the formulas used can

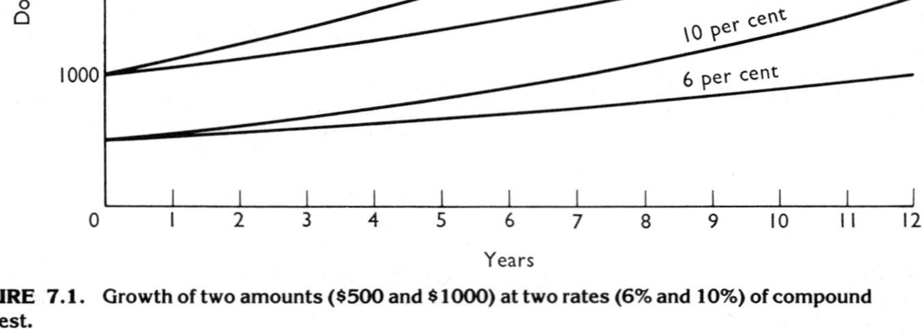

FIGURE 7.1. Growth of two amounts (\$500 and \$1000) at two rates (6% and 10%) of compound interest.

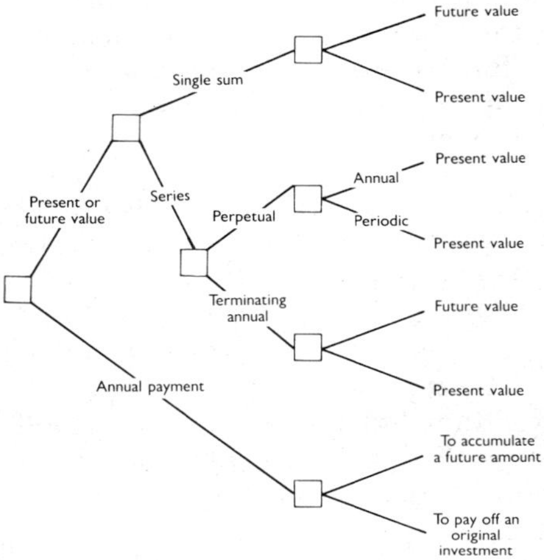

Formula number	Formula	Used to determine
(1)	$FV = PV(1 + i)^n$	Future value of a single sum
(2)	$PV = FV\left[\frac{1}{(1 + i)^n}\right]$	Present value of a single sum
(3)	$PV = \frac{A_p}{i}$	Present value of a perpetual annual series: (capitalized value)
(4)	$PV = A_p\left[\frac{1}{(1 + i)^r - 1}\right]$	Present Value of a perpetual periodic series
(5)	$FV = A\left[\frac{(1 + i)^n - 1}{i}\right]$	Future Value of a terminating annual series
(6)	$PV = A\left[\frac{(1 + 1)^n - 1}{i(1 + i)^n}\right]$	Present value of a terminating annual series
(7)	$A = FV\left[\frac{i}{(1 + i)^n - 1}\right]$	Sinking fund formula, or annuity
(8)	$A = PV\left[\frac{i(1 + i)^n}{(1 + i)^n - 1}\right]$	Installment payment, or capital recovery

Symbols

A = Uniform series of annual (end of year) payments
A_p = Uniform series of perpetual periodic payments
i = Interest rate
n = Number of interest periods (usually years)
r = Number of years between periodic payments
FV = Future value
PV = Present value

FIGURE 7.2. Decision tree for using compound interest formulas. ***Source:*** **Adapted from John E. Gunter and Harry L. Haney, Jr., "A Decision Tree for Compound Interest Formulas." Copyright 1978 by** ***Southern Journal of Applied Forestry*****, Vol. 2, No. 3 (August 1978): 107. Used by permission.**

find excellent presentations in available texts.[2] A minimum basis for understanding the following pages, however, is familiarity with the symbols to be used, and access to the eight formulas of Figure 7.2.[3] Because there is often as much difficulty in selecting the appropriate formula as there is in making the calculations, the formulas are given in the "decision tree" format (Figure 7.2) developed by Gunter and Haney (1978).

In most forestry calculations the number of periods *(n)* is stated in years, but in many other calculations interest may be compounded (or payments made) semiannually, quarterly, or monthly. In such cases the *rate (i)* must be adjusted accordingly. If, for example, interest is being charged at a 1% *monthly* rate, the transaction should be described as one bearing a *nominal* 12% per year, compounded monthly. A two-year loan made under these terms would be calculated

[2]Three of these are: Grant, Ireson, and Leavenworth (1976), Gunter and Haney (1978), and Clutter (1983).

[3]The "end-of-period" convention will be used throughout this book unless otherwise noted. This means that all series amounts *(A)* and *(A_p)* and all future values *(FV)* occur at the end of the period or periods in question.

using $n = 24$ and $i = 1\%$.[4] If compounded quarterly, n would equal 8 and i equal 3%. It is important to realize that a *nominal* annual rate of (say) 12% compounded monthly is *not* identical to a 12% *annual* rate; the differences are usually small, but if rates are high and the total period long the difference can be significant. A 25% nominal rate compounded monthly, for example, gives an "effective" rate (the yearly rate that yields the same result) of slightly over 28%.

Banks occasionally use daily compounding for saving accounts, which gives a result very close to *continuous* compounding. The latter, although rarely used in financial transactions, can be useful when one wishes to treat costs or revenues as being spread throughout a year. Given any PV, the FV resulting from continuous compounding can be calculated from the formula

$$FV_c = (PV)e^{in}$$

where e is the base for natural (Naperian) logarithms and has the value of 2.71828 +, i the interest rate, and n the period in years. The single payment present value formula is, as would be expected, the reciprocal of the above, or

$$PV_c = (PV)e^{-in}$$

For example, a bank advertises that it pays 8% on certificates of deposit and that it is continuously compounded. Compared to 8% compounded annually over a ten-year period, how much of an advantage will this prove to be?

If compounded annually:	$\$1000\ (1.08^{10})$	$= \$2,158.92$
If compounded continuously:	$\$1000\ [e^{(10 \times .08)}]$	$= \$2,225.54$

So you would gain $66.62 over the 10 years from the continuous compounding.

These relationships are particularly useful in theoretical expositions or developments since only continuous functions can be differentiated.[5]

An Example

Using the appropriate formulas, we can work through an example of profit calculation by a timber enterprise—in this case, the problem facing an owner of a 25-acre woodlot near a pulp mill. The owner has just been advised by the company forester to consider planting hybrid poplar. "Fifteen years after planting," the forester says, "we promise to buy the 30 cord per acre yield at a price of no less than $20 per cord." Further inquiry reveals that the woodlot can be prepared for planting at a cost of $15 per acre, nursery stock and planting will cost a total of $60 per acre, while protection and property tax will be a constant (it is hoped) $3 per acre each year. And because poplar is so intolerant of shade, a "weeding" cost of $6 per acre should be included at the end of the second and fourth years.

[4]A formula for future values under compounding periods other than a year is

$$FV = PV\left(1 + \frac{i}{m}\right)^{nm}$$

where m = the number of compounding periods per year. The two-year loan (above) at a 12%/year nominal interest, compounded monthly, would be

$$FV = PV\left(1 + \frac{.12}{12}\right)^{2 \cdot 12} = PV(1.01)^{24}$$

[5]This is an easy function to differentiate, since it is its own derivative—a fact making calculation of generalized marginal cost functions very simple.

To decide whether or not to plant, the owner would want to calculate the present value of prospective costs and revenues. After deciding he should use a 9% discount rate, his calculations should approximate the following:

Costs in first year (present value)	
Land preparation	\$ 15.00
Planting and nursery stock	60.00

Protection and property tax constitute an annual payment for the 15 years. Using formula 6—to determine the value of a terminating annual series:

$$\frac{3\,(1.09)^{15} - 1}{.09\,(1.09)^{15}} = \quad \$\,24.18$$

Each of the two weedings must also be discounted, using formula 2.

First weeding $6/1.09^2$	5.05
Second weeding $6/1.09^4$	4.25
Total present value of costs	\$108.48

His revenue of 30 × \$20 per acre would be realized only after 15 years, so this would also have to be discounted. Again using formula 2,

$$30 \times \$20 = \$600 \qquad \text{and} \qquad \$600\,(1/1.09^{15}) = \$164.72$$

On a per-acre basis, the *present net value* (the *PNV*) of the project would be \$164.72 − \$108.48 = \$56.24, and for the 25 acres the total *PNV* would be \$56.24 × 25 = \$1,406.00.

Does this mean that the investment of \$108.48 (the *PV* of all costs) per acre generates a per acre yield of \$56.24 over the 15-year period? Not exactly: it means that over the fifteen years the investor is gaining a 9% return on every penny spent—and will, at the end of the period, gain an additional \$204.85 per acre (the present value of which, at 9% discount, is \$56.24).

After looking over the owner's calculations, the forester might suggest that this first harvest 15 years after planting would probably be the first of many such operations—that this should really be regarded as the initiation of a permanent operation. To evaluate this possibility we can use the net value already at hand—every 15 years the operator will receive \$204.85 per acre. Using this in formula 4, we would have

$$\$204.85\left[\frac{1}{(1.09)^{15} - 1}\right] = \$41.05$$

which is the present value of a periodic income of \$204.85 that will be received every 15 years, forever, and beginning 15 years from now (a perpetual periodic series). And, as before, this does not include the 9% return being earned on all the expenses.

Many questions can be raised about this problem. Wouldn't the poplar sprout so no replanting would be needed? Did the owner use the "proper" discount rate? Is this the best alternative the owner has? What *is* the actual rate of return on his investment? Should the owner have included his cost of land, or is there some other way of taking this into account? What are the risks inherent in such an undertaking? What if the price is higher—should he sell earlier or later—or, for that matter, is 15 years the best rotation under the given circumstances? We will eventually deal with these questions, but before complicating the picture any further we should become more familiar with some of the concepts underlying "interest."

Time Preference

Individual time preference is the rate at which an individual discounts the value of future income to make it comparable to present income.[6] Individual time preference will therefore be directly affected by an individual's present and expected income, his or her present and expected needs, attitude toward saving versus consumption, age and general outlook on life, and so on. For example, if you were willing to take $1100 a year from now instead of the $1000 your aunt offered today (on the first page of this chapter), then your individual time preference at that particular moment was 10% or less. The old story of the ant and the grasshopper could be explained in terms of the individual time preferences of the two creatures of the fable. The grasshopper had a very high time preference rate, heavily discounting all future income in favor of present high living. On the other hand, the ant displayed all the virtues humanity expects of fellow humans, working all summer long, slaving and saving for the hard winter he could foresee in his future. The ant clearly had a relatively low time preference.[7]

But far more than personal attitude affects individual time preference. An individual's income level is especially important: people with high incomes tend to discount future values at a low rate because their need for additional income today is slight (the marginal utility of additional income today is small). Future values tend to be more nearly equivalent to present values, hence time preference rates are low. Those with very low incomes may need their income today just to keep their family alive, so they discount future values at high rates. They have a high time preference rate.

Social time preference is the rate at which society (rather than the individual) discounts future values as compared to present values. A high time preference discourages acceptance of undertakings that are long in maturing, while lower time preferences favor such investments. Generally (at least before the bomb, cruise missiles, acid rain, and the CO_2 effect), society is thought to have a much longer life expectancy than individuals, and will take a "long view" toward the utilization of resources. As a consequence, social time preference rates tend to be low relative to those of individuals.

However, this is only true in the industrialized regions of the world, where public forest agencies are well established and income levels are sufficiently high to meet most basic social needs. In developing regions, where present incomes are low and government tenures are often insecure, time preference rates will be high and traditional forestry will find little acceptance by rural inhabitants. Their needs are immediate, and benefits that accrue in 20 to 50 years—no matter how large—are not likely to be considered important. Similar conclusions might be reached by low income farmers or small mill operators in our own nation. Yet public agencies, more representative of society as a whole, might view these same forest investments as being quite attractive.

---

[6]The rate of individual time preference may more precisely be defined as "a ratio between the marginal utility of . . . money in more distant future intervals and the present marginal utility of the same amount of money in intervals nearer the present and reduced by unity" (Ciriacy-Wantrup, 1952).

[7]It should be noted, however, that traditional moralistic interpretations of this fable may well have to undergo some changes in our affluent society. Consider that: (1) it seems quite possible that the grasshopper had enough fun during the summer to more than compensate for his having to beg from the ant; (2) the economics of private thrift may well be questioned in times when social security and Medicare are established facts and negative income taxes are being given serious consideration.

Corporate time preference undoubtedly exists, but there is little data one might generalize from. Large, well-established companies—the solid "blue-chip" corporations—would seem likely to take a longer view (have a lower time preference) than the aggressive, fast-growing, younger corporations, but there are many exceptions. Corporations are often said to be "deathless," yet they do disappear—from bankruptcy, sale, merger, or even a simple closing. For many owners, especially stockholders, the management of a corporation is judged by the last year's profit. This means emphasizing short-term rather than long-term earnings, and implies a high time preference rate.

The coordination of immediate and long-run objectives for individuals, corporations, and, particularly, society, is not easy. We will return to this question in the discussion of investment decisions in Part 3.

Risk, Uncertainty, and Expectations

A distinction is sometimes drawn between risk and uncertainty. In such cases the term *risk* is reserved for events of a nature such that the probability of their occurrence can be calculated, while *uncertainty* refers to events whose likelihood of occurrence cannot be quantified in probabilistic (stochastic) terms. The probability of losing a stand through fire or wind storm, for example, can be put on an actuarial basis in many parts of the world. The *risk* of loss, then, can be evaluated and (often) reduced to a known cost through insurance. Future product prices, on the other hand, are *uncertain,* and the uncertainty increases as we move further into the future. While it is true that timber is about the only commodity that has shown an upward price trend for over 150 years, wood product prices have often dropped—sometimes drastically—and forecasts of their actual level even six months ahead are properly greeted with considerable skepticism.

The availability of lumber and other wood-product futures on the commodity market and the ready access to future prices tends to mitigate the uncertainty to some degree and opens the possibility of lessening the effect of (short-term) price fluctuation through "hedging"—buying or selling lumber futures in the commodities market to protect against a price rise or fall. This, however, is much more likely to be practiced by wholesalers of wood products than by manufacturers.

Risk is important in determining appropriate interest rates, since a high risk venture must pay a correspondingly high rate of return if it is to attract investment capital. Relative to their possible rates of return, uranium stocks tend to be lower priced than those of blue chip corporations. Timber production, like forestry itself, is also subject to risks, and the interest charged for forest investment capital (or the return required from forestry) is increased correspondingly. Any increase in risk means an increase in effective interest rates—in forestry as well as in other businesses. Uncertainty has a similar effect on interest rates: the greater the uncertainty the higher the allowance made. The difference here is that since, by definition, the probability cannot be determined, the adjustment made for an uncertain event is more subjective.

Business executives do, however, hold *expectations* with respect to future events. For example, an individual's expectation may be that prices will fall over the next five-year period, then rise substantially. An expectation is "how one feels" about a future event and it depends on a large complex of factors—the present situation, past and present trends, the individual's interpretation of their significance, and even how he or she happens to view life on a particular day.

The effect of expectations on interest rates is somewhat similar to that of risk, but expectations behave in a more volatile fashion. If producers expect business to be good, prices to rise, profit margins to increase, and so on, they will move to increase production levels. On the other hand, a decline in their expectations will cause them to cut back plans for new production or even to slow down existing rates. Neither of these effects is necessarily serious, but such actions tend to be cumulative, and, at times, contagious. Once shaken, confidence in business may take months or even years to become reestablished. The "elasticity of expectations" (Hicks, 1946) changes quickly and can be high; a drop in prices may make producers expect a further drop, which leads to a still further drop, and so on. Expectations are believed to be an important element in the psychology of business booms and recessions, and the concept is given considerable attention in studies of dynamic price models. *Consumer expectations* seem to play a very important role in inflation.

"Pure," "Market," and "Prime" Rates of Interest

The "pure" rate of interest is, essentially, a risk-free rate. In the early post-World War II literature this was often taken as the rate on U.S. Government long-term bonds and was usually thought to be around $2\frac{1}{4}$ to $2\frac{1}{2}$%. Since that time both long- and short-term rates have fluctuated widely—from 4% up to 21% and (by 1987) back down to an almost normal $5\frac{1}{2}$%. This experience, and the recurring inflationary pressures on the economy, has made the whole idea of a "pure" rate much less acceptable. Some economists have pointed out that if one reduces to an annual rate the inflation that has plagued the United States (and much of the world) since 1950, and adds this to the $2\frac{1}{4}$% that was formerly taken as the "pure" rate, the result is not far from the average rate on long-term government bonds.[8] This implies that the general public simply adds to the "pure" rate a factor to allow for the current rate of inflation and their expectation of future inflation rates, and that the factor is derived from past experience. If this is really so—if an inflationary trend is to be anticipated—the idea of a "pure" rate becomes little more than a convenient synonym for risk free, and it will be used in this sense here.

The term market rate of interest usually means the rate charged by lending institutions in the particular "market" being considered. The difficulty with this is that there are far too many markets for the term to carry any real meaning, and even within a specific market the rate will vary according to the duration, amount, risk, cost of supervision, type of collateral, and so on.

The "prime" rate is only slightly less ambiguous. This is usually defined as the rate major banks are charging their "preferred" customers. It would be more accurate, however, to define this as the rate *announced by* particular banks as the one they are charging preferred customers. And although some customers are "more preferred" than others (for some the banks are reported to "shave" the quoted rate), the prime rate announced by major banks, particularly those of New York, is carefully watched by financial institutions everywhere.[9] Changes in the rate are especially newsworthy because interest rates heavily influence prices of both common stocks and bonds.

[8]This was not true during parts of the 1981–1985 period, when the difference between the prime rate and the rate of inflation was often 7 or 8%.

[9]In April 1984 there were an estimated 50 lawsuits by borrowers claiming banks overcharged them "because the prime rate their loans were tied to wasn't the real prime, or best, rate." One such suit, in Atlanta had been settled out of court for a payment reported to be as much as $12\frac{1}{2}$ million dollars. It seemed likely that the definition of "prime rate" would change. Morgan Guaranty Trust began defining the prime as "the rate of interest publicly announced by the bank in New York City from time to time as its prime rate" (*Wall Street Journal*, April 3, 1984, p. 35).

INTEREST, MONEY, AND BANKS

Foresters involved with management activities necessarily become familiar with the use of compound interest. If they are to *understand* how interest rates are formed, why they change, and their importance to our economy, they must also know something of the relationships between money, its supply, our banking system, and interest rates. Earlier in this chapter, interest was described as a measure of money's time value, as the price of money, and as the rate of growth of capital. But what is money?

To be sure, money is the currency we use for day-to-day purchases. But it also includes the sum of all *demand deposits*—checking accounts, NOW accounts, available money market funds, and so on.[10] This is the general definition of the Federal Reserve's M_1+. The Federal Reserve also keeps track of several other categories important to their purposes but not to ours. We should also realize that the definition of money constantly changes as new forms of credit are created. Money, after all, is whatever is acceptable as a medium of exchange for goods and services, or that can be used to service debts.

If interest is to be regarded as the price of money, that price must either be the result of the interaction of the demand for and supply of money or be fixed by some central authority, as many prices are in some nations. In the United States the supply of money is the result of many interactions, and most of these take time to produce an effect. In the very short run the supply can be regarded as fixed, but it most certainly can change over time. In the eyes of many people, including more than a few members of Congress, control of the interest rate through manipulation of the money supply is in the hands of the Board of Governors of our Federal Reserve System.[11] Does "the Fed" control our money supply, and thereby the interest rate? Let's find out.

The supply of money can be influenced by Federal Reserve Board actions in three ways: (1) through changing the reserve deposits required of every member bank; (2) by altering the discount rate on loans made to member banks; and (3) by buying, or selling, government securities in the open market.

Changing the reserve requirements directly affects the money supply. Demand deposit reserve ratios have varied from 18 to 26% (the ratio, incidently, also varies by bank size class). Any increase in the reserve requirement means that affected banks have an immediate reduction in the total value of loans they can make, and this reduces the potential of money supply growth. It also decreases the limits of the money supply system. It is a very effective, but seldom used, instrument.

The second means is through changing the Federal Reserve discount rate. Member banks can borrow from their Federal Reserve Bank to add to their own

[10] What about credit cards—our "plastic money"? These accounts are not part of the money supply, for a credit card is only a temporary substitute for money. Unfortunately, one has to pay, eventually, for all credit card purchases—usually by sending a check to the credit card company. The unpaid portion of any credit card account *does* represent a temporary creation of new money, but it is replaced each month by a transfer from some demand deposit. The total of such accounts, along with other trade credit, seldom varies much from month to month and would be almost impossible to measure.

[11] The Federal Reserve System, established by Congress in 1913, consists of 12 District Banks governed by a 7-member Board of Governors, each appointed by the President for a 14-year term. These District Banks are "Banker's banks": that is, they cannot be used by individuals or corporations, but are used only by other banks. Each District Bank serves as a Central Bank for all Member Banks in the District. Almost all large U.S. banks are members. The independence of the Board of Governors has been a hotly debated topic for many years.

reserve balance, thereby increasing their own capacity to grant commercial loans and, hence, increasing the potential money supply. But because they must pay interest to the Federal Reserve Bank on such a loan, any increase in the discount rate makes it less likely that member banks will find it desirable to borrow. An increase in the discount rate therefore acts as a brake on money supply. Additionally, it signals the whole economy that the Federal Reserve would like to see interest rates increase and the money supply either decrease or slow down its rate of increase. The effect on total money supply is limited because borrowing by member banks is held within strict boundaries, but the signal itself is of substantial importance. Other interest rates tend to follow any changes in the Fed's discount rate, but the effect is indirect.

Open market operations are the most frequently used device for influencing the nation's money supply—and the least understood. If the Reserve buys government securities in the open market, it pays for them by a check on some Federal Reserve Bank. When deposited in the bank by the dealer from whom the security was purchased, this check will be cleared to the Federal Reserve Bank and credited to the member bank's account. Hence this latter bank has increased its reserve, although no other *commercial* bank has had its reserves depleted, and the money supply can thereby be increased. A sale of government bonds by the Fed acts in an exactly opposite manner—if the Fed sells securities it *lowers* the Federal Reserve account of member banks and money tends to become "tight." If it *buys* them it "eases" the money supply.

So—does the Fed control the supply of money in the United States? Control?? Hardly! Influence it? Well . . . yes . . . let's say sometimes. It can and does influence it, but the Fed's ability to raise or lower money supply depends heavily on many other factors, most specifically on the excess reserve balance held by member banks. If the reserve balances are large, actions of the Fed have relatively little effect: if money is already tight, then it can have a far greater immediate influence. In either event the action taken must have time to work its way through the banking system until the actual users of money are affected. And the quantitative effect on supply will still depend on the extent to which available funds are actually borrowed by users.

Now, what about the demand for money? Usually at least two kinds of demand are recognized: *transactions* demand and *financial* demand.[12] Transactions demand is essentially the demand for money used to buy commodities of one kind or another—for food and clothing, entertainment, to pay the rent, buy a car, and so on. As one would expect, the transactions demand curve slopes downward (D_t in Figure 7.3): if the price of money (the interest rate) is high, fewer people take out loans for buying TVs, VCRs, personal computers, or feel able to assume a mortgage. *Financial demand* (D_f) is the demand for funds to use in purchasing securities, pay for new machinery, or make any other type of investment. It also includes the demand for funds individuals or corporations want to keep readily available for subsequent use or to protect themselves against unanticipated or threatened emergencies.

Both demand curves are shown in Figure 7.3 and are added to form the demand curve for money (D_m). The point of intersection with the money supply curve (S_m) determines the interest rate, r. As noted, the supply curve is essentially fixed over short time periods, but it can and does shift—sometimes in response to

[12]"*Financial*" demand as used here includes both the *precautionary* and *speculative* demands recognized by many economists.

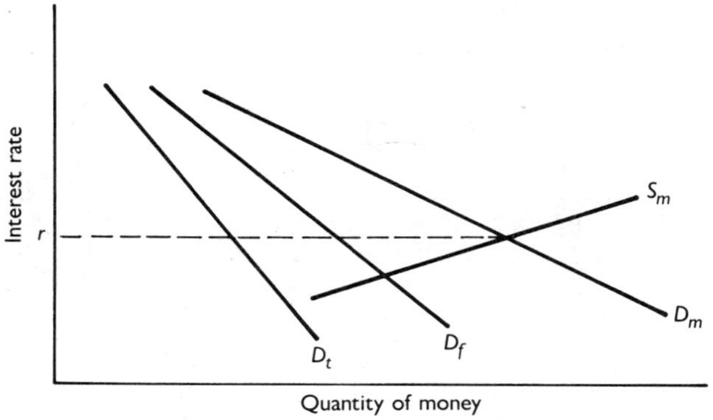

FIGURE 7.3. Aggregation of transactions demand and financial demand forms the demand for money. Combined with the supply curve of money, the interest rate is determined.

actions taken by the Federal Reserve Board, but sometimes not—for reasons already given. However, the figure does show that fewer and fewer investment opportunities will qualify as being desirable. Furthermore, higher interest rates make it more profitable to remain "invested in cash"—to keep assets liquid. Expectations also change rapidly, and can substantially influence the response to changing rates, either offsetting or augmenting action taken by the Fed.

LENGTH OF PRODUCTION PERIOD

Because production periods for timber are relatively long, interest rates have a particular significance to forestry. To illustrate: suppose you have inherited a 25-acre plot and decide to make this your first investment in forestry. You calculate that it will cost $40 an acre to prepare the area for planting and to purchase and plant the seedlings. This amounts to a $1000 investment in your forestry enterprise. The growth of this initial cost at five levels of compound interest and for four different time periods is shown in Table 7.2.

Clearly, few investments can be expected to pay high rates of return for 50 years or more. If one is going to make investments in forestry that will be long in maturing, then funds must be available at low interest rates. Fortunately, interest rates do tend to decrease as the length of the loan period increases, since costs of loan management are decreased and reinvestment costs avoided.

The other side of the coin is illustrated by Table 7.3. If 10% is the appropriate

TABLE 7.2
Growth of $1000 over Four Time Periods at Five Interest Rates

Rate	Time Period			
(in percent)	10 Years	25 Years	50 Years	100 Years
2	1,218.99	1,640.61	2,691.58	7,244.65
6	1,790.85	4,291.87	18,420.15	339,302.08
10	2,593.74	10,834.71	117,390.85	13,780,612.34
15	4,045.56	32,918.95	1,083,657.44	1 billion +
20	6,191.74	95,396.22	9,100,438.15	82 billion +

TABLE 7.3
Present Value of an Income of $1000 When Discounted at Various Rates for Specified Time Periods

Rate (in percent)	Time Period: 10 Years	25 Years	50 Years	100 Years
2	820.35	609.35	371.53	138.03
6	558.39	233.00	54.29	2.95
10	385.54	92.30	8.52	0.07
20	161.51	10.48	0.11	—

discount rate, a per-acre return of $1000 coming 50 years from now warrants spending no more than $8.52 on each acre today. Few silvicultural treatments would qualify. But at 6% for 10 or 25 years the picture is very different. Obviously, both the rate and the time period involved are important to forestry decisions, but high interest rates are certainly not conducive to long-term investments of any kind.

INTEREST AND RESOURCE ALLOCATION

In a market economy such as ours the most important role of interest is that of allocating investment capital. Markets for investment funds in industrialized nations are highly organized and closely integrated. Any change that might affect the return from an investment—risk, expectations, evaluation of uncertainties, or time preferences—will be reflected rapidly in interest rate levels. People having funds to invest balance the opportunities they see in forestry and timber production against all other investment opportunities available to them; they weigh the rate of return from timber, recreation, and other forest production possibilities against their "alternative rate of return"—their "*opportunity cost*" of capital. Conceptually, each investor establishes a schedule of investment opportunities, using the rate of return from each alternative to rank them in desirability. The result can be generalized as a curve formed by plotting the rate of return over the accumulating capital required for investing in successive opportunities (Figure 7.4). If the individual has investment capital (funds available for investing) amounting to q_1, he will invest in those opportunities promising a return of r_1 or higher.

Economists refer to this as a curve showing the marginal efficiency of capital. When used by a corporation this scheduling is called capital budgeting. Some corporations restrict the investment possibilities considered to those within traditional corporate boundaries: a paper corporation, for example, might consider pulpwood production from several alternative areas, along with investment in a new plant, a new paper machine, and so on. More and more, however, the traditional corporate lines are disappearing. Corporations holding extensive forest areas now consider sawlog production along with pulpwood production, and production of timber along with use of their lands for residential development, vacation home sites, campgrounds, or for hunting preserves. Even more broadly oriented are the so-called conglomerates, whose management may view chemicals, oil, transportation, or fertilizer along with recreation, timber production, and cattle when making up their list of alternative investments.

Public investments must also be made according to some schedule of priorities, for public funds are not unlimited. Throughout the world benefit/cost analy-

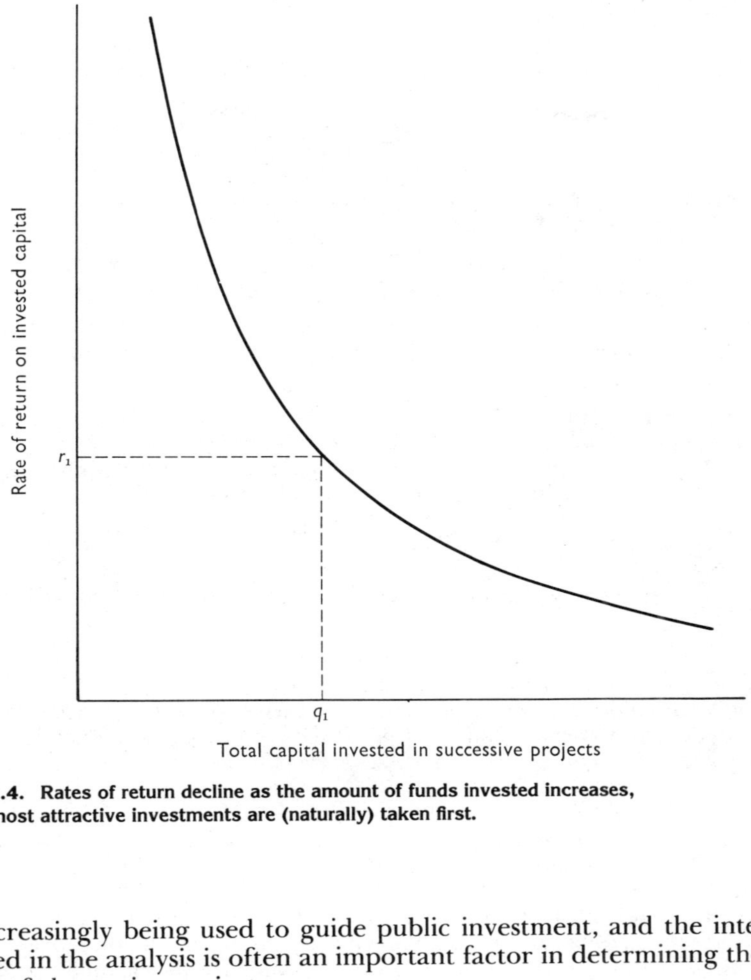

FIGURE 7.4. Rates of return decline as the amount of funds invested increases, since the most attractive investments are (naturally) taken first.

sis is increasingly being used to guide public investment, and the interest rate employed in the analysis is often an important factor in determining the relative ranking of alternative projects.

A FOOTNOTE TO THIS CHAPTER

In 1981 the U.S. banking system began changing in a major fashion: the system began to be "deregulated." Limits on interest rates—which were under great pressure because of inflation—were gradually lifted. Savings and loan institutions were given much more latitude in their activities and banks were able to undertake new functions (e.g, they could establish their own money market funds and also act as stock brokers). Probably most important, banks (as long as their state laws permitted) were able to expand by establishing branch banks in states other than that of the parent bank. Very large banking corporations thus became possible and a wave of bank mergers and acquisitions began. By the end of the 1980s it seemed that our banking structure was moving toward a system consisting of a few, perhaps no more than 10 or 12, regional banks, each with very large assets, hundreds of branches, millions of depositors, and possessing great financial power. How these changes will affect interest rates, the availability of capital to both individuals and corporations, and to forestry, has yet to be determined.

QUESTIONS AND PROBLEMS

1. If $2000 is invested at 8% compound interest today, how much will be accumulated in six years?
2. How much would you have to invest at 10% interest today to accumulate to $4500 in three years?
3. How much of a payment would be needed today to equal a payment of $3500 fifteen years from now at 12% interest?
4. If you were to deposit $5000 now, how much could you withdraw at the end of each year for 15 years and have nothing left if the interest rate is 8%?
5. An equipment manufacturer offers to replace the edger in your sawmill for $16,000. How much will you have to save each year to recover the investment in four years with interest at 7%?
6. A Christmas tree plantation costs $100 per acre to establish. Taxes are $4 per year, and protection costs that begin three years after planting will be $2 per acre annually. Seven hundred trees will be selected and trimmed eight years after planting at a cost of 30¢ per tree. Two years later they must be pruned again, this time at a cost of 40¢ each.

   If the annual interest rate is set at 9%, and all 700 trees can be harvested 12 years after planting:
   - **a.** What price per tree will be necessary if the owner of the plantation is to break even?
   - **b.** What price will be needed if the operation is to net one dollar per tree profit in 12 years?
7. Suppose you have decided to buy a new car under the following terms: Total sale price is $9500, with a down payment of $2000 and the remainder in equal monthly payments over a three-year period. Your bank tells you that you can have the needed money at a *nominal* annual rate of 12%. (A *monthly* rate of 1% is termed a *nominal annual* rate of 12%.)
   - **a.** How much will you be required to pay each month?
   - **b.** What is the "effective annual rate" you are being asked to pay? (The "effective" rate is the *annual* rate that will yield the same return as the monthly rate you are actually paying.)
   - **c.** Suppose you decide to delay the purchase of the car for three years, and instead put the $2000 in a certificate of deposit paying 12% annually over three years, and also arrange to deposit the monthly payment calculated above in a money market fund paying the same nominal rate, and compounded monthly. How much will you have in your account at the end of three years?
8. Recalculate Problem 7 using the price of the car in which you are particularly interested and the going rate for auto loans at your (or the nearest available) bank. Also use whatever rate the bank is paying on CDs or its money market fund when calculating part c.

chapter eight
TAXES*

It is doubtful if any other single institutional factor in the United States is as all-pervading in its effect on our economic behavior as our tax system. City, county, state, and federal government units exist to provide services, and all must be financed. A general understanding of the principles of taxation, combined with a specific knowledge of how taxes affect forest management and its costs and returns, is essential to the modern forester.

After a discussion of taxation *principles*, this chapter will concentrate on four major tax types: the general property tax and a suggested substitute for forest and farm lands (a productivity tax); the yield tax; the income tax (both corporate and personal); and estate and inheritance taxes. Each of these tax types will first be defined and described, then their influence on forestry, and especially on timber production, will be assessed.

PRINCIPLES OF TAXATION

If one is to appraise any tax structure, the goals the particular society has established for its government and its economic system should be known. Social goals are not derived from economic principles, but are basic expressions of the society's value judgments at the time and place those goals are relevant. When those value judgments change, one should expect to see, eventually, corresponding changes in the tax structure.

Since the close of World War II the U.S. economy has taken its general direction from three major goals:

1. Maintenance of a satisfactorily high rate of economic growth.
2. Economic stability (with emphasis on full employment).
3. An income distribution conforming to society's currently held standards of equity.

---

*This chapter draws heavily from the work of Richard and Peggy Musgrave (Musgrave and Musgrave, 1984).

As political administrations change, the priority of these goals may shift, and frequently measures taken to advance one goal prove incompatible with attainment of others. The unprecedented U.S. inflation of the late 1970s and early 1980s, for example, forced a re-thinking of the economic stability goal, and recent decisions to strengthen the U.S. military establishment have led to cutbacks in many programs that were designed with income redistribution goals in mind.

Even though the goals may not always be compatible with one another, all must be considered when deriving taxation principles. Three generally accepted *principles of taxation* are:

1. *Neutrality*: A tax should interfere as little as possible with attainment of an *optimum allocation and use of resources.* Whenever possible, it should assist in reaching this optimum allocation.
2. *Equity*: A tax should be *equitable*; the distribution of the tax load should help attain the desired pattern of income distribution.
3. *Efficiency*: The real *costs of collecting* the tax should be a minimum—and real costs include inconvenience to the taxpayer.

The relation of these principles to the goals of society should be kept in mind when specific tax types are being considered.

Neutrality (Resource Allocation)

Taxes levied on specific goods usually increase the price of that good, thus reducing its consumption relative to pretax levels. In most cases this will alter the allocation of resources. Sometimes the change may be deemed socially desirable: taxes on tobacco and alcoholic beverages, for example, are usually justified in this way, although the additional tax revenues seem to be given at least as much consideration as deterrent effects when lawmakers set "merit tax" levels. Other tax effects may prove harder to assess. A sizable airport tax on tourists entering or leaving a country, for example, might provide funds for airport operation, but also could have unintended effects on the number of visitors to the country's national parks or forest recreation areas. Similarly, taxes on diesel fuel might affect the type of tractors and trucks used in timber harvesting and indirectly increase the price of lumber, thus reducing lumber exports and cutting back domestic residential construction activities.

Equity

Equity problems call for special comment since they may take on several dimensions. From our general commitment to, and belief in, the market system comes one quite generally accepted approach to equity in taxation: a tax system should be devised so that the burden is proportional to the benefits received. A second approach stems from the belief that the tax system should provide equal treatment to all individuals receiving equal incomes[1] and also that persons of different circumstances should receive the "desired" *relative* treatment. Both beliefs are largely satisfied if the tax burden is made proportional to an individual's ability to pay, which in practice is interpreted as (roughly) proportional to income.

Neither the **ability to pay** nor the **benefits received** criteria can be applied generally, and the two are frequently in direct conflict. Most governmental func-

[1]The feeling exists that the "equal treatment for equal incomes" equity guide requires qualification by differentiating between incomes subject to different degrees of risk. Those with incomes subject to high risk should have a relatively lighter tax load.

tions are in the public sector because the benefits accrue to society as a whole rather than to individuals. Hence the benefit criterion, the basis for commercial activities in the private sector, cannot be applied. Furthermore, strict application of the benefit approach would often result in those with low incomes bearing the same per-capita burden as the rich—a violation of the "ability to pay" criterion. Yet some public activities can be financed on a benefit basis without adverse results. Provision and maintenance of public roads through taxes on motor fuel and by license fees are classic examples; creating forest recreation facilities by user charges or financing wildlife management through hunting license fees are common examples from forestry.

The "ability to pay" criterion has no real economic basis. It stems directly from a value judgment of our current society and is in accord with the general ideas of desirable income distribution. Since ability to pay is interpreted in terms of measurable economic well-being, this approach to equity has the very real advantage of being "workable." It should be observed that taxes meeting this criterion (progressive individual income taxes, for example) may work against attaining the desired economic growth through their effect on incentives. At the same time, such a tax may play an important role in attaining stability. This tradeoff between economic stability and economic growth creates a major policy issue whenever attempts are made to raise or lower income tax rates.

Efficiency (Costs of Collection)

The principle that costs of tax collection should be minimized points to the fact that resources used for collecting a tax add nothing useful to the national production of goods or services. Tax legislation should be clearly understandable to both taxpayer and collecting agency, and should be as simple as possible. Generally speaking, a marginal approach is the proper guide for determining the level of expenditure for collection purposes; expenditures should be increased until the additional cost is just matched by the added tax revenue thereby attained. Equity considerations, however, impose a real constraint on collection costs. Obviously, if some individuals can avoid payment, an inequity has been introduced, and those best able to avoid payment are often those able (but not willing) to pay the most.

PROPERTY TAXATION

During the early part of the 1800s taxes on land, houses, stores, factories, carriages, inventories of goods, and similar items were grouped together and called the "general property tax." In 1982 this *ad valorem* tax (levied "according to value") supplied 76.0% of the tax revenue for local governments (Stat. Abs., 1985)—those of the counties, cities, and towns of the nation. Property taxes are particularly important to school districts accounting (in 1982) for nearly 97% of their tax revenue. Broadly speaking, property taxes apply to two classes of goods:

1. *Real* property. Land, buildings, and permanent improvements. Usually real property has a fixed location.
2. *Personal* property. Usually movable, personal property is itself of two kinds:
   a. *Tangible*: Autos, farm machinery, personal effects such as clothing, jewelry, and so on.
   b. *Intangible*: Claims against tangible wealth, such as stocks, bonds, bank deposits, and so on.

Real property is by far the more important of the two, accounting in 1982 for about 87% of the total assessed valuation of all taxable property in the United States. As a result, the property tax today is essentially a tax on real estate. Since forested land subject to the property tax is fully one-fifth of the total U.S. land area, the issues arising from forest land taxes are significant, even though undeveloped land accounted for only 10% of the total assessed property value in the United States.

Administration of the Property Tax

Administration of the property tax requires five steps: assessment, review or appeal, determination of the tax rate, collection and distribution, and provision for delinquency. Each step can have a bearing on the ultimate effect of the tax.

Assessment

Assessment is the process of placing a dollar value on the property concerned. In the United States assessment is based on market value, with property sales in the jurisdiction providing a guide.[2] Reassessment is required as market values change but the period between reassessments varies. The rapid rise in residential dwelling values and corresponding increases in taxes led to strong resentment against property taxes in the late 1970s and to numerous "tax revolts" (typified by California's Proposition 13).[3]

Review or Appeal

The taxpayer has the right to request review of an assessment by a local board, often the county commissioners. If unsatisfied with the results, the taxpayer may contest the assessment in court. Reviews are infrequently requested, however, since knowledge of other property assessments is seldom at hand, assessments are typically at less than market value, and individuals fear that an appeal might result in the raising of a neighbor's assessment rather than in a lowering of their own.

Determination of Tax Rate

In theory, the county, township, school district, or other authority first totals the individual assessments in the jurisdiction, then determines the amount needed for performing all planned functions, subtracts this amount from the estimated revenues other sources will yield, and divides the remainder by the assessment total. In practice, of course, the process tends to be less simple. There may be legal or political limits on the allowable rate level and, if the planned expenditures exceed these limits, either services must be reduced or new revenue sources found. Neither problem has any easy solution.

Collection and Distribution

The individual tax bills are calculated by summing the rates that apply to each taxable unit, then multiplying the result by the assessed value of the property. For example, with a school tax of 8 mills (a mill is one-tenth of a cent), a county tax of 4 mills, and a city tax of 12 mills, the owner of a property assessed at $50,000 will pay $\$50,000 \times 0.024 = \$1200$. Taxes are paid once or twice each year to a designated central office that distributes the collected funds to the proper taxing jurisdiction.

---

[2]In England the basis for property taxation is its rent. Since (theoretically) the earning power of property would be capitalized into its market value, either base should yield essentially the same result, but in practice this is seldom true.

[3]As a matter of fact, while property tax revenues quadrupled over the 1960 to 1980 period, other local taxes were up by a multiple of 10, and the average *effective* rate on single family dwellings dropped from 1.98% in 1971 to 1.28% in 1980. In California where the tax revolt began, the effective rate had *dropped* from 2.48% to 0.98% (Musgrave and Musgrave, 1984, p. 475).

Provision for Delinquency

Following the tax due date, a period of grace is usually allowed during which the taxes, together with accumulated interest (and often a penalty), may be paid. After this period has passed, the property may be sold at auction and the realized funds used to pay the tax. The buyer usually receives a temporary deed that may be redeemed by the delinquent owner at any time within some specified period by paying the overdue taxes and penalties. In some states forest land reverts to state ownership rather than being subject to sale.

Deficiencies of the Property Tax

In general terms there are two reasons for criticizing the property tax as a means of raising public funds: (1) its administration is often poor, and (2) it does not meet the usual criteria for distributing the tax load.

The major administrative deficiency arises from the assessment process. In the past, and even today in some parts of the country, the assessor is an elected official, often without adequate training for what is inherently a difficult task. Poor assessment is most frequent in those places where the assessor is elected on a township basis, a unit scarcely large enough to permit hiring fulltime personnel.

There are more specific grounds for criticism. (1) The general property tax can and often does interfere with an optimum allocation and use of resources. Assessed unit values often vary by size of property and by use, are not necessarily related to market value, and the tax rates vary from one jurisdiction to another. To avoid some of these difficulties most states have procedures for "equalizing" assessments that overcome some inequalities within the state and serve, at least in part, to prevent jurisdictions from competitively lowering property tax rates as an inducement to new industries. Such variation influences the use of resources in ways not conducive to economic efficiency. (2) Property ownership is not necessarily related to an ability to pay, nor is it an accurate indication of benefits received; by either approach to equity problems the property tax is found wanting. (3) Costs of collection would be large if the tax were to be administered at all equitably. All property would have to be identified and assessed each tax period, and at present the means of assessment, of rate determination, and even of disbursement are seldom well known to the average taxpayer.

Forest properties do not escape these drawbacks of property taxation. Indeed, many of the common faults are magnified in the case of forest land. Assessment of timber land is difficult at best. Very few assessors are able to appraise forest land values either accurately or consistently, and the frequent reappraisal that would be needed to achieve something approaching an accurate tax base could scarcely be expected.[4] Additionally, lands held by absentee owners or large corporations tend to be assessed at rates higher than similar properties that are locally owned.

The Time Bias

These factors all tend to make the property tax as usually administered somewhat inimical to forest land management. But the largest drawback of property taxation when applied to forest land stems from a different cause—the long time period involved in stumpage production. The property tax must be paid annually (in some cases, semiannually). A forested acre, however, rarely yields an income every year.

---

[4]This would require initial assessment on the basis of an accurate inventory, then annual adjustment (or reinventory) for all growth, all harvest, and all natural mortality.

Hence the property tax represents an annual cost that accumulates until the harvest cut, and for financial success the eventual harvest cut must yield an amount at least sufficiently large to meet these costs together with the accumulated interest charges. An example, purely fictitious, will illustrate the effect of this "time bias."

Barbara has just become heir to a recently harvested 100-acre hardwood forest that has been managed for timber production for some time. She receives an offer of $20,000 "for cutting rights to all timber on the property" and decides to investigate the alternatives. She learns that the area has been under an uneven-aged silvicultural system with a 10-year cutting cycle. The site is reasonably good and annual growth averages 300 board feet per acre. The last cut left 7000 board feet per acre as growing stock. Three thousand board feet per acre should be harvested in 10 years, at an expected price of $50 per thousand.

Inquiry about taxes reveals that rising demands for forest properties has resulted in a recent reappraisal of county properties. Cutover areas, she is told, are assessed according to the residual timber volume, valued at $25 per thousand; with 7000 board feet her property will be valued at $175 per acre. The tax rate is 40 mills, so the annual tax will be $\$175 \times .040 \times 100 = \$700$. Other total annual costs amount to only $2.00 per acre ($200 per year), so total annual costs are estimated to be around $900. Interest rates are currently at 8%.

Barbara has at least three choices:

1. Continue to manage the forest indefinitely. If this alternative were selected, the present value of the costs would be

$$\$900/.08 = \$11{,}250$$

and the *PV* of the revenues ($15,000 received every 10 years, beginning 10 years from now):

$$\$15{,}000\ [\frac{1}{(1.08)^{10} - 1}] = \$12{,}943.03$$

Hence, the net present value *(NPV)* would be

$$\$12{,}943.03 - \$11{,}250 = \$1{,}693.03$$

2. Sell the timber cutting rights today, for which she would receive $20,000. She would also have 100 acres of cut-over land—which we will assume would have no further value to her unless management expenses were incurred.
3. She can pay the taxes and other costs for the next 10 years, then sell the 10,000 board feet of timber per acre for (she estimates) an average of $50 per thousand. Under this option she would receive $50,000 in 10 years, which has a present net value of

$$\frac{\$50{,}000}{(1.08)^{10}} - \frac{900\ (1.08)^{10} - 1}{.08\ (1.08)^{10}} = \$17{,}120.61$$

And again, she would still have the cut-over land.

Of the choices, alternative 2 is clearly Barbara's best decision. This does involve liquidating the forest, but she could then forget about taxes, land management problems, and the possibility that in 10 years stumpage prices might be lower than she estimates. Besides, someone might offer to buy the cut-over land and, if not, she can let it revert to the state as tax delinquent, or might even give it as an addition to the nearby State Forest and take a state tax deduction.

Fortunately for forestry, the illustration is exaggerated. In most states today the property tax is made bearable by assessments far below the timber's actual market value. But assessments do change, and in some areas reassessment problems are becoming acute. Normally, land is assessed at the value resulting from its "highest and best use," that use bringing the greatest value. Prime recreational land, for example, or land suited to summer home construction usually has a high per-acre value. Practically speaking, the example just given is well suited to show the effect of imposing high assessed values on forest land; $175 per acre is low for second home use, but it would effectively rule out forestry. Zoning laws can help prevent fragmentation of management units, but over time those areas yielding higher returns from nontimber uses will produce timber only as a secondary product, if at all.

The possibility of clear-cutting a forest, then letting the land revert as tax delinquent, may sound farfetched, but faced with a high tax situation and if only timber values are involved, this is what many owners would do under the situation illustrated above. This happened in the northern Lake states during 1910 to 1940, where so much forest land reverted to state ownership that it was called "the new public domain."

Basically, the problem exists because the land and timber are both real property. Land has value because of its income-producing potential. Agricultural land rises in value as its productivity increases, and there is a close relationship between the annual income produced and the market value of agricultural land. Put directly, the value of land is the capitalized value of the net income stream it produces. If forest land values were calculated similarly,[5] they would offer no particular problem, but because of the peculiarities of stumpage production they are not so calculated by assessors, and they *do* offer particular problems. The time period required for production is only the most obvious difficulty in applying the general property tax to forested areas. The dual nature of standing timber and its one-way flexibility of production and marketing paves the way to liquidation.

The problem created by tax delinquent forest lands became so acute in the 1930s that the Congress appointed a study commission to investigate "The Forest Tax Problem." The report of this group, the Fairchild Commission, is still the classic study of its kind. The report proposed:

> *That the property tax be modified in one of three ways in order to remove the excess burden inherent in that form of tax. The first is to reduce the current property tax on the deferred-yield forest property by an amount proportioned to the extent of income deferment in each individual case. The second is to defer the payment of property taxes until income is received. The third is to apply a flat-rate reduction to the property tax based on the degree of income deferment which may be considered typical or average for the State!* (Hall, 1935)

Each of the three alternatives involves a conceptual separation of the land and the timber and suggests modifying the property tax so that it would "impose a burden about equal to that of a tax on income or net yield at a rate equivalent to the property tax rate" (Fairchild, 1935). None of the alternatives has ever been tried in this country, probably because the suggested methods are believed too complex to replace the relatively simple general property tax.

The many drawbacks associated with the property tax might well lead one to ask why this tax has been permitted to survive so long. Why not do away with it

[5]Of course, they can readily be calculated in this same fashion. The calculation requires, however, that future yields and prices be estimated—and future yields are directly dependent on levels of management and levels of growing stock.

entirely? The answer is purely pragmatic. The property tax has survived because it yields a large, stable, and predictable income to the taxing units; it is easily, although often badly, administered; and long use has made it familiar to the American public. Moreover, any drastic revision would result in large windfall gains to property owners, since this tax has often been capitalized into property values.[6] The first factor (its large and predictable yield) is of great importance to local governmental units, which must somehow be able to plan on future revenues. School buildings, sewer and water systems, and fire stations must be built, and payments for them are generally scheduled far in advance. The predictability and dependability of the property tax have become very real assets in planning, even though these same characteristics may also be a serious liability when it comes to promoting economic equity or aiding full employment.

PRODUCTIVITY TAXES

A forest productivity tax would apply the local property tax rate "to a calculated productivity value which varies with land quality but not with timber stocking" (Klemperer, 1983). In 1982, at least seven states made this type of tax either optional or mandatory (Carlan et al., 1982). The most common approach to a forest productivity tax is to "capitalize the net value of the mean annual increment as if it occurred as annual income, given some assumed rotation age" (Klemperer, 1983). The property tax rate is then applied to the per-acre value as calculated for various timber types and site classes, but not for different stocking levels.

The (unmodified) productivity tax burden for a landowner establishing a plantation on bare land would be greater than that under an *ad valorem* property tax that was recalculated each year according to market values (Klemperer, 1983). The reason for this lies in the timing of the payments. An (unadjusted) productivity tax is levied on the capitalized *average* net value per acre over the assumed rotation. The same base value is used each year. Timber values for a plantation, however, begin with zero for the early years, then increase until harvest. For this reason the productivity tax shifts the tax payments toward the present—which increases the present value of the total tax bill (future payments have a much lower present value than present payments).

Calculated productivity values are very sensitive to the interest rate used for capitalization. Under a forest productivity tax one would expect timber owners to argue for high capitalization rates, while nontimber owners would want a lower rate: lower rates would shift more of the tax burden to timber properties. If the capitalization rate is tied to some index that varies with inflation, then an additional problem can be encountered: with high rates the timber values would be undervalued (Gregersen, 1975). For these reasons forest productivity taxes have usually been modified—for example, by lowering estimates of growth or values.

THE YIELD TAX

The Fairchild Report did not recommend the yield tax as a substitute for the general property tax, yet this tax type is the one most widely accepted as overcoming the deficiencies of the property tax. A yield tax is exactly what its name implies, a tax imposed on the yield of some particular operation. In forestry it is normally

[6] That is, the market value of the land has been reduced by an amount that would yield, at market interest rates, an annual income just equal to the amount of the tax. For example, a tax of $4 per acre, at 8%, would reduce the property value by $50/acre if capitalized into the land value. How often this occurs with forest properties is uncertain.

administered as (like the property tax) an ad valorem tax, a tax according to value; hence it is a tax on the value of the forest yield. Because payment is usually made at the time of harvest, yield taxes effectively eliminate the "time bias" of the property tax when it is applied to timber-bearing properties.

Typically, yield taxes in the United States have required a conceptual separation of land and timber, even though land takes on a value only because it has a productive capability. The land is then subjected to the prevailing property tax while the timber values are, essentially, ignored until harvest. In a very real sense this represents a compromise between the general tax and an income tax. As Marquis observed over three decades ago (Marquis, 1952), the yield tax is a compromise "between two contradictory desires: the postponement of tax payments by the owner and the need for current revenue by government units." In addition, the yield tax has frequently been made a vehicle for regulating cutting practices. Clauses specifying some minimum level of forest practice before owners can qualify their lands for classification are often made part of the law.

Yield taxes are sometimes distinguished from severance taxes on the basis of their intent. A yield tax is one intended to replace the general property tax, while a severance tax is a revenue measure levied in addition to the property tax. The difference between severance and yield taxes is really much deeper. Severance taxes are most often applied to mineral or other nonrenewable resources that, once removed, are gone forever. They reflect the quite general notion that such resources are part of society's heritage. A severance tax is a kind of payment to society for the privilege of depleting the resource. Such a tax is almost always fixed on a unit basis—so much per ton of ore, barrel of oil, and so on. A yield tax, however, is usually calculated as some percentage of the product value of a *renewable* resource. Moreover, timber yield taxes are often optional. If the landowner so elects, he or she may have the property placed under the yield tax option and thus removed from liability for property tax. The owner is under no compulsion to do so; it's a personal decision, based on what the owner believes is best financially. In contrast, severance taxes are always compulsory.

At first glance the yield tax would seem to offer a logical means for eliminating some of the difficulties imposed on timber production by the property tax. And in fact it does, but new problems are created, with the most important being the obvious difficulty imposed on the tax-receiving units. A yield tax by its nature simply cannot provide a predictable, dependable source of income. If the tax rate were completely fixed, the income would still vary by the amount of timber cut during the year and by its price. Conceivably, even under a "sustained yield" program, the yield tax income to a local community might be very high during a year when the lands in the county or township were scheduled for harvest, but almost nonexistent during the following years when cuts were scheduled in forest areas outside the taxing unit. To counter this difficulty, some legislation specifies that the yield tax be paid directly to the state, while the state makes an annual payment "in lieu of taxes" (usually fixed on a per-acre basis) directly to the local taxing unit. The yield tax thus becomes a payment made at infrequent and irregular intervals into a revolving fund, out of which *annual* payments are made to local taxing units. Clearly, to inaugurate such a plan almost always requires that a substantial initial grant be made to cover payments in the earlier years of the program.

The administration of a timber yield or severance tax requires the administrator to know either the total revenue from the harvest or the total volume and the (average) price. This can be required from a private seller, but what about the integrated wood product firm cutting its own timber? Here, as Jackson (1980) points out, no market transaction is available for tax audit. Self-reporting may be deemed highly desirable by producers, but it does raise questions of equity since any underpayment means a shifting of tax burden to other taxpayers. Spot checks are expensive and seldom yield adequate proof for legal action.

Yield tax proponents often express surprise at the comparatively small percentage of eligible landowners who have taken advantage of available yield tax options. Some of the more important reasons for the apparent lack of interest on the part of forest owners are: lack of knowledge; cumbersome methods of inspection, application, and approval (too much red tape); restrictions on forest practices, and the threat of future restrictions; lack of sympathy or downright opposition on the part of public officers; excessively high eligibility requirements; and an absence of any clear tax advantage. In a very real sense, however, the small number of acres grossly understates the effectiveness of these laws. Whenever such options exist, the landowner can, if it is believed to be financially advantageous, escape overly heavy property tax assessments, and this possibility is fully recognized by local assessors. Many owners who have not directly used the yield tax option still regard the legislation more favorably. It is a potent "club" to have in hand in case some overly ambitious assessor comes into office.

Changing from one tax system to another can also pose transition problems. A forest owner who has recently harvested his property could avoid taxes on his timber for some time by switching from an ad valorem property tax to the yield tax. In contrast, an owner anticipating a harvest in the near future would probably face a higher tax payment if she were to switch from an ad valorem tax to a yield tax.

EFFECT OF PROPERTY AND YIELD TAXES

The discouraging effect the property tax can have on an anticipated investment in forest land should be quite clear from the example already given. The discouragement will increase with an increase in interest rates, in tax rate, or in the interval between investment and harvest. It will become less as these same items are reduced. Some of the other effects, particularly the effect on management practices, are less easily seen.

Between 1970 and 1984 methods for economic analysis made substantial advances, not the least of which has been the increasing use of mathematical modeling. Several such models have been developed and used to examine the effects of property, yield, income, and other taxes on profitability, rotation length, management intensity, long-term timber supply, and so on. Among these studies are those by Hyde (1980), Jackson (1980), Chang (1983), and Graham-Tomasi (1984). With these analyses, as well as with the graphic analysis presented here, one needs to keep in mind the assumptions—both implicit and explicit—underlying the model, particularly those regarding the form of the production function.

A Comparison

We can begin by taking the "ideal" forest—a fully established, operating, going forest concern—and ask "What effect would a property tax or a yield tax have on the *production* decisions of the manager or owner of a commercial forest?" Figure 8.1*a* gives a total cost curve for such an operating forest, with both cost and output increasing as the intensity of management (and output) is increased. Note that the figure represents the entire operating forest area involved; it is *not* a per-acre figure. At the given unit price, *P*, and under the cost/output relationships shown, the optimum annual output would be *Q*.

Now introduce a tax of *T* dollars that is levied on the forest *property*, Figure

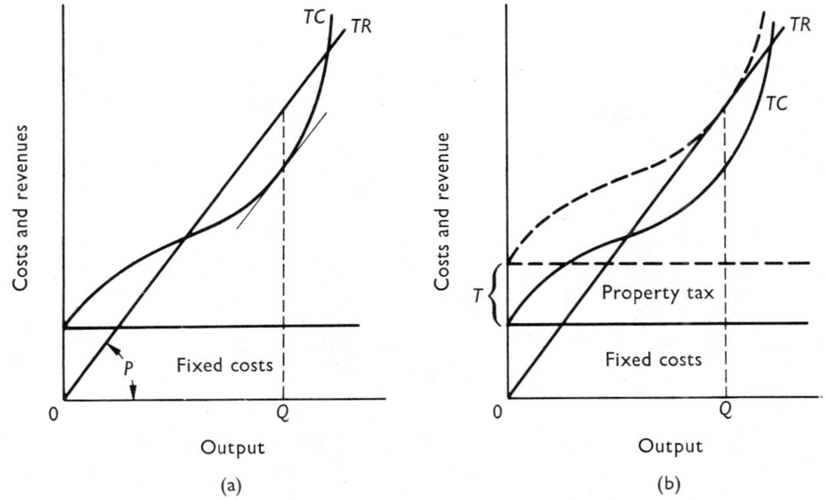

FIGURE 8.1. Effect of property tax on optimum output levels for a fully established "ideal" forest property.

8.1*b*. Since the tax is not directly dependent on the productivity of the property or on its present or anticipated output it will have no effect on the production decisions of the owner or manager.[7] Net revenue is decreased—in fact, it disappears—but all costs are still covered and the optimum output level remains unchanged by the tax. Were the tax to be increased further, it is clear that the long-run outlook for the operation would not be very promising, but from the standpoint of normal production decisions regarding output levels, a property tax of *any* level on this kind of forest property would have no effect: it is a fixed cost, unvarying with output. If the tax were to be increased *above* the level indicated *(T)*, the long-term continuation of the firm would be jeopardized.

Compare the situation of Figure 8.1 with that of Figure 8.2*a* and *b*. The yield tax does have an impact on the production decisions, for a yield tax is a tax on value produced. One can view this tax as reducing the selling price of the product, as in *a*, or as a variable cost *b*. If, for example, stumpage price were \$10 per cord and the yield tax 10% of the stumpage value, the effect of the tax would be to reduce the price to \$9 per cord. As a consequence, the total revenue curve would swing down (from *TR* to *TR′*), its slope decreased (by 10%), and the optimum level of output reduced from *Q* to *Q′*.

Exactly the same result is obtained if one views this tax as a cost. Under the stated conditions the tax amounts to \$1 per cord. Hence the total amount to be paid varies directly with output and the tax is a *variable* cost. In *b* this is reflected by an "upward attenuation" of the total cost curve; the total cost curve is stretched upward and its slope quite obviously increased at every point. The net result is the same as that shown in *a*; optimum annual output is reduced, along with, of course, total net revenue.

[7]If the property were reassessed every year and the tax levied on this new value, then the tax level would depend on the productivity of the property and would affect production decisions.

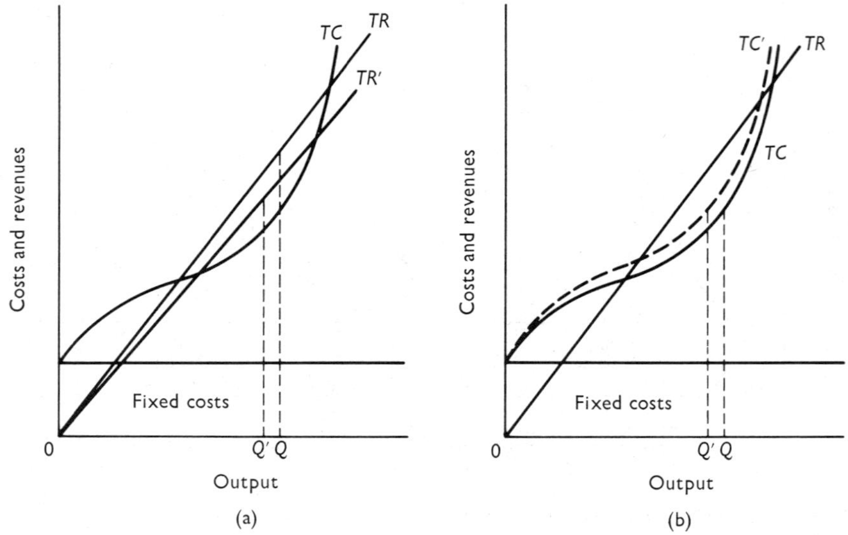

FIGURE 8.2. **The effect of a yield tax can be treated *(a)* as a reduction in revenue, or *(b)* as a variable cost.**

A going forest concern with a steady annual income is an unusual forest in this or in any country. Would the conclusion reached regarding the property tax from Figure 8.1 apply to a farm woodlot or a 500-acre plot of timber owned by a retired postmistress (and managed by a consultant)? It would if the property yielded an annual income to pay the property tax, but this is not the usual situation. The tax must be paid every year, as in the example given for the northern hardwood stand, and interest charges accumulate constantly. The longer the period between harvest cuts the more burdensome the property tax becomes and the more attractive the yield tax appears. The effect of either of these taxes is dependent not only on the rate but on the cutting cycle, or rotation, and even on the kind of silvicultural system employed. One can even argue that the property tax favors all-aged management as compared with even-aged management and, since most of our commercial coniferous species are presently being handled on an even-aged system, the property tax, through this mechanism, has an unfavorable impact on timber supplies (Yoho, 1965). By far the more important influence on supply, however, is the direct effect of the "time bias."

From the foregoing discussion one can draw several conclusions:

1. The effect of either a property or a yield tax will vary with the tax *rate*. When property taxes, for example, are low almost no one bothers about their effect on timber production. When rates rise forest owners begin to think anew about "the forest tax problem."
2. The burden of a property tax increases as the income from the property is moved into the future, or as the periodic incomes from the property become less frequent. With a fully regulated forest having a steady annual income a property tax will affect income *levels* but not input or output *decisions*. Under such conditions a yield tax offers no particular benefit (assuming that effective rates are comparable) and from a societal standpoint may in fact lead to a lower intensity of management.

3. With virgin forests, as presently exist in many of the developing countries, a property tax may be a very useful device to force utilization of presently standing timber, for a property tax cannot be avoided by nonutilization.[8] Under such situations (in developing countries where large areas are held under private ownership) the property tax can aid in achieving the goal of economic growth. And in these situations it may also aid in attaining a better distribution of income. A severance tax or a yield tax, however, can be avoided by postponing the cut. The latter type of tax does nothing to encourage utilization and may in fact inhibit use.
4. Since a yield tax has the same effect as a lowering of the stumpage price it would seem to make some marginal stands inoperable. However, since a yield tax is a tax on the stumpage price, and prices for stumpage approach zero on truly marginal properties, the effect would be small.

It should also be recognized that any tax on privately produced timber that is directly on or can be shifted back to the stumpage producer produces a bias in favor of public forestry, for the latter pays no tax. One possibility that has received little attention is that of a timber "sales tax," levied on the purchaser, *not the producer*, of the timber (Jackson, 1980). Such a tax could, and probably would, be shifted back to the producer in the form of a lower stumpage price, but would avoid the bias toward public production. It should be reasonably easy for a state to administer a timber sales tax, although distribution of the tax yield to the county or township where it originated might cause problems. Such a tax would also eliminate the problem of self-reporting encountered with yield or severance taxes.

THE INCOME TAX

The year 1913 is a memorable one in U.S. tax history. In that year the Sixteenth Amendment to the Constitution was ratified, providing that "the Congress shall have power to lay and collect taxes on incomes, from whatever source derived, without apportionment among the several states, and without regard to any census or enumeration."[9] Since then taxes on both corporate and personal income have increased in relative importance and are now the primary source of federal tax revenue. They are also of great importance to state and even city governments as well. In 1985 the personal income tax provided 44.74% of all federal budget tax receipts, while corporate income taxes accounted for another 9.1% (down from a 1950 high of 27.1%). Thus, 53.75% of federal budget receipts came from the two income taxes. The role income taxation now plays in federal and state financing, together with its absolute size (personal and corporation federal income taxes totaled $396.1 billion in 1985) makes the income tax of primary importance to corporations and to citizens of the United States.[10]

---

[8]But since in developing countries most legislators are property owners—often very large ones—the likelihood of securing passage of property taxation is not high.

[9]An income tax was levied during the Civil War and in 1872 was held to be an indirect tax and hence valid. But parts of the 1894 personal income tax were held to be direct and therefore unconstitutional. The Sixteenth Amendment was the result, although the corporate income tax was enacted in 1909, four years ahead of the ratification of the Sixteenth Amendment.

[10]Other receipts: payroll taxes, 28% (in 1982, 35%); sales and excise taxes, 4.3%; estate and gift taxes, 1.2%; customs duties and miscellaneous receipts, 2.5%. The percentages given here and above are from Musgrave and Musgrave, 1984, Table 2.2, p. 30.

The Tax Reform Act of 1986[11]

In 1985 President Reagen proposed major changes in the federal income tax law, and by early 1986 potential revisions were being debated by the House Congressional Committee. Prospects for genuine reform, however, seemed bleak: in April the *Wall Street Journal* reported that the original targets—elimination of tax breaks for special interest groups—would apparently be retained, including "existing tax breaks for the oil, timber and mining industries."

To everyone's amazement, a proposal of the Chairman, Senator Packwood, was cleared by the Committee, passed by both Houses of Congress, and signed by the President. The new bill was widely regarded as effecting the most sweeping change in our federal tax structure during the 76 years following enactment of the federal income tax and is generally referred to as the Tax Reform Act of 1986.

There were indeed many changes. To mention just a few: most (but not all) the special tax breaks were abolished, the exemption levels and the rate structures of both the individual and corporate taxes were altered, many depreciation provisions were modified or eliminated, and the favored treatment accorded capital gains was repealed. Those of specific importance to forestry will be treated at appropriate points in this chapter, although with so many changes in both corporate and individual tax law it will be several years before the effects of these changes will be clear. Moreover, one can be quite sure that during the coming two or three years other changes will be made; special interest groups were as surprised as anyone by the acceptance of the reform, but strong efforts will be made to reestablish lost advantages. Over the next decade it will probably be even more necessary than before to review annually the latest issue of the Forest Service "Guide to Federal Income Tax for Timber Owners," Agriculture Handbook Number 596 or its successor, or one of the more detailed treatments of timber taxation, and to keep abreast of tax changes that could influence forest investment decisions.

State Income Taxes

State income taxes vary greatly and (like the federal tax) change from year to year. In 1982 all but seven of the 50 states had some form of income tax: in 37 states these taxes were similar in structure to that of the federal government. As with the federal tax, the greatest impact on forestry was on income from timber sales. Capital gains were generally treated in much the same fashion as by the federal system, but seven states taxed 100% of the gain. In 1982, state income tax rates varied from 1.9% on all taxable income in Indiana to West Virginia's complicated system which ranged (in 24 steps) from 2.1% of the first $4000 of taxable income to 9.6% of that over $400,000 for married persons filing jointly (McGee et al., 1984). Deductions, tax levels on capital gains, methods of allowing for expenses—all vary between states. Because of this wide variation, attention from this point on will be confined to the federal income tax.

Personal Income Tax

To discuss income taxation there must be an agreement as to what constitutes income. The basic definition used here is the one now accepted by most fiscal theorists (Musgrave and Musgrave, 1984): income is equal to consumption during

[11]Much of the material concerning the Tax Reform Act of 1986 followed from a telephone call to William Siegel of the Southern Forest Experiment Station, New Orleans, La. He is probably best known as the author of Agricultured Handbook No. 596—the Forest Service "Guide to Federal Income Tax for Timber Growers."

the given period, plus the increase in net worth. As a practical expedient, however, tax legislation in the United States interprets the concept of income as a *realized* flow of wealth during the specified period. Although the difference may at first appear slight, it is important. To illustrate: if the house I own increases in value from $30,000 to $60,000, my net worth has similarly increased, and quite clearly my potential spending power for consumption has risen by $30,000. But until I actually sell the house—until the income is *realized*—it is not taxable. There is no recognized flow of wealth until an exchange transaction of some kind occurs.

This approach is a practical necessity because it would be virtually impossible to revalue all assets every tax period. It is not, however, a logical necessity, and the basic definition must be kept in mind when looking at income taxes from a policy and from an equity standpoint. And the practical approach does introduce inequities. For example, the rent I save by owning the house in which I live is tax free; the use value I receive is not counted as income. But a tax would have to be paid on any part of my salary used to rent a house. Many other such inequalities exist—the value of services rendered by family members, home-produced food and clothing, and the value received from do-it-yourself projects are illustrations.

The fact that the tax is levied on income as received, along with the requirement that income determinations be made on a yearly basis, are two of the most important features of the U.S. income tax. The third is its progressive nature: the tax rate increases as the income base increases. Many of the problems associated with this tax, in forestry as well as in other areas, stem from these features.

Progressive Rate Structure

In previous years the progressive rate structure was usually given top billing as the most obvious, and to some the most painful, feature of the federal personal income tax. There were 14 different rates of income taxation ranging from 11% to 50% in 1986; in 1988 there will be only 2 rates—15% and 28%. During the late 1950s and early 1960s the maximum marginal rate was 91%, a level at which many believed the consequent reduction in incentive was a hinderence to economic growth: individuals in the top income bracket received only 9% of the marginal return earned by their efforts. In 1965 this maximum was reduced to 70%, in 1980 to 50%, and under the tax reform of 1986 the top rate will be 28%. One might argue about the social desirability of lowering the maximum rates, but there can be little doubt of its positive effect on incentives to work and to invest when compared to the previous rate structure.

It should be recognized that these rates were, and are, *marginal*, applying only to that part one's income *above* specified levels. The *average* tax is much lower than the rate structure indicates. In 1986, for example, a married couple with two children, and whose entire income of $15,000 was from wages, filing a joint return and not itemizing deductions, would have had a tax liability of $867—about 5.78% of their total income. The *marginal* rate (the rate on the last $50 of income) was 14%. At an annual salary of $40,000 and with all other conditions unchanged, tax payments would have been $6123, or 15.31% of total income and fully 28% of the last $50 earned. The maximum rate (50%) applied only to amounts in excess of $175,000.

Agreement with Taxation Principles

Personal income taxes are reasonably in accord with the three major principles stated earlier. A tax on personal income does affect individual freedom of choice and probably has some affect on the choice between work and leisure. Such a tax, however, is generally held to be relatively free from direct interference with the allocation and use of resources.

The largest impact of a personal income tax comes through its effect on income distribution. Because of its progressive rate structure the income tax has favorable effects on the social objectives of income redistribution and equity (although this effect was decidedly reduced by the tax reform of 1986). At the same time, progressive income taxes may unfavorably affect the rate of capital formation, since richer people pay heavier taxes and they are also the ones most likely to save. This in turn may adversely affect the rate of growth of the economy—depending, of course, on how the federal government chooses to spend the funds it receives.

In general, the personal income tax is regarded as the most equitable means of apportioning the costs of government among the population. The distribution of the tax load conforms with the income distribution pattern approved by society and helps attain this desired pattern. The real costs of collection seem within reason (less than 1%) though one might hope for considerable reductions in the cost of compliance, which may be excessive. Because of the large sums involved, the personal income tax has become an important tool of fiscal policy in maintaining economic stability, especially since the federal (and most state) tax on salaries and wages is collected at the source.

Corporate Income Tax

A corporation is a legal entity, quite distinct from its owners. It receives income, but does not necessarily pay this income out to the people who own the corporation. Corporate earnings not paid out as dividends cannot be reached by a personal income tax except to the extent they are reflected in taxes on capital gain. Without some sort of parallel tax treatment the corporation would become a major device allowing profits to be reinvested tax free—hence, the corporation income tax. Yet the same rate structure should not apply to corporations as to individuals. Ability to pay has little meaning in a corporate context, while the use of progressive rates would be highly discriminatory against stockholders in large companies as contrasted to smaller ones.

Rate Structure

As with the personal tax, rates for the corporate tax vary from year to year. In 1986 (and prior to July 1, 1987) there were five rates for ordinary income: the first \$25,000 of *net* income was taxed at 15%, the second \$25,000 at 18%, the third at 30%, the fourth at 40%, and all over \$100,000 at 46%. After July 1, 1987 the first \$50,000 will be taxed at 15%, the next \$25,000 at 25%, from \$75,000 to \$100,000 at 34%, from \$100,000 to \$335,000 at 39%, and all over \$335,000 at 34%. Although the rate progression is sometimes defended as a means of encouraging growth of new or small businesses (reinvestment of earnings is a major source of expansion funds), this objective would be better satisfied by a tax based on size, not net returns. Steps believed to support small firms, however, are politically popular even though they may in fact foster or support inefficiencies.

Both personal and corporate tax laws recognize that losses as well as gains can result from business activities. Like other tax specifics, the loss limits and carryover privileges change from time to time.

Three items require some further development before we can look into the effect of personal and corporation income taxes on forestry: capital gains, depreciation and depletion.

Capital Gains

Both corporate and personal income tax laws recognize two types of gain and loss: ordinary and capital. Realized capital gains are "gains which result from the sale of assets other than those held in the ordinary conduct of business" (Musgrave and Musgrave, 1984, p. 337). For example, if one sells a forested area for a higher price than was originally paid, the difference is a *capital gain*. In the United States the major source of capital gains (in dollar terms) is common stock of corporations, with real estate a poor second. Population growth, urbanization, speculation, the reinvestment of corporate earnings, a rising general price level, a decline in the market rate of interest—all these and many other factors produce capital gains.

There are some who argue strongly that capital gains should not be taxed. Canada views such gains more leniently than the United States does, and until 1962 Great Britain did not tax capital gains at all. Several other countries still regard such gains as nontaxable income. Part of the argument relates to the fact that an individual's income may be used to enhance future earnings rather than for consumption. Certainly, if one changes the definition of income from "potential spending power for consumption" to "potential earning power" a different type of tax might result, especially if equity principles are revised or ignored. To illustrate: a drop in interest rates would normally increase security prices, while increasing interest rates cause a fall. If interest rates were halved, for example, prices of newly issued bonds would approximately double; a "perpetual" bond yielding $8 annually that sold for $1000 when interest rates were around 8% would increase in value to around $2000 when interest rates dropped to 4%. From the standpoint of earning power there has been no gain in capital at all, for it now takes $2000 to earn what $1000 earned before the decline in interest rates. If one sold the bond for $2000, however, the $1000 gain would be just as available for spending as $1000 earned as salary or wages. Naturally, the extent of any increase in price due to a lowering of interest rates depends on the time to bond maturity.

Similarly, if one buys a piece of forest property for $5000 and sells it for $10,000, the $5000 gain is as much income (potential spending power for consumption) to that person as any salary or consulting fee might be. To ignore this income would mean a failure to treat people receiving equal incomes in the same fashion. And since capital gains are much more important to high- than to low-income groups, failure to tax such income would violate the ability to pay criterion as well as being contrary to usual equity considerations.

Problems of Capital Gains

The difficulty created by capital gains arise primarily from characteristics of the income tax: progressive rates, and tax levied on income as realized, and income being segregated into definite yearly periods. If capital gains could be taxed as they accrued, whether realized or not, most of the argument for special treatment would disappear. But this is an administrative impossibility, and the established treatment constitutes a compromise. The alternative of permitting capital gains to go untaxed, however, would produce inequities that most people are apparently unwilling to accept (other than on gains accrued until death on assets bequeathed to heirs).

The situation for both individuals and corporations was particularly difficult when the *rate* applied to capital gains was lower than that on ordinary income. In 1986 a taxpayer included 40% of any long-term capital gain as ordinary income, with the remaining 60% free of tax.[12] The top rate paid by anyone on such long-term capital gains was therefore 50% of 40%, or 20%. In 1988 all such gains will be taxed as ordinary income, but with a maximum rate of 28%. (As an interim year, 1987 has a somewhat peculiar rate structure—a phasing in of the new rates.) Capital gains for corporations will be taxed as ordinary income—a decided change from previous practice.

For individuals in upper-income brackets, or for corporations with over $100,000 profit, there was a real advantage in having gains classified under the capital gain alternative. In 1986 it meant that any profit a corporation could classify as a capital gain would, after taxes, be one-third larger that if it were ordinary net income. For an individual at the top levels, it meant receiving 80% of the gain instead of 50%. In 1988 there still can be differences, for the top rate on capital gains will be 28% (except for joint filers receiving from $71,900 to $149,250 and single taxpayers in the $43,150 to $89,560 bracket, who will pay 33%).

From the standpoint of the general economy the most serious criticism of special treatment for capital gains stems from the so-called "locked in" problem. Investors are encouraged to hold on to securities or other properties on which gains have been made rather than selling them and reinvesting in some other venture. The problem becomes most severe in the case of older people, since capital gains on property held until death are not taxed. The significance of this problem is difficult to assess: when the maximum rate of the tax was 20% the rational investor would not be greatly influenced if the opportunity for greater gain was clear—but there is no denying the fact that the *margin* was altered.

Depreciation and Depletion

Depreciation and depletion problems are also traceable to the time element, but arise from the nature of capital expenditures rather than gains or losses. When a firm buys land, or buildings, or equipment, or invests capital in road construction, bridges, a fire tower, and so on, the firm is not allowed to charge off the entire sum as an expense in the year during which the purchase or expenditure was made as long as the asset has a life expectancy of over one year. Yet common sense demands that the expense be allowed some way, for these costs are just as necessary to the conduct of a business as those incurred for gasoline, coal, or wages. Such expenses are recovered by the deduction of an annual depreciation allowance, designed to permit the total recovery of the entire cost (less salvage value) over the life of the asset. Should it be evenly divided? Should the original cost or the replacement cost be the basis of the total amount recovered? How does one determine the life of a piece of equipment before it is used, and so on. To avoid confusion, these questions have been somewhat rigidly answered by the Internal Revenue Code. Before 1981 the cost of any asset was deducted over its useful life, and three methods of allocating the costs over time were allowed. Most individuals used the straight-line method, recovering the purchase price over the expected life of the asset in equal annual amounts. The two alternatives were the diminishing balance method and the sum-of-years-digits method.

In 1981 the Economic Recovery Tax Act (ERTA) replaced previous methods with a new system, the accelerated cost recovery system (the ACRS). This new system made no reference to useful life but instead placed the asset in one of four

[12]In June 1984 the holding period for a long-term capital gain was reduced from one year to six months. For properties or items purchased after January 1, 1988, the term will again be one year.

cost recovery classes, each applying certain percentages to the original cost basis to determine the allowable annual depletion. Compared to previous methods, the ACRS allowed a much faster tax writeoff of a capital asset—a way of stimulating new capital investment and, through this, increasing production—the reason behind the change.

The 1986 tax reform made several important changes in depreciation. There are no longer special allowances for the various assets used in timber operations—fences, buildings, and so on. Now the same general rules for other depreciable assets apply. The Accelerated Cost Recovery System was modified in several ways. A 7-year and a 20-year recovery class were added to the 3-, 5-, 10-, and 15-year classes. There are options for the depreciation method—a 200% declining balance with a shift to the straight-line method at a predetermined point, and straight line on a class-by-class and year-by-year basis. Some items can still be depreciated under a method not based on years—temporary logging roads and bridges, for example, can be depreciated over the life of the logging operation or by the unit of production as timber is harvested.

A depletion allowance at first sight appears similar to that for depreciation, but for many years was primarily a preferential treatment for those investing in extractive industries. The oil, gas, and mining industries were the recipients of special favors in this respect. Firms in these industries were permitted (1) to expense rather than depreciate outlays for exploration cost, and (2) to deduct a specified percentage of gross receipts for tax purposes, as long as the well or mine was operated. Percentage depletion was criticized for years by almost all students of taxation, and finally, in the Revenue Act of 1969, the maximum rates allowed were reduced from 27.5% to 22%. The Tax Reduction Act of 1975 eliminated percentage depletion for the oil and gas industries except for small producers, for whom the allowance was to be gradually reduced to 15%. Other natural resources have retained these allowances, with rates varying up to a maximum of 22%.

It is emphasized that timber does not share in such favored treatment. The timber depletion allowance is much the same as the depreciation permitted other industries and makes possible the recovery of an initial investment.

Forestry and Income Taxation

What has all this to do with the problems of timber production? Quite a lot. Because of the peculiarities of producing timber, forestry has been a particularly difficult production process to fit into our income tax structure. Timberland is often held for substantial periods of time before the timber is harvested—certainly it would not be unusual to have a period of over a year between the original purchase and the harvest cut. Should the resulting gain from such cutting be counted as a capital gain or as ordinary income? If I buy stumpage for some future cutting I can recover my capital expenditure when that stumpage is harvested. But what if I hold it five years and stumpage prices happen to double? What if the timber *grows* between the times I buy and sell the timber? If I operate on a clear-cut and plant system of pulpwood production, can I subtract the planting costs as a regular expense? What about regular silvicultural treatment costs? What about fertilizing or spraying against insect attacks? These are just a few of the questions that arise.

The following discussion is at best a compromise. The intention has been to concentrate on the principles that are involved; the actual working out of particular cases at specific times is left to others.[13]

[13]See, for example, *A Guide to Federal Income Tax for Timber Owners*. Agr. Handbook No. 596, current edition or supplement.

Capital Gains on Timber

Prior to 1944, providing standing timber had been owned for a year or more, an owner could sell the timber outright and claim the difference between the original purchase cost and the final sale price as a capital gain. If the timber had been acquired before 1913, the difference between its estimated value in 1913 and the time of sale could be claimed as a capital gain. If timber were bought, then sold, with six months or less intervening, any difference was treated as ordinary income. Clear enough! But what if a farmer cut his own timber and sold logs? Or what if a lumber company cut timber that it had held for 20 years and sawed it into lumber? Could the increase in value (or the loss, if any) then be treated as a capital gain? Before World War II the answer was a definite NO. To qualify as a capital gain the timber had to be sold outright and neither of the above cases constitutes a sale of *timber*. Then in 1944, Section 117(k) was added to the Internal Revenue Code, an addition now known (in the Internal Revenue Code of 1954, as amended) as Section 631(a) and (b). Section 631(a) states that the cutting of timber for sale or use by a taxpayer who owns or has a contract right to cut the timber *shall be considered as a sale or exchange* of the timber cut, providing the individual has owned the timber or held the cutting contract for a period of more than (in 1986) six months (after 1988, one year). Section 631(b) gives the same right to any owner who "retains an economic interest" in the timber, a step made necessary by the fact that stumpage payments are often made periodically as the cutting progresses and may take place a considerable time after the sale itself.

These changes made it possible for *any* timber owner to count as capital gains any change between the purchase price of his timber and its market value, provided he had held the stumpage (or retained an economic interest in it) for the required interval—one year unless acquired between June 22, 1984, and January 1, 1988. Sale value, for tax purposes, is "the fair market value as of the first day of the taxable year in which the timber is sold." As stated earlier, beginning with 1987 for corporations (1988 for individuals), capital gains of any kind are to be treated as ordinary income. For 1987 the *personal* income tax places a maximum rate of 28% on capital gains, while the rate on ordinary income is higher for those in the middle- and high-income brackets.

It should be emphasized that it is still important to separate capital gains from other types of income. Capital gain remains a concept of tax law; Sections 631a and b still exist and are applicable, and one can subtract a bigger proportion of any large capital *loss* instead of carrying it forward. Moreover, individual owners may be able to avoid paying Social Security taxes on long-term timber gains, for capital gains are not presently subject to the Social Security tax.

Depletion and Capital Expenditures

Two difficulties of timber income taxation are peculiar to timber production: the need to define and allow for timber depletion, and the annoying but necessary division between *capital* expenditures (which are recovered through depreciation) and *operating* expenses (which are deducted directly from gross income).

Depletion

When timberland is purchased, part of the purchase price goes for land and the rest for whatever timber exists at the time. For tax purposes the two must be kept separate, and the "timber account" will usually require three subaccounts: (1) merchantable timber, (2) young growth (trees of premerchantable size), and (3) plantations, or deferred reforestation (includes site preparation, seed or seedlings, and planting). Of the original purchase cost, only that portion attributable to merchantable timber is recoverable through the depletion allowance. However, if any appreciable time elapses between purchase and harvest, the original volume will have changed, and because of the dual nature of standing timber there is no practical means of separating the original from the newly added product.

To overcome this problem, the Code specifies a procedure for calculating a "depletion unit," usually expressed in dollars per thousand board feet or cords. Included in this depletion unit are the various carrying charges and capital expenditures—the latter encompassing tree planting expenses if any have been incurred. Successive cuttings require successive adjustments of the depletion unit. The procedure is somewhat involved, and there is always a variety of items that seem not to fit the standard case. Still, no basic problems of principles are involved in the calculation of the depletion allowance under current income tax legislation. The details are well covered in the excellent handbook prepared by the Forest Service (USDA Agr. Handbook No 596).

Capital Expenditures

A capital expenditure is an expenditure made for equipment, buildings, trucks, machinery, office furniture, copyrights, or patents or other property with a useful life of more than one year. Such expenditures can be recovered through depreciation over the useful life of the property. Operating expenses, on the other hand, can be deducted directly from gross income. Almost every business has expenditures of both kinds, and in nearly all operations "borderline" expenditures exist that are difficult to classify. Because of the time period involved in timber production, forestry enterprises raise special problems, many of which have been settled by administrative ruling.

The reforestation sub-account carries some special advantages. Due to a retroactive statement in ERTA, the Economic Recovery Act of 1981, up to $10,000 of qualified reforestation expenditures made after December 31, 1979, including site preparation, depletion of tools, nursery stock and planting costs, can be amortized over an 84-month period; any excess must be capitalized. One-fourteenth can be taken the first year, $\frac{1}{7}$ each of the next six, and the remainder during the eighth year. Additionally, since the planted trees have an expected life of more than seven years, the owner can claim 10% of the reforestation costs as an investment tax credit. The act was slightly modified by the Tax Equity and Fiscal Responsibility Act of 1982 (TEFRA), which required that the total amount amortized be reduced by one-half of the tax credit taken the first year. (For details, see Haney, u.d., or Clutter et al., 1983).

If, for example, reforestation costs averaged $80 per acre for planting a 100-acre area, the owner could take an $800 tax credit the year of planting, then amortize $8000 − 400 = $7600. His amortization schedule would be:

$\frac{1}{14}$ of $7600, or $542.86 the first year

$\frac{1}{7}$ of $7600, or $1085.71 the next six years

The remainder, or $542.85, the eighth (last) year.

The impact of this deduction on the owner's income tax would depend on his or her taxable income for the year and the rates then in effect. If the owner's taxable income *before* the deduction was $25,000, the saving (in 1986) would be $1121. Because no more than $10,000 can be amortized in this way, the tax saving is important only to relatively small forest owners; a corporation planting 20,000 acres per year at $80 per acre would have reforestation costs of $1.6 million. Probably the $10,000 would still be amortized over the 84 months, but the saving would appear almost negligible.

The 1986 Tax Reform made no change in amortization provisions.

Effect on Production Decisions

To see the effect of an ordinary income tax on individual or corporate production decisions[14] of a going timber growing enterprise (excluding the capital gains provision) the usual set of cost curves proves useful (Fig. 8.3).

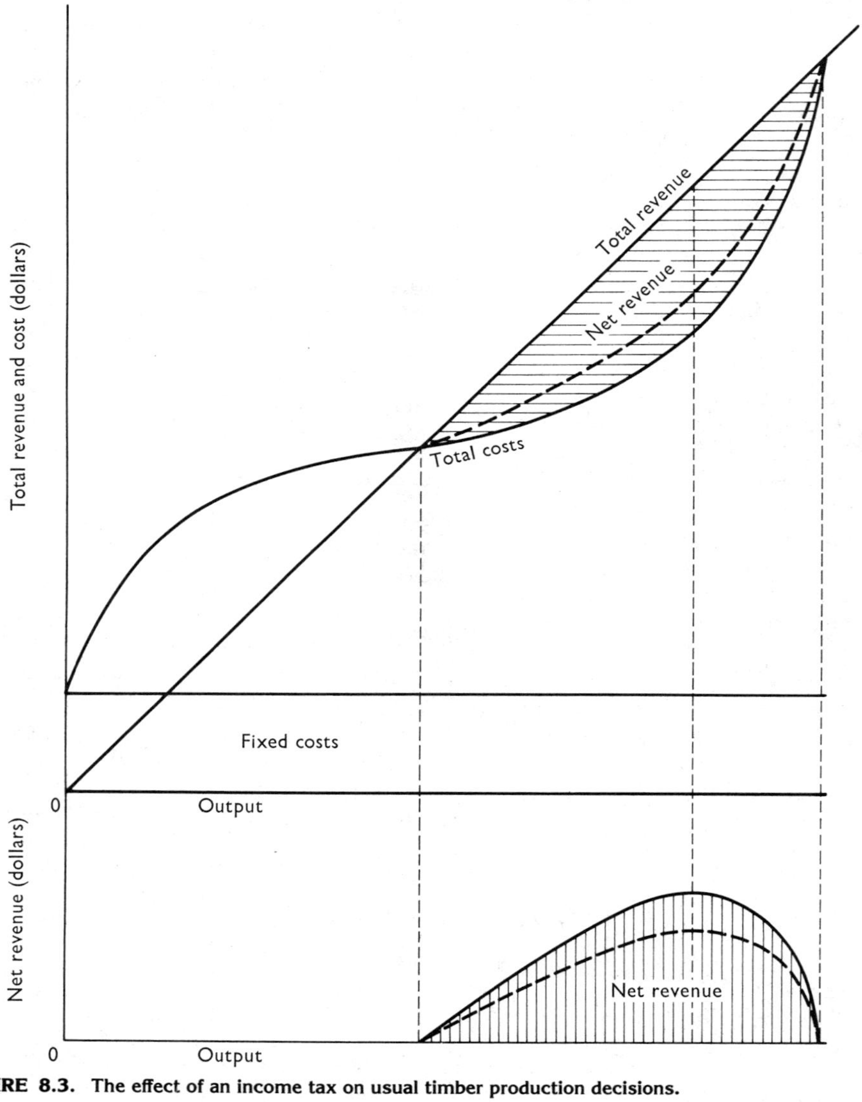

FIGURE 8.3. **The effect of an income tax on usual timber production decisions.**

[14]Here, as with property tax, the "production" decisions meant are those involving *how much* to produce from a given forest property through changing the inputs or the intensity of management.

Because the tax is on *net* income it will be levied on some vertical "slice" of the shaded area of the upper diagram—the area between the total cost and total revenue curve—since this represents net revenue. To facilitate seeing the effect of the tax, this net revenue is reproduced directly below the total cost-revenue diagram, with the output (horizontal) scale maintained in the same units, but with the dollar units converted to net revenue.

The income tax always represents a *percentage* reduction of the net revenue. Since all output (under our present assumptions) would qualify for capital gains treatment, the tax in 1986 would be no more than 20% of net revenue. In 1987 for corporations the timber income would be lumped with other income. The optimum output level will be the output maximizing net revenue—where the slopes of the two curves (cost and revenue) are equal. And on the lower diagram this same point will be where the slope of the net revenue curve is zero (where it is perfectly horizontal). As long as the income tax is a percentage reduction of net revenue there will be no effect on timber production decisions. The point of optimum operation *before* application of the tax will be exactly the same as the optimum output *after* the tax. Hence, a tax on ordinary income is *neutral* (Dennis, 1985). The dashed line in the lower diagram of net revenue is drawn to indicate a 20% reduction, as is the similar line in the upper net revenue diagram.

The same reasoning holds for all levels of the personal income tax. Production decisions will not be affected as long as the *marginal* tax rate is less than 100%, for then the after tax income will increase whenever net revenue increases.

Any special tax treatment, however, can change management intensity and so may indirectly alter the optimal rotation age (Dennis, 1985). Further discussion of this point is deferred until rotation determination has been discussed (Chapter 14).

It is important to emphasize here that we have been discussing *timber* production. Once nontimber values are brought into the analysis the picture changes: any advantage given timber by special tax treatment favors timber production whenever choices are made between timber and nontimber products, and such choices are being made with increasing frequency. The statement that "anything that favors timber production favors forestry" is not always supportable.

Effect on Investment Decisions

Any tax affecting the absolute size of the net return will have a definite influence on investment decisions. Suppose, for example, we consider the problem facing the board of directors of a paper corporation when deciding whether or not to embark on a land acquisition and planting program. If the rate of return on the planned investment worked out at 12% *before* tax, it would have approached 6% *after* tax under the 1986 corporate income tax, and from July 1, 1987, on should be about 8%. Compared to a no tax situation, the incentive for business investment is decidedly reduced—one of the major criticisms of the corporate income tax. If all businesses are treated in exactly the same way, however, resource allocation between industries will be unaffected.

Before 1987 the capital gain provision gave forestry a decided advantage over many industries. In the example above, the return before tax was 12%, but only slightly over 6% after tax. If the return was treated as a capital gain, however, the program would return nearly 9.5%. As a consequence, timber production would have moved up in the capital budget of the firm. When choosing between two projects differing only in that profit from one, but not from the other, could be treated as a capital gain, the capital gain possibility would be selected. The 1986 Tax Reform eliminated this possibility.

Section 117(k) and its successor, Sections 631(a) and (b), made timber production one of the favored industries and also eliminated a very undesirable situation. Prior to enactment of these Sections there was a strong tax incentive to break up a forest property, to sell the land and timber together rather than manage the stand, holding the land, and selling only the timber. There is no doubt that Sections 631(a) and (b) provided a decided encouragement to invest in timber-growing enterprises, especially for large firms.[15] The rapid increase in investment in timber growing following World War II has been cited by many as proof of the efficacy of the provisions (Briggs, 1962; Fortson and Hargreaves, 1974). How much of this increase is really due to the change in the Code and how much to other factors is a moot question, but there is no doubt about the increase itself, nor of the investment incentive the changes provided. How much forestry investment will decline after removal of capital gain status is something tax scholars will be watching carefully in 1987 and thereafter.

Basic Policy Issues

Given the need for amending the previously undesirable situation, and accepting that timber production investments are encouraged by the present tax treatment, we can move on to two somewhat more difficult questions: "*Should* timber growing be thus favored, and is this particular kind of treatment the best means of attaining the desired ends?"

Many factors may be cited to support the present treatment. Forests yield social benefits for which owners are not recompensed, and there are many reasons why a national government should encourage forest ownership. A tax advantage is one way of offering this encouragement. Stumpage production is a long, time-consuming process that extends over many years. Tree growth and the aging of wine have been the classic examples for illustrating long-term investment problems; it makes little sense to tax all income from the growth of many years in any single year just because the harvest happened to come at that time. If other industries are given generous treatment, forestry deserves to be treated equally.

Offsetting these positive items is the knowledge that any interference or alteration in resource allocation by a tax is presumed undesirable unless the alteration can be proved to promote a closer attainment to the optimum. Is timber production more important to the nation than, say, automobile manufacture, or newspaper publishing, or educational institutions? The fact that some resources are given unwarranted benefits should be no real excuse for enlarging the area of mistake! An adequate system of averaging returns over several years would take care of most difficulties caused by harvesting accumulated growth.

One major criticism of Sections 631(a) and (b) came from the lack of any direct connection between forest management practice and benefits received from the tax. From 1944 through 1986 any owner, regardless of management or lack of it, could claim the difference between purchase cost and the market value at the beginning of the tax year as a capital gain. Purchasers of national forest stumpage could benefit as well as an owner who had carefully managed his or her stand for several years, as could owners who had not managed their forest at all.[16]

Most foresters and industrial operators have expressed strong opinion that

[15] In 1971 five companies received about 40% of the benefit to all individuals and corporations from the special tax treatment of timber gains (Sunley, 1976).

[16] Presumably, any really careful manager would stand to benefit more than owners who did nothing to their stands, since the values present in the forest should be higher in the former case.

Sections 631(a) and (b) have been highly beneficial to forestry, and that the general public therefore benefited. (U.S. Congress, 1963). Everyone agrees that forestry has gained. That the general public has gained, however, is far more difficult to substantiate.[17]

Value Added Tax (the VAT)

This kind of tax is common in Europe but still rare in the United States. It has been discussed several times as a possible substitute for all, or part of, an income tax—especially in countries where income taxes are high. Value added means just what it says: it is the dollar measure of the value added by a firm, industry, or individual to the total worth of a product. A value added tax, therefore, is a tax on the *cost* added in each stage of production and distribution. The idea of the value added tax, as well as much of its theoretical development, came from the United States out of a desire to have a *neutral* tax that would provide large revenues without altering the efficiency of the free market system.

With such a tax a person or firm paying for goods or services *pays* a tax thereon and *collects* a tax on sales. In that sense the VAT is a "sales" tax. Value added—the base of the tax—measures the value of the goods and services claimed by the taxpayer as his or her contribution to the value of the final product. Taxes paid are credited against taxes collected, and only the balance is paid to the taxing jurisdiction.

Because it is (usually) rebated on exports and imposed on imports, a VAT can be of substantial assistance in balance of payments problems—another reason for U.S. interest. It is usually assumed that value added taxes will be passed on to consumers.

Like all other taxes, however, it is difficult to determine just who really bears the tax burden. Most taxes on produced goods are incorporated into the prices paid and the ultimate consumer bears this cost. But because the final price is thereby increased, the supply curve has been "pushed up" and the total quantity purchased reduced. Thus it is not correct to say the final consumer pays the *total* cost, for the entire economy—producers, labor, owners of capital, and the entrepreneurs—have shared in the payment. Any tax received by government comes from the combined efforts of the productive factors—including those managing natural resources.

Estate and Inheritance Taxes

Generally known as "death duties" or "death taxes," estate and inheritance taxes are imposed only once, at the death of the owner. They differ in that estate taxes are levied on the entire estate as a unit, while inheritance taxes are imposed on the share left to each heir. A few people have argued that since it represents an increase in wealth to the recipient, an inheritance should be included in regular income rather than being given separate treatment. Most people believe that separate treatment is required since very frequently—as in the case of an inheritance by a widow or children—the receipt of the property does not represent any real increase in well-being because the beneficiaries have already had full use of the property before the death of the owner. An additional reason for separate treat-

[17]As examples of the dispute about the efficacy of the capital gains timber tax see (for a favorable, though scarcely analytical, treatment) Fortson and Hargreaves (1974). This may be contrasted with Sunley's article (1976), which reviews the advantages usually cited for capital gain treatment of timber but finds few of these to be well founded. The report of a Society of American Foresters Study Group on Forest Taxation is in a special section of the *Journal of Forestry* (S.A.F., 1980).

ment is that death itself, although inevitable, is highly irregular, and so, consequently, is an inheritance. Most often estate and inheritance taxes represent a deliberate policy aimed at avoiding excessive or undue concentration of wealth.

The U.S. federal government requires (in 1986) an estate tax report to be filed for all estates whose gross value exceeds $600,000. Rates (which are applied to the estate value after all debts are paid) are lower than in some other countries but are steeply progressive. Under the Economic Recovery Act of 1981 rates were changed substantially. During the 1970s the exemption equivalent was $60,000. This was increased to $175,000 in 1981, to $325,000 in 1984, and to $600,000 in 1986. Formerly, rates ranged from 18 to 70% according to final estate value (after all debts are paid); they now range from 18 to 50%, with the top rate applied to estates over $2.5 million. The limit to marital deductions was removed in 1982; now any amount may be left, tax free, to a surviving wife or husband.

A 1926 law permitting 80% of any state inheritance tax to be credited against the federal estate tax (although relatively unimportant today because the 80% applies only to federal liabilities under the 1926 rate schedule) resulted in allowing all states having at least the minimum tax level to pick up this amount. In 1980 there were 29 states with inheritance taxes going beyond this, and 10 more with estate taxes. Rates vary considerably between states and range from 8 to 21% as well as by the closeness of the heir to the deceased. Exemptions were considerably lower than those granted by the federal tax.

Possible Effects on Forestry

The effect such taxes *can* have on forestry is most easily seen by an example. Suppose an uncle were to die, leaving you a forest property of 10,000 acres, which was duly appraised at $300 per acre. About the time you recovered from the shock of inheriting a three-million dollar property you might be somewhat disconcerted to learn that (at 1986 rates) the federal estate tax could be as much as 1.2 million dollars. You might be surprised by the lack of sympathy your friends would provide at your inheritance being cut to 1.8 million, but your real problem will be "Where can I raise $1,200,000 to pay the tax?" Probably you would find it necessary to sell off part of the estate, thus obtaining the required amount—but it may prove difficult to find a buyer. It should also be noted that under the described circumstances the law would permit the estate tax to be paid in installments over a 10-year period. If you could pay "only" $125,000 per year, you might be able to keep the operation going.[18]

But suppose your uncle had been building up a real sustained yield forestry concern with a modern sawmill designed to cut and finish the annual output of the forest, an all-weather road system, fire suppression equipment, and so on. If you still find it necessary to sell part of the forest, your property is no longer the well-balanced operation that your beloved uncle succeeded in constructing. Quite possibly you will decide it will be to your benefit to sell the whole property to a much larger corporation, exchanging it for shares of the company's stock at a favorable price. The tax could then be paid by selling whatever stock might be necessary, but keeping the forest property intact.

It may also be pointed out that this situation is made to order for anyone holding spare cash and looking for a way to make a fast dollar. Such a person might well offer you a *cash* price considerably below the market value of the property—

[18]There are several ways to avoid full payment of estate taxes through trust arrangements. For establishing such a trust, see your lawyer!

say two million dollars — that you might take in order to get part of the estate. Six months to a year later the new owner could liquidate the entire operation, taking the difference between the purchase price and the highest value he or she could establish for the stumpage as of January 1 of the year purchased. Costs could be recovered through depreciation, the mill sold, profit taken on the timber, and the land retained.

Death taxes impose a decided disadvantage on individual forest management as compared to corporate forestry, for a corporation is essentially deathless. Such taxes also tend to stimulate mergers and thereby increase large forest ownerships. Since large ownerships tend on the average to be more efficient than small ones, the estate tax may act to produce a more efficient allocation of resources, although some might deem it less desirable from a social standpoint. Consequently, the estate tax can have a serious effect on forest properties by forcing dismemberment of an efficient operation. It could also act as a deterrent to investment if the potential investor decides to anticipate such a likelihood (Yoho, u.d.). On the other hand, it might not have either such effect, and it could even be beneficial.

Death taxes in Great Britain offer an interesting contrast to such possibilities, through a sort of "double reverse" of the impact. English death taxes are large, and they have forced dismemberment of some estates. But standing timber is not subject to death taxes; that is, the value of standing timber is not included when evaluating an estate. Hence one finds wealthy estate owners devoting large areas to forestry as a hedge against death duties. The timber can be cut and sold after the owner dies, and used to pay whatever tax is levied on the estate.

Empirical evidence on the actual effect of the U.S. estate tax is neither abundant nor convincing. There is an undoubted effect on liquidity, probably the most serious factor so far as forestry is concerned. Because the tax is levied only once, and then at the very outset of the initiation of management by any single owner, the effect on management decisions would usually be slight. However, the decisions taken to *avoid* payment of estate and inheritance taxes can at times be very substantial, as witness the case in Great Britain.

From an equity standpoint the federal estate tax seems quite justified. The recipient of a bequest or inheritance has certainly been made better off and this increase can justifiably be taxed. The number of people receiving significant additions to their income from inheritances is so small that those who experience such gains should (from an equity standpoint) make some additional tax payment. Perhaps one of the best reasons for such taxes today is that they provide a means of reaching a large sum that otherwise escapes taxation altogether. Present legislation permits any capital gain *on property transferred at death* to escape income taxation, and death duties place at least some tax on these large amounts.

Estate and inheritance taxes imposed by a state can substantially increase the difficulties faced by an inheritor of forest land. Some type of death tax is imposed by all states, but the form and the burden of these taxes vary greatly, as do the possibilities of taking special tax saving measures (Olsen et al., 1981).

QUESTIONS AND PROBLEMS

1. What are the three major principles of taxation?
2. How would you compare the property tax and income tax from the standpoint of the degree to which these taxes are in accord with taxation principles?

3. How are property tax rates determined?
4. Why have most foresters felt that property taxes have an adverse effect on forestry? Deal specifically with the time bias said to be present when forest holdings are subject to property taxes.
5. To what extent does a severance tax differ from the property tax in its effect on forest investment and management?
6. It has been suggested that a sales tax levied on the purchaser of stumpage but not the *producer* (sometimes called a use tax) might beneficially be substituted for either the property or yield tax. Comment on this proposition.
7. What are the principle characteristics of the personal income tax as presently levied in the United States?
8. Discuss the pros and cons of having progressive tax rates as part of the corporate income tax structure.
9. What is a capital gain, and why is it peculiarly important to forestry?
10. What is a depletion allowance, and why is it a complicating factor in tax liability calculations?
11. What is the effect of an income tax on the day-to-day (or year-to-year) management decisions on an operating private forest? What effect does the present tax structure have on forest investment decisions?
12. Sections 631a and b are sometimes termed milestones for private forestry practice and management in the United States. Discuss these sections in terms of (a) effects on forest *management,* (b) in terms of forest *investment,* and (c) in terms of the major taxation principles.
13. How do estate and inheritance taxes effect private forest management (industrial and nonindustrial ownerships) in this country?
14. Compare corporate and personal income taxes of the current year with those of 1986 as presented here. Did the Tax Reform Act of 1986 hurt forestry? Did it affect timber production? Did it harm forestry more than it did other industries? Be sure you can defend your answers.

chapter nine
TIMBER DEMAND, SUPPLY, AND PRICE

The demand for major wood products in the United States was the subject of Chapter 3, where demand was interpreted broadly. For most wood-based products demand is not direct, but derived from the sale or expected sale of finished houses, furniture, boxes or toys, or of newspapers, magazines, books, and grocery bags, or of any or all the hundreds of items into which lumber, plywood, and paper are fabricated. Although seldom given more than passing attention in introductory courses, derived demand is of singular importance to anyone who would understand markets for natural resources. It is essential for an understanding of stumpage prices — which is the reason for treating it now. Once the principles are in hand they will be used to develop an understanding of the demand, supply, and pricing of standing timber.

DERIVED DEMAND

The classic approach to derived demand may be illustrated as follows. In Figure 9.1*a* assume D_h to be the demand curve, and S_h the supply curve, for new residential housing in the United States, measured in terms of square feet of floor space. Q_h and P_h are, respectively, the equilibrium quantity and price for residential housing floor space in the nation. S_o is the aggregate supply curve for all factors of housing production *other than lumber* — factors such as paint, glass, bricks, labor, concrete, capital, nails, and so on.

The demand curve for lumber used in residential construction may now be "derived" by subtracting this supply curve for all other productive factors (S_o) from the housing demand curve (D_h). The result of this subtraction is shown in Figure 9.1*b* and is the demand for housing lumber (D_L) (the shaded portions of the two diagrams are equal).

As a concept, derived demand is easily understood. However, when stated in terms of dollars per billion square feet of floor space the demand and supply for residential housing lumber are not easily interpreted by a sawmill owner — or even a forester. So, let's work a simplified example through, step by step.

For variety, we'll start with a demand curve for newsprint, from which we

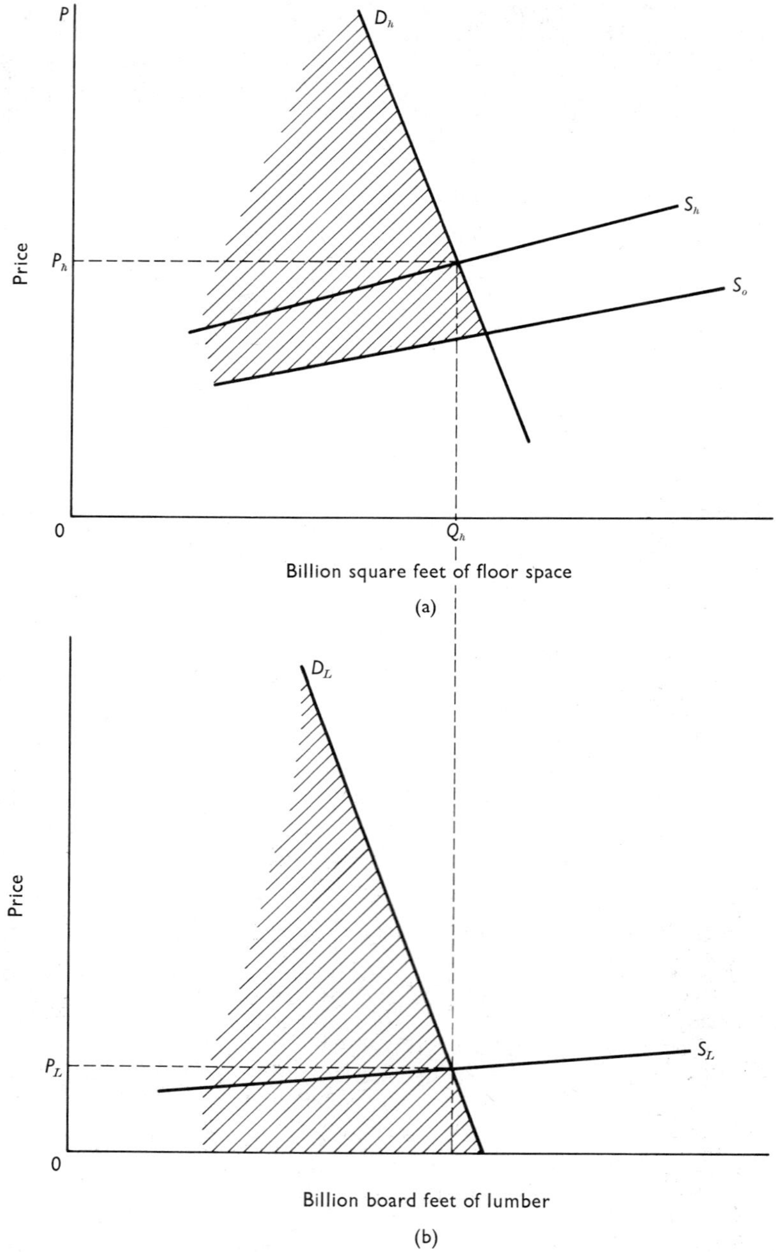

FIGURE 9.1. Demand for lumber derived from the demand of new residential construction (measured in sq. ft. of floor space).

wish to derive the demand for pulpwood. We'll assume our technology is such that each ton of newsprint requires $2\frac{1}{2}$ cords of pulpwood.

1. Let P_n^d be the demand price for newsprint in dollars per ton, and Q_n the quantity of newsprint in hundreds of tons. The demand curve for newsprint,

$$P_n^d = 450 - 0.35Q_n \tag{1}$$

and the supply curve,

$$P_n^s = 100 + 0.14Q_n \tag{2}$$

are both illustrated in Figure 9.2*a*.

2. Let P_o^s be the supply price for all nonwood production factors (*o*ther factors), expressed in dollars per ton of newsprint. The assumed supply curve for all these nonwood factors (Figure 9.2*a*) is

$$P_o^s = 80 + 0.06Q_n \tag{3}$$

3. Subtracting (3) from (1) yields the equation for the demand price of pulpwood in dollars per ton of newsprint (P_{pwn}^d), with quantities still expressed in newsprint terms (Q_n).

$$P_{pwn}^d = 370 - 0.410Q_n \tag{4}$$

However, we want the demand curve for pulpwood to be in pulpwood terms—in cords and in dollars per cord.

4. Since each unit of newsprint (100 tons) requires using 250 cords of pulpwood, we can change the horizontal scale of Figure 9.2*a* into cords simply by multiplying each scale unit by 250 (Figure 9.2*b*). The function must be similarly changed, but to maintain the relationship we must divide the coefficient of equation (4) by 250 (each single 100 ton unit must be replaced by 250 cords). This gives us

$$P_{pwn}^d = 370 - \frac{0.41}{250}Q_{cd} \qquad \text{or} \qquad P_{pwn}^d = 370 - 0.00164Q_{cd} \tag{5}$$

To rescale the graph in dollars per cord instead of per ton of newsprint we need only divide each unit of newsprint by 2.5 (Figure 9.2*b*). The same can be done with the equation, remembering that if we divide one side of an equation by a number we must also divide the other. So, dividing equation (5) through by 2.5 yields

$$P_{pw}^d = 148 - 0.00065Q_{cd} \tag{6}$$

as the demand curve for pulpwood—derived from that for newsprint. Now, with the supply curve for pulpwood given as

$$P_{pw}^s = 8 + 0.000118Q_{cd} \tag{7}$$

the equilibrium quantity and price of pulpwood can be found by solving the demand and supply equations [Equations (6) and (7)] simultaneously:

$$\begin{aligned} P_{pwd}^d &= 148 - 0.000656\ Q_{cd} \\ -P_{pwd}^s &= \quad 8 + 0.000128\ Q_{cd} \\ \hline 0 &= 140 - 0.000784\ Q_{cd} \\ Q_{cd} &= 178{,}571.43 \text{ cords} \end{aligned}$$

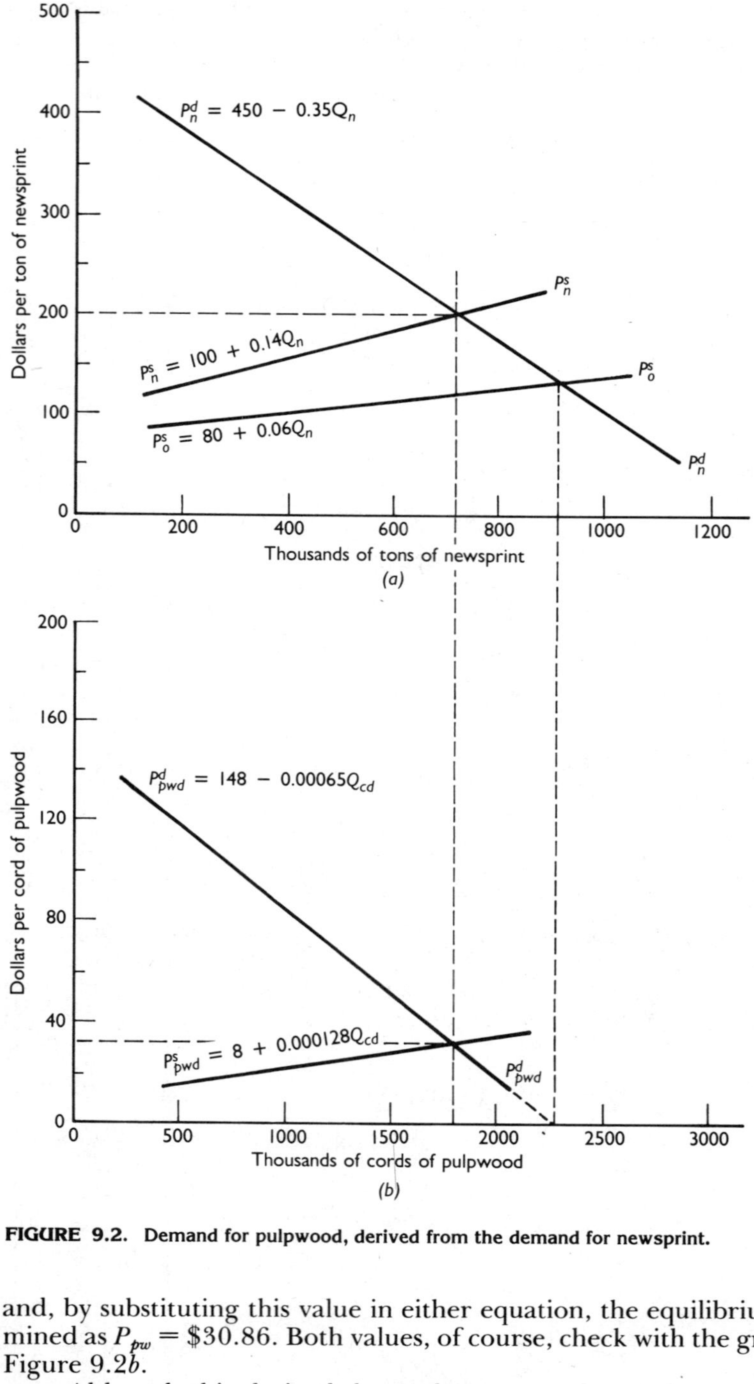

FIGURE 9.2. Demand for pulpwood, derived from the demand for newsprint.

and, by substituting this value in either equation, the equilibrium price is determined as $P_{pw} = \$30.86$. Both values, of course, check with the graphic solution of Figure 9.2*b*.

Although this derived demand curve can be used to estimate how prices might change as pulpwood prices are altered, it is not precisely the same as the normal Marshallian demand curves. Always before, the *ceteris paribus* assumption was present; with derived demand, however, any movement along the curve implies that the prices of all other productive factors will change as indicated by the supply curve of other productive factors. Moreover, a derived demand analysis has a built-in assumption of fixed proportions of all input factors. In the newsprint-pulpwood illustration this means that newsprint is assumed to require certain fixed proportions of chemicals, pulpwood, machinery, labor, capital, and so on. In the short term this may be a reasonable assumption, but it would scarcely be realistic over longer time periods.

Marshall gives four propositions respecting the elasticity of derived demand curves (Marshall, 1947). The derived demand for a product will be *more inelastic*:

1. the more essential the factor being examined;
2. the more inelastic the demand for the final product;
3. the smaller the fraction of total cost that goes into the factor;
4. the more inelastic the supply curve of the other factors.

Samuelson, long ago, added a fifth that could be important in forest operations.

5. when used by firms with monopolistic power—firms able to follow pricing policies permitting them to maintain a virtually constant production rate even though factor prices change (Samuelson, 1951).

Citing these propositions, some authors have concluded that the demand for lumber (for example) must indeed be quite inelastic. An examination of the points as they apply to lumber would certainly lead to agreement; with current techniques of construction, lumber is quite essential to houses; the demand for the final product (housing) is probably inelastic at given income and population levels; lumber makes up no more than 15 to 20% of the total cost of a house; the supply curve for other productive factors (labor, capital, paint, glass, etc.) could be no more than moderately elastic.

What is frequently overlooked is that most of the conclusions regarding the elasticity of lumber demand stem from analyses treating lumber as a single commodity.[1] The findings are analogous to those one would expect from an analysis of the demand for food. Certainly in the United States the demand for food would be highly inelastic; after all, the human stomach is of limited capacity. Yet a different conclusion about demand elasticity should be anticipated if one were to analyze the demand for frying chickens instead of the demand for food. Other meats are readily substituted for chicken in our diet, and a relatively small change in fryer prices can and does have a decided effect on sales. A similar case occurs with the demand for a specific wood product produced in a given region, for example, redwood siding, or southern pine box lumber, or hardwood plywood paneling, or any of the thousands of wood product items.

The demand for siding, for example, would be derived from the demand for housing and would probably be quite inelastic; it is difficult to build a house without some sort of siding. But there are many siding materials—plywood of several types, cedar shingles, brick, aluminum, pine and fir materials, red cedar, redwood, and so on. At any given time each of these materials will have a specific price. Given the choice of siding materials that exists, a relatively small change in the price of, say, redwood siding lumber *could* have a very substantial effect on the

[1]For a derived demand-based analysis of regional lumber demand using market linkages to estimate "elasticities of price transmission" (defined as equal to % change in product price/% change in factor price, see Haynes (1977).

amount of redwood siding sold. The demand for many lumber products may well be quite *elastic* when viewed from a specific product standpoint. An overreliance on an inelastic derived demand for lumber could have very serious effects for the producer of almost any specific lumber product.

DEMAND FOR STUMPAGE

Production of stumpage is the traditional objective of forestry, and to this day the sale of timber remains the major source of revenue to most forest landowners. Stumpage price is a key factor to the successful practice of forestry by most wood-based industries and to many nonindustrial private forest owners.

In this price-forming process all the physical and economic factors of timber production and management and of wood product consumption are brought together within the political and economic framework of the nation. Small wonder that this subject has a strong fascination for anyone concerned with forest economics.

The demand for stumpage is derived from the demand for wood products. Stumpage is required if lumber, pulp, plywood, or any wood product is to be manufactured. People demand housing for which lumber is needed; lumber requires sawlogs, and sawlogs must come from stumpage. A similar chain stretches from paper to pulp to pulpwood to stumpage and, if an understanding of stumpage price formation is sought, none of these steps can be bypassed. Since lumber is still the most important single wood product, our development of stumpage demand will be in terms of sawlog stumpage, although pulpwood, plywood, or any other wood product would have served equally well.

Derived Demand for Sawlogs

Because there is virtually no substitute for sawlogs in manufacturing lumber, some writers have concluded that the demand for sawlogs must be inelastic (Mead, 1966).[2] The fact that log demands may be derived from paper and plywood as well as from lumber does not alter this conclusion as long as one approaches sawlog demand as if there were a national sawlog market. But national markets for sawlogs do not exist! As soon as one begins thinking about stumpage, questions of "What stumpage?" and "Where?" come to mind. Because of the immobility of standing timber, the **location** and **accessibility** of stumpage are highly important aspects of its demand. Furthermore, we know that **species** and **quality** are important factors in the demand for any specific timber stand—and only specific areas are sold. Finally, because the location of the tract might limit either the number of possible buyers or the alternative sources of stumpage supply to the buyer, or both, the **structure of the stumpage market** itself might prove to be a significant factor. Even though usual price theory treatment assumes these factors to be constant, a theory of stumpage price must take these factors into account if it is to be at all realistic.

Sawlog Demand for a Single Mill

The starting point for our analysis will be the demand of a single mill for sawlogs. As a specific example, assume the mill to be an efficient-sized southern pine sawmill producing a variety of grades and sizes, and selling on the national market. With annual sales of about 50 million board feet, its output accounts for only 0.45% of

[2]Unfortunately, Mead did not distinguish between demand for logs and demand for stumpage.

total southern pine sales, which in turn must compete with other species from other regions of the country.

One would expect the demand curve for this firm's product to be quite elastic within the usual range of output, with departures from perfect elasticity probably insignificant in most cases, arising primarily because of customer loyalty.

To derive the sawlog demand curve for this mill requires not only knowledge of its product demand, but also of the supply curve for all productive factors other than logs used by the mill. Even though substantial variation can be expected in this industry, some useful generalizations are possible.

If "normal capacity" is defined as the annual output resulting from a single shift working a 5-day week, 8 hours a day, then output can be reduced by (1) instituting a shorter work week (fewer hours per day, or a 4-day week, for example), or (2) closing down the mill for longer periods. The possibility of reducing output by such means is decreased during periods of abundant alternative employment opportunities for unskilled labor or by any employment of specialized labor. Output can be increased by adding overtime or increasing the average log diameter, the latter a possibility that will be ignored for now. Within a single shift operation, the nontimber marginal costs of this sawmill should be nearly constant up to the normal capacity output and would then increase quite steeply, due to increased costs of overtime labor (S_1S_1 in Figure 9.3).

Significant increases in output could be achieved by operating on a two- or three-shift schedule. The second shift would have somewhat higher costs (S_2S_2), for the new labor would presumably be a little less efficient and the needed supervisory staff less skilled. Both factors would tend to be more pronounced for a third shift, thus raising the supply curve for nonlog factors even higher (S_3S_3).

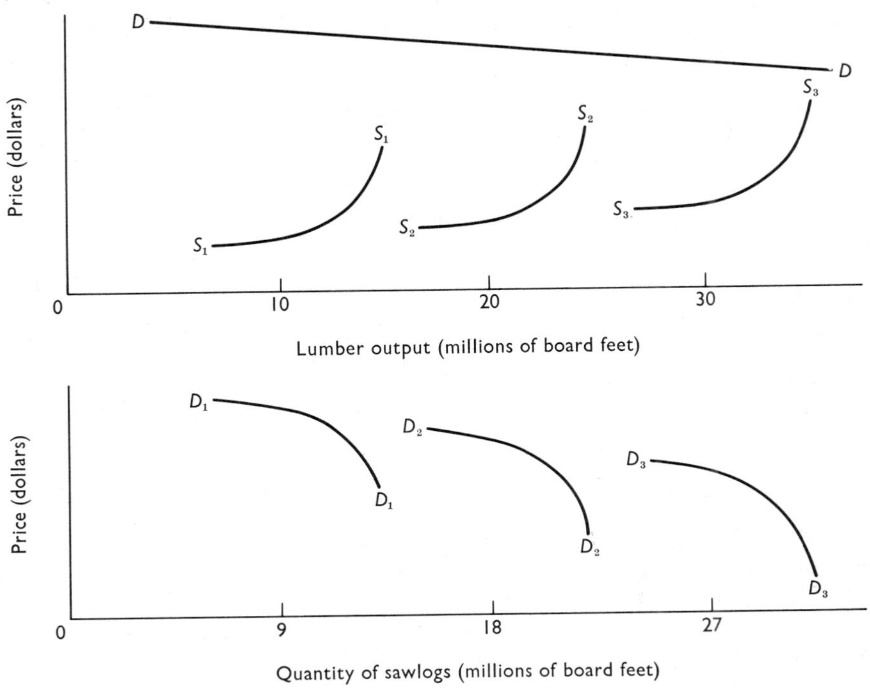

FIGURE 9.3. Derivation of sawlog demand for a single mill.

The relation between the product (lumber) demand for this one mill and the three supply curves for nonlog factors is given in Figure 9.3. The area between the relevant segments of the supply curves and the firm's marginal revenue curve (here assumed to be the demand curve) is plotted directly below, yielding the three derived demand curves for sawlogs.[3] The two quantity scales differ because of the characteristic "overrun" of most sawmills. As shown here, the demand is elastic over part of each shift and inelastic over other portions, but conditions leading to different results could easily be encountered, especially under nonpure competition.

Sawlog Demand for a Group of Mills

According to usual price theory, the *market* demand for sawlogs would be determined by aggregating the sawlog demand curves of all mills in the sawlog market area. A summation of the supply curves for all nonlog productive factors is thereby implied, but would be acceptable only if the sawmills were independent in their demands for all productive factors. The traditional assumption of independence between cost curves of individual firms, however, is clearly lacking here.

The lumber demand curve (D_1), together with the supply curve for all nonlog productive factors (*SS*) is shown for a group of mills in Figure 9.4. Product demand could still be regarded as being perfectly elastic in most instances. The supply curve for other factors (*SS*) would probably tend to be *in*elastic, primarily because a general expansion of the labor force would be needed if a group of mills were to increase output; such a general expansion would be difficult. The discontinuities noticed in the illustration of the single mill (Figure 9.3) would be less pronounced since each mill in the group would have a different production schedule for adding shifts and for hiring new workers or for going on overtime. As a consequence of

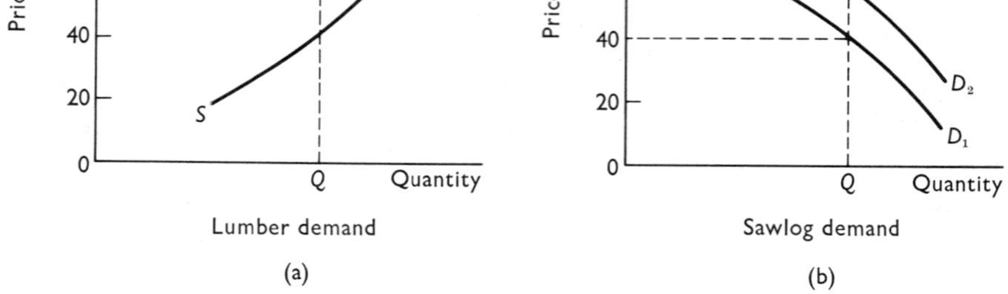

FIGURE 9.4. Derivation of sawlog demand for a group of mills.

[3]Practically speaking, a second or third shift is not normally instituted unless the mill's demand curve has "shifted" outward far enough to assure coverage of these added costs. Such shifts do occur, sometimes as a result of increased construction activity in the locality where the mill's output is normally sold or because of decreased availability of lumber from other areas due to inclement weather, shipping difficulties, strikes, or a variety of other reasons.

these two factors, the derived demand curve for sawlogs would tend to be smoother, and probably considerably less elastic, than that for a single mill.

If the demand for lumber were to shift sufficiently to cause a price increase of \$15 (from D_1 to D_2), and if the supply of all other productive factors remained unchanged, then the demand for sawlogs would also shift up by \$15. The effect on sawlog prices could be quite dramatic; at the indicated quantity of Q, a \$15 increase in lumber price (15%) produces a $37\frac{1}{2}\%$ increase in the price of sawlogs *with no change in the output whatever*.

By aggregating all such sawlog demand curves for a given species over an entire region (say, for Douglas fir sawlogs in the Pacific Northwest) a regional demand curve for sawlogs might be constructed. Then, by aggregating these regional curves for each species, one might similarly obtain a "national" demand curve for sawlogs. This demand, however, would involve the same problems encountered in obtaining a demand curve for pets: one must add cats and dogs, canaries, turtles, goldfish, and so on. The result would have no more, and no less, significance.

Three further practical complications must be mentioned. The one concerning species has already been alluded to; each species group, and often individual species, must be treated separately since the end product (and, therefore, the sawlog price) often varies by species. The second concerns quality; high-quality sawlogs are those yielding increased proportions of lumber in the more valuable grades. Because the demand curve for such lumber is at a higher level than that for lower grades, the demand for high-quality sawlogs is "pulled up." Sawlog demand might be regarded as a "family" of curves, with each particular curve appropriate to a given sawlog grade or quality range.

The third complication concerns time. Lumber stocks in retail yards absorb sudden changes in construction activity just as mill stocks help smooth out shipments and production after orders have been received at the mill. But plans and contracts for log delivery need to be made well before product orders are received. The effective demand for logs is really derived from an anticipated demand for lumber, and this is surely influenced by the mill owners' expectations of construction activity. Since expectations are difficult to quantify, this influence will reduce the accuracy of the estimated demand, and because expectations are generally more volatile than actual occurrences one should anticipate substantial fluctuations in log demand—perhaps even more than would occur in the lumber market.

SUPPLY

Earlier chapters emphasized that an industry supply curve is an aggregation of marginal cost curves and that a supply curve serves as a producer response curve by showing the quantitative response made by producers when the price offered for their product is changed. The supply curve for sawlogs must similarly reflect the marginal costs of sawlogs and must also show how producers will respond quantitatively to changes in sawlog prices. Construction of the required sawlog supply curve will be aided by considering the cost of delivered sawlogs as being composed of two parts. The cost of transforming standing timber into logs on the mill deck will be called the **cost of availability**. The second part is the cost of standing timber, or stumpage, which will be related to (but is not identical to) its **reservation price**. Both cost of availability and reservation price require explaining.

Cost of Availability

With no topographic variation and a homogeneous timber stand, log production costs of felling, limbing, bucking, skidding, and loading can be regarded as con-

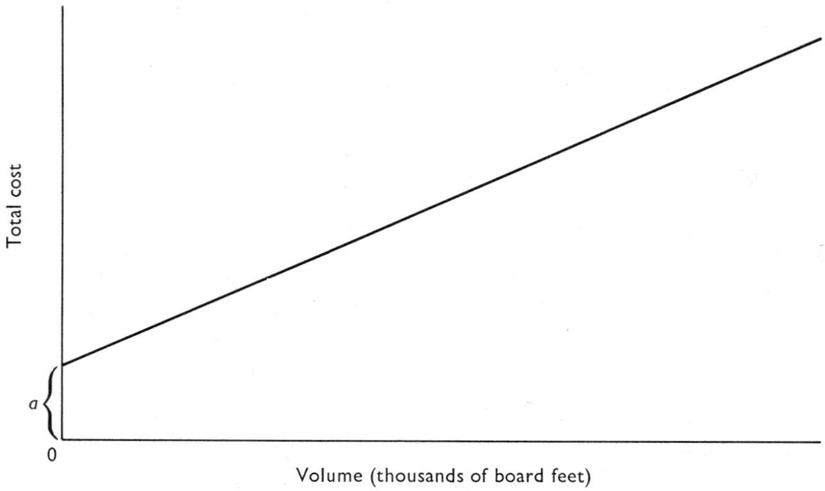

FIGURE 9.5. Total log production costs as a function of volume.

stant per thousand board feet, except for the costs of getting workers and equipment to the logging site. For any given area, total log production costs will increase proportionately to volume removed (Figure 9.5), and costs of getting workers and equipment to and from work—*a* in Figure 9.5—will vary with location.[4]

The other significant cost variable is that of log transportation. With any given method, hauling costs per thousand board feet tend to be linear when expressed as a function of distance, with the unit cost directly dependent on the method used. Generally speaking, the more elaborate, the larger, or the faster the means of transportation, the higher the fixed charges and the smaller the cost per additional mile. The general pattern of transport cost curves consequently tends to resemble that of Figure 9.6. Only four types of equipment are illustrated, but, if one considers the possible variables and combinations, it becomes realistic to regard the transport function as an envelope curve of the "family" of linear relationships, much as shown in the figure, with costs of loading and unloading indicated by the intercept on the vertical axis. When used to estimate transport costs, such a curve implies that for any particular distance the most efficient equipment is used.

If all other costs of log production (felling, bucking, skidding, etc.) are added to the transportation cost, the transport function will be shifted directly upward as in Figure 9.7.[5] And, finally, this logging cost curve *(LC)* should be increased by the rate of return regarded as normal by logging contractors. In Figure 9.7 this rate is 10%, yielding the *cost of availability curve (CA)*.[6]

---

[4]If logging crews report for work at the logging site, this cost may not be borne by the logging company. The cost curve would then begin at the origin. Such situations are not uncommon in the South, the Lake states, or in many nations other than the United States.

[5]The cost of getting crews to the logging site also increases with distance. This cost is not shown in Figure 9.7, but its effect is readily seen.

[6]Ordinarily, the return on investment would be treated no differently from any other cost; it is simply one of the costs of logging. For convenience at a later point, however, the separation is made here. It will also be noted that in this diagram the rate of return has been assessed against all costs, both fixed and variable. This may be debated, but is believed more realistic than confining the return to fixed investment. Moreover, it is in accord with modern concepts of cash flow.

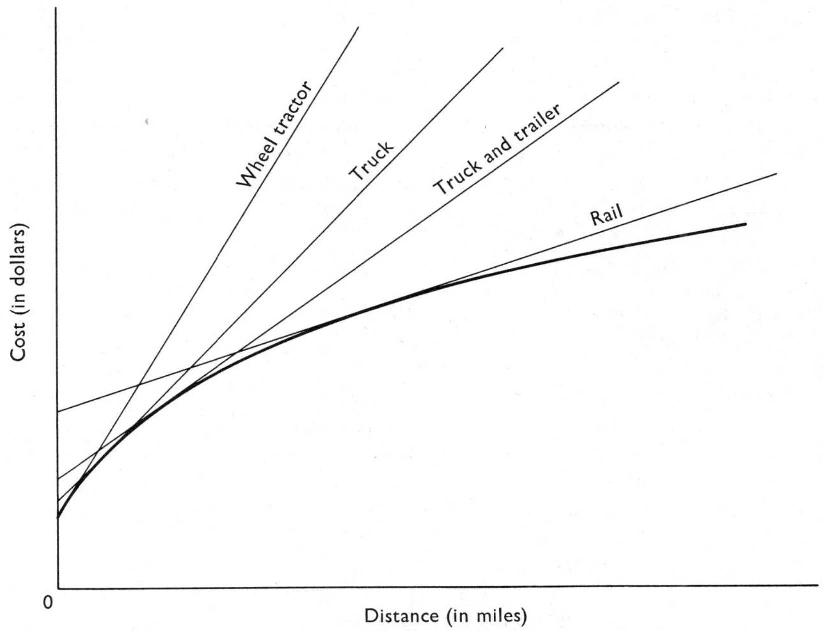

FIGURE 9.6. The usual form of a transport function (including transfer costs).

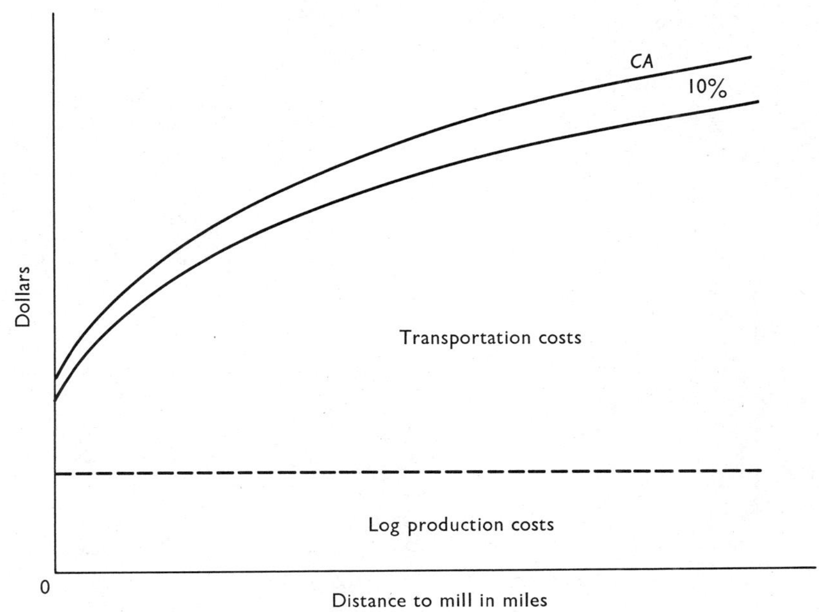

FIGURE 9.7. Derivation of the cost of availability.

From this curve the cost of availability for any stand within the flat plain being used in the present illustration can be determined. Forests, however, are rarely on flat plains, and both topography and the established transportation facilities distort the cost of availability curve. Cost of availability should directly reflect not only distance, but all other factors influencing transportation and log production costs. At some points within this chapter it will become useful to make the cost of availability curve more generally applicable, and in these instances the term *cost distance* will be used—the distance that corresponds to a cost of X dollars on the cost of availability curve.

Several important aspects of the CA curve should be noted from Figure 9.7:

1. Any reduction in log production costs—the use of log harvesters, whole tree logging, of labor training, and so on—may increase the distance over which logging can be conducted profitably.
2. A lowering of transport costs will also increase the profitable logging radius. Here, however, the shape of the curve as well as its position might alter. For example, if the slope of the transport function were to be decreased (from CA_1 to CA_2 in Figure 9.8), perhaps by use of larger or faster hauling equipment, the result could be an upward shift of the CA curve near the mills (between 0 and a in Figure 9.8) but a downward shift at greater distances (beyond a in the figure).
3. The logging radius has a nonlinear relationship to the area that can be logged. For example, a cost reduction that would permit increasing the profitable logging radius from 30 to 40 miles could increase the potential timber supply area by almost 80%.
4. A reduction in the rate of return on invested logging capital could, for similar reasons, produce a substantial increase in the potential timber supply area.

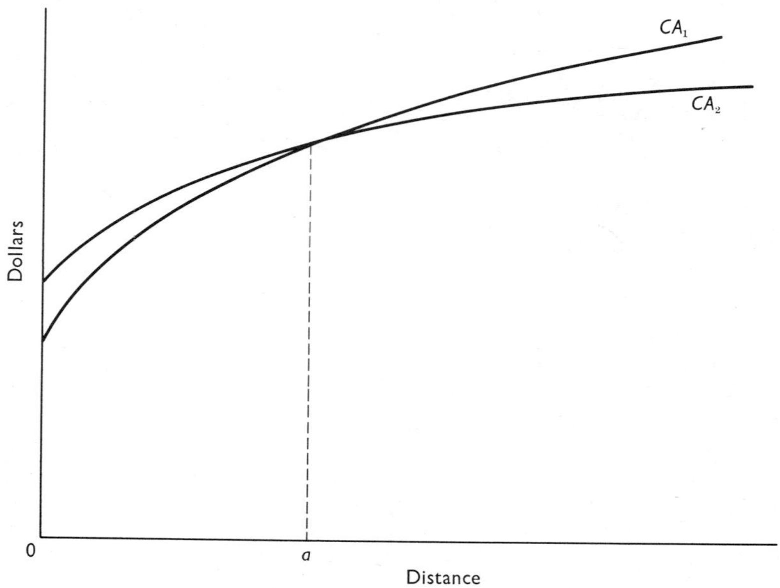

FIGURE 9.8. Effect on cost of availability of changing equipment used for transportation.

Reservation Price

In any region with a reasonably dispersed forest land ownership pattern some owners will be actively seeking to sell all or part of their timber, others would be willing to sell if the price were sufficiently attractive, and there may be a few who are simply not in the market. This variability has many reasons—owners hold different price expectations and different time preferences, and the objectives for which forest land and timber is held are extremely varied.

Owners who expect stumpage prices to rise in the near future will tend to hold their stumpage today. People with low time preference (those who value future incomes highly in comparison with income received today) will also tend to keep their timber, preferring to have their income in the future. And individuals who purchase or hold forest land for recreation, wildlife, or other nontimber purposes will be less interested in selling stumpage if its sale impedes realization of their major objective. The length of time a stand has been under management may also influence an owner's time preference.

Most owners who consider selling their timber will make an appraisal of some sort—a more or less formal estimate of the stumpage value. Once made, this will have an important bearing on the price of their timber, for an appraisal influences (and sometimes determines) the lowest price an owner will accept. An appraisal requires estimating present or future volume and quality of the stand, and then its value. The appraisal of publicly owned stumpage, which is usually by some variation of the conversion return approach, will be described later in this chapter. Many industrial owners, and some larger private owners, also use this method, but other approaches are common.

The aim of an appraisal is to establish a **fair market value**, usually defined as one established by negotiation between a willing buyer and a willing seller, with neither under any compulsion. If both buyer and seller are knowledgeable of the product and the market, the price established by the bargaining process will be essentially the same as determined by the intersection of the market's supply and demand curves.

Transactions evidence provides another basis for estimating stumpage values. For the nonindustrial owner this evidence is usually obtained by questioning previous timber sellers or others knowledgeable about timber prices in the area. Stand location and timber quality must also be part of the comparison. (As used in court trials, the term also implies ability to prove the price at which specific transactions were made.)

Both these approaches are used for immature as well as mature timber. For immature stands, however, many owners take a more sophisticated approach, one of estimating the anticipated income the timber would yield and its timing, then discounting this to its present value.

These factors, and others, all serve to form an owner's **reservation price**.[7] The reservation price for a particular volume or area of timber is that price necessary to persuade the owner to sell the volume or area. It is the owner's estimate of his opportunity cost of selling now, the highest present value of any future expected timber price—the lowest price at which the owner would be willing to sell. At any price below this reservation price the owner will "reserve" his timber.

[7]Many of the ideas presented in this chapter have grown out of discussions with students and other colleagues. The general concept of reservation price came from Dr. Henry Vaux of the University of California.

Individuals have many reasons for reserving timber, and not all are necessarily financial. Sentimental, political, social, aesthetic, or any reason people have for owning forests may be sufficient to keep an owner from selling. Corporations wishing to protect future supplies, or a society wishing to preserve flexibility in its resource base or primarily interested in nontimber products and acting through state or national governments, may also reserve timber from present sale. Most private forest owners, however, have some price at which they would be willing to sell. And, occasionally, the reservation price may even be negative; should an owner want to clear land for pasture, for example, she might be willing to pay to have the timber removed. Colonization schemes in developing countries sometimes produce similar results.

Whatever the reason, it is the price fixed in the owner's mind that is important. Since only owners have the final word on whether to sell or not to sell at whatever price is offered, it follows that one of the keys to timber supply is its ownership. Information on U.S. timber ownership was given earlier; here we need only the reminder that so far as timber production and sale are concerned, the United States is a "mixed economy." Part of our standing timber is owned by private industry operating under competitive market forces and part is publicly owned and managed. And since few people care whether the wood in the houses they own or the paper they read was sold by the Forest Service or by Anderson's Tree Farm, both private and public timber must be brought together in the supply picture.

The short-term supply function is the logical place for this marriage, but both bride and groom must first be identified. We will first develop the short-term supply curve for sawlogs from privately owned areas and then discuss the supply curve for sawlogs from public timberland and to some extent compare it with that for the private sector. The two will then be combined to produce a sawlog supply model which, together with the model of sawlog demand just presented, will allow us to examine the factors influencing sawlog prices. From this point it is a short step to the development of a "theory of stumpage price."

Sawlog Supply from Private Timber

Any timber sale is necessarily made from existing forest growing stock. One might sell all the growing stock of a stand, the trees above (or below) some stated diameter or quality, individual trees as marked, or make almost any other kind of full or partial sale. It will aid initial development of a sawlog supply curve, however, to assume that a stand can only be sold outright.

Elements of Supply Curve

Suppose now that the reservation price had been determined for every stand in a given market area. If to the *reservation price* (per M) of each stand one now adds its *cost of availability* (per M), the resulting total will be the minimum price any buyer must pay if sawlogs are to be put on his log deck from the specific stand. This total is called the *reservation cost* (not *price*) of the stand. If all stands are now ranked according to their increasing reservation cost, and these increasing costs plotted over the accumulating total volume of sawlogs, the resulting cost curve will be the short-term sawlog supply curve for logs delivered to the mill.

A Numerical Example

The approach is most easily demonstrated by constructing two related tables, such as Tables 9.1 and 9.2. With these as formats, one could (at least conceptually) enter the data from each timber stand in the area—the stand volume, the owner's

TABLE 9.1
Conceptual Procedure for Constructing a Sawlog Supply Curve (the Data)

1 Reservation Price (dollars per M)	2 Cost of Availability (dollars per M)	3 Reservation Cost (dollars per M)	4 Volume (million of board feet)	5 Rank (no.)
52.00	56.00	108.00	1.200	10
34.00	43.00	77.00	1.600	3
60.00	24.00	84.00	.800	7
20.00	62.00	82.00	.400	6
52.00	36.00	88.00	.400	8
72.00	24.00	96.00	1.200	9
44.00	29.00	73.00	1.000	1
46.00	34.00	80.00	2.200	5
30.00	45.00	75.00	.400	2
60.00	18.00	78.00	.800	4

reservation price, and its cost of availability—and then derive the sawlog supply curve.

For example, in Table 9.1 the reservation price (column 1) and the cost of availability (column 2) are added to produce the reservation cost (column 3). Stand volume is shown in column 4, and in column 5 the stands are assigned a rank according to their reservation cost, with the stand of lowest cost being ranked first. In Table 9.2 the data have been regrouped according to their ranked order (columns 6, 7, and 8). The accumulated volume (column 9) has been calculated by adding the volumes stand by stand as they are listed in column 7.

The supply curve for sawlogs (*SS* in Figure 9.9) is now easily constructed by plotting the reservation cost (column 8) over the accumulated volume (column 9).

TABLE 9.2
Conceptual Procedure for Constructing a Sawlog Supply Curve (the Data after Ranking, and with Accumulated Volumes)

6 Rank (no.)	7 Volume (millions of board feet)	8 Reservation Cost (dollars per M)	9 Accumulated Volume (millions of board feet)
1	1.000	73	1.000
2	.400	75	1.400
3	1.600	77	3.000
4	.800	78	3.800
5	2.200	80	6.000
6	.400	82	6.400
7	.800	84	7.200
8	1.600	88	8.800
9	1.200	96	10.000
10	1.200	108	11.200

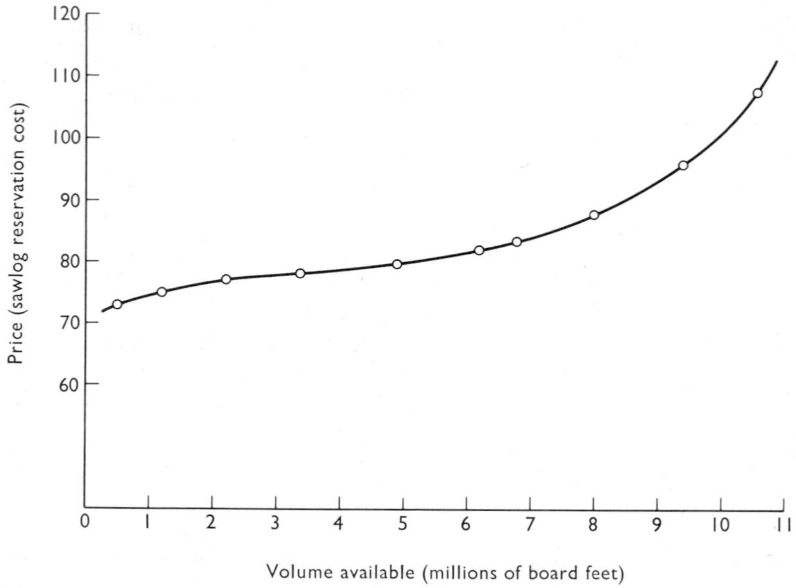

FIGURE 9.9. A short-run sawlog supply curve.

As with all marginal cost curves plotted from discrete data, one should use the midpoint of the added output for plotting. With the data of Table 9.2 the result is the short-term sawlog supply curve of Figure 9.9. In most market areas of this country there would be many stands and a smooth curve would normally result. For ease of illustration, only ten stands are utilized here.

Each stand in the market area has now been entered in the supply curve at its sawlog reservation cost per 1000 board feet of sawlogs, so the sawlog supply curve shows how much an additional 1000 board feet of sawlogs would cost the buyer at any possible quantitative level of production. This reservation cost curve of sawlogs is an aggregate marginal cost curve, just like any other short-term supply curve, and it shows the response of sawlog producers to changing sawlog prices.

Aggregation over Distance

The sawlog supply curve requires that volumes be aggregated over distance. Figure 9.10 may help to clarify this aspect. Here, 49 separate stands have been ranked according to their reservation cost and the supply curve, *SS*, plotted for the market area. The volume of each stand is indicated by the length of the interval occupied on the volume scale at the figure's base (at a zero price level). This same aggregate of volumes is "disaggregated" on the lower part of the diagram, where the volume of each stand is plotted at its distance from the market, with the distance being measured in cost miles (see page 108). When the volumes are all added together, the total volume on the quantity axis is completely accounted for. Since the stands were ranked according to their sawlog reservation cost, the curve formed is necessarily a rising one. Buyers look at every stand, regardless of its location, from the viewpoint of "What will the logs cost me at my mill site?" The supply curve for a market area must include all timber within the timbershed.

There would be cases where reservation price would not be the same for all possible sale volumes from the same stand—where different volumes could be removed from the stand at different reservation prices. In such cases the stand being considered must be represented by a *curve* of reservation cost over volume (rather than as a single point) and the curves then aggregated to produce the short-term sawlog supply curve. The process is identical to that of aggregating the marginal cost curves of firms in an industry. The difference is not in the process of aggregation, but in the fact that each such curve has its origin at its own (ranked) reservation cost.

Sawlog Supply from Public Timber

A thorough discussion of public timber supply would require a complete review of the policies of multiple use, of sustained yield-even flow, of appraisal, of bidding, and, perhaps most important, of the procedures and policies underlying the deter-

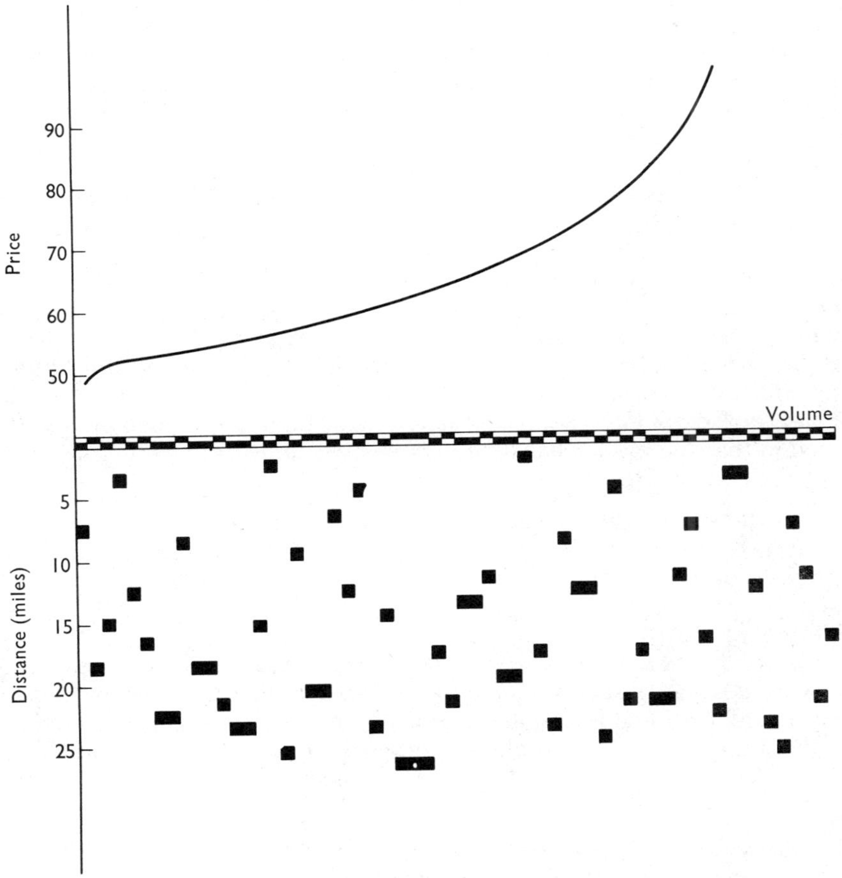

FIGURE 9.10. The sawlog supply curve requires that volumes be aggregated over distance. The buyer is not interested in the stand's location, per se, but rather in the cost of logs delivered to the mill site. Stands are therefore ranks according to their reservation cost, regardless of location.

mination of the so-called "allowable cut" from forest service management units.[8] Fortunately, the background of federal timber policy need not be reviewed to develop the supply relationships, although this particular subject area provides an excellent illustration of how economics and policy are, or should be, related. Very briefly, the significant facts of public land timber supply are the following.

Institutional and Administrative Factors

1. Under the Organic Administration Act of 1897, national forests were established for watershed protection and "to furnish a continuous supply of timber for the use and necessities of citizens of the United States." This established the policy of sustained yield, which was reaffirmed by the Multiple Use Act of 1960, then redefined by the Renewable Resources Planning Act of 1974, the National Forest Management Act of 1976, and the Forest and Rangeland Renewable Resources Research Act of 1978. The Forest Management Act limits "the sale of timber from each national forest *to a quantity which can be removed from such forest annually in perpetuity on a sustained-yield basis*: *Provided*, that in order to meet overall multiple-use objectives, the Secretary may establish an allowable sale quantity for any decade which departs from the projected long-term average sale quantity that would otherwise be established." The end result of the changes was to reaffirm the policy of "even flow-sustained yield," but did provide a possible departure.
2. An "allowable cut" is determined for each forest. Since the early 1970s the Forest Service has increasingly used computer models, often employing linear programming, for allowable cut determination. Timber RAM, the first such model to be widely applied, now permits most managerial options to be evaluated.[9] At least through 1986, however, market price had not directly entered such calculations.
3. The timber is advertised for sale at "not less than the appraised value." Bidding may be either by submission of sealed bids or by oral bidding at scheduled auctions.
4. The contract for harvesting is awarded to the highest qualified bidder. (Qualified in that the bidder has demonstrated an ability and willingness to carry out the contract terms.)

From the standpoint of short-term supply curve formation, the two most important factors are those involved in determining (1) the allowable cut, and (2) the appraised price.

The Allowable Cut

The actual volume of mature timber is predetermined, although the measurement of these volumes may pose problems and can also produce variation in results. Sale volumes on any National Forest are normally limited "to a quantity which can be removed from such forest annually in perpetuity on a sustained yield basis," although the Secretary of Agriculture "may establish an allowable sale quantity for

---

[8]The discussion here is entirely in terms of the U.S. Forest Service, since 87% of all federal (66% of all public) commercial forest land is under the administration of this agency.

[9]Timber RAM was initially described by David Navon (1967). Also see the articles by Neff (1973), and by Chappelle et al. (1976). This last article states: "It appears that undue reliance has been put on the optimizing features of Timber RAM without seriously pondering the more important question of what should I optimize?" The model most currently used is a planning model called FORPLAN. It is much more flexible—and complicated.

any decade which departs from the . . . sale quantity that would otherwise be established." This is the so-called even flow-sustained yield policy, now modified slightly to the nondeclining-even flow approach. Combined with the objective of maximizing timber yield by cutting at the culmination of mean annual increment, this policy rejects objectives and goals derived from an economic base. The rotation age is fixed by the size of tree desired and the expected growth rate. Growth on immature stands is determined by site factors, and to a lesser extent by the expected level of management. Growth on the entire forest is also influenced by the area of immature timber it includes. Under present procedures, market price has no direct influence on the allowable cut volume. Once the timber is sold, there is some flexibility in the time of cutting. Usually the amount of timber sold but uncut is around 2 to $2\frac{1}{2}$ times the annual allowable cut, but due to the required road building and other exigencies, there is less flexibility here than might be expected.

Allowable Cut Effect

The allowable cut effect has been defined as "the immediate increase in today's allowable cut which is attributable to expected future increases in yields."[10] It can arise anytime a forest with excess inventory (usually old growth timber) is under some form of volume regulation and is also subject to some constraint on the distribution of the cut over time. Many—probably most—of the national forests in the Pacific Northwest are (or, until recently, were) in this category.

To appreciate this effect, consider a forest in which a major part of the area is old growth timber, with the remainder consisting of recently burned timberland now in need of reforesting. The forest is under volume regulation and is subject to the even flow constraint cited in the Forest Management Act, limiting "the sale of timber from each national forest to a quantity which can be removed from such forest annually in perpetuity on a sustained-yield basis." The management plan calls for a rotation of 70 years, and the allowable cut has accordingly been calculated by taking the total volume of timber that is over rotation age and dividing this by the anticipated rotation, then adding the mean annual increment of the entire forest.

$$\text{Allowable cut} = \frac{\text{Volume of mature timber over rotation age}}{\text{Rotation length}} + \text{Mean annual increment}$$

Now suppose a 3000-acre plantation project having an estimated yield of 28,000 board feet per acre after 70 years is proposed for the forest. The financial analysis of the project, however, yields a slightly negative present value when using the approved discount rate—the project fails as a viable economic investment.

But then someone observes that if the proposed plantation project is implemented, the mean annual increment will be substantially increased. In fact, with project implementation the annual allowable cut can immediately be increased by $(28{,}000 \times 3000)/70 = 1{,}200{,}000$ board feet. At \$100 net per thousand this would mean an instant income increase of \$120,000 per year—and when this is added to the discounted present value of the plantation the B/C ratio looks unusually good.

---

[10]The term was apparently first used by Schweitzer, Sassaman, and Schallau (1972). Several others, including Flora in 1966, Pearse in 1965, and Ervasti in 1970, had commented at length on the influence this effect might have on harvest cuts and forest investments and were cited by Schweitzer et al. (1972).

Is this a proper way of treating this "bonus"? Try the with and without test. With the project, allowable cut is increased 1.2 million board feet per year; without it, no increase! What's wrong?

To attribute the extra $120,000 per year to the *plantation* is not correct. The money is coming from the old growth and could be obtained either with or without the plantation — except for the institutional requirement constraining the cut and the particular method used in its calculation. But if one is looking at this purely from an investment standpoint, then the with and without test comes through with flying colors — it gives the potential investor the correct answer. The investor does the right thing, but for the wrong reason!

Clearly, under these conditions investment funds may be misallocated. There are probably many other areas which, if planted, would yield a better return than the 3000 acres in our example, but lack the old growth forest nearby (Teeguarden, 1973). A somewhat different sort of misallocation was observed in British Columbia (Pearse, 1965; Forster, u.d.). Here the allowable cut from large areas of old growth timber was calculated much as illustrated above. By purchasing private cutover land and adding the area to a sustained yield management unit (a tree farm license), a corporation could gain an immediate increase in its annual harvest. The benefit was not the anticipated yield from the purchased land, but the increased harvest from the public area.

One should bear in mind that neither the even flow policy nor the nondeclining even flow policy are derived from an economic base. Binkley observed that these policies stem from efforts — in large part successful — to stem the indiscriminate cutting of virgin forests at a time when they were regarded as a gift of nature, available to anyone who would put the timber to a useful purpose. The policy has been perpetuated, at least in part, because it serves a useful political end by providing a guide for "steering the Forest Service among competing interest groups" (Binkley, 1980).

Given the existing legal constraints on national forests, if the objective is to maximize present net worth, then their net benefits should be calculated with the allowable cut effect. One should, however, be aware that the benefits stemming from the ACE result from the constraints, not the practice that gives rise to the effect. And if timber per se has a value to society over and above that measured in the market, then even though the net financial benefits are negative, social welfare may increase.

The Appraisal Procedure

Current appraisal procedures will be reviewed at the end of this chapter. For present purposes it is sufficient to know that (1) Forest Service timber appraisals involve a residual calculation; (2) for sawtimber the starting point is lumber value, from which all costs of production other than stumpage are subtracted; and (3) this "conversion surplus" is then divided (not necessarily equally) into two parts: (a) an allowance for "profit and risk" and (b) a remainder for "stumpage return." Within the framework of sawlog supply already established in this chapter one can regard the appraisal procedure as requiring, for each stand, an estimation of the cost of availability which is then subtracted from the (estimated) sawlog value.

Constructing the Short-term Supply Curve

The short-term sawlog supply curve for public timber is constructed in the same manner as that for private timber — by adding together, for each stand, its cost of availability and its reservation price, with the latter now being the *appraised* price. Unlike private timber, however, there is a positive, administratively fixed minimum. If it is assumed that all appraisal estimations and calculations have been perfectly performed, the **sawlog reservation cost** for each stand in the present example will be identical.[11] The sawlog supply curve for timber from federal lands will therefore be perfectly elastic up to the amount of the allowable cut, then become perfectly inelastic. Such a supply curve is illustrated in Figure 9.11, with Q_a being the amount of the allowable cut.

While both public and private short-term sawlog supply curves are based on sawlog reservation cost, there are three significant differences between the two curves. The first is obvious: the calculation of an allowable cut makes the sawlog supply curve for public timber perfectly inelastic at the allowable cut quantity, but perfectly elastic for all quantities up to this amount. The second difference is only apparent if one attempts to quantify the sawlog supply curve for a particular area. Publication of the appraised prices of stands gives one the sawlog reservation costs directly, while the task of determining these costs for private timber can be difficult. The third difference deals with the stability of the curves. The sawlog reservation cost of private timber varies with the owner's expectations, while that for public timber is not directly influenced by expectations.

Sawlog Supply from Mixed Ownerships

If the land ownership pattern in the area is either all public or all private, then the sawlog supply curves shown in Figures 9.10 and 9.11 are representative of what might be expected in typical situations. In many parts of the United States, how-

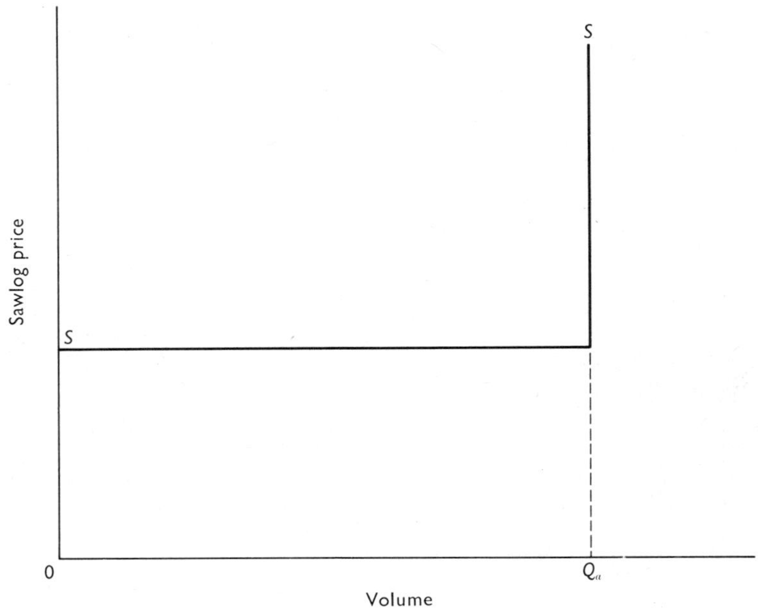

FIGURE 9.11. Sawlog supply curve for timber from public lands.

[11]This is necessarily true because of the assumption of (1) a perfectly homogeneous forest on (2) a flat plain and (3) a single group of mills located at its center. Hence sawlog value less cost of availability equals appraised price.

ever, forest land ownership is decidedly mixed,[12] especially in parts of the West, and in such instances the two supply curves can simply be added together horizontally. Figure 9.12 gives the result of such an addition, using the two curves previously developed as an illustration. The "break" in the curve caused by the inflexibility of allowable cut calculations on public lands is readily apparent, as is the perfectly elastic portion.

PRICE FORMATION—SAWLOGS

Given the short-term demand and supply curves already developed, there is no particular problem in price formation for sawlogs from either private or public timberland. With these two functions, and assuming a competitive market structure, the intersection of the two curves will indicate the equilibrium price and quantity. Figure 9.13*a* and *b* illustrates for both cases how an equilibrium price (P_m) and quantity (q_m) would be determined. In the case of logs from private timber sources, Figure 9.13*a*, a shift of the sawlog demand curve outward, perhaps to D_2, would cause sawlog prices to increase (to P_2), thereby bringing in more sawlogs from private lands. With logs from only public sources, however (Figure 9.13*b*), there are two possible outcomes if demand increases.

1. If the demand curve intersects the elastic portion of the curve (as does D_1) any place short of the "corner," then the full allowable cut will not be sold.
2. Only if the intersection occurs at or above this corner—if it intersects the inelastic portion of the supply curve—will the full allowable cut be sold. The price effect will be transmitted backward to stumpage prices, which will be examined shortly.

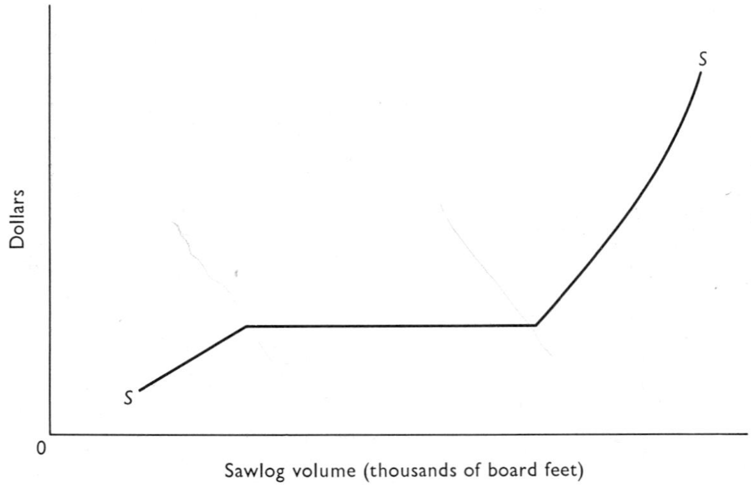

FIGURE 9.12. Sawlog supply curve for timber from mixed ownerships.

[12]The extreme in this regard is that found in some Western states, where the public domain was used to encourage railroad construction by granting alternate sections to certain corporations, thus creating a checkerboard of public and private forest land ownership.

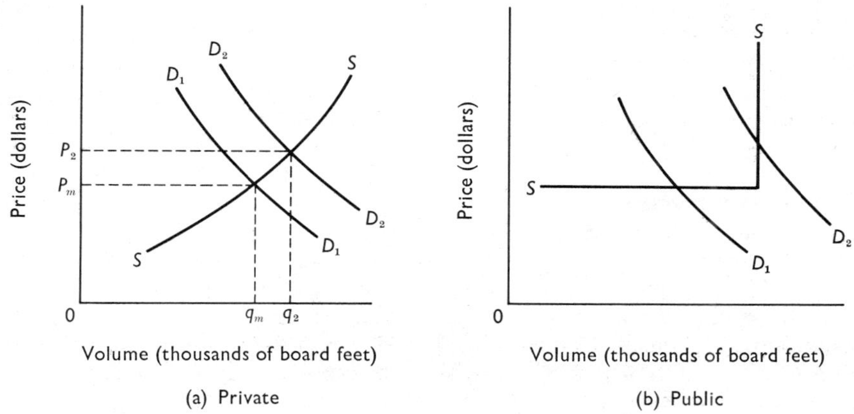

FIGURE 9.13. Price formation for sawlogs.

The most interesting case is that of Figure 9.14, wherein sawlogs are available from either (or both) public or private lands. Here the three possible situations are shown by the three demand curves, D_1, D_2, and D_3. The results in terms of sawlog prices and quantities, and in terms of who sells and who does not, are sufficiently clear to require no further comment.

PRICE FORMATION: STUMPAGE

In a perfectly competitive market the sawlog stumpage price at any specific location will equal the sawlog price at the mill minus the cost of availability. But forming stumpage demand and supply curves presents a peculiar problem because the

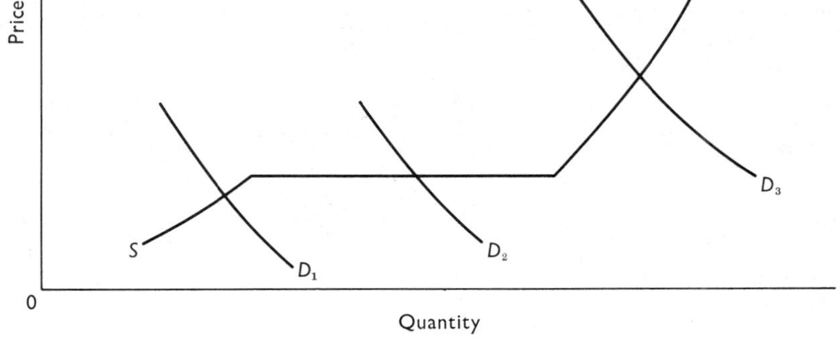

FIGURE 9.14. Sawlog price determination when sawlogs may be obtained from either public or private sources.

demand for stumpage originates at one point in space and the supply in another. The buyer (mill operator) views each possible stand from *his* market point, while the seller (the timber owner-producer) views it from his own very different market point. Both buyer and seller are aware of the difference, but this effectively prevents stumpage supply and demand curves, at least of the usual variety, from being constructed.

Significance of Location

The significance of location, and the way it distorts *stumpage* demand and supply, may be clarified by reexamining a supply curve for sawlogs. Figure 9.15 is the same sawlog supply curve shown in Figure 9.9, constructed from the data of Table 9.2. Suppose the sawlog market price is at \$92, delivered at mill site—that is, assume the demand for sawlogs is elastic at a level of \$92. Now from the given *supply* curve, we can identify exactly which stands will be purchased and can also pinpoint their cost of availability, for this cost is part of the data used to construct the supply curve. The cost of availability is the information needed to derive the stumpage demand curve, for if we subtract it from the market price the result will be the derived demand curve for stumpage, derived from a perfectly elastic demand curve for sawlogs. (There is no real need to assume perfect elasticity but the

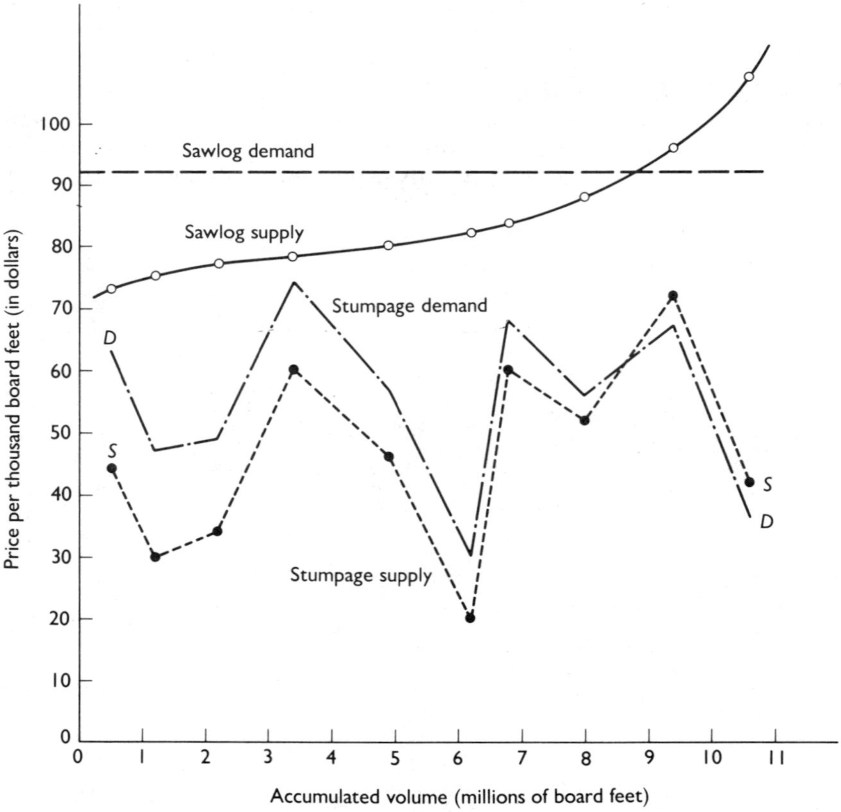

FIGURE 9.15. A demand and supply curve for stumpage—derived from the demand and supply curves for sawlogs.

illustration encompasses only 10 stands, hardly enough to make up the total supply of an area.)

Price for Individual Stands

The result of this subtraction is shown by the broken line, *DD*, in Figure 9.15 and represents the stumpage price that will be paid for each stand. This very irregular line is in fact a perfectly elastic demand curve for stumpage. The reservation prices are also shown in the figure, connected by the dashed line, *SS*. In a very real sense this is the supply curve of stumpage (although not really continuous), while the demand curve is that of *DD*, the dashed line. The intersection of the stumpage demand and supply curves identifies the total volume of stumpage that will be purchased, which of the stands will be taken, and their prices. But the intersection does not indicate *the* stumpage price, for this price differs with every stand. The dashed line of stumpage price lies above the reservation price (supply curve) for all stands taken, and below for all remaining stands, but the price level of the intersection has no general meaning.

A More General Model

For most purposes a simpler model that stops short of determining specific demand and supply curves for stumpage can be used. In Figure 9.16*b* the sawlog demand and supply situation (supply from private lands only) is shown as establishing the sawlog price, P_m, and the quantity, Q_m, at the selected market point. Part *a* of the diagram shows the same market price (P_m) that is established in part *b*, but plotted over a horizontal axis scaled in distance instead of volume units. The cost of availability at each distance has been subtracted from P_m, and the result plotted over the appropriate point of the lower (distance) scale. The cost of availability curve is reproduced, but now inverted, and with the origin transposed to P_m. The new curve, intersecting the price axis below the indicated sawlog price by the cost of loading and unloading logs, declines with distance from the market, thus tracing out stumpage prices as the distance between the market and the stand is increased.

A Model for Mechanics

For mechanically minded individuals, the model of Figure 9.17 may prove helpful. Suppose the cost of availability curve were traced on a piece of transparent plastic,

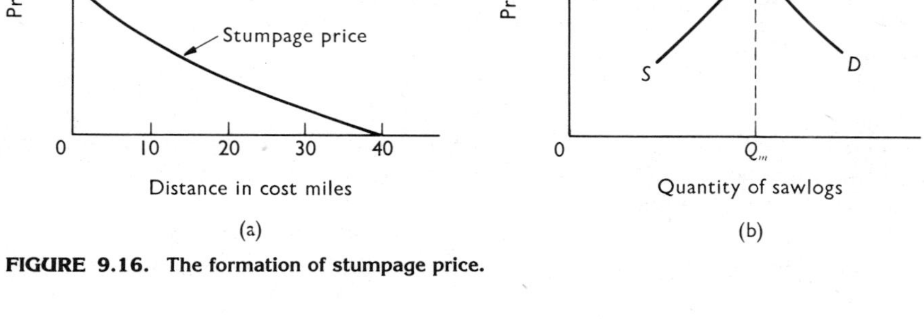

FIGURE 9.16. The formation of stumpage price.

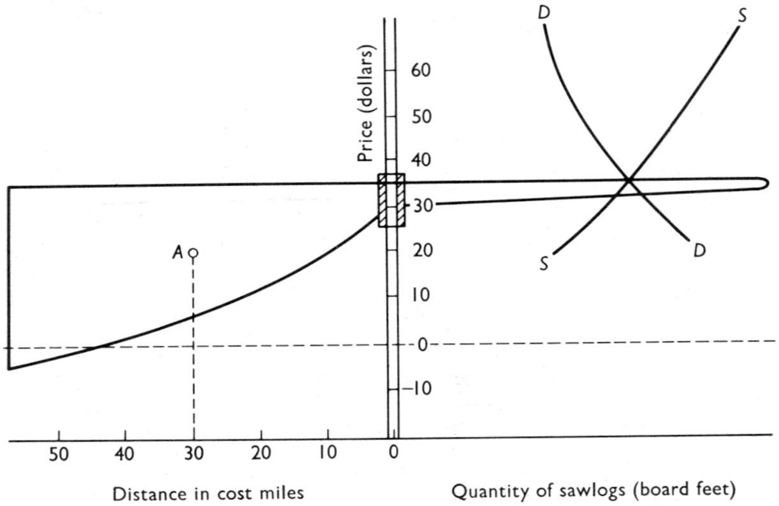

FIGURE 9.17. Stumpage price determination.

then cut out with a horizontally projecting arm (as in Fig. 9.17) and attached to a vertical rod or pin so it might be slipped up or down as desired. The rod is scaled in dollar units. The right side of the model has as its components the sawlog supply and demand curves, with the vertical rod serving as the price scale and the horizontal axis scaled in timber volume units (board feet). A shifting of either the supply or demand curves raises or lowers the intersection, and hence the height, of the plastic indicator, thus keeping the top of the indicator always at the market price of sawlogs, while the lower (curved) portion indicates stumpage price at any cost distance from the market—on the left side of the diagram. For example, in Figure 9.17, with the demand and supply curves determining a price of $35 for sawlogs at the mill site, a timber stand located 10 cost miles from the market will have a stumpage price of $20, while at 45 cost miles the stumpage price would be zero.

Note carefully the role of reservation price in this example. A stand with a reservation price of $20 that is located 30 cost miles from the mill (*A* in Fig. 9.17) will not be sold under the prevailing conditions. Only those stands with a reservation price equal to or below the indicated stumpage price will be sold. But the sawlog supply curve is made up by a rank ordering of the reservation cost (reservation price plus the cost of availability) for all stands in the market, and exactly enough timber will be available at the indicated set of stumpage prices to supply the mills with the (equilibrium) quantity of logs they demand.

The mechanical approach can be carried considerably further, and there are advantages to discussing at least one additional step. Suppose a map of the entire timbershed—the flat plain of homogeneous timber—was available, drawn to a scale identical to that used in making the stumpage price curve of Figure 9.17. Suppose further that the vertical rod, together with the attached indicator carrying the inverted cost of availability curve—the whole apparatus of Figure 9.17—was then pinned into the map at the location of the group of mills. As the indicator, set at the market price for sawlogs and carrying the stumpage price curve, is swung about the rod it will describe a pointed, parasol-like surface over the entire timbershed. The height of this surface directly over any location on the map will indicate the stumpage price for that location. Figure 9.18 conveys at least an idea of how the surface might appear under the hypothesized conditions.

Note how the surface must change if it is to represent the real world rather than a flat, homogeneous forest. Instead of the symmetrical surface of the figure, the stumpage price surface over a forest area (and using real miles instead of cost miles) will be full of indentations and convolutions. Rough topography increases logging costs, so the stumpage price surface would dip down over any such area; a high-speed highway lowers transport costs, so here the surface would be further above the map. But the basic elements would still remain. Under competitive conditions the stumpage price for any stand in the forest will be the price of sawlogs at the delivery point *less* the cost of availability, and unless the owner's reservation price is at or below this price the stand will remain uncut.[13]

One can see that mills with different cost structures, or at different locations, could have different offering prices for any particular stand. But as each mill's stumpage price curve swings over its timbershed—whether in competition with other mills or not—it will sweep up just enough stumpage to meet its demand for sawlogs at its own equilibrium price. And note also that it is in no way necessary that every mill will have the same equilibrium price for sawlogs.

Short-term reservation prices are particularly prone to change, and anything affecting future price expectations tends to shift the sawlog supply curve, changing timber as well as sawlog prices. Although there is no clear-cut line for distinguishing between short- and long-term factors in sawlog or stumpage pricing, and to some extent the division is an artificial one, it will be necessary to look at the more important factors that serve to shift stumpage prices under particular conditions and to see the effect of relaxing at least some of the more stringent assumptions. It will be especially important to examine the impact of a few of the important institutional factors. Before proceeding to this analysis (Chapter 10), we'll review other approaches to stumpage price formation.

The Residual Approach to Stumpage Price

To many, the determination of stumpage price seems quite simple. Stumpage is a raw material. If it has a value at all, it is because some firm or some individual

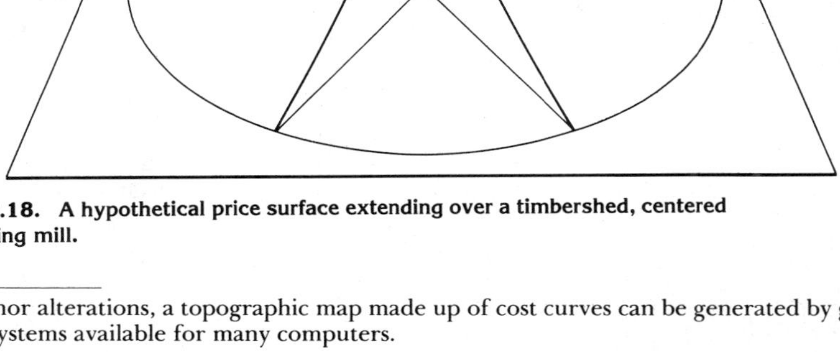

FIGURE 9.18. A hypothetical price surface extending over a timbershed, centered on the buying mill.

[13]With minor alterations, a topographic map made up of cost curves can be generated by geographic mapping systems available for many computers.

believes this raw material can be fashioned into a finished or semifinished product that can be sold profitably. It follows that the starting point for determining stumpage price should be the selling price of the product or products that are to be produced from the raw material. By subtracting all costs from the product's final sale price—from felling costs through costs of manufacturing—and including whatever amount is necessary to repay the entrepreneur for his or her contribution, one derives a residual value. This approach is the time-honored technique of timber appraisal, well developed in textbooks of forest management and forest valuation. As far as it goes, it is a direct application of derived demand: all nonwood costs are subtracted from the (given) market (demand) price of lumber.

Stumpage Price and Rent Theory

Most economists regard trees simply as another natural resource. Standing timber is included in the general concept of land, and the return to stumpage is a form of rent. As a topic in rent theory one might illustrate determination of stumpage price with a diagram such as that of Figure 9.19*a* and *b*. Average total costs of lumber production (ATC_1) and the corresponding marginal costs (MC) are shown in their usual form. The average total cost curve depicts all production costs, including whatever return on investment is necessary to keep the lumber producer content and in business. At a price of P_1 the firm will be in equilibrium, just covering all production costs, including the indicated stumpage payment, s_1. If one were to aggregate all the marginal cost curves of all the firms producing lumber for this market, the resulting supply curve would be that of SS in Figure 9.19*b*, which intersects the demand curve for lumber (DD) at a price of P_1.

Now suppose mortgage rates fall and residential building activity increases, shifting the lumber demand curve out to the new position, D_1D_1. As a result lumber prices rise to P_2, and all producers in the industry will increase their output until once again their marginal cost will be equated to the new price, P_2. At this price the firm will make a substantial profit, its size indicated by the shaded rectangle of the figure and which will average $(P_2 - P_1)$ per 1000 board feet.

But now assume the timber from which the lumber is being produced is being purchased from another party who controls the supply. Previous to the lumber

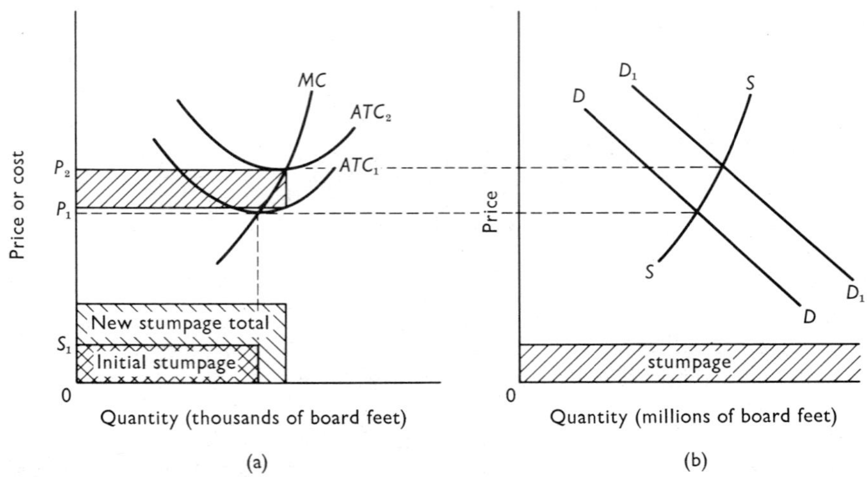

FIGURE 9.19. Stumpage as a rent.

price increase, the stumpage owner could charge only S_1 for the timber. Now, however, he can make an additional charge of exactly $P_2 - P_1$ per 1000 board feet. At the new stumpage price the lumber producer will be just as well off as at the outset—he is no worse off than before the demand curve shifted. His average total cost curve will simply "slide up" the marginal cost curve, eventually reaching the position shown in Figure 9.19*a*. This is the classical case of Ricardian increasing rent referred to in Chapter 4, and it fits the residual approach to stumpage pricing on which the appraisal process rests; P_2 equals the selling price of the lumber, and $P_1 - S_1$ equals the average cost of production *plus the margin for profit and risk* that is normally included in the total cost curve.

Note that the rent theory approach is not opposed to the supply-demand model. The differences stem from the assumptions underlying the two perspectives. One might say that the rent approach to stumpage price determination proceeds entirely from the manufacturer's side. All the elements of the analysis concern production of the wood product—lumber, pulp, or plywood, as the case might be.[14] The derived demand approach also begins with an end product such as lumber, from which *all* nonwood production costs are subtracted. The difference lies in the treatment of supply. Neither rent theory nor the related appraisal procedure involve consideration of stumpage production costs or the possibility of a reservation price on the part of forest owners. It is implicitly assumed that the supply of all productive factors other than timber is perfectly elastic, that the supply of timber offered is perfectly inelastic with respect to price, and that the forest owner is in a position to extract the rent. With these assumptions, the rent-based appraisal model of stumpage pricing provides little assistance for analyzing the effect of changing timber supplies and/or production costs on either stumpage prices or on product prices. These assumptions would have to be changed. This is possible, but it alters the nature of the theoretical rent structure.

In a stable, short-term period an appraisal, if accurately performed, can yield a valid estimate of the dollar amount available for stumpage, profit, and risk. The purpose of this book, however, is to develop an understanding of how forestry fits into our economy, and how important economic forces, national and international, influence possible and probable returns to timber production and ownership. For this task the demand-supply model seems to be the better tool.

QUESTIONS AND PROBLEMS

1. Assume that the demand curve for reproductions of antique furniture of bird's-eye maple can be represented by the equation

$$P_a^d = 600 - 0.01Q_a^d$$

and the supply curve as

$$P_a^s = 200 + 0.00333Q_a^s$$

[14] This itself has posed a severe conceptual problem to residual theory ever since pulpwood production became important to timber producers. Should one begin with the value of pulp or of paper? What is a fair profit margin on pulp or paper manufacture? And how accurate, or even significant, is the calculation of a residual that amounts to less than 10%—and often as little as 2%—of the initial sum?

where P_a^d = demand price per unit of antiques, P_a^s = supply price per unit of antiques, while Q_a^s and Q_a^d are the numbers of antique units produced and/or sold.

If P_{of}^s is the supply price for all productive factors other than bird's-eye plywood, the supply curve for nonplywood productive factors (we will assume) is expressed by

$$P_{of}^s = 150 + .001667Q_a$$

where P_{of}^s is the price of all nonplywood factors, and Q_a is units of furniture produced. Each unit of furniture requires 64 sq. ft. of plywood.

a. Plot the equations. (Note that the quantities will be several thousand.) Then, using a graphic approach, derive the demand curve for bird's-eye maple plywood to be used in antique furniture. Replot the derived demand curve, with the scales showing quantities demanded in terms of 4 × 8 foot plywood sheets, and price per sheet.

b. Check your work by calculating the equations for the demand and supply for bird's-eye plywood sheets for antique furniture, in terms of numbers and prices of plywood sheets.

c. How much bird's-eye maple plywood will be sold for antique furniture?

d. What do you think might happen to producers of bird's-eye maple plywood if the price of the antique reproductions were to drop to $250? What would happen to the producers of the reproductions?

2. Would you expect the derived demand for lumber to be more, or less, elastic than the demand for housing in the United States? Why?

3. Explain the two concepts, cost of availability and reservation price, as they pertain to stumpage price.

4. Why would the supply curve for stumpage in a particular National Forest tend to be horizontal?

5. Explain how the stumpage supply curve in an area where forests are of mixed (public and private) ownership might have three segments, first rising at low prices, then a horizontal portion, and (at still higher prices) another rising part.

6. Stumpage price is most often termed a residual. Explain.

7. Economists usually refer to stumpage returns as rent. Show why.

chapter ten

TIMBER PRICES IN A MARKET ECONOMY

The interplay of product and factor prices with consumer and producer actions was stressed when reviewing the role of prices in a free enterprise economy (Chapter 3). In this chapter specific relationships of stumpage price to product price, institutional factors, and land use will be examined. One of the most significant of these is the connection between timber prices and final product prices. And since discussion of the rent-appraisal procedure provides an excellent introductory background, the relationship between product and timber prices is a convenient starting point.

RELATIONS BETWEEN TIMBER AND PRODUCT PRICES

Suppose bank lending rates were suddenly reduced. With money available at lower interest rates the demand curve for new housing could shift outward, as illustrated in *a* of Figure 10.1. As previously seen, the resulting increase in construction activity would then produce an outward shift in the demand for lumber, from *DD* and *D′D′* (b in Figure 10.1). The increase in lumber price produced by this shift would be reflected by a corresponding outward and upward shift in the derived demand curve for sawlogs at the mills being supplied by the forest area under study (*c* in Figure 10.1). This increased demand would, in turn, raise the entire schedule of stumpage prices to the new position shown in *d*.

The changes in lumber, sawlog and stumpage prices would not necessarily occur simultaneously. Major producers, however, pay very close attention to mortgage rates, housing starts, and so on, and to factors affecting them. Consequently, lumber price changes are often anticipated by suppliers.

Short-term Effects on Private Stumpage

Such an increase in prices would have two short-term effects; within the zone of previous positive stumpage values (from point *a* to *M*) a number of stands would now be sold that were not previously available because of the high reservation prices held by their owners. Logs from these newly available, higher priced stands would tend to shift the supply curve for sawlogs outward at least slightly, perhaps enough to somewhat lower both sawlog and stumpage prices from the first newly

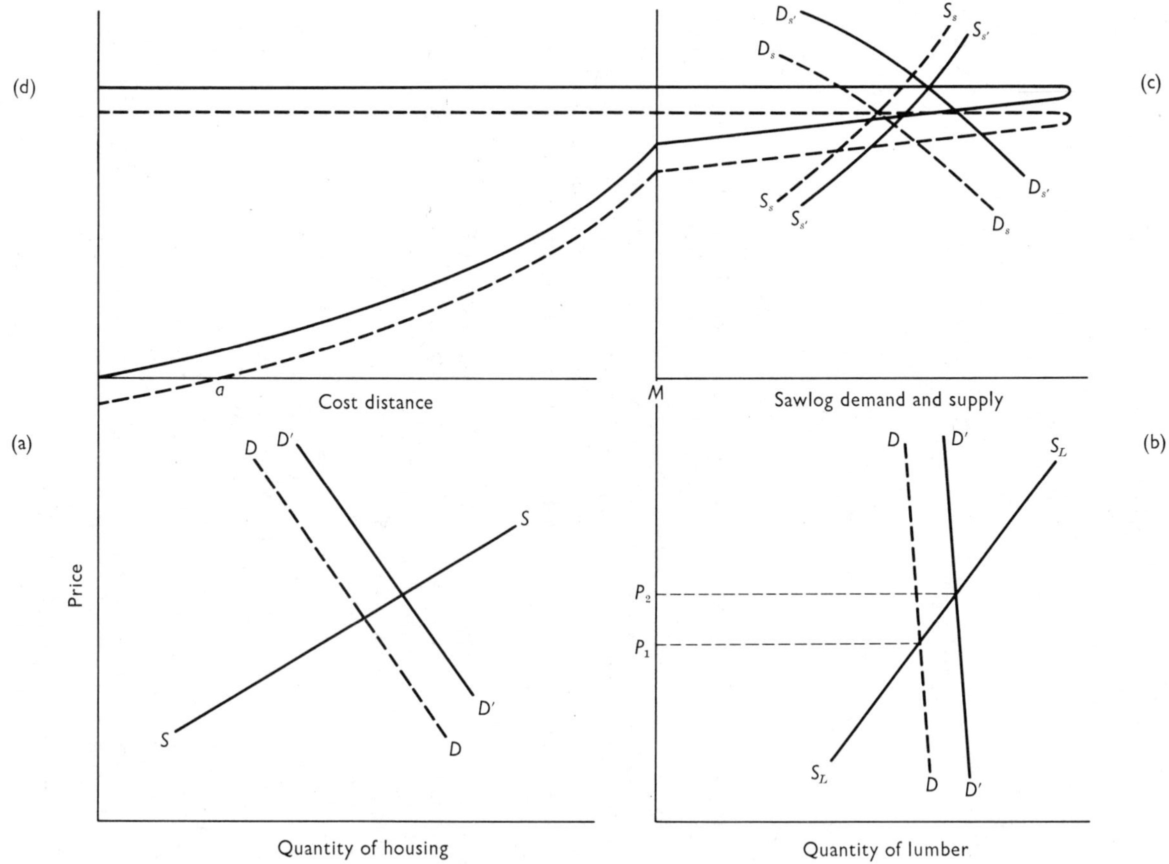

FIGURE 10.1. The relation between lumber and stumpage price (short term).

established high point. But this new stumpage price schedule would also increase the distance over which logs might profitably be transported to the mills, thus making new timber areas available as supply sources. With these new supplies the sawlog supply curve would again shift outward, resulting in a further drop in sawlog and stumpage prices. Eventually, the equilibrium price of timber would settle at a higher than original level, with some new areas sending timber into the mills. If owners' stumpage price expectations rise as a result of these shifts, a further round of adjustments would take place with the end result depending on the relative shifts of the demand and supply curves. But note that no particular increase in growth or in sustained yield capacity is necessarily needed. An increase in timber price will normally (under conditions of private land ownership) result in more timber being sold over the short term.

Effect on Stumpage if Ownership Is Mixed

The same increase in residential construction activity could have a quite different effect on sawlog and stumpage prices where sawlog stumpage was being obtained from both public and private sources. The sawlog demand and supply curves for public and private timber can be simultaneously treated by using back-to-back supply and demand diagrams, as in Figure 10.2. Unfortunately, such an illustration does not lend itself to depiciting the stumpage price curve directly, but by now this extension should be readily envisioned. With sawlog supply from private sources on the right side and from publicly managed areas on the left, the quantity taken by lumber manufacturers is represented by the distance between the points of intersection of the demand curve(s). (The fact that two demand curves are used does not mean there are two separate sources of demand; the same demand draws from the two supply sources.) With demand at D_1, for example, sawlog price will be at P_1 and q_1 will represent the quantity of sawlogs delivered to the mills. At this level of demand *all* logs will be supplied by private timberlands; the reservation price of

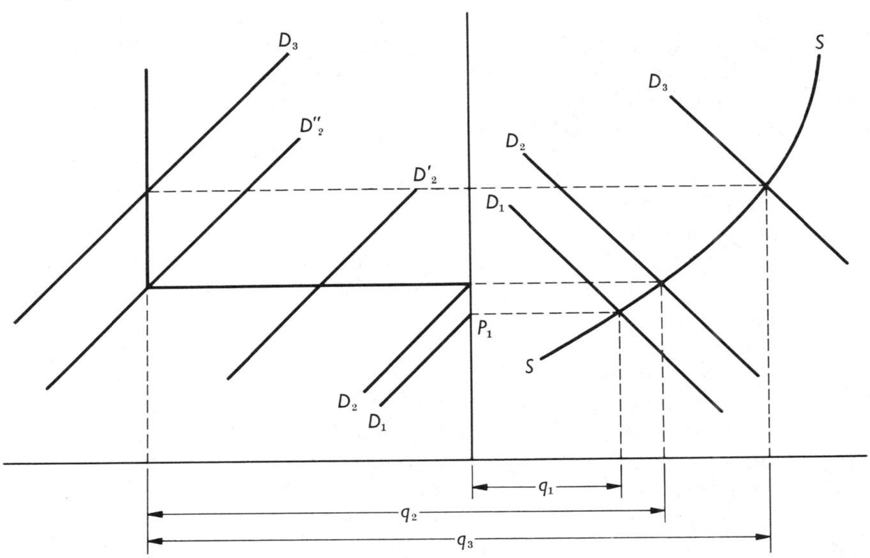

FIGURE 10.2. Effect of shifting demand on sources of supply.

public timber (the appraised price) reserves it from sale. Under current conditions such a situation is unlikely, since use of market prices in the appraisal process would usually lower the appraised price of public timber to the P_1 level. During the 1930 depression, however, when lumber prices were extremely depressed and public timber was withheld from the market as an aid to private timber owners, the situation was not unlike that just described.

As demand shifts outward to D_2 (other things being equal), supplies from public lands will begin to become economically available, and under the conditions assumed in Figure 10.2 the demand for sawlogs can continue to shift—through D_2' as far as D_2''—until the full allowable cut is being purchased before any further rise in price will take place. Normally, such a large increase would be of a long-term nature. New mills would be required and expansion of the old mills would undoubtedly take place. In the meantime the supply curves for both public and private timber would undergo considerable change.

But suppose demand for sawlogs expanded beyond that indicated by D_2'', for example, D_3 (on both sides of the diagram). With this level of demand, stumpage prices will increase *for both public and private timber*. So far as public timber sale is concerned, two events can be predicted with confidence: (1) bids for public timber will begin to exceed the minimum appraised price, first by small amounts, and then, as sawlog demand shifts further, the bidding may far exceed appraised prices; and (2) lumber manufacturers will start agitating for an increase in the allowable cut from public forests. And in part they will be correct, for if the allowable cut is increased, the supply curve will shift outward and, if the shift is large enough, stumpage prices may begin to come down.

If future demands do not decline, however, increasing the present allowable cut may force a greater reduction at some future date. Such a reduction is especially likely if the increased cut reduces the growing stock, unless the increased incomes from larger cuts lead to intensified management, thereby offsetting this reduction. Under the impact of competitive bidding for the limited short-term supply, and if the area concerned competes with mills that are in a better supply situation, some units may be forced out of business. A similar result may come about if the allowable cut now comes almost entirely from old growth timber, as is the case with many of our Western forests. In this situation, too, increasing the allowable cut can only hasten the day when the old growth will be exhausted and the dependent mills forced to close down.

Discriminatory Pricing

By separating public and private timber supplies the possibility of *discriminatory pricing* can be explored. Suppose, for example, the sawlog supply and demand situation is as shown in Figure 10.3. The entire allowable cut from public lands is being taken, as indicated by the position of the demand curve, D_1D_1, and overbidding of the appraised price is usual.

Now suppose foreign timber manufacturers enter the market, seeking to buy timber at whatever price is possible. Any purchase of new quantities under the conditions illustrated here will raise sawlog prices and either force domestic producers to find supplies elsewhere, meet the competition by paying the price required to bring new supplies into the market, or go out of business. But there is another possibility. If local timber buyers can persuade their elected congressional representatives that it is "wrong" to sell public timber to foreign manufacturers when U.S. mills are themselves seeking new supplies (foreign competition is forcing American laborers out of jobs by closing American sawmills and plywood

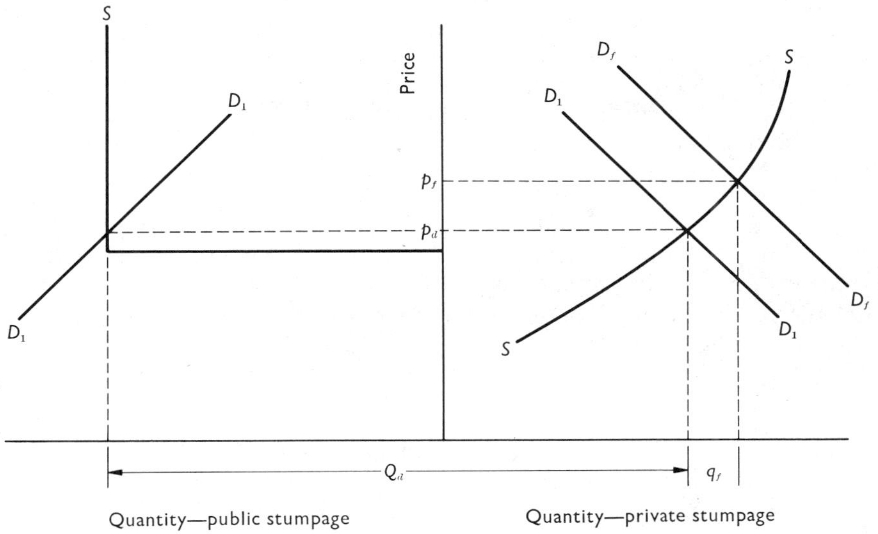

FIGURE 10.3. Discriminatory pricing of export logs.

factories, etc.), then an embargo might be declared on the sale of public timber to foreign buyers. If the embargo is enforced, it is possible (but not probable) that domestic buyers can continue to purchase both public and private timber in the same quantity and price (p_d) while a few private timber owners sell the desired amount (q_f) to foreign manufacturers at the higher price of P_f—and reap the corresponding profits (subject only to capital gains).

Such price discrimination is not rare in international sales, but it is usually possible to maintain price differentials of this type only if the market is firmly controlled. In the above situation, if demand is really active, the new sales will shortly raise the price of all stumpage, including that from public lands, to the new higher price level. Under these circumstances the new supplies will probably be taken from timber formerly reserved for later harvest. This tends to raise price expectations, shifting the supply curve for private timber upward and raising the price to all buyers, domestic and foreign. The log export market to Japan provides an example.

Expectations

So far the discussion has been in terms of the effect that shifting lumber demands might have on stumpage prices. But what happens if sawtimber owners, observing past increases in both lumber and stumpage prices, and reading of the boom in house construction, anticipate a decided rise in lumber and stumpage prices; that is, suppose expectations of stumpage owners rise. This change in expectations would quickly raise the reservation price of owner's timber, making sawlogs more costly by shifting the sawlog supply curve upward and, if the increase is at all widespread in the industry, cutting back production. Because short-term lumber demand is quite inelastic, and because under specified conditions lumber producers are in a seller's market, lumber prices will be increased as stocks are diminished. Such an action would reinforce the expectations of timber owners and again raise their reservation price.[1] A mutually reinforcing spiral may thus be set in motion. Increased timber prices raise lumber prices; builders hurry to lay in stocks to protect their commitments; this permits lumber producers to further raise lumber prices, which encourages timber owners to boost stumpage prices again, which raises lumber prices, and so on. The spiral is aided by the appraisal process, for with higher lumber prices the appraised price of public timber is necessarily increased. Present systems of allowable cut determination and pricing on public timberland have a decided escalator effect on timber prices.

If for some reason the demand curve for lumber shifts *inward* and lumber prices decline, then opposite forces are set in motion, but with one very important difference. A decline in lumber price, unless very prolonged, is unlikely to have a seriously depressing effect on price expectations of timber owners. The history of lumber and stumpage prices since the turn of the century has been one of almost constant advance. Declines have occurred, but for nearly four decades each upward jump has been enough higher than the succeeding decline to result in a net advance of substantial proportions.

The interaction between lumber and stumpage price is particularly important during periods of sustained price disequilibrium. In the years immediately following World War II, for example, many lumber manufacturers were unconcerned about stumpage prices since any resulting cost increase was willingly absorbed by lumber retailers (and in turn by consumers). As a consequence, public land sawtimber sale prices often exceeded the "appraised" value by very large percentages, a phenomenon difficult to explain under any strict interpretation of residual theory. The same difficulty occurs in reverse during periods of declining lumber prices. Confronted with falling lumber prices but committed to buy stumpage at prices derived from previously high lumber price levels, mill owners have been placed in very difficult economic positions. The situation with respect to federal timber sales was greatly eased when the Forest Service instituted a "review" procedure, permitting limited renegotiation after price changes of given relative amounts.

The steep decline in lumber prices accompanying the record interest rates of the early 1980s (and the resultant drop in construction activity) provided an exaggerated example. Substantial amounts of National Forest timber, purchased during the 1979 to 1982 period, were still uncut when lumber prices slumped. Buyers of such stumpage suddenly discovered they were under contract to cut timber costing considerably more than they could expect to gain from the sale of lumber the timber might produce. Some manufacturers claimed—probably correctly—that holding them to the contract price could result in bankruptcy, with the accompanying employment loss, and so on. Congressional hearings resulted in a temporizing action by the Forest Service. The 1984 rise in lumber prices lessened the difficulty,[2] and legislation of 1984–1985 allowed substantial relief in the form of Forest Service "buybacks" of many stumpage purchase contracts signed before the drastic lumber price decline.

---

[1]In this connection the more advanced student may wish to check the concept of elasticity of expectations, the ratio of the percent increase in price to the percent increase in expected price (Hicks, 1946).

[2]Interestingly, while support for treating rising stumpage prices as capital gain has been very strong, the accompanying privilege of treating stumpage price *decreases* as a capital loss has received little comment—perhaps only proving what everyone knows, that profits are preferred to losses. And one should recognize that if a firm is facing closure from declining sales there is little consolation to be gained from claiming a capital loss.

MARKET STRUCTURE AND STUMPAGE PRICES

The competitive structure of the market in which timber is sold is important in determining stumpage price. The point of departure for theoretical development is that of pure competition, the situation assumed in developing the supply-demand model of the previous chapter. It is doubtful, however, if timber markets with many sellers and many buyers should be called characteristic of U.S. timber markets. Real-life conditions probably range from the extreme of pure competition all the way to bilateral monopoly (a single seller and a single buyer), and almost every conceivable combination of buyer and seller conditions could be found someplace in the United States.

Some degree of timber market imperfection is inevitable, because forests occupy extensive land areas and transportation costs per unit of volume are high. Since no two processing plants can occupy the same point in space, *some* location advantage will accrue to *some* firm in most timber buying situations. The question is one of degree; how much price protection is afforded depends on the cost of moving logs to or from the nearest alternative. Spatial monopoly is almost always present to some extent, but it can be of great significance in forestry.

The four combinations to be reviewed here are those involving (1) many sellers, or (2) a single seller of timber, with (3) oligopsony, and (4) monopsony, on the buyer's side of the picture. The review will also be confined to short-term effects; the extension to long-term effects on both price and quantity aspects of timber sales should be made by the student.

Many Stumpage Sellers and a Single Buyer

The major departures from purely competitive conditions brought about by having a single buyer (monopsony) can be seen from the data in Table 10.1 and the corresponding figure (Figure 10.4). Viewing a schedule such as that of the table, the monopsonistic buyer would reason as follows: "I can buy 200,000 cords at an effective price of $20 per cord, or I can buy 225,000 cords at an effective price of $20.50 per cord. The additional 25,000 cords will, therefore, actually cost me $612,500, or $24.50 per cord, since in getting the 25,000 additional cords I must raise the price of pulpwood by 50¢."[3]

Such a buyer will not consider the supply curve of pulpwood as such, but

TABLE 10.1
Schedule of Reservation Costs for Pulpwood

Quantity Available (cords)	Price (dollars)
150,000	19.00
175,000	19.50
200,000	20.00
225,000	20.50
250,000	21.00
275,000	21.50

[3]In this particular example:

225,000 cords @ $20.50 =	$4,612,500
200,000 cords @ $20.00 =	−4,000,000
Cost of additional 25,000 cords	$ 612,500

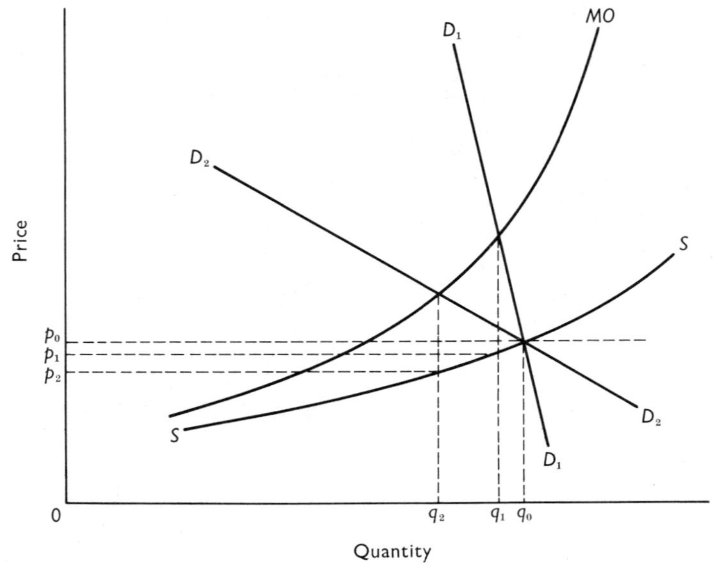

FIGURE 10.4. Price determination under conditions of monopsonistic buying from many owners.

rather the marginal outlay required by the purchase. The relevant curve is the buyer's **marginal outlay curve** (*MO* in Figure 10.4) showing the additional expenditure caused by the purchase of each additional unit. It is the curve marginal to the reservation cost schedule. A rising pulpwood supply curve will act to push up the marginal cost curve of any monopsonistic buyer and will also make it steeper, since each additional unit purchased adds to the production cost.

To aid in evaluating the possible effect of monopsony on, for example, sawlog price and quantity relationships, two possible demand curves are shown in Figure 10.4 One (D_1D_1) represents a very inelastic demand on the part of a monopsonistic buyer, and the other (D_2D_2) the relatively elastic demand that might be generated by a single large sawmill selling on a national and highly competitive market. In appraising the effect of this market structure on timber price and quantity, we must consider the shape and elasticity of both the sawlog demand and supply curves. If the supply curve is perfectly elastic, monopsony will have no effect on either price or quantity. The marginal outlay curve will be identical to the supply curve and will be horizontal. But if the supply curve is rising (as illustrated in Figure 10.4, and the more probable situation under reservation cost conditions), the effect of monopsony will be to reduce the quantity of sawlogs purchased and lower the purchase price. The more *elastic* the demand curve, and the more *inelastic* the supply curve, the greater the effective reduction of both price and quantity.

In the figure, for example, purely competitive conditions on both sides would indicate a price of p_o and a quantity of q_o under either of the demand situations shown. But the monopsonist with the demand curve D_1D_1 would purchase the quantity q_1 at the price p_1, while the monopsonist with the demand curve D_2D_2 would purchase q_2 at a price of p_2. From the broad, national viewpoint one can say that a monopsonistic buyer purchasing in a purely competitive market would always result in fewer resources being devoted to producing and using sawlogs (and hence sawlog stumpage) and would usually mean (1) lower returns to timber owners, (2) higher profits for the buying industry, and, perhaps (3) higher prices for consumers, depending on the market conditions under which the final product is sold.

In attempting to evaluate the actual effect of monopsony on stumpage markets, one must consider an important practical qualification. Because of its derived nature, an elastic demand curve for timber ordinarily implies an even more elastic demand for the final product. If product demand is elastic, then competitive pressure will normally force firms to construct highly specialized plants. Such plants tend to have a distinctly limited output range, especially with processes such as chemical pulp production. This leads to an inelastic demand for raw materials such as pulpwood, which also limits the degree to which price and quantity can be affected by a buyer.

Studies of national forest sawlog stumpage prices have indicated that sales made to single bidders often bring lower prices than those involving several bidders, indicating the possibility of monopsonistic price influence. Usually other factors are involved, and careful study is needed before concluding that monopsony has had any serious effect on U.S. timber prices.[4] Yet common sense would tell one that whenever timber must be sold to a single buyer if it is to be sold at all, then prices would at least tend to be lower than if the sale were made under competitive pressure. This seems to be the situation in many localities, although no studies of a conclusive nature can be cited.

Under monopsonistic conditions, with many sellers, one must also recognize the probability of discriminatory pricing. A corporation, for example, paying a higher price for all pulpwood harvested more than 60 miles from the plant, is not necessarily acting out of goodness of heart to cover the extra cost involved. The fact that the added supply permits paying a lower price for timber purchased within the 60 mile limit might also be a factor.

Many Timber Sellers and Few Buyers

If there are only a few buyers in an area with many timber sellers, the theoretical effect on timber price can vary greatly. At one extreme oligopsony could have the same effect as that already described under monopsony, for the possibility of either collusive oligopsony or development of price leadership (where one buyer sets a low price with all other buyers following) must be recognized. This possibility is tempered by three facts:

1. The actual effect on price will depend on the elasticity of both the demand and supply curves for the timber. With a perfectly elastic supply curve there would be no effect whatever on stumpage prices.
2. Any oligopsonistic buyer, when considering a buying price reduction, would be limited by the knowledge that other buyers, even though operating under similar circumstances, might promptly bid up the price to limit output or to undermine confidence in those from whom they are attempting to buy.
3. Since buying price agreements are illegal, there is little hope of enforcing them, even if they are made.

Whether oligopsonistic buying from competitive sellers has in fact depressed timber prices below those that might have been obtained under competitive conditions cannot be claimed without careful study. In the single such example known to

[4]Public timber sales have other constraints that limit the effect of monopsony. See Chapter 16.

the author (a pulpwood case), it seemed clear that prices were actually higher, not lower, than the level to be expected under purely competitive conditions, apparently because buyers wanted to maintain good supply relations.

Single Seller and Single Buyer

The possible influence of bilateral monopoly on both selling price and volume of standing timber sold can be assessed through Figure 10.5. As before, a rising sawlog supply curve has been used, based primarily on the assumption that a single owner would almost surely tend to increase prices as more and more available timber stands were sold.

A monopsonistic buyer, observing his marginal outlay curve *(MO)* and with the demand situation as shown *(DD)*, would want to purchase Q_{ms} stumpage at a price of P_{ms}. Compared with the outcome under pure competition, this implies on the buyer's part a reduction in quantity purchased and in the price paid for sawlogs—and timber. The extent of the price reduction and, in fact, the degree to which the monopsonistic power can be effective, will depend primarily on the elasticity of the supply curve.

But with bilateral monopoly the buyer is not the only source of market imperfection. To a monopolistic owner the demand curve is his average revenue curve, and to take full advantage of his position he would try to equate *marginal* revenue to his supply curve. In the case illustrated, this would mean selling Q_{mp} at a price of P_{mp}, well above price levels in a competitive market and so reducing the quantity of wood products available to consumers. Because of the relationship between raw material and final product prices, the prices would be increased as long as the amount produced constituted a significant portion of the industry

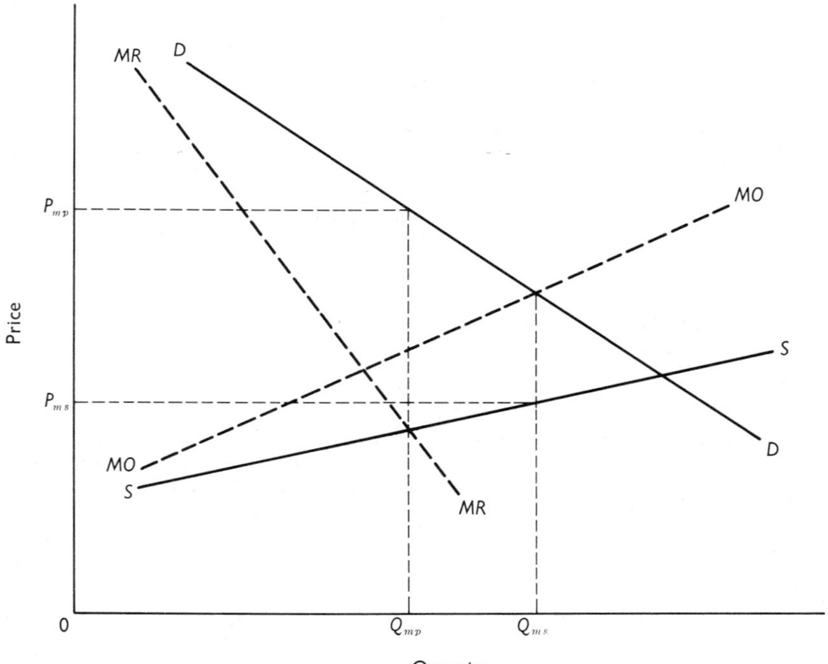

FIGURE 10.5. Price effects of bilateral monopoly.

output.[5] The elasticity of demand for sawlogs in the monopolized area will be significant in determining possible price effects. With a given supply curve, the more inelastic the demand the greater the possible impact on price and resource allocation.

The outcome of bilateral monopoly is indeterminate. Any one of several results is possible, since the monopsonist buyer will strive for a low purchase price and a restricted output, while the monopolist seller will want a high price and also a restricted (but different) output. The final result can fall at either extreme if one party holds a dominating bargaining position; otherwise it can fall anyplace in between. If it happens that the bargaining power is equally divided, then price will probably fall somewhere near the middle and output will approach the competitive level. It is also theoretically possible, but definitely illegal, for the two parties to get together, maximize their combined return, and divide the loot.

Single Stumpage Seller and Few Buyers

The illustration used for bilateral monopoly (Figure 10.5) can also be used for the one seller-few buyers situation, for the limiting case (where the few buyers unite) is that of bilateral monopoly. The possible price effects of the buying group, however, are limited by the elasticity of the supply curve, by legal constraints, and by the individual buyer's knowledge that his or her own actions may influence other buyers in their price offers.

The practical importance of monopoly on U.S. timber prices is difficult to assess. There are areas in this country where virtually all standing timber is owned or controlled by a single individual, firm, or agency. Most large private ownerships are held by wood-using corporations, where monopoly of timber is probably a by-product of the raw material security such ownership affords. Price effects could arise only if other manufacturers attempted to buy in the monopolized area, a possible situation, but not of great importance in the United States. With timberland, monopolistic power is more likely to be directed toward restricting entry into the monopolist's market area and toward maximizing any tax advantages arising from timber ownership.

The prime example of standing timber monopoly in this country is public land holdings. The Forest Service and the Bureau of Land Management provide examples of local public timber monopoly, especially in the West, where large forest areas were withdrawn from the public domain and made part of the national forest system or where railroad land grants were returned to public ownership. In 1977, 62.6% of all softwood sawtimber inventory in the United States was under public ownership. In the three Pacific Coast states over 70% was publicly owned, with 55% in National Forests (USDA For. Ser., 1982).

Constraints on Public Timber Sales

Timber sales from U.S. public agencies are made under policy constraints that would seem to prevent any direct economic exploitation of their monopoly position. With the U.S. Forest Service, two constraints are particularly significant: (1) the allowable cut is determined almost entirely by biological rather than market criteria; and (2) minimum prices are established by appraisal, using costs and returns from operators of average efficiency. Both were discussed in the previous chapter. The first constraint means than the inelastic part of the supply curve from

[5]If the product is sold in a perfectly competitive market, then product prices could not be affected. Here, however, we are discussing market *imperfections*.

public lands (Figure 10.6) is established at an output *(a)* that is unrelated to the economic monopoly position. The second means that the level of the completely elastic part of the curve is independent from, and usually decidedly lower than, the level that would be determined under monopolistic pricing.

This does not mean, however, that the public monopoly of sawlog supplies has no influence on price. Sawlog (and hence stumpage) prices are definitely affected by the quantity offered for sale by the agencies, and if the quantity demanded by mills exceeds that established by the allowable cut policy—if the demand curve for sawlogs intersects the supply curve in its inelastic portion as does D_1D_1 in Figure 10.6—then the appraised price is primarily useful as a starting point for the auction. It has little relevance to the final sale price. The *objective* of the restricted output level may not be that of acquiring monopoly gains, but the *result* so far as price and the availability of wood products to consumers is concerned may be very similar. If one further considers the interaction between stumpage and lumber prices over time, especially when the escalating effect of appraisal is included, there can be little doubt that the public monopoly of sawtimber has and does influence the price of stumpage, of sawlogs, and of lumber. Whether the influence has been beneficial or otherwise is a different question and much depends on the criteria chosen. At this time we are primarily interested in establishing the economic framework and in exploring some of its implications.

STUMPAGE PRICES IN OTHER NATIONS

In developing nations the landholding pattern sometimes leads to monopolistic timber pricing even under conditions of strict bilateral monopoly. In Mexico, for

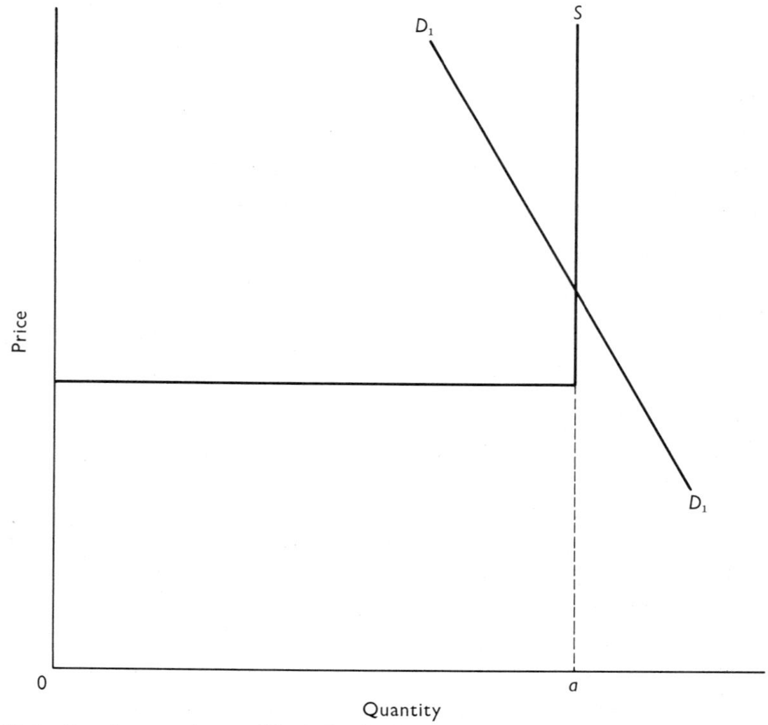

FIGURE 10.6. Supply curve from public timber lands.

example, large areas of forest land have been used to establish new *ejidos*, a form of communal land ownership.[6] Lumber and pulp producers, even though holding an exclusive right to purchase any timber sold in a specified area including one or more *ejidos*, were sometimes forced to close down because of excessive stumpage prices set by the *ejidatarios*.

The more usual case in developing countries is monopsony. To aid the industrialization process governments often grant timber concessions (exclusive rights to harvest timber over a specified area of publicly owned forest). Such concessions are made to individual concerns, either domestic or foreign. If private holdings are intermixed with public lands, the only market for the private timber may be controlled by the concession holder, and monopsonistic pricing may result.

Stumpage pricing, combined with sales procedures, can be a powerful tool to promote efficient timber utilization and to further the desired pattern of industrial development (Gray, 1983).

Pricing methods can be particularly important where public forests are dominant. In past years such countries have frequently sold timber at a flat rate per unit, regardless of access costs and sometimes even ignoring species value differences. Such practices usually result in "creaming" or "high-grading" the forest (cutting only the highly valued timber) and nearly always in taking the easily accessible timber first.

The long-run effects, however, can be even more important. Gradually, as costs of accessibility rise and as species values drop, the firms favored by the unit price rate become unprofitable. Worse, the remaining forest may no longer be attractive to industry unless prices are substantially decreased. Provincial forests of Ontario and Quebec in Canada, where pulpwood was uniformly priced at $2 per 100 cu. ft., offer examples of this, as do those in many developing nations.

Where prices are administered or where they are determined by "negotiation," other problems arise. Prices that favor industrial wood use may work against the well-being of local inhabitants who presently rely on wood for fuel, fodder, or other necessities. Moreover, negotiation between a public official and a corporate buyer may offer opportunities for corruption. The subject of concession pricing alone offers interesting opportunities for research.

EFFECTS OF TAX STRUCTURE

Any tax on timber paid by a stumpage owner necessarily reduces income, consequently tends to lessen the owner's financial interest in timber production, and in the long run must act to increase prices of wood products (Chapter 8). One can scarcely claim that the property tax "discourages" timber production, for timber will grow whether taxed or not. But anything that reduces income from timber production tends to reduce the owner's interest in forest management.

Very often the indirect effect of a tax is greater than the direct effect. A tax levied on wood products, for example, or one tending to shift the supply curve of other productive factors upward (such as a tax on gasoline or diesel fuel that would increase transportation cost) will either depress timber prices, raise lumber prices, or do both. Such a tax is almost always shifted. Although levied on transportation

[6]Of ancient origin, *ejidos* are of great political importance in Mexico. Because most of the land suited to agriculture was granted earlier, forest land has been taken to establish new *ejidos* for the expanding rural population. Unfortunately for Mexican forestry, Mexico's wood-using industries, and the nation, this has resulted in rapid destruction of forests, with erosion and desert the foreseeable end. Mexico cannot afford to waste its resources; a major change in policy must come about if real disaster is to be avoided.

and paid directly by the logging industry, it is ultimately passed along and paid by stumpage owners or by lumber consumers, or both.

Capital Gains

Probably of more concern from the standpoint of stumpage price formation is the possible effect of capital gains treatment of timber. This has always been a delicate subject in corporate circles, and data to prove or disprove various hypotheses is understandably difficult to obtain. Sections 631 (a) and (b) referred especially to timber (see Chapter 8); the effects of these special provisions are still of interest with reference to stumpage prices; even though capital gains are (in 1987–1988) to be treated as ordinary income.

Two possible influences of these provisions on stumpage price need examination. The first applied to firms that owned all or part of their timber supply. It was necessary to make it possible for such firms to profit from both timber and manufacturing operations. A problem arose, however, because profit from timber was taxed as a capital gain (in 1984, for example, at 28%) while that from manufacturing and sale of the product was taxed as ordinary corporate net income (in 1984 at 46%). Thus every net dollar gain from timber yielded 72¢ after-tax profit, while the corresponding return from a manufacturing dollar was only 54¢. A company's timber costs were a matter of record, but if both forest and manufacturing operations were under control of the same unit no actual sale of timber occurred. Clearly, the firm would like to have as much of every total profit dollar attributed to timber as possible. What value should such a company place on its timber harvest? Price guides were available if the company had purchased timber from nearby private growers or sold to others from its own holdings. Purchases from public timber sales could also be used to assist in establishing a price, and consultants were often engaged. To assure that the company's reported transfer price (the "price" paid its timber division by its mill or logging division) was a "fair market value," the IRS required comparative sales data from an outside third source (such as a reputable consultant) to support the timber values submitted.

The second possibility came from wood-based firms who bought timber from either private or public sources. Any such sale, of course, was a matter of record. As the timber was a factor of wood product production, one might presume any profit could come only from product sales and, as such, would be taxed as ordinary profit (at a rate of 46% in 1984). However, Section 631(a) provided (see Chapter 8) that the cutting of timber for sale *or use* by a taxpayer who owned *or had a contract right to cut the timber* should be considered as a sale or exchange, while Section 631(b) gave this same right to any owner who retained an economic interest in the timber, in both cases provided (after 1984) that the timber had been held for over 6 months. In such cases, the increase in value of the timber concerned could be treated as a capital gain. Any increase in timber value shifted profit from manufacturing to profit from holding the timber, and thereby increased the after-tax return from every such dollar to 72¢ instead of 54¢—an increase of $33\frac{1}{3}$%. In effect, such a firm was in the same position as the one with its own producing forest, except the former's timber *cost* was fixed by its purchase price.

The key to the questions raised in both cases lies in determining how value might be placed on corporation-held timber, whether the timber had been owned for a long period of time or for a period less than 6 months. Various methods were used, but the most usual involved transactions evidence. If the company could show that similar forest timber sold for $150 per M the court would often accept the price.[7] Therefore, it would pay the company to bid up the price on a few timber sales—enough to establish the higher price if called on to do so—since for every dollar-per-thousand increase in timber value established, their profit per thousand increased by 33⅓¢. There were severe problems in using evidence from a few purchases to establish a general price rise in stumpage values, especially if those purchases had been made by the company immediately concerned. There is no question, however, that there was a short-term incentive for timber-owning corporations, or for those holding cutting contracts for six months or more, to bid up timber prices.

There was also an important secondary effect. Any general increase in stumpage price was eventually reflected in the price of the final product. And with the usual public land appraisal techniques, higher product prices meant a larger residual was available for stumpage, which tended to raise timber prices further. From this standpoint Sections 631(a) and (b) tended to act as built-in escalators of stumpage price, although it must be remembered that the appraised price was primarily important as a floor whenever the demand for sawlogs was at a level such that the intersection with the supply curve from public timber fell in the inelastic portion.

That stumpage prices were pushed up by the corporate income tax structure, and especially by Sections 631(a) and (b), was never completely demonstrated. In fact, Mead's study of the Douglas fir industry showed quite the opposite (Mead, 1966). Data to prove that bidding practices tended to increase sales prices on either public or private timber were very hard to obtain and were not in the interest of any company to divulge. Executives of large timberland-owning corporations, however, are not naive. Opportunities for increasing the profitability of an operation do not remain unexploited for long. Over time, increased stumpage prices would tend to force up lumber prices, reduce the quantity of lumber sold and, at least until the excess production capacity was eliminated, decrease the opportunity for profit. Whether this would most affect large corporations that were better able to take advantage of capital gains and of scale economies or smaller companies that normally benefit little from capital gains provisions is a point that has not received careful analysis.

EFFECTS OF OTHER FOREST LAND USES

The demand for forest *land* is only partially dependent on the demand for timber-based products: if demands for nontimber products increase, fewer acres will be available for stumpage production. The effect of this reduction on stumpage price could be considerable, but perhaps even more important is the effect on forest management for timber production.

In general, as wilderness areas, national parks, roadside scenic strips, and other nontimber uses remove timberland from the market, reservation prices for timber will shift upward, thus raising stumpage price. It also is possible that reservation prices will move up as a result of owner's anticipations of these other demands. A long-term upward shift in reservation prices, however, would probably be limited by the knowledge that intensified management practices such as use of improved, high-yielding strains, the application of fertilization, and so on, and particularly the widespread establishment of man-made forests, can bring about large increases in timber production per unit of area.

[7]The usual procedure, as with any income tax dispute, would be first to make the return using the $150 timber value. The Internal Revenue Service might accept or dispute the price. If the Internal Revenue Service insisted that $150 was too high, the case would eventually be brought to court, with the court deciding the issue. It is here that transactions evidence became important.

Natural and Planted Forests

Figure 10.7 illustrates the probable short-term relationship of natural to man-made forests. The demand curve *(DD)* is an aggregate of demand for sawlogs, pulpwood, veneer bolts, and so on.[8] Two supply curves are shown, with S_n representing the supply curve for material from natural forests and S_p the supply curve for material from plantations. Both curves are reservation cost schedules. The curve S_p includes the reservation prices of all plantation owners, primarily influenced by their expectations for *timber* prices and rising opportunity costs for land; and S_n includes the reservation prices of owners of all natural forests in the market area. These will be influenced primarily by expectations of demand for forest land to be used for camping, wilderness areas, watershed protection, esthetics, summer home sites, fishing camps, and so on, and also by timber price expectations. Both supply curves are necessarily rising, and may be added together to form the single supply curve, S_t, shown in *a*. The discontinuity in this latter curve arises because man-made forests are not likely to be established unless price expectations are above some minimal level, while at least some natural stands will be available to supply timber at relatively low prices.

The intersection of S_t with *DD* indicates the total quantity of timber that will be taken at the market price, P_t. At this price the amount Q_n will be produced from natural stands and Q_p from plantations. To facilitate the analysis that follows, the same two supply curves are shown in Figure 10.7*b* as a set of "back-to-back" supply curves. Here, at any price above P_x, the total quantity of timber provided will be indicated by the horizontal distance between the two supply curves.

Figure 10.8 is an extension of this same diagram, constructed by incorporating time as the third dimension. At the right of the figure *(a)* are two conventional aggregated short-term demand and supply curves for pulpwood and sawlogs, with the lower one representing the situation as it might appear today (essentially the same as that of the previous figure) and the upper showing the supply-demand

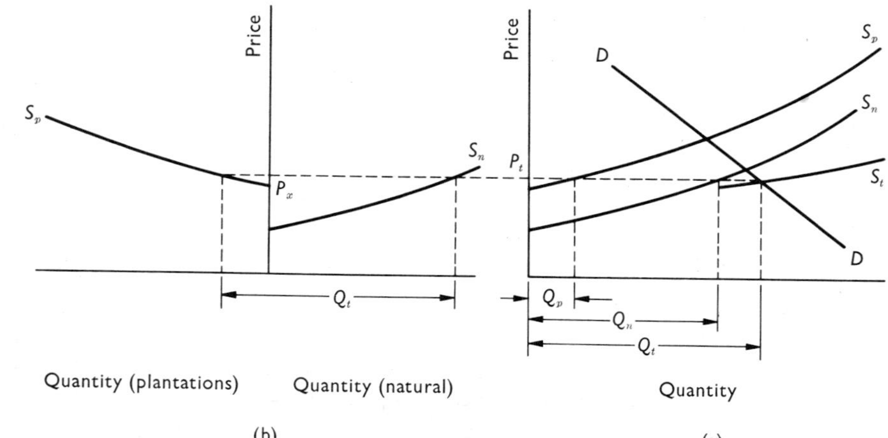

FIGURE 10.7. Short-term relationships between natural and man-made forests.

[8]Demands for raw materials such as pulpwood, sawlogs, and so on may be aggregated in the same fashion as the derived demands for lumber used in shipping were aggregated in Chapter 9. All the demand must pertain to the same kind of material in the same market.

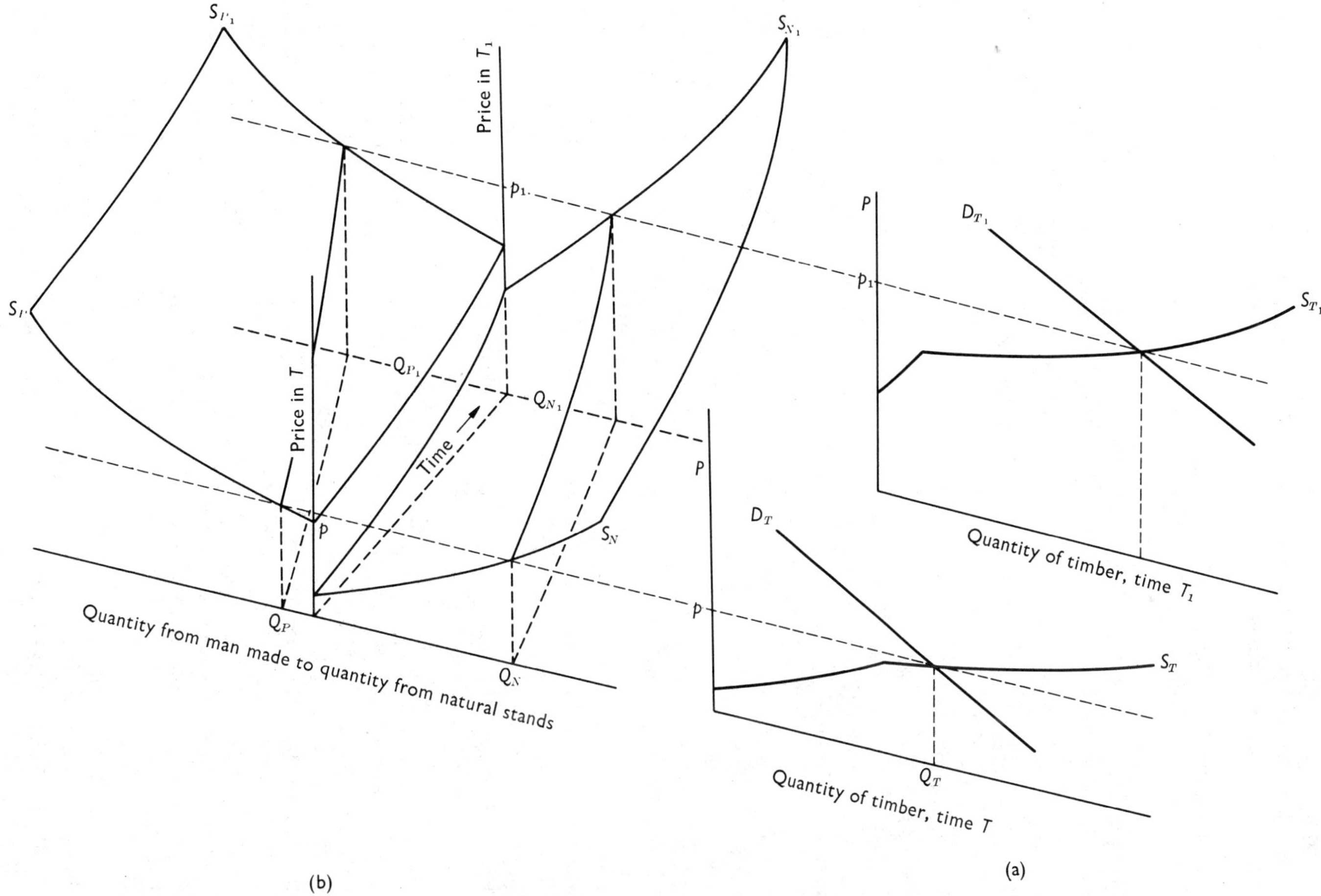

FIGURE 10.8. Three-dimensional representation of the changing relationships over time between natural and man-made forests.

picture at some future period, say, for the year T_1. The supply curves of the two-dimensional figures on the right have been produced by adding together the two timber supply curves—the reservation cost schedules—of both natural and man-made forest areas in a given market. The differing levels and shapes of the two sets of curves indicate possible shifts that might occur as price expectations rise over time.

The same two supply curves are shown individually on the front and back of a three-dimensional diagram (Fig. 10.8*b*), this time using "back-to-back" demand and supply curves as in Figure 10.7, but with time added as a third axis. As the two supply curves shift over time, they produce two supply surfaces—which are related to the demand aspects through price. The total quantity of timber available at any given price level is now the sum of the horizontal distances between the price axis and the two supply surfaces. If price is sufficiently high, then plantation wood will be available and the total quantity will be the horizontal distance between the two surfaces.

Rising Price Expectations

In Figure 10.8, the demand is shifting outward. Timber demand is increasing over time: more and more wood is wanted—by increasing populations, by newly discovered processes or products, and so on. Available forest land, however, and consequently supplies of the timber this land can produce, are expected to decrease over time. If no other response were generated, the quantity of timber available for wood products would decrease substantially, to Q_{N_1}. But there may well be another response. Under the spur of higher price expectations, more capital and other inputs may be devoted to plantation establishment and intensive forest management. Because large increases in productivity are possible from such measures, and because it seems likely that demand for such intensively managed areas by nontimber uses would be relatively low, the reservation cost schedule for these stands should rise less sharply than that for the remaining natural stands. The end result could be as illustrated—a large increase in production from man-made forests, perhaps more than enough to compensate for the loss of production from natural stands. Timber output in T_1 is shown to be substantially increased over that at time T, with well over half the output coming from plantation sources. At the same time, prices are seen to be rising sharply, from p to p_1.

The interaction between the two supply surfaces is probably the most intriguing part of this model. Price expectations of forest land owners result from many factors not connected with timber markets per se. Alternative uses for forest land influence price expectations of forest landowners, not only for the land, but also for the timber on its surface. Yet the price of timber over the long term is subject to very real constraints. As price expectations rise, man-made forests can and will be established, and their production should serve to hold down timber prices. Additionally, management on existing forests will be—or should be—intensified. For if the quantity of stumpage available in the not-too-distant future can be substantially increased through intensive management, *higher cuts can be made on present stands.*

Here again the institutional factor of public land ownership takes on far-reaching significance. If allowable cut policies on public timberlands are determined by applying biological formulas that utilize long rotations; if these formulas take no consideration of changing economic conditions, such as shifting demands; if no cognizance is given to the possible increased output under intensified management; then this constraint to rising prices may be prevented from having any effect.

Under such conditions both stumpage and lumber and veneer prices may rise drastically.[9]

At the same time it must be emphasized that no claim for quantitative exactness or for correctness can be made regarding the relative importance of production from natural and man-made forests for any particular time. The direction of the change, and the influence of the two possible kinds of forest production, however, are clearly indicated. In fact, a strong argument can be made that the model (not the specific curves) is appropriate to forestry in both the United States and Western Europe today. In both areas a clear tendency toward reliance on plantation forestry for timber production is evident and also toward diversion of natural forests to nontimber uses.

Rapid advances in forest genetics and in silvicultural treatments, with the latter including aerial application of fertilizers and pesticides, promise dramatic increases in yields—yields that are not just doubled, but multiplied by three, four, or even more times. Such yield increases could result both directly and indirectly in *lower* unit wood costs—directly because of the increased volumes per acre and indirectly because smaller areas would be needed to supply manufacturing plants, thereby permitting more efficient mill location, lower transport costs, and so on. Additionally, technological changes are enabling industry to use woods of lower quality and smaller sizes as they increasingly take advantage of the "pulverization principle" (Duerr, 1986). In the not-so-long view—perhaps a generation or so—the supply curve for industrial wood may shift out, and down, over time, bringing substantially lower wood supply costs to wood-using industries. Such developments, coupled with the increasing public demand for natural areas, *may make natural forests too valuable to cut.*

QUESTIONS AND PROBLEMS

1. Trace the probable effects of an outward shift in the demand for housing on the demand for sawlogs. Can you say anything about the effect of such a shift on the demand for pulpwood?
2. What would be the likely effect of this shift on the price of stumpage if it is privately owned? What if it is stumpage in a Western national forest? What about stumpage in one of the Eastern national forests?
3. Under what conditions might discriminatory pricing be employed in timber markets? (Think of situations other than those involving export logs!)
4. Why is it believed that price expectations play a significant role in stumpage price levels? List the factors you think might raise price expectations. Are any of these likely to be playing an important role today?
5. What is a marginal outlay curve and why is it important where one buyer (or a very few buyers) provide the only market for timber growers? Why are natural resources, especially timber, peculiarly liable to monopsonistic markets?

[9] This entire sequence—the failure of public forestry officials to recognize economic signals or to pay attention to market prices—is the subject of discussion by many economists. See, for example, the books by Jackson (op. cit.), by Hyde (op. cit.), and several articles by Marion Clawson, among them those in *Science* (1976, 1979).

6. Would you expect sawlog markets, or pulpwood markets, to be more likely under monopsonistic or oligopsonistic influence? Why?

7. What is the likely price result if a single timber buyer has only one owner from whom he or she may buy? Where would you think the advantage might lie if there is only one possible buyer for timber from a particular national forest? Why?

8. What constraints are particularly significant in sales of national forest timber?

9. Are timber concessions made by developing nations to foreign corporations "good" or "bad" for the developing nation? Discuss.

10. Does the present capital gain structure of our income tax have any effect on stumpage prices? Could it have?

11. As one would anticipate, the use of forest lands for wilderness, recreation, roadside scenic strips, national and state parks, and so on, has, and will have, an effect on stumpage prices for both private and public timber. What effect will improved management have? What about the potential effect of genetically improved strains of timber trees? Will timber plantations established today influence current stumpage prices?

PART 3

Economic Guides to Investment Decisions

Using the concepts developed in previous chapters, this part presents economic guidelines for forest investment decision making. After a discussion of management objectives for forest owners, Chapter 11 introduces the concept of an investment project. This is followed by a presentation of discounted cash flow, internal rate of return, and the payout period — three tools often employed by corporations when making financial investment analyses. A pulpwood plantation is used as an example of calculating after tax discounted cash flow.

The necessary background for public investment decision making is presented in Chapter 12. Concepts from modern welfare economics are reviewed and illustrated, including those of preference, utility, indifference curves, and Pareto optimality.

The principle guide to public investment decision making — benefit/cost analysis — is treated in Chapter 13. New concepts basic to an understanding of this important analytic tool — those of willingness to pay and of consumer's surplus — are described in some detail. Problems arising from projects with multiple objectives are also discussed. An example of a public project, however, will be delayed until nontimber uses have been treated.

The final chapter of the part deals with making choices among several available investment opportunities. Here also the problem of determining an optimum harvest rotation — a classic problem of forestry — is treated.

chapter eleven
TIMBER INVESTMENT OBJECTIVES AND CRITERIA

Whether owned by a corporation, a farmer, a school teacher or the general public, if a forest is managed, some investment will be required. Investing in forestry is not confined to buying stock in Weyerhaeuser, Boise Cascade, or Champion International: many alternatives of forest management can, and should, be regarded as investment alternatives. The decision whether to plant or to rely on natural regeneration is an investment decision, as is the choice of rotation, the degree of fire protection desirable, the number and quality of roads, firelines, or campgrounds to build, or the selection of species to favor in a regeneration cut.

Investments are generally made with some, or perhaps several, objectives more or less clearly in mind. Not all objectives are of a financial nature, and the first task of this chapter is to examine objectives for the different kinds of forest landowners. The second task is to define and explain the concept of an investment project. Once these have been accomplished, the criteria for financial investment (or project) selection will be discussed and illustrated, along with factors influencing the choice of interest rates.

OBJECTIVES

Objectives of private forest landowners, especially industrial owners, tend to be less complex than those for public forests. Industrial objectives are seldom stated in purely financial terms, although these are much more easily quantified. Nevertheless, most firms require a careful financial analysis before making investment decisions. Public forest programs must also be appraised in financial terms, for funds must be obtained, spent, and accounted for. These reasons make it desirable to deal first with the objectives and criteria for investment decisions most often attributed to industrial firms. Once these objectives are in hand, criteria used for deciding among projects are more easily understood. Objectives and criteria for nonindustrial private forest owners and for public forestry can then be treated.

Large Industrial Holdings

Forest industries owned 68.8 million acres of commercial forest land of the United States in 1977—about 14% of the U.S. total (U.S.D.A., 1982).[1] The 20 largest forest product companies owned on average enough timber to assure over 40% of their needed long-term wood supply (Clephane, 1978). None of the top 31 paper industry companies owned less than 360,000 acres of timberland (O'Laughlin and Ellefson, 1982b).

Why do they maintain these large timber holdings? The most important reason cited was to maintain a sure raw material supply for their mills, since a papermill must operate continuously at full capacity if efficient production is to be achieved. Other reasons were summarized as follows.

1. Timberland provides insurance against both raw material shortages and short-term price fluctuations.
2. It is a low-risk investment, providing liquidity (ownership decision is readily reversible) and a hedge against inflation (both timber and land hold their value over the long term).
3. Acquisition is becoming increasingly difficult.
4. Timberland affords income taxation advantages under capital gains provisions.

It was also observed that "boards of directors simply like to own a large piece of the Earth."

There are additional strategic reasons for timberland ownership. Such ownership assures a low-cost wood supply; it can be close to production facilities; it permits integration of processing and marketing; the pattern of ownership can yield competitive advantages; and both nationally and regionally it improves a corporation's competitive position (O'Laughlin and Ellefson, 1982b). To some degree, owning forest land would also seem to discourage price increases on the part of private suppliers. It is a definite hedge against potential timber shortages resulting from urbanization or the withdrawal of public forests from timber production.

Given these many objectives for timberland ownership by wood-based corporations, what objectives guide investment in forest management? When 20 companies were asked this question, only six replied that they attempted to maximize revenues from their land use. Five stated their objective as meeting production requirements, while another five stated they tried to maximize productive capability of the land (from Enk, 1975, in O'Laughlin and Ellefson, 1982a). Two had no definite strategy, and the remaining two strove to develop "long-term highest dollar value of land."

But corporate strategy changes with conditions. By 1983 profits in the larger elements of the timber industry had fallen over 50% from the 1979 level and as a result several Pacific Northwest companies had closed or curtailed operations at various mills. Among the larger firms, Diamond International Corporation had been liquidated, and St. Regis Paper Company—one of the older manufacturers—was bought by Champion International, itself a combination of what a few years before had been two strong but independent corporations. More significant to forestry, several firms, including long-established forest landowners such as ITT,

[1]According to O'Laughlin and Ellefson (1982a), International Paper had the largest ownership, but Weyerhaeuser holdings were more valuable. The holdings of either company are equivalent in size to the state of Maryland.

Rayonier, and Crown Zellerbach, were trying to sell part or all of their timber holdings. Some corporations have apparently realized that under certain conditions forest land can serve as a source of needed cash, either through outright sale or as collateral for a needed loan.

Satisficing, and Galbraith's Technostructure

Economists have been critical of profit maximization as an objective, and two have been particularly persuasive. Herbert Simon introduced the idea that firms strive for a "satisfactory" return, not maximum profit (Simon, 1979), while Galbraith claimed decisions in large U.S. corporations are made by the only people who can run a modern firm, the "technostructure." This is composed of the groups, formal and informal, of individuals whose combined knowledge makes modern business possible. Their goal is not profit maximization, but survival and growth.

According to Simon (1979), firms will try to earn a return sufficient to keep stockholders reasonably happy, permit management to handle sales promotion, advertising, increase market shares without too much interference, meet labor requirements for wage adjustment, and so on.

More specifically, Simon points out that any theory of rational choice (on which profit maximization depends) requires knowledge of all alternatives open to the decision-maker, certainly in the evaluation of consequences, and the ability to compare consequences in terms of some "consistent measure of utility." Yet virtually all business decisions are made under conditions of incomplete knowledge and uncertain futures. Hence (he says), decisions are made by a process of "bounded rationality": when making a decision, but lacking omniscience, the decision-maker searches for an alternative with the aspiration of finding one that is good enough—not one that will maximize profits, but one that will suffice under the conditions that prevail. This process of selection Simon called "satisficing."[2]

Galbraith carries this line of thought further (Galbraith, 1967), arguing that one can scarcely rationalize profit maximization as an objective for large corporations since the owners of these corporations, the stockholders, know little about the corporation, and management doesn't know the stockholders. The most basic goal of the technostructure—the decision-making group in a large corporation—is survival, while the second goal is corporate growth. To achieve these the corporation must earn a satisfactory rate of return, but maximizing profit would probably reduce the rate of growth, which is measured in terms of sales. Sales, however, are substantially within control of the large corporation: it establishes its own pricing policy, and (so Galbraith says) to such firms capital has ceased to be a limiting factor of production, for the modern large firm generates its own capital through retained earnings.

A corporation necessarily plans its output and its purchases of raw materials far in advance. This ability to make plans, and the power to enforce them, says Galbraith, are the two most significant corporate facts of life today. The large corporation controls the quantity of products produced, the prices at which these products are sold, and the prices at which raw materials are purchased. Through advertising, consumers are induced to conform to the carefully structured plans of the corporation. This combination of planning and control results in the most efficient production system ever developed. The thoughtful student may question

[2]The best presentation of satisficing is in the lecture Simon gave in 1978 at Stockholm, when awarded the Nobel Prize in Economic Science (Simon, 1979).

the worth of this efficiency if the cost is that of individualism, but the production efficiency per se cannot be seriously doubted.[3]

Neither Simon's satisficing nor Galbraith's industrial state have been fully accepted. In most industries, including those that are forest based, large corporations are not dominant. With the possible exception of some regional prices, few forest-based companies control their product price to any significant degree, and almost all rely on stock or bond issues or on direct borrowing for major capital needs.

Simon's satisficing may be a more accurate description of industrial financial objectives than is profit maximization, and the process of bounded rationalization seems a better fit to decision making. To date, however, satisficing has not been as profitable an assumption for economic analysis as has profit maximization, but it serves to reemphasize the need for interpreting the terms maximize and optimize broadly.

Traditional Forestry Objectives

Two rather special objectives have been devised as peculiarly suited to forestry: (1) maximization of the land (or soil) expectation value and (2) maximization of forest rent. Both of these will be treated in some detail, but since they are most often applied to problems of rotation determination it seems more appropriate to reserve discussion until that somewhat special case is treated in Chapter 14. For now it will be sufficient to observe that the first of these — land rent, or "Le" (for land expectation) — is simply the present value of a periodic income that is repeated for an infinite time period. Forest rent deals only with a total forest property; maximizing forest rent means maximizing the *average* net revenue *per acre* from the total forest property.

Nonindustrial Private Forests

In 1977 some 278 million acres of commercial forest land were privately owned by 7 million individuals or groups with no wood-producing facilities. These are the NIPFs — the Nonindustrial Private Forests. Because they include 58% of the nation's commercial forest — all the private forest land capable of growing 20 cubic feet per acre except that owned by wood-using industries — their biological potential for timber production is large. Nearly 91% of the NIPFs are in the East, where they constitute almost three-fourths of the commercial forest land. In 1977 they produced three-fourths of the nation's hardwood roundwood and sawtimber supplies, and a little over one-third of its softwoods (For. Res. Rep. 23, 1982). About 74% of the NIPF area is in units of over 100 acres, held by more than 600,000 individual owners. Some of these units are quite large: more than 1000 owners hold over 10,000 acres, and nearly 40% of the total area is in units of 500 acres or more held by 63,000 owners. There are also 1.2 million owners holding between 10 and 50 acres each, and 5.5 million owning less than 10 acres. But these 6.7 million owners control only about one-eighth of the total nonindustrial private forest land (Crowell, 1983).[4] More than half of the owners are professionals, teachers, businessmen, clericals — white-collar workers — and less than 30% are farmers. Most are over 50 years old and have held their land for more than 10 years (Sizemore, 1977).

---

[3]The system may not be efficient if viewed from other than production aspects. It is to this latter point that Galbraith addresses his most critical, and caustic, comments.

[4]It seems likely that Crowell's figures include all forest land, while data from Forest Resource Report No. 23 and from Sizemore's paper refer to commercial forest land.

The series of Forest Service timber situation reports beginning in 1920 had predicted an eventual timber shortage and identified the NIPFs as the best prospect for increasing U. S. timber yields. Many of these areas could be highly productive and were located relatively near population centers. But according to the conventional wisdom of the time, the typical NIPF was poorly stocked, overgrazed, overcut, and badly managed. Numerous studies cited reasons for the situation: the long time period required to grow timber ("the typical forest land owner can produce grandchildren more quickly than wood in commercial quantities"), holdings too small to be managed or harvested properly (diseconomies of small scale), owners lacking market information and the technical skills needed for management, prices and rates of return too low, and, in too many cases, owners who seemed uninterested in improving forest holdings (Raup, 1978).

In the mid-1970s a new wave of workshops, conferences, and studies of the NIPFs began.[5] Clawson observed that many of the NIPFs were nonindustrial "simply because no timber processing firm was willing to offer an attractive price to the owner." Moreover, by comparing forest lands of the same region, he showed that the NIPFs compared favorably to nearby industrial forests in fiber productivity and were superior to that of regionally similar National Forests (Clawson, 1977). Management of the NIPFs was defended by several (Vardaman, 1978; Zivunska, 1978; Sedgo and Ostermeier, 1978), who observed that owners of NIPFs might have objectives that gave timber production a low priority. If NIPFs were to make a sizable contribution to the national wood supply, owner objectives would have to become at least supportive of timber production. Raup pointed out that small size could be an advantage: capital carrying charges are a major cost of timber production, but with NIPFs the opportunity cost is only the *differential* in financial yield between forestry and alternative land uses. If a 100-acre forested area will pay 4% on its capital value of $100 per acre, and the alternative use pays 8%, the owner is giving up just $400 a year for his or her 100-acre forest. And the owner might well be willing to pay this for the enjoyment obtained from the ownership, while an owner with a large holding can't afford a 4% loss. From many standpoints, owners of nonindustrial private forests should be considered *consumers*, not *producers* (Raup, 1978).

Objectives of ownership determine how and for what a forest will be managed, and owners of larger areas are much more likely to be interested in timber production. Ownership categories are changing. In 1977 somewhat less than 42% of the NIPF area was owned by farmers, down from almost 60% in 1952. The average size of holding is increasing at the large end of the ownership size scale—there is an evident trend toward consolidation—while the opposite holds for the other end: small ownerships are becoming more fragmented (Yoho, 1985). Studies of ownership objectives are not lacking, but with over 7 million owners no single objective can be identified.[6] Where stumpage markets are active, as in the Southern states, many owners look on their forestland as a "savings account"—a means of accumulating cash for some future need such as college education for their children, a new car, a house, or for expanding their farm area. Others, probably

---

[5]Between 1977 and 1986 at least five major conferences or workshops were held to discuss problems of nonindustrial private forests. The first, at Arlie, Virginia, was reported by Sedjo and Ostermeier (1978). Two were held at Duke University (Royer and Convery (1980) and Royer and Risbrudt (1983), a fourth in St. Louis (U.S.D.A. For. Ser., State and Private, 1983), and the fifth in Michigan (Mich. Dept. Nat. Res. For. Man. Div., 1985). Additionally, a Society of American Foresters Task Force issued its report on "Improving Outputs from Non-Industrial Private Forests" (1979).

[6]For a selected bibliography of ownership studies, see p. 256 in Forest Resource Report No. 23 (1982).

more frequently those in the Northeast or Lake states, still regard their forested area as a source of fuelwood for home use or for sale, and a place for squirrels and birds. Small holdings are frequently residential sites, where amenity values usually preclude timber sales.

Basing his categories on "investment strategies," Yoho identified five groups of NIPF owners (Yoho, 1985):

1. *Custodial investors.* No investment strategy. These are accidental owners of forest lands—inheritors, part owners of a yet-undivided estate, those waiting to sell, and so on.
2. *Sideline investors.* Also accidental owners, these own their forests as part of a farm, or as a residential or recreational property. They pay minimal attention to the forest.
3. *Speculators.* These owners deliberately purchased the forest land they own, but plan no further expenditures. Holding the land for its expected value increase, the owners maintain costs at a minimum.
4. *Hobby investors.* Typically these owners live on retirement incomes and the forest is their major investment. Ownerships in this group will be quite stable.
5. *True investors.* These are the most important owners from a timber supply standpoint, reacting to changing prices, and other conditions. They may be building an estate, or striving to increase or maintain current cash flows, or both. Forest land lends itself to such purposes: tree growth is tax free to the present owner (it will be taxed when passed to heirs), tax credits are available, and gains from timber sales are treated as capital gains.

Since investment was interpreted as having timber as the product, the classification is also useful as a guide to prospective timber output. In general, the size of forest holding increases from the smallest (custodial investors) to the largest (true investors). Likelihood of timber production also increases with size of holding. If social benefits rather than financial return had been the measure, a somewhat different order would probably have resulted.

Federal and state assistance programs for the NIPFs have also been designed with increased timber production as the objective. These programs are of three major types: **direct fiscal** (primarily tax programs), **nonfiscal** (subsidy programs), and **indirect** programs (technical assistance, information, and education) (Skok and Gregersen, 1975). Tax advantages granted timber have been discussed in previous chapters and are illustrated in the example at the end of this chapter.

The nonfiscal programs began during the dust bowl period in 1936. The first was the Agricultural Conservation Program (ACP) and from a forestry standpoint it concentrated on tree planting and timber stand improvement. In recent years this program, now the Rural Environment Assistance Program (REAP), has provided some cost-sharing funds for forest practices. It was essentially replaced in 1973 by FIP, the Forest Incentives Program, which made cost-sharing funds directly available for forestry practices. The primary goal of FIP is timber production, protection, and production of nontimber forest outputs. Although the pros and cons of the program have been vigorously debated, it would appear that when combined with vigorous parallel state programs of technical assistance, education, and cost sharing, the output of timber and other forest products has been increased at a modest public expenditure. If choices must be made, technical assistance is likely to bring the better result. The key to real success may lie in effective coordination of the several agencies that must work together and in carefully targeting owners who would be receptive. Most studies of such owners indicate that areas under 100 acres are unlikely to be cost effective in use of cost-sharing funds. Of the several states having successful programs, those of Georgia (Cubbage et al., 1985) and Virginia (Hall and Starr, 1985) are excellent examples.[7]

Virtually all the studies of NIPFs have taken timber volume or timber value as the output measure. Almost all have assumed that maximizing either the present value of financial returns, or the internal rate of return, would be the financial objective of owners if land area were sufficient—probably over 50 acres. Smaller areas were generally assumed to have other objectives.

Small Corporations

Little more need be said about holdings of small or medium-sized corporations. While theoretically smaller corporations may be considered deathless, in practice they often seem to operate on rather short term horizons. Their objectives appear similar to those of any small business concern and subject to much the same set of influences. Maximization of net present value, with land value established by its opportunity cost to the concern in question, would seem an appropriate financial objective.

THE PROJECT CONCEPT

One of the important concepts of this chapter is that of a project. This term will be used to indicate any planned investment undertaking or proposal that can reasonably be analyzed or evaluated as an independent unit.[8] A project can consist of (for example) planting a 300,000 acre pulpwood forest over a 20-year period, plus building the pulp and paper plant needed to convert the wood, constructing the road system and acquiring the equipment for harvesting the wood, and so on. Or it might consist only of thinning one acre in a farm woodlot. The criterion is not size or cost, but rather a positive answer to the question "Does it make sense to evaluate this as an independent undertaking?"

This definition does at times raise questions. Is a sawmill independent of its timber supply? Not completely, but it often makes sense to build a sawmill to use timber purchased from other owners, as most U.S. sawmills do today. Is a timber producer independent of his market? Not completely, but many timber producers are able to sell their product to independently owned mills. Should road construction be part of a large plantation project? Here the question is simply "Can the project be completed and used without road construction activities?" Clearly, any project planned for a "new" area—one in which there have been no forestry activities—will necessarily include some items unneeded in places where forestry operations are common.

The project approach has the overriding advantage of providing a general framework for making forestry decisions that have economic content or implications—and this covers most forestry decisions. Moreover, the guidelines are applicable to both private and public forestry decisions, particularly those of management, permitting ready comparison of decisions on public forests with those made for private forest management, and thereby offering a possibility of identifying the reasons for existing differences.

---

[7]Sources for this paragraph include the following: Wallace and Silver (1983), Boyd (1983), de Steiguer (1984), Mills and Cain (1976), Michigan Dept. of Natural Resources Forest Management Division (1985).

[8]Little and Mirrlees use a similar definition in their book, *Project Appraisal and Planning for Developing Countries* (1974).

A complete project report for larger projects is often called a feasibility study or feasibility report. This chapter deals primarily with financial and economic analysis, but one should recognize that this is only part of such a study. A complete feasibility report will usually include, in addition to the economic aspects:

1. Any technical details needed to justify statements made or implied about project inputs and outputs.
2. Assumptions made about the economy in general. For example: What will be the growth rate of GNP? Will inflation be increasing, decreasing, or steady? What about employment? Will population changes be of significance?
3. Any special risks to the project that should be known in advance.
4. A discussion of managerial needs.
5. An indication of the alternatives considered and why they were rejected. This applies especially to alternative sites, technologies, and scale.
6. Any changes—political, legislative, or technical—that will be needed to make the project successful, or that might interfere with the project's success.
7. A discussion of markets for the project production: what sort of market analysis was made, the period covered, probable competitors, possibility of price competition, etc.
8. The expected source of financing.
9. Potential tax changes.
10. Environmental impacts, if any.

There is one other important qualification: no matter how large or small the project may be, the worth of the project analysis will depend directly on the accuracy of the technical information, the human and managerial techniques to be applied, and the accuracy of the economic data used. Each is worth examining in a forestry context.

Technical Information

Forestry projects usually require substantial amounts of biological data. To analyze a thinning project, for example, requires that the analyst have information concerning the increase in growth that will follow the thinning, the quantity of wood to be removed, the time before the harvest cut (or next thinning), the quality of the wood to be removed and of the wood remaining and that will be added, whether or not any mortality will result in the residual stand—in fact, all the silvicultural data.

Human and Managerial Information

This is difficult to assess, but important. As an example, think of a 10-year planting program proposed by a State Forestry Department. Will the planting crews be properly trained, the equipment well maintained and efficiently operated, the planting stock properly selected, graded, and handled? Will the project manager properly schedule procurement and planting of the stock, the clearing of the planting site, and so on? Will supervisors manage their crews so the work will be done well and on time? Probably the most common error in this category is in the matter of estimating the time needed to complete particular jobs—especially those involving construction.

Economic Information

This is usually the direct responsibility of the analyst, but unfortunately this fact does little to solve the problem. Material costs and product prices can be determined for the current period, but what about prices for things needed next year—or 10 years from now? Will diesel fuel be more, or less, expensive in five years? Will labor costs be rising? What about workers' compensation? And how does one estimate the level of income taxes, property taxes, or interest rates over the coming years?

This is where sensitivity analysis plays an important role. A sensitivity analysis systematically tests how the earnings of a project change when actual events differ from the estimates made about them. Because prices, wages, taxes, and interest rates can and do change over time, the possible effect of these changes should be evaluated. Modern computers make it relatively easy to test different prices, interest rates, and so on, and see how project earnings are thereby changed. In this way the sensitivity of the project to changing factor values is determined. Sensitivity analysis differs from risk analysis, which requires the assignment of statistical probabilities to the likelihood of specific levels of change in the project elements. Both types of analyses *aid* the decision-maker: they do not make the decision.

Project Decisions and Objectives

Some projects can be formulated and evaluated on the back of an envelope; a few calculations made by someone completely familiar with the possible problems may be all that is needed. In such cases, it is clear that most decisions will be made by the project designer. At the other extreme are the large projects—building a pulp mill or an integrated manufacturing plant, or establishing a plantation in a developing nation to provide foreign exchange through exporting either raw materials or finished products, or designing and building a facility to produce a new product such as oriented strand board. Such projects may require several years of work by a sizable team of experts. Although decisions made by team members during project design have much to do with its success or failure, the final decision of whether to invest or not will probably be made by the owner, the forest supervisor, the Chairman of the Board, Congress—someone bearing responsibility for the final outcome. Any suitable guidelines must be formulated with the objectives of this final decision-maker in mind.

FINANCIAL ANALYSIS

Of the various types of financial analysis developed over the years, three are of particular significance at this point: discounted cash flow, or DCF; internal rate of return, or IRR; and the payout period. The first two—discounted cash flow and the internal rate of return—are related, but different. In most cases, both criteria will be used by the decision-maker.

Discounted Cash Flow

This procedure is well suited to the analysis of almost all forest investments. Even for public projects, where decisions may be based on social or political reasons, a financial analysis of some type should be made, for if a project is to be implemented it must be financed and the funds obtained from some source. A DCF analysis lets the agency's financial officer know how much must be on hand to meet project expenditures and when the money must be there.

A discounted cash flow analysis requires four major steps:

1. All expected inputs and outputs must be specified quantitatively.
2. Each input and output must be scheduled. That is, the time of making each input or output must be specified.
3. A value must be placed on each input and output.
4. Finally, these data are combined into some measure of profitability.

This means that *all payments received from the sale of project outputs for each year* during the life of the project must be quantitatively estimated. This includes the sale of anything left at the end of the project—the salvage value of all equipment, sale value of any buildings, and (in forestry, of special importance) the value of any land and any timber remaining at the end of the project. It also means that all payments made for goods or services used by the project, and the year during which the payment is made, must be determined. This includes all capital costs whether for initial equipment or replacement, and all current costs such as property taxes, insurance, fire protection charges, and so on.

A discounted cash flow analysis differs from usual accounting practices in several ways. First, it is based on cash received and paid out, not the values of goods and services produced and received; hence the name: cash flow. Second, normal accounting practice usually shows financial liabilities with respect to interest and taxes, not payments. This can cause big differences in timing. Finally, normal accounting practice includes an allowance for depreciation and obsolescence of capital. With DCF there are no such allowances, but the replacement cost of any item is included when the expenditure is made. As will be seen, this can somewhat complicate the picture when calculating profit after tax, since tax law treats depreciation differently.

Discounting

The effect of time on both costs and revenues is (as indicated by its name) accounted for by discounting. The choice of the discount rate is of great importance and will be considered at some length after the mechanics of the analysis are well in hand. In most cases, however, the rate (or rates) to be used are given to the analyst.

The discounting procedure itself is simple: the sum of all costs and of all revenues for each period (normally a year, but not necessarily so) is discounted to its present value. In most cases a single discount rate is used, although any anticipated rate changes can be taken into account with little difficulty. For example, suppose the discount rate over a five-year period was expected to change in the following sequence:

Year 1	8%
Year 2	8
Year 3	10
Year 4	9
Year 5	8

The appropriate discount for each period would be:

$$\text{Year } 1 = 1/(1+.08) = 0.9259$$

$$\text{Year } 2 = 1/(1+.08)\,(1+.08) = 0.85734$$

$$\text{Year } 3 = 1/(1+.08)\,(1+.08)\,(1+.10) = 0.7794$$

$$\text{Year } 4 = 1/(1+.08)\,(1+.08)\,(1+.10)\,(1+.09) = 0.71504$$

$$\text{Year } 5 = 1/(1+.08)\,(1+.08)\,(1+.10)\,(1+.09)\,(1+.08) = 0.66208$$

Once the sums of all costs and all revenues have been calculated for each year (period) and appropriately discounted, the total discounted cost (the sum over all years) is subtracted from the total discounted revenue to give a single figure—the discounted net cash flow stemming from the project.

As indicated in Chapter 7, both the level of the costs and revenues and their timing are important, along with the discount rate itself. In fact, it is probably the prominence given to the timing of both revenues and expenditures that has made this procedure so popular. As a simple illustration, consider the following.

A logging superintendent has been asked to review the road system in a newly purchased timber tract. Convinced that a substantial rebuilding program is needed, he prepares a prospective schedule of construction costs, together with an estimate of the savings that would result if his plan is accepted (Table 11.1) and presents these at the staff meeting. "With savings fully $460,000 above costs, this is clearly a profitable undertaking" concludes the superintendent.

The company accountant, however, has been busy with his calculator. "How come you have a saving of 300,000 in year five if it's only 250,000 for the remaining years?" he asks. The logging superintendent replies, "I figure we can get about 50,000 for the used equipment we'll have left—so that's been added to the 250,000 saved in year five, making 300,000 in that year." "Fine" says the accountant, "but I'm afraid you're forgetting that these costs and savings come at different times. Our company figures every investment should earn at least 12%. If my calculations are correct, the *discounted* figures would show that the present value of your estimated annual costs is $1,076,130, while the cash income has a present value of only $970,400. In my book this is not a good investment! I suggest we think about this and take it up next week."

Before the next meeting the economist asks the road supervisor if a somewhat slower construction schedule wouldn't work just as well. "Sure—but it will cost more" he is told. Together they work out a new set of estimates for the staff meeting (Table 11.2).

In their presentation they point out that, due to the stretch-out of construction work, the program's cost is now $275,000 higher than the earlier estimate, while the savings go up only $20,000. The project, however, now earns a profit with a present value of $21,440 as well as paying a full 12% return on all costs. As expected, the new project is approved.

The two graphs of Figure 11.1 help one to see why these changes occur. Clearly, if costs can be deferred to a later date they become less significant, while moving incomes forward also makes a project more desirable.

TABLE 11.1
Road Building Costs and Resultant Savings (in thousands of dollars)

Year	Cost	Savings
0	150	20
1	300	40
2	400	100
3	400	150
4	50	200
5	10	300
6	10	250
7	10	250
8	10	250
9	10	250
	$1350	$1810

TABLE 11.2
Costs and Savings of Road Plan (in thousands of dollars), with NPV at Several Rates

	Original		Stretch-Out	
Year	Cost	Savings	Cost	Savings
0	150.00	20.00	75.00	50.00
1	300.00	40.00	150.00	80.00
2	400.00	100.00	180.00	120.00
3	400.00	150.00	180.00	180.00
4	50.00	200.00	180.00	190.00
5	10.00	300.00	180.00	210.00
6	10.00	250.00	180.00	230.00
7	10.00	250.00	180.00	240.00
8	10.00	250.00	160.00	280.00
9	10.00	250.00	160.00	250.00
Total Cost	1350.00	1810.00	1625.00	1830.00
NPV				
8%	1154.35	1176.57	1150.85	1213.74
10	1113.87	1066.64	1066.54	1106.94
12	1076.13	970.40	992.01	1013.45
14	1040.86	885.83	925.87	931.30
16	1007.85	811.25	866.96	858.85
15	1024.09	847.39	895.57	893.95
14.8	1027.40	854.88	901.50	901.24

In this regard it is important to recognize that the level of the discount rate can make a big difference. For example, if a discount rate of 8% had been used in the road construction project the results would change significantly. Both project plans would be profitable in that they would earn at least 8%, although the stretch-out plan is more advantageous, earning nearly \$64,000 profit as compared to only \$22,220 for the first. Bear in mind, however, that taxes, depreciation, and depletion have not yet been brought into the picture. This example only introduces the idea of a DCF analysis and gives enough of an understanding to permit making comparisons between DCF analysis and other approaches.

Internal Rate of Return

A project's internal rate of return (or IRR) is the rate earned on all project costs by the anticipated revenues. It has several names: the profitability index, the project rate of return, the ROI or return on investment, but is probably best known as the IRR. This particular rate will make the sum of all discounted project costs exactly equal to the sum of all discounted project revenues, which means it is the discount rate that makes the net present value of a project equal zero. Programs for computing internal rates of return are available at almost any computing center, and

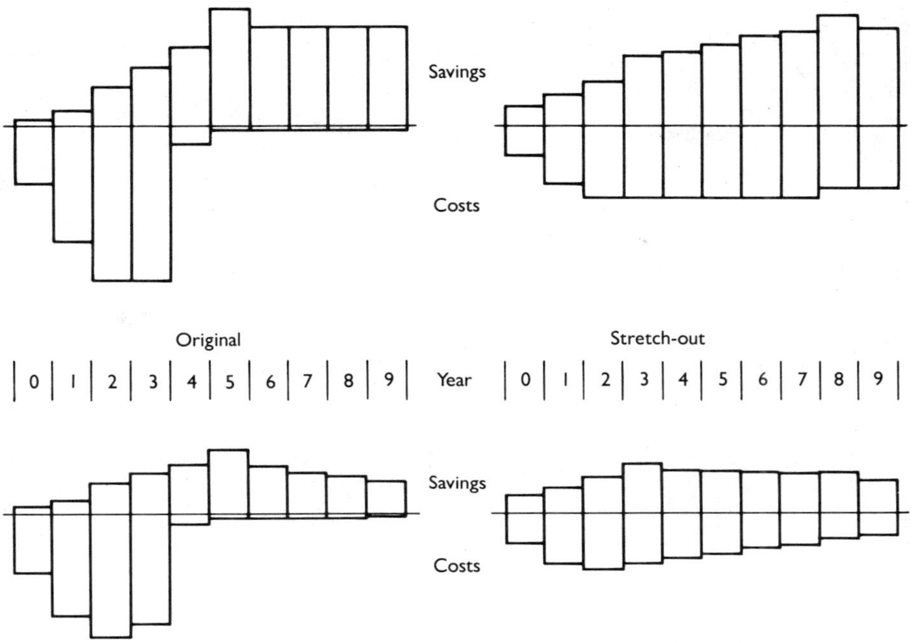

FIGURE 11.1. Costs and savings of original and stretch-out plans. The upper graph shows timing of costs and savings. The lower graph shows present values of costs and savings at discount rate of 12%.

many of the electronic business calculators are preprogrammed for this calculation.[9]

A convenient method for approximating the IRR is a simple graphic procedure that also has the advantage of making obvious the nature of an internal rate of return. Using the road construction project as an example, and the net present value of the project at 12% (1013.45 − 992.01 − 21.44), we might next calculate the NPV using (say) a 15% discount rate, which yields a value of 893.95 − 895.57, or −1.62. Now, plotting present value over discount rates (Figure 11.2), we can extend the line connecting the two points (*PV* at 12% and at 15%) so it crosses the *PV* = 0 line. If we wish, we could take the interest rate thus identified—the rate that would apparently make the NPV = 0—and calculate the NPV at this discount rate. If the calculation yields a positive NPV we could try using a somewhat higher rate; if the result is negative we could lower the rate. Successive trials such as this could be repeated until the desired degree of accuracy has been attained—in our case, approximately 14.8%.

[9]One formula for estimating IRR, based on the equality of discounted costs and revenues, is

$$\sum_{t=0}^{n}\left[C_t\frac{1}{(1+i)}^t\right]=\sum_{t=0}^{n}\left[R_t\frac{1}{(1+i)}^t\right]$$

where C = Cost, R = Revenue, i = internal rate of return, n = length of the project in years, and t = the time, or year, in which cost or revenue is incurred. By using alternative values for i (much as in the graphic method given above), the value of i that makes the two sides of the equation equal (the true IRR) can be estimated as precisely as desired.

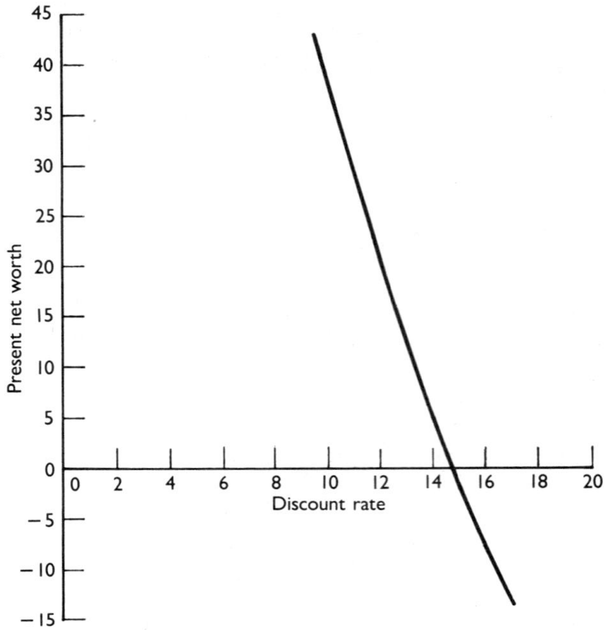

FIGURE 11.2. Estimating the internal rate of return for the road-building project.

In calculations of this type one should remember that nearly always the costs and revenues are estimates, often with substantial error margins. Under these conditions, calculation of an IRR much beyond a single percentage point becomes quite unnecessary and may create an illusion of precision.

From the standpoint of project evaluation the internal rate of return is in a somewhat different category from other criteria. It is a discount rate, but completely independent from market rates, alternative rates, and so on. Its value depends entirely on the amount and timing of the project's costs and returns. In fact, this is one of its greatest advantages, for its use permits making an analysis without having to decide just what interest rate is appropriate—an advantage that will be more obvious after the discussion of how an appropriate interest rate is determined.

So far as project criteria are concerned, the IRR does just what it purports to do—it tells one the rate of return earned by investment in this specific project. It does not, however, always provide an infallible guide to the best one of two or more projects—the subject of Chapter 14.

Other Project Criteria

Payout Period

A decade or so ago the "payout" (or payback) period was used by many companies as an investment guide: if the investment could be completely recovered in (say) five years, then the investment would be made, but not if it would take longer. Today only some of the smaller companies use this sort of guide, especially those with weak financial resources, in which case the required payout period might well be as little as two years.

The major drawback to such a guide is that it tends to remove from consideration many projects that would be far more profitable. Use of the payout period is particularly damaging when applied to decisions regarding new, or replacement, equipment, since it tends to keep obsolete or inefficient machinery and methods in use.[10] In the road-building project cited earlier, the project as developed by the logging superintendent had a payout period of a little over 7 years, while the stretched-out version's payout period was 8 years.

The above does not mean that the length of the payout period is unimportant. Certainly, if choosing between two projects that are essentially similar in all other respects, one that would pay out in three years and the other in six, the one with the shorter payout period would be selected. Flexibility is much increased if cash is not tied up.

INTEREST RATES FOR FOREST INVESTMENTS

Of the criteria mentioned for private industrial projects, maximizing the discounted cash flow is favored by most economists, but the IRR is probably most used by business executives. Use of the discounted cash flow—the NPV criteria—requires that someone choose the appropriate interest rate. General principles underlying market rates of interest were discussed in Chapter 7; now a closer look is needed to see if there is some way of determining the rate that might be applicable to particular problems and owners.

Impact of Inflation

The unprecedented inflationary period triggered by the energy crisis of 1973 changed traditional interest rate patterns and, combined with changing expectations regarding economic stability, seriously eroded investor confidence. Prior to that time an inflation rate of 4% was considered disastrously high; one might hear of such rates in less developed countries, but certainly not in the economically mature, industrialized nations that had acquired the economic know-how permitting them to control such aberrations. So much for conventional wisdom! When the United States, along with most of the industrialized world (except Switzerland and, for a while, West Germany) began experiencing double digit inflation, previous investment guides were often worse than useless. Conservative investors who had kept their savings in an account paying the (then) usual 4 to 4½% suddenly discovered that a 6 or 7% annual inflation rate meant their real income was dropping and their savings accounts were rapidly losing value.

As inflation rates continued to rise, interest rates took on a completely new meaning. Corporations, and many individuals, began to factor inflation into their investment decisions, usually by adding their expectation of inflation to whatever rate of return they felt necessary. Interest rates responded as one would anticipate: in 1982 the prime rate soared above 20%, while inflation was running at an annual rate of over 12%.

Forestry was hit in several ways—by the effect of high rates on construction

[10]Techniques of calculating when or if a particular piece of equipment should be retired or replaced can become quite involved, since depreciation methods, potential for more efficiency from new machinery (or personnel), or for declining efficiency from use of old, tax considerations, and so on, all can have an influence on the result. See Grant, et al., 1976, Chapter 17.

(see Chapter 3), by the uncertainty inflation induced everywhere, and of course by the knowledge that the relatively long-term investments often needed in forestry simply could not produce returns of 18 to 24%.[11]

Inflation was not a new phenomenon to economic historians, even though it was a new experience to most U.S. citizens of the 1970s. The question to economists was not "Will it stop?" but instead "How long will it continue?" By the end of 1984 the annual inflation rate had dropped to less than 5% and the rate for the entire year of 1986 was only 1.17—the lowest annual inflation rate in 25 years. Investors, however, tend to have long memories, especially of events that have proved costly. Through 1985 and into 1986 expectations of future interest rates still reflected fear of a possible return to high inflation—a fear bolstered by high federal expenditures and record budget deficits. As a result, the mid-1980s witnessed a strange reversal: long-term interest rates were higher than short-term ones. Even in early 1986, if one subtracted the current rate of inflation from the rate on short-term U.S. Treasury notes, the result would be a "real" rate of 7% to 8%—at least three or four points higher than anything experienced in the twentieth century prior to 1970. In early 1987, however, the *Wall Street Journal* quoted the rate on 30-day treasury bills at 5.23%, 30 year notes (due 2016) were at 7½%, the prime rate 7½%, and worries about inflation and the national debt, although still alive, were much reduced. This recent experience with high and fluctuating interest rates makes a discussion of rates appropriate to either private or public investment in forestry considerably less specific than desirable, but it in no way reduces the importance of discount rates to forest management decisions.

There is another problem with interest rates; asked to suggest an interest rate applicable to forestry enterprises, for example, most business executives would probably say, "The market rate, of course." But which market rate? At any given time there are a whole host of them. And even if one were to specify the borrower, his or her purpose, the size of the loan, its duration, and the method of repayment, the interest rate to be charged would vary with when and where the loan was to be made (because the supply and demand of investment funds change with time and place).

Earlier forestry practice—especially in textbooks—was to select a rate that was (1) sufficiently low to assure demonstrating that "good" forestry would pay, and (2) sufficiently high not to be rejected immediately by students primarily familiar with interest rates paid on saving accounts—generally rates within the 4 to 7% range. The 1979–1984 period changed this drastically. Seventeen percent mortgages and widespread acceptance of the ubiquitous credit card with its 18% charge for unpaid balances made almost everyone aware that higher rates were all too possible. It has also made the task of selecting a rate for long-term forestry projects much more difficult.

One must recognize at the outset that ownership necessarily plays an important role. Public and private opportunities should certainly be separated and at least the small woodland owner from the large corporation. In the following discussion, keep in mind that these are the extremes, selected for the purpose of illustrating the range of applicable rates and the consequences of their use. Not all woodland owners, or all corporations, or even all public agencies, behave alike.

[11]As a matter of fact, few, if any, long-term investments can be expected to return such rates. Generally long-term securities (15 years and over) are issued at lower rates than medium- and short-term securities, and are often callable—which means they can be redeemed either at some specified time or whenever the issuer feels it is to his or her advantage, which usually means whenever funds can be borrowed at a significantly lower rate.

The answers are frequently intertwined with the market structure of the organization, the age and income of the owner, or the nature and objective of the agency in question.

Nonindustrial Private Owners

A direct approach to interest rate levels applicable to particular borrowers is that of determining the rate regularly paid by that group for loans. For example, loans are sought by farmers for a variety of purposes, with one of the most frequent being for financing new equipment. Farm equipment loans, like those for automobiles, are often handled directly by the manufacturer through a subsidiary corporation. Bank loans for such purchases can be arranged, with rates depending on the individual and on the collateral put up for the loan (most often the equipment itself). The rate charged a farmer or other nonindustrial private forest owner buying an automobile or similar equipment would be no less than the usual bank rate (in 1987 around 8%) and one would normally expect it to be substantially higher.

With loans to individuals, time preference plays an important role in determining an acceptable rate. As discussed in Chapter 7, time preference rates are generally correlated inversely with income—low-income groups have high time preferences and high income groups the reverse. Farmers probably have a relatively high time preference—especially in 1987, when many farmers are facing foreclosure forced by loans obtained, often for equipment, when rates were much higher. One would expect a farm woodlot owner to use a high discount rate—probably at least 12% or more—if he or she could be interested at all in a forest investment. Yet this would not necessarily rule out timber production: both the rate and the amount of the cost are important in time problems, and a considerable part of the necessary investment in farm woodlot operations is in terms of labor and nonspecialized equipment. The opportunity costs of labor and equipment must be used in analyzing farm projects, and in many instances these costs will be low.

Compared to farmers, nonfarm, nonindustrial owners hold more forest land, probably have a much wider income range, higher incomes, and perhaps easier access to bank loans—the last because they are more likely to have collateral familiar to bankers. These owners, however, would probably face higher costs—hired labor and hired equipment, plus higher supervision charges. Home mortgage rates might indicate something of the rate appropriate to these nonfarmer, nonindustrial private forest landowners, and in early 1987 these were quoted at about 9%.[12]

One would expect to find relatively few owners of small, understocked woodlots, or woodlots with a high proportion of low grade timber species, that could be interested in timberstand improvement investments aimed at sawtimber production. On the other hand, suggestions to establish plantations of fast-growing pulpwood on areas that would interfere little with other land uses might find ready acceptance, especially if these activities are subsidized through free planting stock, soil conservation payments, and other such measures.

The Large Corporation

As with any investor, alternative rates of return are significant to the large corporation. Because the capital requirements of large corporations are met to a consider-

[12]Federal Home Loan Mortgage Corporation, conventional fixed rate 30-year mortgage, quoted in the *Wall Street Journal*, January 23, 1987.

able extent from retained earnings, the guiding alternatives are those available within the corporation itself. The desire for corporate growth takes second place only to the desire for long-range security, and most large corporations today are simultaneously pushing expansion plans on several fronts. In a few cases their alternatives lie within the range offered by a single product type—paper, or chemicals, or automobiles—but most of the large conglomerate corporations produce an amazing array of products. This, together with the expansion of the U.S. economy, has resulted in average rates of return to industry over the past three decades of from 10 to over 20% *after taxes*. The significance of the range of investment opportunities to the interest rate appropriate to large timber-based industries is direct. If, for example, a new oriented strand board plant or particle-board plant can earn, say, 20% on the required investment, or a new paper machine could earn a similar rate of return, investments yielding lesser rates will be taken only if other reasons offset the difference.

Timber-based industries are not an exception to the general run of U.S. corporations. In 1984—a year of recovery for most industries but a poor year for forest products—the median return on stockholder's equity for the *paper, fiber, and wood products* group of the Fortune 500 was 12.4%,[13] while the median for all 500 largest firms was 13.6%. The 15 largest wood-product firms (all those among the 200 largest U.S. industrial corporations for 1984) included only 2 firms heavily involved in producing products other than those derived from wood. Over the ten year period 1974–1984 the median total return to investors[14] for the paper, fiber, and wood products group was 19.56%, while the median for all 500 firms was 18.15%

From the above figures it would seem that alternative rates of return, after taxes, for large timber-producing corporations in the United States must be no less than 15, and in some cases, probably 25%. This would mean a conservative estimate of returns on invested capital of around 20% *before* taxes. If this strikes the reader as being too high, he or she could contrast these rates with those expected from investment in developing countries. Here the *minimum* anticipated return must normally be well over 20% if the investment is to be given serious consideration—and that with a payout period that may be no more than 5 years!

The opportunities for timber growing in the United States that will return 15% after taxes on invested capital are, at best, somewhat limited under current cost and price conditions, especially if returns are calculated according to the usual textbook examples. A 15% interest charge on costs would make many plantations appear decidedly unprofitable, yet U.S. corporations have made large investments in timber lands, plantations, and various forest management activities, and are heavily committed to major timber investment programs. The obvious discrepancy between the nominal rate of return from forest plantations and an alternative rate of return on capital of 15% merits some explanation.

The preceding discussion of modern corporate objectives suggests that the three prime objectives of management in large corporations are:

1. Maintenance of a satisfactory rate of dividend payment.
2. Steady corporate growth as measured by sales.

---

[13]From *The Fortune Directory 1985: The 500 Largest U.S. Industrial Corporations*. Stockholder's equity is the sum of capital stock, surplus, and retained earnings at the company's year end.

[14]Total return to investors includes both price appreciation and dividend yield on stock and assumes reinvestment of all dividends including stock splits and dividends.

3. The avoidance of foreseeable risk of either a product price collapse or a raw material price rise.

Clearly, paper machines could not earn from 15 to 20% on their investment costs (and therefore dividend payments could not be made at a satisfactory rate) unless fed with pulp. Steady corporate growth cannot be anticipated unless new supplies of raw material can be foreseen. The surest means of preventing a disastrous price rise in wood supplies comes through owning or otherwise controlling a sufficiently high proportion of the required wood to prevent any combination of producers from restricting the needed supply.

What rate of return does a modern corporation really earn from its forests? The rate earned depends on the value these objectives are accorded by management. Their attainment may not be assured by forest ownership, but it is certainly enhanced. If investments are made in forest plantations when the alternative rate of return on available capital is over 15%, one can be sure that management believes the forest investment will pay no less!

Cost of Capital, or COC

For a discounted cash flow analysis the company should use its opportunity cost of investment capital—its real cost of capital—but this is difficult to quantify. Some use a weighted average cost of capital calculated by estimating the cost of capital obtained by issuing stock (equity capital) and the cost of borrowing (debt), and weighting these by the percentage of each currently outstanding (Grant et al., 1976). Others have used the price of the company's common stock or its P/E ratio (the price/earnings ratio), sometimes averaged over time, or adjusted in another way. All these methods should include the cost of retained earnings—the source of a substantial part of most corporate investments. These items, and their treatment, vary from company to company. Some of them, such as stock prices and price/earnings ratios, change from day to day.

Results from any of these approaches are highly dependent on accounting procedures used by the company, how it maintains its book value (the historic cost of items), and how tax records are maintained: after-tax earnings should be used if earnings enter the calculations. And since the company is trying to pin down its opportunity cost of investment capital, it should also take into account the alternative investments available.

From our standpoint, the thing to hold in mind is that the discount rate appropriate to any investor is the investor's opportunity cost of the needed capital, and this is seldom either an easy or precise calculation.

We will be returning to discounted cash flow and internal rates of return later in this section. Before looking more deeply into public production and its economic background—the subject of Chapter 12—it seems highly desirable to work through a somewhat more complete example of financial analysis in the context of a proposal for a pulpwood plantation. Only the financial analysis will be given, and it will be compressed, but the treatment and effects of depletion and the income tax are illustrated.

EXAMPLE: A Proposed Plantation

A Southern-based investment company has purchased a distressed and run-down cattle farm of 2000 acres for $200,000. One alternative the company decides to investigate is that of planting the area with loblolly pine and producing pulpwood for sale to a nearby paper mill. As part of the feasibility study a financial project analysis is to be made.

The company engages the services of a forestry consultant who prepares an estimate of potential yields under his proposed management regime. Anticipated yields are given in Figure 11.3 while the needed numerical data are as follows.

Age of Stand (years)	Volume before Thinning (cords)
10	8.6
15	(5.3 cds removed)
20	34.0
22	(7 cds removed)
25	46.0
30	60.5

The first thinning, to be made at age 15 after the growing season, will remove 5.3 cords of pulpwood. A second thinning of 7 cords will be made at age 22. A 30-year rotation is anticipated, with pulpwood the only product. Site preparation and planting are to be done by a forestry and logging contractor with whom the

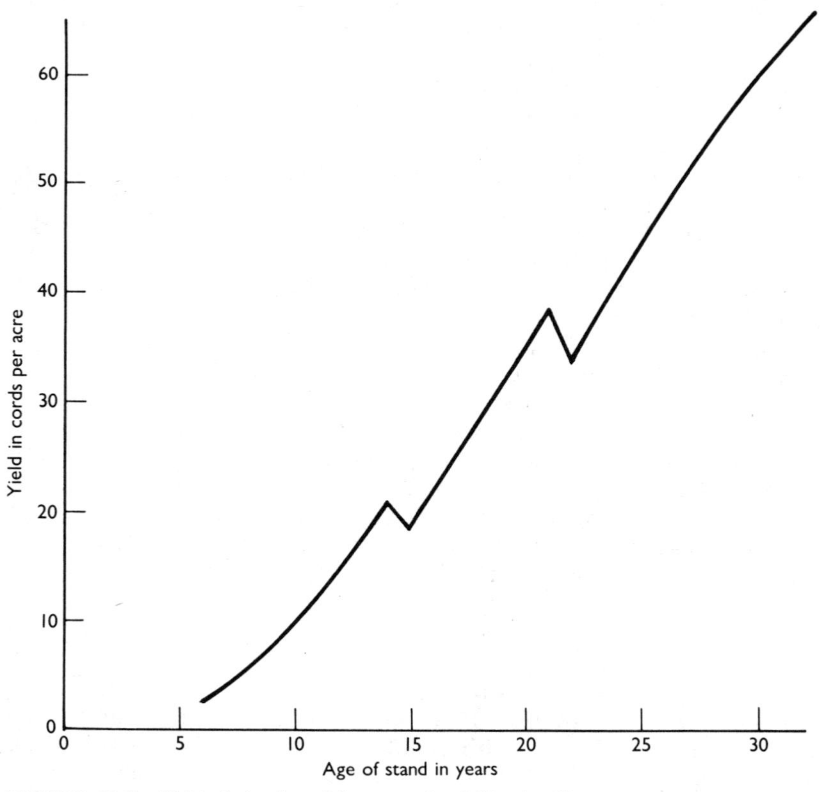

FIGURE 11.3. Yield of stand used for example of Chapter 11.

company has had previous experience. Current costs per acre are estimated to be as follows.

Site preparation and planting	60.00
Annual costs of administration and maintenance, including fire prevention and suppression and a retaining fee for consulting	1.50
Property tax (also an annual cost)	.80
Herbicide treatment in year 7 (*Note*: this treatment is eligible for cost sharing with Dept. of Agriculture, which contributes 50%)	15.00
Sale preparation: first thinning	4.00
second thinning	4.00
final harvest	2.00
Broadcast burn at age 25	2.00
Broadcast burn at age 28	2.00

Company economists report that real costs and property taxes are not expected to rise over the time periods involved. In fact, they expect that costs (other than taxes) may drop because of improving technology, but recommend using current cost levels as a conservative measure. The economists also recommend that an 8% discount rate be used for this operation since interest rates have dropped below 9% and are still going down.

Revenues

Prices for pulpwood stumpage have been rising slowly in real terms. Company economists estimate that real stumpage prices from the area (within 10 miles of the mill) will continue to rise at about 1% per year. The following stumpage price estimates therefore include the expected increases of 1% annually from now through the next 30 years.

Pulpwood stumpage per cord, first thinning current price $13.00	$15.09
Pulpwood stumpage, second thinning current price $16.00	19.92
Pulpwood stumpage, harvest (30 years) current price, $19.00 per cord	25.61

Differentials in *current* pulpwood prices arise from two causes: (1) larger trees have more usable volume per stem, and (2) harvesting costs per cord are lower with increasing tree size.

Taxes

Present per-acre property taxes of 80¢ per acre per year are assumed to remain constant. Assumptions about income taxes are much less certain: in 1986 a 20% rate was proposed for capital gains, but the existing rate was 28% and before that was 30%. (The Tax Reform Bill of 1986 set the maximum tax on such gains at 28%.) Tax revisions would close most of the presently existing loopholes and exceptions, even to the extent of abolishing capital gains (all such gains would be treated as regular income). If capital gains were to be abolished, income from timber sales could be taxed at anything from 17 to 50%, depending on the final outcome of the tax revision.

For this analysis it was decided to assume that capital gains would continue to be allowed, and at a 25% tax rate. For the ordinary income tax the recommendation was to use a 45% tax rate. No severance tax or yield tax was in force at the time of the analysis, nor was one expected.

Calculation of After-Tax Present Net Worth of Project

Under expected conditions the most important taxes will be those on property and on incomes. This means the accounting procedures used must permit keeping costs and revenues in a way that allows both capital gains and ordinary income to be properly calculated, and for plantations this normally requires at least three separate accounts: (1) a timber account, maintained in both volume and value terms (which permits calculation of timber depletion); (2) a land account, in which the cost of the land and any nondepreciable land improvements are recorded; and (3) an equipment account, so that the proper depreciation schedules can be maintained. Additionally, the regular ongoing expenses must be recorded.

Before December 31, 1979, all planting and site preparation costs had to be carried as capital costs and could not be recovered until the timber was harvested. From then through 1986, up to $10,000 could be amortized over an 84-month period, and 10% taken as a tax credit (see Chapter 8), and this allowance will be accepted in the present example. (This tax credit was abolished for 1987, 1988.) Accordingly, $1000 can be taken as a tax credit, and $10,000 − 1000/2, or $9500, amortized over an 84-month period. This results in the following amortization schedule:

Year 0	$9500/14 = $679
Years 1 through 6	$9500/7 = $1357
Year 7	the remainder, or $679

The amortization is shown in Table 11.3 on line 18. The remaining part of the reforestation costs must be capitalized—carried ahead until some harvest or exchange occurs.

Taxes, administrative costs, charges for fire protection, and other costs can either be treated as deductible expenses or as capital costs, depending on how one elects. If other income is being received, it is usually best to expense such items, which will be done in the example. No equipment account is needed since all work will be contracted.

The remaining part of the reforestation costs (first year, 120,000 − 679 = 119,321) is entered in the timber capitalization account (line 25 in Table 11.3) to be reduced by the amount amortized each year (and increased if additional eligible costs are incurred). These charges are accumulated in the timber account (line 25) and the depletion allowance eventually calculated. This depletion, similar to the depreciation allowed on (for example) a building or a piece of equipment, is made in recognition of the fact that substantial costs may have been incurred before the timber is sold and the owner is entitled to recover these costs.

There is one other complication: planting costs have been incurred for all the trees in the plantation, but different trees in the stand may be sold at different times. In the present case, part of the timber will be sold at age 15, another part at age 22, and the final sale at age 30. The depletion balance that has accrued to the stand at any time must be allocated over the entire stand, and only a proportion of the total depletion allowance can be used for any intermediate harvest. The depletion account for a plantation is begun at the time of planting, and by law must go to zero with the final harvest.

To determine the allowable depletion, the cost basis of the timber to be cut must be established. This normally includes the original purchase price of the timber (not including the land), site preparation, planting costs, and other items of the timber producing process—with one significant exception: interest can only be included as a cost if it has actually been paid.

The capitalized costs are accumulated in the timber account until some income is realized. In our example the first harvest is the thinning at age 15, and at that time the timber account includes the following:

Site preparation and planting	\$120,000
Less amortized amount	− 9,500
Herbicide treatment in year 7 50% of \$15 × 2,000	+15,000
Total timber account costs to end of year 15	\$125,500

Sale preparation costs, along with any carrying charges during the year of harvest, will be deducted from the revenues received from the sale.

Depletion Unit

The total of these capitalized costs is next divided by the timber volume in the merchantable timber account (the amount on January 1 of the year concerned before the cut), giving the cost per unit of volume (the depletion unit referred to in Chapter 8). In the present instance,

$$\$125{,}500/48{,}000 = \$2.615 \text{ per cord}$$

Multiplying this depletion unit by the quantity being removed gives the depletion allowed for the current harvest,

$$\$2.615 \times 10{,}600 = \$27{,}715$$

which is subtracted from the revenue received along with the other cost items in the timber account. The remainder (line 36 in Table 11.3) is the income subject to taxation—in this case as a capital gain.

10,600 cords at \$15.09 per cord	\$159,954
Less sale preparation costs at \$4/acre	− 8,000
Less expenses in current year	− 4,600
Less depletion	−27,715
Taxable as a capital gain	119,639

The remainder of the capitalized costs—those *not* allocated to the current harvest—must be carried forward to the next harvest period.

Capital Gains Tax

At a 25% rate, the capital gains tax on this amount would be

$$\$119{,}639 \times .25 = \$29{,}910$$

Subtracting this tax from the taxable income received during the year (total income minus operating charges, depletion, and sale preparation costs) gives the after-tax net income from the first thinning:

$$\$119{,}639 - 29{,}910 = \$89{,}729$$

TABLE 11.3
Computer Printout for DCF Analysis of Plantation Problem

I	A	B	C	D	E
1	Year	0	1	2	3
2					
3	Timber (cost) acc't				
4	Site prep	40000			
5	Planting	80000			
6	Herbicide				
7	Fertilizer				
8	Adsshnistration	3000	3000	3000	3000
9	Property tax	1600	1600	1600	1600
10	Harvest prep				
11	Burning				
12	Cost cash flow	124600	4600	4600	4600
13					
14	Total expenditure	124600	4600	4600	4600
15	Reforestation exp.	120000			
16	Operating costs	4600	4600	4600	4600
17	Capitalized exp.	120000	0	0	0
18	Amortization	679	1357	1357	1357
19					
20	Timber (vol) account				
21	Volume, cords				
22	Volume harvested				
23	Depletion unit				
24	Depletion allowed				
25	Remaining timb. acct.	119321	117964	116607	115250
26					
27	Revenues				
28	Cost share payment				
29	Stumpage sales				
30	Total revenue	0	0	0	0
31	Deductable expenses				
32	Harvest prep				
33	Operating expenses				
34	Depletion allowed				
35	Total deductions				
36	Taxable gain	0	0	0	0
37	Capital gains tax	0	0	0	0
38	After-tax net income				
39	After-tax positive cash flow				
40					
41	Tax saving				
42	Oper. exp	4600	4600	4600	4600
43	Amortization	679	1357	1357	1357
44	Inv tax credit	1000			
45	Total tax saving	3376	2681	2681	2681
46					
47	ATCF incld tax saving	3376	2681	2681	2681
48	Land value	-200000			
49	Total net ATCF	-321224	-1919	-1919	-1919
50					
51	Discounted cash flow at 6% including land values				209804
52	Discounted cash flow at 8% including land values				-12404
53	Discounted cash flow at 10% including land values				-140357

F II	G II	H II	I II	J II	K II	L II
4	5	6	7	8	9	10
			15000			
3000	3000	3000	3000	3000	3000	3000
1600	1600	1600	1600	1600	1600	1600
4600	4600	4600	19600	4600	4600	4600
4600	4600	4600	19600	4600	4600	4600
4600	4600	4600	4600	4600	4600	4600
0	0	0	15000	0	0	0
1357	1357	1357	679	0	0	
	3000	5000	8000	11200	15000	19200
113893	112536	111179	125500	125500	125500	125500
			15000			
0	0	0	15000	0	0	0
			4600			
0	0	0	10400	0	0	0
0	0	0		0	0	0
			5720			
			5720			
4600	4600	4600	4600	4600	4600	4600
1357	1357	1357	679	0	0	0
2681	2681	2681	2376	2070	2070	2070
2681	2681	2681	8096	2070	2070	2070
-1919	-1919	-1919	-11504	-2530	2530	-2530

TABLE 11.3 *(continued)*
Computer Printout for DCF Analysis of Plantation Problem

		M II	N II	O II	P II
1	Year	11	12	13	14
2					
3	Timber (cost) acc't				
4	Site prep				
5	Planting				
6	Herbicide				
7	Fertilizer				
8	Adsshnistration	3000	3000	3000	3000
9	Property tax	1600	1600	1600	1600
10	Harvest prep				
11	Burning				
12	Cost cash flow	4600	4600	4600	4600
13					
14	Total expenditure	4600	4600	4600	4600
15	Reforestation exp.				
16	Operating costs	4600	4600	4600	4600
17	Capitalized exp.	0	0	0	0
18	Amortization				
19					
20	Timber (vol) account				
21	Volume, cords	24200	29600	35400	41600
22	Volume harvested				
23	Depletion unit				
24	Depletion allowed				
25	Remaining timb. acct.	125500	125500	125500	125500
26					
27	Revenues				
28	Cost share payment				
29	Stumpage sales				
30	Total revenue	0	0	0	0
31	Deductable expenses				
32	Harvest prep				
33	Operating expenses				
34	Depletion allowed				
35	Total deductions				
36	Taxable gain	0	0	0	0
37	Capital gains tax	0	0	0	0
38	After-tax net income				
39	After-tax positive cash flow				
40					
41	Tax saving				
42	Oper. exp	4600	4600	4600	4600
43	Amortization				
44	Inv tax credit				
45	Total tax saving	2070	2070	2070	2070
46					
47	ATCF incld tax saving	2070	2070	2070	2070
48	Land value				
49	Total net ATCF	-2530	-2530	-2530	-2530
50					
51	Discounted cash flow at 6% including land values				
52	Discounted cash flow at 8% including land values				
53	Discounted cash flow at 10% including land values				

Q II	R II	S II	T II	U II	V II	W II
15	16	17	18	19	20	21
3000	3000	3000	3000	3000	3000	3000
1600	1600	1600	1600	1600	1600	1600
8000						
	4600	4600	4600	4600	4600	4600
12600	4600	4600	4600	4600	.4600	4600
12600	4600	4600	4600	4600	4600	4600
0	0	0	0	0	0	0
48000	44400	50600	57000	63600	70200	76600
10600						
2.61						
27715						
97785	97785	97785	97785	97785	97785	97785
159954						
159954	0	0	0	0	0	0
8000						
4600						
27715						
40315						
119639	0	0	0	0	0	0
29910	0	0	0	0	0	0
89730						
117444						
12600	4600	4600	4600	4600	4600	4600
5670	2070	2070	2070	2070	2070	2070
123114	2070	2070	2070	2070	2070	2070
123114	-2530	-2530	-2530	-2530	2530	-2530

TABLE 11.3 *(continued)*
Computer Printout for DCF Analysis of Plantation Problem

		X II	Y II	Z II	AA II
1	Year	22	23	24	25
2					
3	Timber (cost) acc't				
4	Site prep				
5	Planting				
6	Herbicide				
7	Fertilizer				
8	Adsshnistration	3000	3000	3000	3000
9	Property tax	1600	1600	1600	1600
10	Harvest prep	8000			
11	Burning				4000
12	Cost cash flow		4600	4600	4600
13					
14	Total expenditure	12600	4600	4600	8600
15	Reforestation exp.				
16	Operating costs	4600	4600	4600	8600
17	Capitalized exp.	0	0	0	4000
18	Amortization				
19					
20	Timber (vol) account				
21	Volume, cords	90600	75400	82800	90000
22	Volume harvested	14000			
23	Depletion unit	1.28			
24	Depletion allowed	17872			
25	Remaining timb. acct.	79913	79913	79913	83913
26					
27	Revenues				
28	Cost share payment				
29	Stumpage sales	278880			
30	Total revenue	278880	0	0	0
31	Deductable expenses				
32	Harvest prep	8000			
33	Operating expenses	4600			
34	Depletion allowed	17872			
35	Total deductions	30472			
36	Taxable gain	248408	0	0	0
37	Capital gains tax	62102	0	0	0
38	After-tax net income	186306			
39	After-tax positive cash flow		204178		
40					
41	Tax saving				
42	Oper. exp	4600	4600	4600	8600
43	Amortization				
44	Inv tax credit				
45	Total tax saving	2070	2070	2070	3870
46					
47	ATCF incld tax saving	206248	2070	2070	3870
48	Land value				
49	Total net ATCF	206248	-2530	-2530	-730
50					
51	Discounted cash flow at 6% including land values				
52	Discounted cash flow at 8% including land values				
53	Discounted cash flow at 10% including land values				

AB II	AC II	AD II	AE II	AF I
26	27	28	29	30
3000	3000	3000	3000	3000
1600	1600	1600	1600	1600
				4000
		4000		
4600	4600	4600	4600	
4600	4600	8600	4600	8600
4600	4600	8600	4600	4600
0	0	4000	0	0
96800	103400	109800	115200	121000
				121000
				87913
83913	83913	87913	87913	
				3098810
0	0	0	0	3098810
				4000
				4600
				87913
				96513
0	0	0	0	3002297
0	0	0	0	750574
				2251722
				2339636
4600	4600	8600	4600	4600
2070	2070	3870	2070	2070
2070	2070	3870	2070	2341706
Land value after capital gain tax				284465
-2530	-2530	-730	-2530	2626171

and to determine the after-tax cash flow, or ATCF, requires only that the depletion allowed be added to this after tax net income.

$$\text{ATFC} = \$89{,}729 + 27{,}715 = \$117{,}444$$

Calculating After-tax Income from Second Thinning

Adjusting the Depletion Unit

As earlier, before the tax can be calculated, the allowable depletion must be determined. This involves (a) determining the amount remaining in the depletion account after the most recent harvest, and (b) adding to this the additional costs incurred during the period between the first and second harvest. Because operating costs are being subtracted from other income and there are no additional costs:

Remaining in timber account after previous cut	
$125,500 − 27,715 = $97,785	$97,785
Additional costs during 6 years	—
Total allowable depletion	$97,785

Volume per acre on January 1 in year of second thinning (age 22) is 76,000 cords, so the depletion unit (line 23) will be

$$\$97{,}785/76{,}600 = \$1.277$$

and the depletion allowed will be $1.277 × 14,000 = $17,872

The net revenue from the thinning will be

14,000 cds × $19.92 per cd. = $278,880	$278,800
Depletion allowance	−17,872
Operating expenses	− 4,600
Harvesting preparation	− 8,000
Income subject to capital gains tax	$248,408

The capital gains tax due in year 22 will be

$$.25\ (\$248{,}408) = \$62{,}102$$

The after-tax net income (line 38) will be

$$\$248{,}408 - \$62{,}102 = \$186{,}306$$

and the ATCF

$$\$186{,}306 + \$17{,}872 = \$204{,}178$$

Calculating After-tax Income from Final Harvest

Since the entire stand is to be harvested, the remainder of the depletion account will be charged against the revenue.

Revenue at final harvest:	
121,000 cds × $25.61 = $3,098,810	$3,098,810
Remaining in timber account (includes 2 burns)	87,913
Operating expenses	4,600
Harvest preparation	4,000
Taxable capital gain	$3,002,297

The tax on capital gain will be

$$\$3,002,297 \times .25 = \$750,574$$

and the after-tax net income is

$$\$3,002,297 - 750,574 = \$2,251,723$$

while the after-tax cash flow will be

$$\$2,251,723 + \$87,913 = \$2,339,635$$

Tax Savings

Providing that income from sources other than this project is being received, the plantation offers three opportunities for tax savings, all of which should be credited to the project. These three are: (1) operating costs, since these can be charged against other income; (2) the \$10,000 to be amortized—which also reduces income; and (3) the tax credit of 10% of the first \$10,000 of reforestation cost. The first two yield a tax saving of the ordinary income tax amount, which in our case is 45% of taxable income. The third saving—the tax credit—is subtracted directly from the tax itself, so is a \$1000 saving. These savings are shown in lines 42, 43, and 44, and their total on line 45. Some of these savings occur each year and are added to the after-tax positive cash flow (line 39) to determine the total ATCF from the project (line 47).

The land account for the project keeps track of payments made for the land itself and for all depreciable and nondepreciable land improvements. Depreciable costs include bridges, fences, and nonpermanent land improvements; most are depreciable over a 15-year life span under the accelerated cost recovery system. Nondepreciable land improvements are those of a permanent nature: leveling, clearing, grading, and so on. Costs for such improvements can only be recovered at the sale of the property.

In the example used here no additional costs were involved, so the original cost of the land (\$200,000) and its final value less the capital gain tax (net value of \$234,465) are shown on line 48. Land prices have been rising for many years, and increased at a particularly rapid rate during the 1950–1982 period. In the example, land prices are estimated to continue rising, but at a much reduced rate of 1.5% annually. This would make the land worth \$156.31 per acre (or \$312,620) at the end of the 30 years. On this amount the capital gain tax would be

$$(\$312,620 - \$200,000).25 = \$28,155$$

and the estimated after-tax value (the salvage value) of the land would be \$284,465 (line 48). The original land cost is shown as negative and is added directly to the positive cash flow as a calculating convenience rather than including it in the cost stream (Table 11.3).

Calculation of Discounted Cash Flow

With the costs and after-tax cash flows at hand, calculation of the discounted cash flow offers no difficulty. In the present example the after-tax discounted cash flow from the plantation at an 8% discount rate is −\$12,404, which means that after allowing for a return of 8% on all expenditures, including the original purchase price of the land, there would be a loss of \$12,404 from the project.

One would expect almost any corporation today to request the analyst to indicate the sensitivity of the calculated cash flows to changes in several variables

—probably in the discount rate, the tax rate, and certainly in the timber yields and the product prices. Additionally, they would want to know the internal rate of return from the project. All these items are easily calculated; one needs only to change a data entry on the computer setup and the new figures appear. (See problem 17 at the end of the chapter and the computer printout of Table 11.3.) In the present case, for example, with all data as given above including land at \$100 per acre, the IRR is around 7.9%. Note that in some cases the figures in Table 11.3 do not precisely match those in the text; the text figures were calculated with two-decimal accuracy while the computer carried many more.

Depreciation

Since the corporation is hiring an outside firm to do all the planting and harvesting, there are no equipment or buildings to depreciate. Depreciation, however, has essentially the same purpose as depletion: it is simply a way of permitting a firm or individual to recover the cost of items purchased for the ordinary conduct of the particular operation.

Like depletion, depreciation serves to reduce the "taxable income," and so reduces either the regular income tax or the capital gains tax, depending on whether the income is long or short term.

QUESTIONS AND PROBLEMS

1. The objective of maximizing the net present value of an investment undertaking is usually taken as a rational financial objective for a corporation, either large or small. And it is almost universally agreed that this maximization is to be taken in a general sense rather than as a precise mathematical concept. Simon, however, claims that it is not maximization, but satisficing that is the objective of larger corporations. What does he mean by this term, and in what ways does it differ from the more traditional one of maximizing profit or maximizing net present value?
2. Galbraith carries the idea even further. What are the major characteristics of his criticism of profit maximization? What is your opinion regarding his approach?
3. Forestry has developed a couple of objectives of its own: soil rent (or soil expectation) and forest rent. Define each of these and show how they differ from maximization of net present value—if they do differ.
4. Are any of these objectives appropriate to farm owners of forests? How about owners of private nonindustrial forests?
5. As used in this book (as throughout much of the business world, government agencies, and by economists), the word project has a specific meaning. Carefully define this term.
6. What is a feasibility study (or feasibility report)?
7. A project analysis is heavily dependent on technical information, on the human and managerial techniques to be used, and on the accuracy and availability of economic data. What does this mean with regard to forestry projects?
8. Of the various types of financial analyses developed over the years, three are of particular significance today: discounted cash flow, internal rate of return, and the payout period. Describe and discuss the relative merits and drawbacks of each.

9. What are the four major steps in a DCF analysis, and what does each imply with respect to a forestry project?

10. Define the IRR, or internal rate of return.

11. As an analyst you have prepared a table of expected cash flows over the life of a proposed five-year project, given below. Determine the net present value of the project using three different discount rates: 5, 10, and 15%. Also calculate the internal rate of return. What else would you need to know before making a definite recommendation?

Year	Cost	Return
0	800,000	50,000
1	400,000	300,000
2	200,000	400,000
3	100,000	600,000
4	100,000	900,000

12. What interest rate would you recommend that a farmer use for calculating the net discounted cash flow from a forest plantation? What would you want to know about the farmer?

13. What interest rate would you recommend to a company such as Georgia-Pacific? If you were working for them, how would you know what rate to use?

14. What interest rate would you recommend that a nonindustrial owner of 2000 acres of bare land in Southeastern United States use in a discounted cash flow analysis and what would you want to know about her?

15. If you were an economic consultant to forest companies, some large, some small, and were asked to do a financial analysis of a forestry project covering 20 years, what would you do about taxes? How would you handle prices of both products and necessary inputs?

16. What is depletion, and how is it handled in a discounted cash flow analysis?

17. Using the information given in the plantation example, determine the after-tax discounted cash flow from the project if prices were to rise an additional 50%. What if yields were to be 25% less than planned?

chapter twelve

PUBLIC PRODUCTION AND WELFARE

Until now the discussion of timber and timber products has been based on the assumption that production is both responsive to individuals' desires for goods (consumer sovereignty) and privately controlled. In the United States, however, a significant portion of the forest land is under public ownership and management, and a substantial part of the forest products produced are not sold to individuals through organized markets. Although essentially identical methods are used to evaluate both public and private forest investments, most people would agree that purely financial objectives are not suited to public management. But if this is so, then what should these objectives be? If there are no organized markets for some products, does it necessarily follow that these goods should be produced in the public sector? And if a product is not priced through market transactions, what can be said about the quantity that should be produced? These are questions that need answers—or at least discussion—before public forest investment decisions can meaningfully be studied. In this chapter we will take a closer look at these related problems, first by reviewing some of the contributions of welfare economics, then by looking directly at public production.

WELFARE ECONOMICS

Production and price theory are parts of what economists call positive economics, which deals with understanding and predicting economic behavior. Welfare economics is normative economics. It deals with problems of the total well-being, or welfare, of individuals and of society. While not specifying what ought to be, a major emphasis of welfare economics is on how society might move closer toward its specified economic objectives. As Schumpeter has commented, the purpose of positive economics is to explain, while the purpose of welfare economics is to prescribe (Schumpeter, 1954).

Few would claim that the framework employed in modern welfare economics is completely satisfactory. In fact, some critics believe the major contribution of welfare economists has been to prove there is no way to deal quantitatively with welfare problems. But in many cases welfare economics can provide partial answers to important questions of production and resource allocation and, as Scitovsky wrote more than 30 years ago, "to provide partial answers to vital problems is at least as important as it is to provide complete answers to lesser questions" (Scitovsky, 1951).

Our look at welfare economics will be brief.[1] However, we will examine with reasonable care those aspects that are basic to understanding the analytic framework of public investment decision making. The beginning point will be the concept of "utility," from which we will go directly to individual preference systems, Pareto optimality, the compensation principle, and social welfare functions. A discussion of public production will conclude the chapter, but welfare economics will be encountered again in Chapter 13, for it is the basis of benefit/cost analysis.

Utility

"Utility" is the term used in welfare economics to mean happiness, or satisfaction, and an individual's welfare is represented by his or her utility level. The prime difficulty of welfare economics stems from the fact that utility levels cannot be observed directly. This isn't as severe a drawback as one might at first expect, for the basic theories of how consumers react to price or income changes can all be derived from an ordinal system — one set up by *ranking* alternatives. When dealing with public choices, however, it would be very helpful if one could say (for example) that an afternoon spent duck hunting was enjoyed twice as much as an afternoon on the beach. It would be still more helpful if one could then measure the utility of duck hunting and use these same units to measure the utility of other goods and services. In order to make social investment decisions, however, one would also require some method for weighting the units so they could reflect *social*, rather than individual, preferences. But this, as we will see, is wishful thinking.

Individual Preference Systems

While nearly everyone rejects materialism as a guide to a full life ("man does not live by bread alone") we usually accept the general notion that some aggregate of material things roughly indicates a person's wealth, and that wealth is at least a partial indicator of welfare. However, this approach raises serious problems when attempting to compare the "well-offness" of people with different tastes, cultures, or living standards or those living in different time periods. In fact, a direct interpersonal comparison of welfare (or utility) is not possible even between two individuals living in the same community at the same time, since no person can predict the criteria used by another to judge welfare or utility levels, or precisely interpret the degree of satisfaction experienced by another.

Not only do people differ, but so do the options and choices open to them. Most of our theories regarding preference and decision making in economic matters assume that individual consumers are rational beings — rational in the sense that they will make decisions in a manner they believe will result in the most benefit (or least harm) to themselves. Rationality, however, requires that an individual's preferences be consistent (if I rank A, B, and C in that order today I will similarly rank them another time), and transitive (if I prefer A to B and B to C, I will prefer A to C). Yet both these requirements (which underlie even such fundamental economic tools as the negatively sloping demand curve) are being seriously ques-

[1] There has been much theoretical work in welfare economics since the 1950s. For an introductory sample, see Arrow's seminal work on social and individual values (1951), Paul Samuelson (1956), Willig (1976), Randall and Stoll (1980), Krutilla (1980), and for a short review, Herfindahl and Kneese (1974). One of the better thorough treatments is that of Just, Hueth, and Schmitz (1982).

tioned. Carefully conducted experiments indicate that preferences are often reversed and that they are often not transitive (Grether and Plott, 1979; Heiner, 1983; Schoemaker, 1982).

The assumption that people choose so as to maximize their own satisfaction is critical because it permits construction of a preference system such as that of Figure 12.1. Each of the curves (such as *aa* or *bb*) is an indifference curve, showing all combinations of X and Y between which the individual is indifferent (as indicated by points P, P_1, or P_2 on *aa*). Assuming that more of the good is always preferred to less, the individual will always *prefer* a combination on *bb* to any on *aa*, because the combinations on *bb* always represent more of either X or Y. Note that the *quantity*, or *degree*, of satisfaction *does not have to be measured or known*. The system is *ordinal* rather than *cardinal*: it is a more than and less than system, but not a this much more than or this much less than system. Any movement (on Fig. 12.1) toward the upper right-hand corner therefore represents an increase in satisfaction, but one cannot say by how much satisfaction is increased.

The Pareto Optimum

In welfare economics, as in most aspects of public forestry, the concern is much less with the individual than with society as a whole. Unfortunately, additional problems frequently arise when attempting to measure the welfare of a group. To understand these difficulties, suppose we begin by dealing with a group of just two

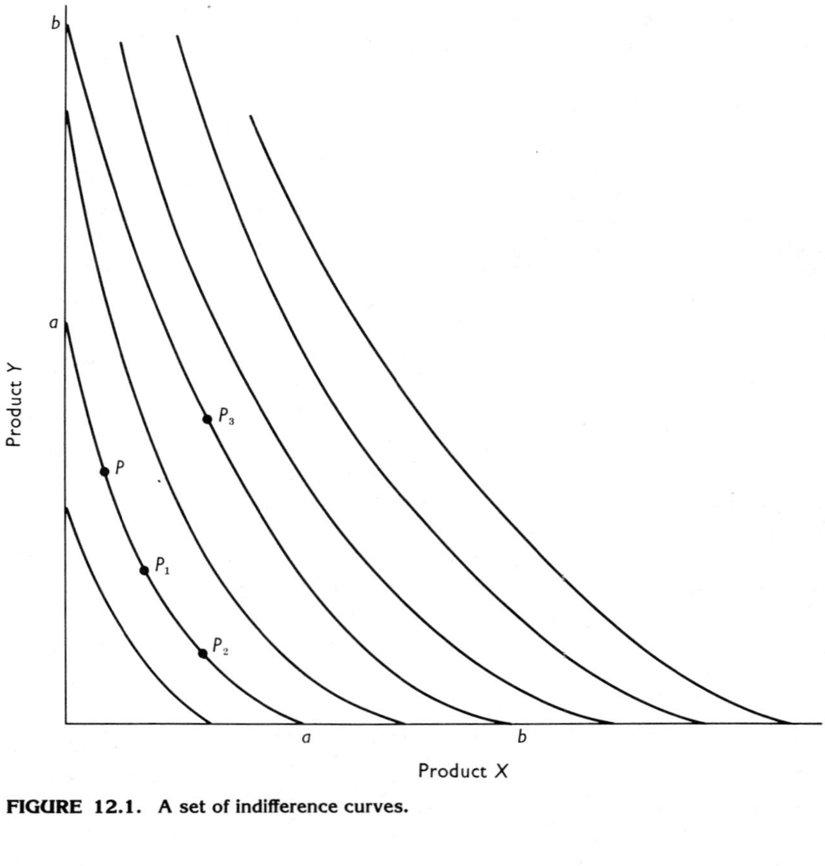

FIGURE 12.1. A set of indifference curves.

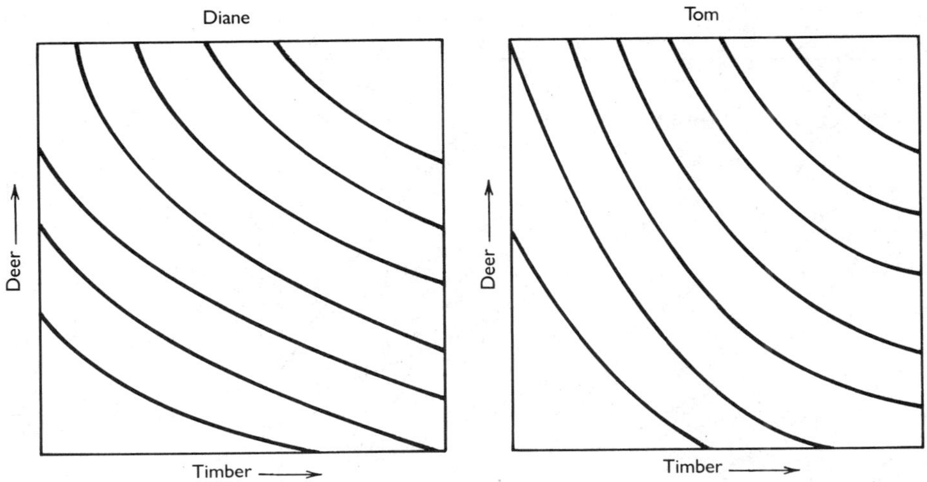

FIGURE 12.2. **Preference maps for two individuals, where the two commodities are limited in amount.**

individuals, Diane and Tom, and two commodities, timber and deer. The preference maps for each might appear as those in Figure 12.2*a* and *b*.

If we assume further that the two commodities are limited in supply, and that the two axes in Figure 12.2 shows the total quantities of deer and timber available, then the diagrams can be combined as in Figure 12.3. Here Diane's preference map is unchanged, while Tom's has been rotated 180 degrees and then superimposed directly over Diane's. Any point within the box now represents some specific allocation of deer and timber between Diane and Tom; at P, for example, Diane has T_d of timber and D_d of deer, while Tom has T_t of timber and D_t of deer.

But the particular division of the two commodities indicated by P is clearly not the best possible! By moving along Diane's indifference curve (i.e., reallocating the quantities of the two resources), Diane's state of satisfaction could be maintained while Tom's is increased and, furthermore, this improvement can be continued *until the point of tangency of the two sets of indifference curves is reached*. Alternatively, the two goods could be reallocated so Tom's satisfaction is held constant (by moving from P along Tom's indifference curve) while Diane's welfare is increased. In fact, the allocation could also be changed so that *both* Tom and Diane are made better off.

Note that the indicated reallocations can be recommended without knowing *how much* the welfare of either person is changed. Any point not on the line connecting the tangency points of the indifference curves indicates an allocation that can be improved: one of the individuals can be made better off while leaving the other at least as well off as before. Once this line is reached, however, the situation is different: any movement *along* this line means one person can gain only if the other loses. Points *on* this line (which for obvious reasons is often called the conflict line) are called **Pareto optimal**, after the Italian originator of the concept (Pareto, 1896).

Pareto optimality was an important conceptual breakthrough, and the concept remains fundamental to modern welfare economics. If it is possible to make at least one person better off without making anyone worse off, that possibility repre-

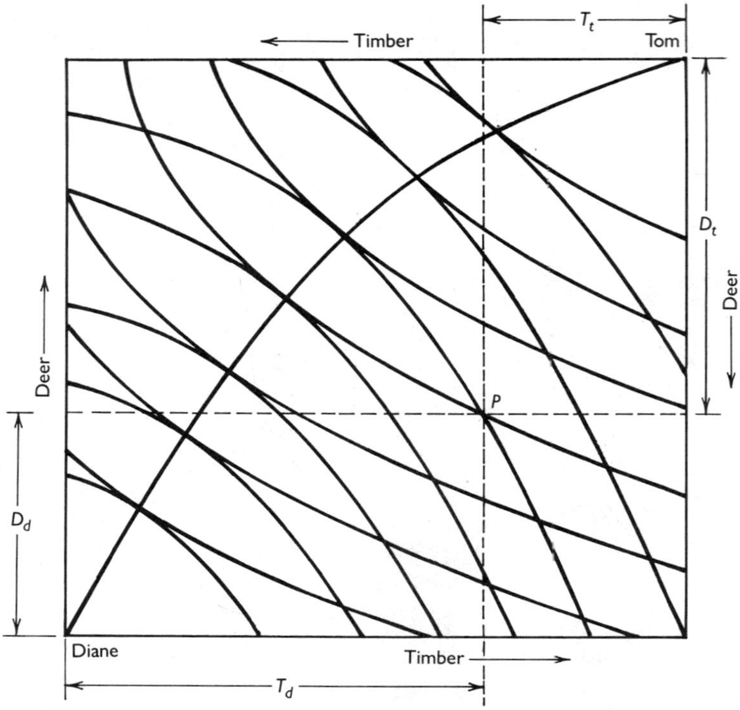

FIGURE 12.3. The two preference maps of Figure 12.2, combined into a single system. Points on the conflict line are Pareto optimal.

sents a Pareto *improvement.* (Making a 100-acre tract of inaccessible timber part of a wilderness area, for example, may make a wilderness lover or a wilderness organization better off without making any timber operator—or anyone else—worse off. This would be a Pareto improvement.) To be considered a Pareto *optimum*, no change can be made that would make some person better off without making another person worse off. As a guide for public investment the concept is therefore of limited usefulness, because almost any recommended policy will create both winners and losers.

The Compensation Principle

The **compensation principle** was a later development, aimed at carrying the concept of Pareto optimality one step further by allowing for losers as well as winners (Kaldor, 1939; Hicks, 1939). According to this principle, a movement from one allocation *(A)* to another *(B)* is preferred *if the gainers can compensate the losers in a way such that everyone will be better off.* Note that the principle does not state that the losers *must* be compensated—only that the *potential* must exist for their compensation. This reformulation of Pareto optimality in terms of *potential* compensation opened the door to a much wider set of applications. Now, as long as it was *possible* for winners to compensate the losers, or (more cynically) to buy off or bribe the losers, the reallocation should be made. More recently, economists have questioned whether the principle of compensation is valid unless compensation is *actually* paid and have pointed out that any recommendation for compensation requires an implicit judgment about the existing income distribution—as does, in fact, a recommendation that compensation should *not* be paid (Just et al., 1982). For our purposes it is sufficient to note that although the topic remains controversial, the criterion of *potential* Pareto optimality has become the standard for modern welfare economics.

Social Welfare Functions

The idea of developing a welfare function for society has long intrigued economists. Essentially, such a function would permit the construction of a set of indifference curves for society, much as we have done for Diane and Tom. This kind of function (sometimes called a Bergsonian welfare function after its originator) would permit solving many problems (Bergson, 1938). Given such a function, presumably measured in some kind of utility unit, one could determine the welfare (or total social utility) produced by (say) timber production in an area, and then similarly ascertain the welfare that would be produced if the area were devoted to wilderness purposes. The use best serving society could then be determined by comparing the two. However, no acceptable common measure of utility has yet been devised, because there is no way to determine *how much more* I like (for example) trout fishing than you do; interpersonal comparisons of utility have no logical basis. Unless we are willing to accept someone's judgment as to how important *our* welfare is, a social welfare function is impossible.[2]

Society, however, still has to make decisions. And although it remains controversial, the rule most often followed is the principle of potential compensation: if the gainers from an action can potentially compensate those who lose, the move should be approved (providing the required funds are available). Sometimes compensation payments are actually made, and sometimes not. Here economists, *as citizens*, can make recommendations as to what they think is best, but the decisions also involve questions important to political scientists, philosophers, lawyers, and many others, including the affected citizens! For the most part, decisions calling for judgments involving distributional issues are delegated to the political arena and made by elected representatives.

PUBLIC BUDGET POLICY OBJECTIVES

Economists concerned with governmental revenue and expenditure problems (with public finance) view a government as shaping its budget with three objectives in mind: (1) to promote the most efficient allocation of resources; (2) to promote a more equitable distribution of income and wealth; and (3) to promote social and economic stability while maintaining growth. These same objectives were necessarily included as objectives of tax policy (Chapter 8). Each has an important bearing on the production of forest products, although attention here will be concentrated primarily on the allocation objective.

Allocation

Resources are used in producing goods and services for the satisfaction of society's wants, and if the most wants are to be satisfied, then resources must be used efficiently. These wants are of two very broad types, **private wants** and **public**

[2]Arrow has stated this in the context of his "General Impossibility Theorem": "If we exclude the possibility of interpersonal comparisons of utility, then the only methods of passing from individual tastes to social preferences which will be satisfactory and which will be defined for a wide range of sets of individual orderings are either imposed or dictatorial" (Arrow, 1963).

wants, and the goods and services that satisfy these wants are appropriately called **private goods** and **public goods**. The private goods bundle is made up of all those goods that are *rival and exclusive in consumption*. For example, the wood I burn in my fireplace is not available to others for heating; if my family lives in a house, that house is not available to other families. Public goods, however, are *nonrival (or nonexclusive) in consumption*. I can enjoy a forest landscape, and my enjoyment does not necessarily interfere with the enjoyment this same landscape provides others. Additionally, noncontributors cannot be *excluded* from the benefits provided by a public good. To illustrate, reductions in air pollution from a power plant are enjoyed by everyone living in an area whether or not their tax dollars actually paid for the scrubbers, filters, and other equipment that brought about the reduction. Similarly, we all enjoy benefits (and bear some costs) from national defense and police protection whether or not we contribute directly through payment of national or local taxes.

The important difference from a production standpoint is that the market, operating through a system of consumer sovereignty, effectively allocates most *private* goods. But the market system fails to allocate *public* goods effectively. If the benefits I receive are not reduced by the fact that you share the same good, and if either (or both) of us can enjoy the benefits whether we contribute to their production or not, then the market system cannot signal to producers the correct quantity of a good that is needed.

Private Goods

In the United States, major reliance is placed on the market system to allocate resources efficiently. Private corporations and firms produce most of the products, which in turn are primarily purchased by private consumers or investors. There are, however, some situations when the government directly intervenes to secure a better resource allocation. Examples have been mentioned in previous chapters: anti-trust laws protect against exercise of monopoly power, public commissions set rates for electric utility and gas companies, minimum prices are established for certain agricultural products, and so on. If externalities are sufficiently important and obvious, public intervention may also be required.

Externalities

An externality is created "where an action of one economic agent affects the utility or production possibilities of another in a way that is not reflected in the marketplace" (Just et al., 1982). Since either producers or consumers can be the cause or the recipient, a four-way classification is often used: producer-producer, producer-consumer, consumer-producer, and consumer-consumer. When logging operations result in silted-over salmon spawning beds, the timber producer has created an externality for the fishing industry—an example of producer-producer externalities. Producer-consumer examples are commonplace: smoke production from slash burning, road congestion and road damage caused by logging trucks, the odor from a sulphate mill, and so on. Consumer-producer externalities are perhaps less publicized, but far from rare: campers and hunters increase fire hazards and reduce freedom of action on industrial forests, regulations to protect endangered species on public forests restrict timber harvesting, bears often damage mature timber in western forests, deer destroy newly planted seedlings, and so on. Consumer-consumer externalities may be the most difficult to control: canoe enthusiasts suffer substantial utility losses from power-boat lovers, and trout fishermen suffer utility losses from both; snowmobile users and trailbike riders reduce the enjoyment of hikers. And there are more subtle utility losses: hunters cause great unhappiness in the minds of many people, and some nature lovers strongly resent the harvest of any remaining virgin timber.

Not all externalities are negative. Logging roads make it possible for recreationists to get to otherwise inaccessible areas, private forests provide shelter and living space for much of our wildlife, the headwaters of virtually all our major rivers are protected by private and national forests, floods are prevented or flood damage reduced because forests are being protected for timber production; many other examples of positive affect abound.

The basic difficulty with externalities should be clear from the examples: private costs may not be total costs; social costs of considerable magnitude may also be incurred, and these are seldom recognized by the marketplace. When these externalities become sufficiently important, public intervention may be required to correct the market deficiencies. The intervention usually takes one of three forms: taxes or subsidies (a tax is imposed, for example, on firms polluting a river), regulation by standards or quotas (stream pollution is not allowed to exceed a given limit of oxygen demand, or burning is not permitted unless atmospheric conditions meet specific standards), or by establishment of property rights (which may take care of the immediate problem but often has substantial income redistribution effects). The economic impact of these interventions varies and measurement always poses problems, for market prices of externalities do not exist. These problems will be discussed further after benefit/cost analysis has been examined (in the next chapter) and again during the discussion of nonmarket determined values (Chapter 15).

Public Goods

Public goods are of two kinds: (1) those satisfying social wants, which are called **social goods**, and (2) those meeting merit wants, called **merit goods**. While not accepted by all students of public finance, the distinction is useful in dealing with public production of forest products. Both kinds of public goods deserve further examination.

Social Wants and Goods

Social wants are determined in accordance with individual preference systems but are seldom revealed directly through existing markets. The desire of millions of people for cleaner air, lakes, and streams provides one example, while the desire for aesthetic enjoyment is another. And in some cases the preferences may be revealed indirectly. Privately marketed camping facilities, fishing ponds, and hunting areas, although differing from those publicly provided, at least partially reveal the values involved, while higher property values in city areas where air pollution is lowest provide an indication of willingness to pay for clear air.[3]

If social wants are to be satisfied through government action, it is necessary to devise a government revenue-expenditure system to provide the money needed to produce or maintain these social goods. At times the government may also decide to collect some or all the production costs from individuals in accordance with their expressed desire. However, user fees and the like are seldom set with any intention of making revenues equal to the production costs incurred.[4]

---

[3]Methods for estimating nonmarket determined values are discussed in Chapter 16.

[4]Note that this in no way means that the government should *produce* the goods; the goods could as well be produced by private means, but paid for and distributed by the government.

Identifying social goods or services and their values may be a greater problem than that of measurement (Driver, 1985). For example, if one were to enumerate the social values produced by a National Forest in the Appalachian Mountains, aesthetic values would certainly be listed along with those of wildlife, stream fishing, recreation, and so on. These are "use" values. But it may well be that the concept of intrinsic value, which is sometimes separated into option value, and existence value, may be of at least equal importance to society.

The concept of **option value** has been undergoing substantial refinement since its introduction by Weisbrod (1964). Originally **option demand** was essentially what the words imply: the value an individual places on the opportunity to make a specific choice at some future time. For example, at this moment I have no plans to watch Alaskan brown bears fish salmon in their native habitat. But I may be willing to pay something to preserve an area where, if, and when I so desire, I can watch those bears fish. I'm willing to pay to keep that option open—to preserve that choice. Uncertainty was later made a part of the definition; that is, option demand is "a willingness to pay for the preservation of an area by an uncertain user" (Howe, 1979). Bishop makes option value a risk premium, an adjustment to monetary measures of welfare to reflect uncertainty about future states of the world (Bishop, 1982)[5] and observes that risk (including also uncertainty) can also be present in supply. Bishop further suggests that option *price* may be a more valuable tool than option *value*. It may result in replacing consumer surplus—a topic of the next chapter—as a measure of the worth of public goods.

Existence values, first proposed by Krutilla (1967), have the distinguishing feature that they stem from altruistic motives. People want their descendants, friends—or even just future generations—to be able to see giant redwoods and grizzly bears and have the opportunity to experience the solitude of a wilderness. And people are willing to pay to assure these opportunities for others. How these values should be integrated into benefit/cost analysis and made comparable to the presently recognized "use" values is a question that is only now being proposed. From the little evidence at hand these values, in monetary terms, may approach or even exceed those of marketed forest products (Brookshire et al., 1983).

Merit Wants and Merit Goods

Merit wants are not determined by an individual's preferences, but rather by the preferences of some "higher" authority. In socialistic systems the higher authority is often the Central Planning Committee; in the United States it is usually a legislative body such as Congress. Merit goods may be produced as private goods, but if so they are not produced in quite the same way or in quite the same amounts: housing, for example, is often produced within the private sector as a private good, but public funds and incentives significantly influence production decisions. Public housing is also provided by government contractors in response to governmental decisions, then made available to particular individuals either through subsidy or through direct grants. Merit goods include things that ought to be available to all citizens, such as basic housing, food, water, and education. Many people also regard wildlife, water, and forest recreation as merit goods. The provision of forest recreation, for example, might be regarded by some as a basic responsibility of the federal government.

[5]The Bishop article (1982) is a good source of references on option values.

Distribution of Income and Wealth

What constitutes an ideal distribution of wealth or of income must, in the last analysis, be a matter of personal judgment. Equal monetary wealth would not yield equal welfare unless everyone placed the same value on the possession of money, and this is certainly not true. There are many who feel equality should be interpreted in terms of opportunity, rather than wealth, but others point out that opportunity is often closely associated with wealth and income. Nevertheless, there are some widely held notions about how income and wealth should be distributed in our nation: people should not be allowed to starve; some level of medical care should be provided to all; children should be fed, clothed, and educated even though their parents cannot bear the associated expense. And, in general, people agree that any redistribution of income should be directed from the rich to the poor.

The significant aspect of distribution here is that the allocation of public goods, whether social or merit, together with the collection of taxes for their production, can influence the distribution of income and of wealth in the nation. Hence, the distribution of public forest goods should be in general accord with the ideas currently held by society regarding a proper income distribution.

Stability and Growth

In this country the primary responsibility for maintaining high employment and reasonably stable prices has been assumed by the federal government. In general, the fiscal means of attaining this objective involves compensatory action. Before 1980 most economists and politicians agreed that if unemployment threatened, action should be taken to increase the level of aggregate demand; if inflation threatened, demand should be lowered. Then, with the advent of so-called supply side economics, there was a shift toward emphasizing supply considerations (encouraging production) under both situations. Of course, monetary and other tools also exist for attaining these objectives and are used in association with those of a fiscal nature. The point here is that stable prices and full employment are important national goals that have both political and economic significance. Those who make decisions for public forest management should recognize, and be responsive to, any changes in these policies.

In other countries, particularly in the so-called developing nations, different objectives may take on great importance. The goal of economic growth, for example, or one of obtaining or maintaining a positive balance of payments, may be given top priority.

PUBLIC PRODUCTION

In the preceding pages a three-way distinction between private goods, social goods, and merit goods has been presented. The distinctions between each are derived from the nature of the want the product is intended to satisfy, and *not* from the production process or whether it was located in the private or public sector. We now need to examine some factors that might influence whether production occurs under public or private control. Remember that publicly produced goods are not necessarily public goods (i.e., they do not necessarily exhibit nonrivalry in, or nonexclusion from, consumption).

Sometimes public production is necessary simply because there is no market for the product being considered. If no one can be excluded from enjoying a good once it is produced, then it is unlikely that private enterprise will undertake its production, for there would be no means of establishing or collecting a price. The beauty enjoyed while traveling a mountain highway provides an example. Sometimes exclusion is possible but the collection of a user charge would be prohibitively costly; small forest campgrounds in wilderness areas are a case in point. Occasionally, private production and marketing are also possible but would be considerably more costly than provision by public means; for example, it is unlikely that wilderness areas would be maintained or provided by private industry.

Public production can also be required because of significant positive or negative externalities (Ayres and Kneese, 1969). If the externalities can be identified and sometimes marketed (as in the case of chemicals reclaimed from waste), there may be no necessity for public production. However, the methodology used in producing a primary product may well affect the kind of externalities that result, and this relationship may provide sufficient justification for public rather than private production. Under some circumstances forestry may offer a good example: the many externalities of forest management—watershed protection, flood control, air purification, provision of scenic beauty and wildlife habitat, and so on—are sufficient in the eyes of some to warrant far greater public ownership of forest land than what presently exists.

A need for public production may also arise from monopoly power or from the threat of such power, and from other market imperfections as well. The possibility of private owners closing off or establishing monopolies over parts of Yellowstone National Park or the valley of Yosemite National Park loomed large in the minds of legislators when they delegated these areas to public control. Occasionally, time lags produce market imperfections that lead to public production; the time required to produce large sawtimber, for example, would make it difficult for private suppliers not already owning old growth forests to react rapidly to an increased demand for large sawlogs. At other times the transaction cost is simply too high: the collection of user fees in small forest campgrounds has already been mentioned.

Some goods are publicly produced because of an unwillingness to accept private market solutions. Such goods could be produced in the private sector but have been turned over to public production because the pattern of consumption under private markets would be undesirable. Our system of elementary schools and high schools provide an everyday example. Urban playgrounds, hunting and fishing privileges, and many forms of outdoor recreation are provided both privately and publicly, although most people feel these should be available to all citizens regardless of their ability to pay. One may observe that different societies decide allocation and production to public or private means very differently. In Scotland, hunting and fishing rights are privately marketed while golf courses are heavily subsidized and provided as public goods, a pattern of consumption that contrasts vividly with that in the United States.

Listing examples of how public production comes about does not of itself provide justification for public production. Public forest ownership, however, is one of the dominating institutional factors in the total U.S. forestry picture today. If we seek an understanding of forestry's role in our present economic society, it is necessary to look a little more closely at how social decisions are made, because social considerations have an important influence on much of forest production.

How Much and What?

The *private* competitive market provides an efficient mechanism for aggregating individual wants into effective demand. Each consumer decides which and how much of the thousands of available products to buy, thereby maximizing his or her own satisfaction within existing income constraints (perhaps supplemented by a VISA or MasterCard credit charge account). Each dollar spent represents a "vote" for the production of some good or service. The market adds up these votes so that, at least ideally, everyone gets as much as he or she is willing and able to pay for. Efficiency in production and in resource allocation is assumed to be assured by competitive forces, which is the reason for looking askance at market imperfections. Since utility isn't measurable, and since interpersonal comparisons are ruled out, no one can prove that this particular pattern of production and consumption will maximize social welfare. But (as stated earlier) no one can prove that any other pattern is better.

On the other hand, earlier parts of this chapter stressed that effective markets do not exist for *public* goods, either social or merit. Government, through purchases of raw materials and finished products and through its ability to tax, can provide a kind of collective action that substitutes for the market by assembling the necessary resources and organizing them to meet anticipated social wants for these publicly provided goods. But how does society decide how much of what should be produced by the public sector? How does society decide to spend dollars on building more campsites instead of more timber access roads, or to take an area out of timber production and declare it a wilderness area?

The private market weights votes by the dollars that are spent, but this weighting system is unavailable when the product is a public good. Arrow concluded that if utility is not cardinally measured, and if interpersonal comparisons have been ruled out, there is no way to aggregate individual wants for public goods; there is no means of weighting the individual wants to derive a picture of social wants (Arrow, 1963). But even though individual wants cannot be aggregated, some problems can be at least partially resolved using the approaches we have already discussed.

Consider, for example, the problem of determining the proper mix of outdoor recreation and timber to be produced from some given public forest area. The production possibilities under two different expenditure levels are shown in Figure 12.4, with curve *P-P* representing the present situation and *P′P′* the possibility under a considerably increased cost. Timber output is measured in cubic feet and recreation in visitor days.

Now suppose one could draw the preference maps representing the attitudes of two different groups of people. One group might be called timber advocates, who fervently believe God and Congress meant forests to "produce sawlogs, not play." Their preference map is shown in Figure 12.5*a*, with the timber and recreation scales kept the same as those of the production possibility curves in the previous figure. Directly besides this, Figure 12.5*b* is the preference map—the set of indifference curves—that might represent some hiking, hunting, or camping group whose members believe forests should have a very different purpose, as the relative slopes of the indifference curves clearly show.

All three sets of curves have been combined in Figure 12.6. Clearly, if the combination presently being produced is P_t, then the timber interests will be happy with the way available funds are being spent (although they may want larger appropriations). But the recreationists will not be content. They would only be satisfied if the combinations were switched to P_r. According to the Pareto criterion, however, no change from the present spending pattern can be recommended unambiguously, for any shift along the production possibility curve that would improve the lot of the recreationists will definitely harm those interested in timber.

There are three possibilities for obtaining cooperation from both groups. If

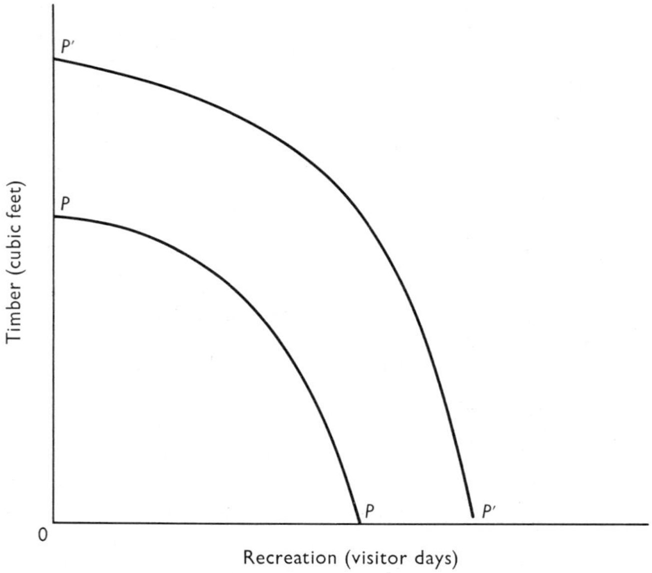

FIGURE 12.4. Production possibility curves for timber and recreation on a public forest.

the complete pattern of possibility curves gave the two optimum expansion paths shown by the dashed lines (Fig. 12.6), one optimum for timber and the other for recreation, then any combination that fell outside the enclosed (shaded) area could be improved; both the timber interests and the recreation interests could be made happier with no increase in cost. Alternatively, either group could be made better off while doing no harm to the interests of the second group.

The second possibility lies in negotiation and compromise. The Pareto criterion provides an excellent argument for maintaining the status quo; almost any change can be shown to reduce someone's satisfaction. But when numerous

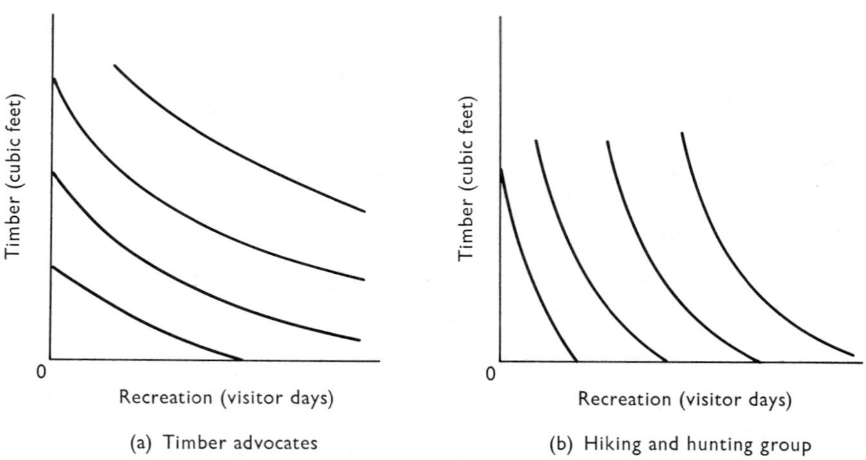

FIGURE 12.5. Preference maps for two groups of people: *(a)* timber advocates, and *(b)* hiking and hunting group.

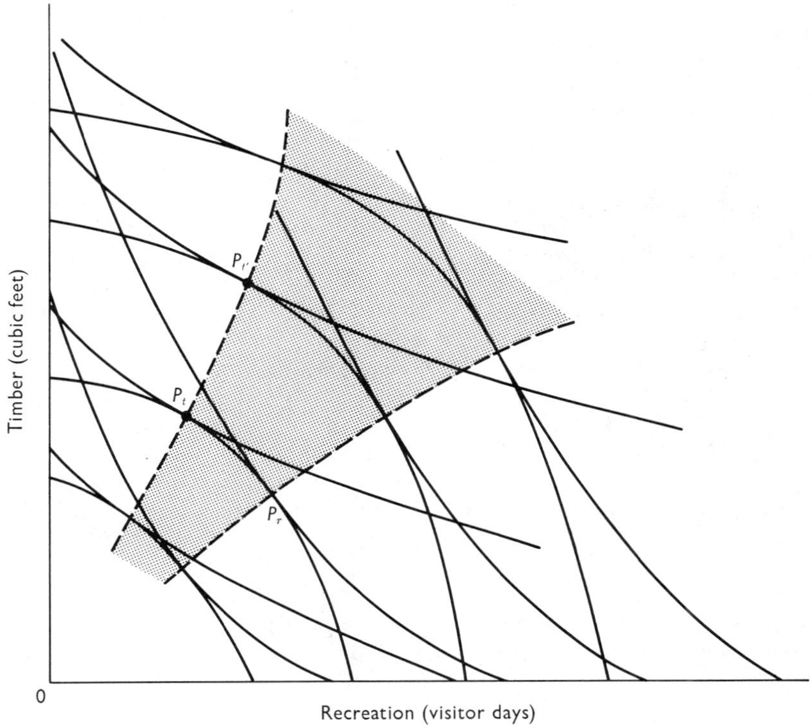

FIGURE 12.6. The differing optimum production paths for the two groups. Any combination lying outside the shaded area can be improved, with both groups benefitting. The shaded area is a region of policy conflict, unless costs are increased.

changes are possible there is often opportunity for negotiating compromises between groups with opposing views. A prime timber area may be designated as a wilderness area in spite of widespread opposition by timber interests—provided the area so designated is substantially reduced in size from earlier proposals, or that allowable cuts from adjoining national forests are increased. National Parks have taken over private timberlands, but often only after substantial compensation settlements have been made. Negotiation and compromise are often possible if alternatives exist.

The third possibility for cooperation is to advocate an increase in expenditure. To take an extreme case, even if the increase were used to move from P_t to $P_{t'}$ (thus keeping an optimum output from the standpoint of the timber interests), it nevertheless results in the provision of more outdoor recreation. Moreover, one would normally expect the agency involved to cooperate willingly with the two groups to obtain larger appropriations.

From an analytic viewpoint, however, no increase in the cost level can be recommended, even though both groups would benefit. Such a recommendation would imply knowledge of the preference map for *all* taxpayers (the basis for a true social welfare function), as well as that for all timber and recreation interests. This knowledge clearly implies both cardinal measurement of utility (or welfare) and interpersonal comparisons of utility, both of which have already been ruled out. It is worth reemphasizing, however, that there has been no reference to the price or to the value of either recreation or timber in the analysis; how much easier and more definite the conclusions could have been if prices for the two products were available and comparable!

Societal Choice Systems

Perhaps the approach of aggregating individual wants to derive social wants is overly narrow, based too much on our concept of private markets and consumer sovereignty.[6] Perhaps we should pay far more attention to the determination and payment of compensation to all "losers" (Thurow, 1980). Without pretense of analyzing such alternative approaches or even reviewing all those currently being considered, we will briefly present two possibilities.

One possibility is to regard individuals as possessing *social* as well as *individual* interests. This leads to the "pluralistic" viewpoint advocated by many sociologists, anthropologists, political scientists, and even some economists, which holds that people have a "plural" set of values and loyalties to family, to club, to employer, to city, to state, to church, to university, and so on. If state decisions are being considered, then the preferences revealed by people's actions with respect to state decisions are appropriate rather than those typifying their choice patterns as individuals. Thus, for example, people may be willing to accept being taxed for things they believe are socially desirable even though the tax paid exceeds the benefits (welfare) they themselves obtain.

A second possibility carries this approach even further, regarding society as a distinct and separate entity. Under this view society possesses its own set of values and preferences and often approves actions to further develop these values through political processes. Society might, for example, decide that big trees are beautiful and clear lumber is important, sufficiently important to warrant production by long rotations. Such a decision would not rule out economic criteria; it would only mean that big timber should be produced—that timber size had become a production constraint. In effect one says, "The decision has been made to provide the public good called big timber. It is up to the agency responsible for timber production to grow these trees at as low a cost as possible." Similarly, if one accepts that society has its own values and preferences, then one can argue that the public can decide to convert an area from commercial timber production to use as a wilderness area, even though this means a reduction in timber supply, an increase in timber costs, and may force some timber businesses to close!

There is no single, universally accepted view of the manner in which society decides either how much or what kinds of public goods should be produced, or even where they should be produced. That a tremendous variety of goods is now being produced in the public sector is accepted as part of our economic system. The full extent of this production is seldom realized, but its significance is highlighted by the factual observation that in only one other country of the world, the USSR, does government control as much production as in the United States. If one makes public production the sole criterion of whether a country is socialistic or capitalistic, then the United States would necessarily be called the second largest socialistic power in the world today. It would also be the largest private enterprise power. With public goods looming this large in our economy it is only to be expected that rather special approaches have been devised for analyzing the need for some of this public expenditure. The most important of these is *benefit/cost analysis*, to which we now return.

---

[6]Here recall also the points made in Chapters 2 and 11 about consumer sovereignty and the objectives of large corporations.

QUESTIONS AND PROBLEMS

1. With reference to welfare economics, what does the term utility mean? As useful as this concept as proved, it still has major drawbacks. What is the most significant drawback?
2. What is meant by the problem of interpersonal comparisons, and why does this spell trouble for measuring the social contribution of a public project?
3. What is the difference between a cardinal system of measurement and one that is ordinal?
4. Define, or describe, a Pareto optimum. Illustrate a Pareto optimum by using indifference curves.
5. What is the Compensation Principle, and why is it considered important?
6. Define and distinguish between the following:

   Private goods
   Public goods
   Private wants
   Public wants
   Merit goods
7. What is an option value?
8. We know public production is very important in, and to, forestry in this country. Under what circumstances is public production justified in what is usually called a market economy? Do you believe public production of forest products is warranted in the United States today? If so, why?
9. Give two possibilities for approaching the problem of determining a social choice system.
10. What is an externality? Describe four types, and give an example of each that might be encountered in forestry.

chapter thirteen
EVALUATING PUBLIC PROJECTS: BENEFIT/COST ANALYSIS

Prior to World War II, students of public finance were concerned primarily with the revenue side of government—with questions of tax incidence, tax effects on incentive, tax equity, and similar problems. Water resource development was the only area of the federal government in which formal efforts to analyze the efficiency of spending proposals were being made. These efforts were largely due to the Flood Control Act of 1936, which required the U. S. Army Corps of Engineers to evaluate the costs and benefits of all water resource projects "to whomsoever they accrue."

As government expenditures grew larger, their impact on the economy became much more significant, and in the early postwar years the expenditure side of government began to attract greater attention. Efficiency in government expenditures, as well as in obtaining revenues, became a matter of concern to an increasing number of economists, legislators, and government executives. Although overly simplistic and heavily criticized, methods developed under the Flood Control Act gradually led to the application of economic logic to budget allocation in a variety of areas, and eventually to the Program Planning and Budgeting System (PPBS) under President Johnson. By the late 1970s questions of federal governmental expenditures, and particularly the size and rate of growth of the federal deficit, had become major campaign issues.

By this time the general framework for analyzing public expenditures was well established. Two major forms, similar in concept but with very different objectives, had been developed: **cost effective analysis** and **benefit/cost analysis**. Essentially, a cost effective analysis seeks to answer the question "Accepting the objective of this project as given, how can it be attained most efficiently?" A benefit/cost analysis, however, asks a much more difficult question, for it purports to be a way of determining what society prefers. This is no small task.

The format now used in benefit/cost analysis stems back to the 1950 report of the Subcommittee on Benefits and Costs to the Interagency River Basin Committee (Subcommittee, 1950), which was primarily concerned with reconciling the widely varying approaches taken by government agencies concerned with river basin projects. These irrigation and flood control projects were often of a size and scope that made usual ways of measuring size meaningless. Consequently, project size was measured by its total cost: each additional million dollars was assumed to be spent on the necessary concrete, generators, transmission lines, engineers, laborers, and so on. Benefits were also measured in dollar terms, since the benefits of most concern were those showing increased agricultural or energy output. Flood control benefits proved difficult to evaluate.

Costs were plotted on one axis and benefits on the other, both expressed in the same dollar units. The resulting figure was similar to that shown in Figure 13.1, the usual presentation of the benefit/cost model. Any benefit level involves a specific cost level, and since the two axes are similarly scaled, a line drawn at a 45 degree angle from the origin identifies all points where benefits and costs are equal. This feature proved to be particularly useful when discussing the optimum project scale and its "justification" (could the project's congressional sponsors justify the required expenditure).

After much discussion it was agreed that the optimum scale for any project is the one that will maximize *net* benefits. This is the marginal solution: marginal benefits must equal marginal costs. In Figure 13.1 net benefits are maximized at an expenditure of 250 million dollars with expected total benefits of 300 million. At this point the slope of the benefit/cost curve is 45 degrees: if both benefits and cost are changing at exactly the same rate the slope of the curve must be 45 degrees. The benefit/cost ratio is 300/250, or 1.2+, but note that this is not the point where the B/C ratio is highest. Net project benefit, not the benefit/cost ratio, is maximized.

Projects were justified if they showed a benefit/cost ratio above one, for, when this is true, social benefits exceed costs and completion of the project increases social welfare. However, this does not mean that all projects with a positive benefit/cost ratio above one should be built! In this regard four concerns are paramount:

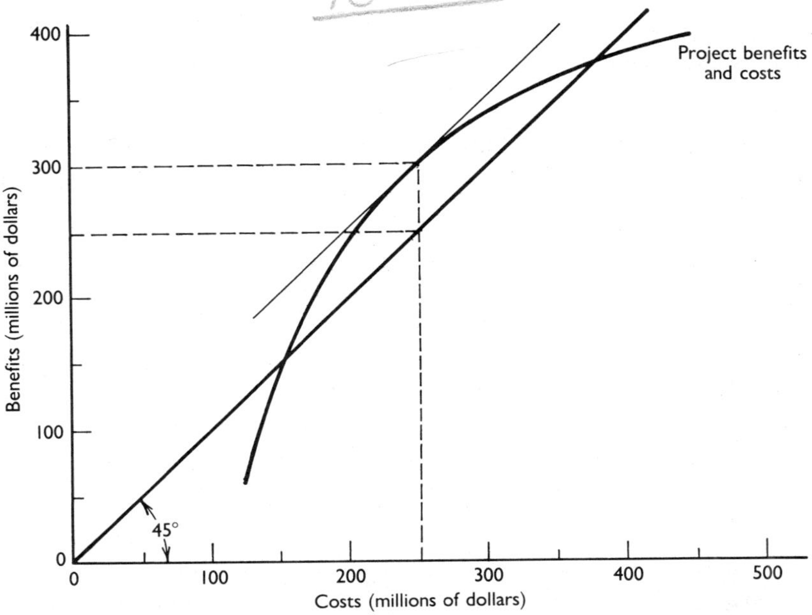

FIGURE 13.1. The usual benefit/cost model.

1. Would other projects be of greater benefit? To assure the best projects would be chosen, each project's costs were estimated by determining the *opportunity cost* of all inputs, this being the value of the input if used in an alternative project.
2. Would available funds permit completing the project? Some means of ranking projects had to be devised (the subject of Chapter 14).
3. Both monetary and nonmonetary benefits and costs must enter the analysis. This problem will be a major part of Part 4—nontimber products.
4. Uncertainty—uncertainty as to the realization of the actual benefits and costs, and of their values as well.

OBJECTIVES OF PUBLIC FOREST PROJECTS

As on privately owned forests, the objective of a project is a principle determinant of the criteria used for judging its worth. Public objectives, however, can differ substantially from those of private owners. Recognizing that exclusive use of monetary measures would be unsuited to public forestry projects, the Multiple Use-Sustained Yield Act (1960) was made to state that national forest resources were to be managed to produce products "in the combination that will best meet the needs of the American people . . . and not necessarily the combination of uses that will give the greatest dollar return or the greatest unit output."

Aggregate Consumption

Underlying virtually all public investment projects is the objective of raising local, regional, or national living standards. Few would claim that a nation's level of living can be accurately measured by any aggregate of material items, yet nearly everyone agrees that consumption is an important component of well-being.

Measuring aggregate consumption poses three major problems: how to convert the heterogeneous bundle of goods and services consumed by a nation into a single measure; how to add the consumption levels of individuals having different tastes and receiving different incomes; and how to compare consumption levels that take place at different time periods. All three are complicated by the necessity to base the measuring unit on the expressed desires and wants of individuals.

The solution to the first problem, how to add up the individual consumption levels, can be visualized as a two-stage process: (1) the total amount of each good or service consumed is first weighted by its market price, then (2) this weighted aggregate consumption is measured by consumers' "willingness to pay," a concept to be defined later in this chapter. The final result is not necessarily equal to the market value of the good, nor is it a true measure of utility. (Goods without established markets require different treatment and will be covered in Chapters 15 and 16.)

The second problem, that people have different tastes and different incomes, is made especially difficult by the relationship between taste and income. Although individuals have different tastes, they express these differences in their buying habits: weighting the quantities purchased by market prices takes taste into account, but only to the extent that all markets are equally accessible to all individuals. The major constraint on market accessibility, however, is income, and public projects can impact income distribution. The problem is not only how to add consumption levels of individuals with different incomes, but also one of deciding how incomes will be affected by the project, and whether a project should or should not be used to redistribute incomes. So far as the aggregate consumption objective is concerned, this problem is often ignored, but when incomes will be affected significantly adjustments are needed. Projects, however, are often the least controversial method available to assist low income groups or regions and are frequently chosen with income redistribution in mind. When this is the case income distribution can be brought into the analysis by giving additional weight to the consumption of the favored group or region. Most analysts would argue that income redistribution should be approached directly, and that to use projects to achieve such goals confuses issues.

The third problem, that of aggregating consumption and costs over time, is handled by discounting future consumption to obtain its present value. Obviously, the choice of the discount rate for public projects (the social discount rate) is important and will come up again.

To summarize the discussion so far: a major objective of social benefit/cost analysis is to determine the change in living standards that would follow implementation of the project being analyzed. That change will be determined by the project's net contribution to aggregate consumption, weighted by market prices and measured by consumer's willingness to pay. We know that consumption is an inadequate measure of welfare, and that both the weighting system (market prices) and the measuring system (willingness to pay) are far from perfect. Benefit/cost analysis can only present a partial picture of what a project may be worth to society. But remember: no one claims that benefit/cost analysis determines what is *best* for society; it only attempts to show society's *preferences*, and this is something a decision-maker should know even though that preference may later be disregarded.

Multiple Objectives

Most private sector projects have a single primary objective. Market values reflect the returns from each product and guide the owner to the allocation maximizing net returns. If the owner wants to change his profit maximizing output mix, he can do so, but will carefully estimate his opportunity costs.

The objectives of public projects are more complex. Regional employment and maintaining the raw material base for existing communities have been important public forestry objectives to the Pacific Northwest since the mid-1970s. They will continue to be important for that region, but will also increasingly conflict with public desire for more wilderness and other nontimber products. Producing timber for the national market has long been given preference in most public forestry projects, but substantial areas of timber have had to be withdrawn to protect (for example) the spotted owl in the Pacific Northwest and the Kirtland's warbler in Michigan. Protecting watersheds, providing forage and range, improving hunting opportunities, satisfying demands for firewood seekers, assuring supplies of timber, wildlife, or natural areas for future generations—all such objectives could be precisely handled by measuring each output, then assigning weights to each product or service that would result in the required allocation of effort. But who will decide the weights?

Weighting Objectives

The value judgments needed to decide (for example) the relative merits of timber production or spotted owl protection, or whether to concentrate planting programs in Utah or Idaho instead of in the Lake states, reflect political considerations. But political leaders find it difficult to make these decisions explicitly. In part this is because politicians have many responsibilities: they must make decisions that will best preserve the national interest, represent and serve their particular electorate, and while doing so act in a way that will enhance the security of their own position. And politicians are only occasionally trained in economic analysis—a thorough understanding of economics has never been a prerequisite for a successful political career in the United States. In fact, logical analysis may even create problems, since (as every "two-handed" politician understands) a certain amount of ambiguity often carries obvious advantages. So economic considerations usually emerge as only one of many types of objectives guiding public sector projects (Dasgupta, Sen, and Marglin, 1972).

The Four Account System of the U.S. Water Resources Council

An important recent attempt to define explicit criteria for multiple objective resource planning is that of the U.S. Water Resources Council, which in 1983 released a report, usually called "Principles and Guidelines" (U.S. Water Resources, 1983). This effort is rooted in the benefit/cost tradition and seeks to encourage consistent planning by federal agencies when evaluating water resource projects.

The major contribution of the Principles and Guidelines is its description of four accounts, to be used in evaluating the effects of alternative project proposals. The first of these is concerned with any project-related changes in the economic value of output. The other three accounts require: (a) the display of information on environmental effects, (b) regional economic and distributional effects, and (c) any other social effects of a proposed project.

This presentation is notable for at least two reasons. First, the information required as part of each account is quite detailed and usually well described. The environmental quality account, for example, must display any anticipated changes in ecological, cultural, or aesthetic attributes, and also present an overall summary of project effects on environmental resources. Second, no attempt is made to collapse the four accounts into a single indicator of project attractiveness. Indeed, the explicit intent of the system is to encourage the use of information about anticipated project effects on diverse social objectives, and to recognize the basic incommensurability of the different project assessment criteria.

This approach holds important implications for the modern forest planner or manager. When used for forestry projects, the approach stresses the significance of the nonfinancial impacts the project may have. As a result, the use of multiple accounts requires the analyst to recognize the impact of forest practices on key ecological indicators, on the health of forest communities, and on the distribution of income and employment opportunities within a state or region. This calls for a sophistication and breadth, both of analysis and of presentation, that would not have been required a decade or two earlier.

Before leaving the discussion of multiple objectives, we should recognize that some analysts regard multiple objectives as myths—they feel there can be only *one* objective and other considerations are simply constraints on the optimal solution. This approach has some merit: when alternatives are identified and choices made explicit, the associated tradeoffs become easier to see. If only one objective can be selected, someone—the analyst, the decision-maker, project formulator, politician, or the public—is forced to make a choice: what does society really want?

But democratic societies are not single-minded. There is no necessary agreement on priorities or on objectives, nor is there a way to force unanimity. Social choices and individual values will not always concur, nor should we expect them to. Many projects will have multiple objectives, and the choices that are made will not necessarily be those derived from financial, or economic, analysis.

UNDERLYING CONCEPTS

Two of the more basic concepts underlying benefit/cost analysis are not specific to the methodology but concern the way public choices are made. The first of these is closely related to our democratic form of government: the belief that *individuals* count. The second concerns the use (or misuse) of benefit/cost analysis by public decision-makers.

The Role of Individuals

In a nation governed by democratic principles, public decisions are normally made through individual participation: individuals must have a voice in the decision. Benefit/cost analysis provides a means for recording individual preferences, usually through the marketplace. If no market prices are available, then the analyst must use some other means for approximating suitably comparable measures.

But should individual preference *always* rule? In most democracies, including the United States, important public decisions have been made in which the individual's opinion was neither sought nor considered.[1] The Vietnam War, military involvement in Central America, registration for a possible military draft, capital punishment, and production of nuclear weapons are well-known examples that might not have been supported by a majority of individuals. In certain situations governments (to the extent their citizens allow) tend to become paternalistic, telling individuals what is "good" for them.

Benefit/Cost in Public Decision Making

The second concept concerns the use of benefit/cost analysis in public decision making. It is best revealed by asking "When should benefit/cost analysis substitute for political decision making?" The answer is NEVER. Any time benefit/cost analysis is allowed to substitute for the political process, it is implied that maximization of societal preferences is "a good thing"—and this would make benefit/cost analysis heavily value-loaded. If we remember that benefit cost analysis is always a *guide*, not a *rule*, then it is far less normative. Yet it seems difficult to avoid the impression that benefit/cost analysis can be used as a "golden rule"—that it is perfectly objective, that its use eliminates the necessity for using judgment. Unfortunately, this is just not so! Benefit/cost analysis can only provide a guide to what society wants, and it is not necessarily true that what society wants will be good for it. Nor does it follow that an analysis, even one carefully made, includes all the factors that a decision-maker should weigh. One reference puts it this way (Dasgupta, Sen, and Marglin, 1972):

> *It is worth remembering that social benefit-cost analysis is not a technique but an approach. It provides a rational framework for project choice using national objectives and values. Projects are judged in terms of their precise impact on the economy, and this impact is evaluated by using parameters reflecting national goals, social objectives and global facts. This is relevant not merely for the evaluation of given projects but also for formulating new ones and implementing the chosen projects.*

[1] A reviewer suggested that this statement is wrong—that opinions are both sought and considered, but these are opinions of the "attentive public"—those individuals and groups who are interested in the issue and express their opinions. "The rest of us are general publics—we don't care enough to get involved. When we got tired (of the Vietnam War) the 'general public' became attentive and changed government policy!" He further suggested that we have a dual representative government—one government of elected officials, and another, "somewhat self-elected," of attentive publics who actively participate in the policy formation process.

"Financial" and "Social" Analysis

As with private industrial projects, every proposed public project requires determination of anticipated cash flows. Agencies have to prepare budgets and budgets require approval; financial officers need to know how much they will be required to spend and when the funds must be available. Legislators may override the efficiency criterion when making public project decisions, but they want to know the monetary costs and returns of projects they are asked to approve. For this reason public projects generally require two distinct types of evaluation. The first is a financial analysis that deals only with monetary costs and benefits. The second, termed an economic analysis, includes *all* costs and revenues whether measured in monetary or other terms.

In *form* the financial analysis of public projects is identical to the discounted cash flow analysis of private industry presented in Chapter 11. It is the *values* that differ, for what are costs or benefits to a private owner may not be so valued by the public. Taxes provide a simple example: federal, state, and local taxes are costs for any private project, but from a public viewpoint taxes are simply a transfer of funds from individuals to the public at large. They have no bearing whatever on public projects. Benefits tend to be somewhat more complex, even those that come from products with well-established markets. To understand how social benefits are treated requires a step back to the basic concepts of utility and preference and the related concept of *consumer's surplus*.

WILLINGNESS TO PAY AND CONSUMER'S SURPLUS

Willingness to Pay

Three basic concepts of welfare economics—utility, preference, and consumer's surplus—underlie **willingness to pay**, the measure of value used in benefit/cost analysis. **Utility** was defined in Chapter 12 as the unobservable measure of welfare. **Preference**, which reflects both wants and tastes, is taken as the indicator of utility. If A *prefers* beer to wine, then beer is more wanted than wine. We generally assume that individuals choose from among competing goods on the basis of the effect on themselves rather than on others—a form, some say, of selfishness.

To relate this to utility the reasoning goes as follows: if beer is preferred to wine, then the utility of beer is perceived to be greater than that of wine. If the individual is indifferent to the outcome of the choice, then the utilities of beer and wine are equal.

Prices, when available, can be used to *rank* preference (but not to measure them). With prices we can say:

1. If an individual prefers beer to wine, then the person will be willing to pay more for beer then for wine.
2. If a price, P, is paid for beer, the individual is willing to pay that price; otherwise she would go without.
3. Hence the price, P, an individual is *willing to pay* is an indication of her preference for that product, *and* of the utility that product has for her.
4. The market permits individuals to choose products. Individuals allocate their incomes in accord with their preferences, and thus their willingness to pay indicates their preferences and their expected utilities.

Now the problem becomes one of how to measure *willingness to pay*.

Consumer's Surplus

Marshall's Solution

Near the beginning of this century Marshall set out to solve the problem of measuring willingness to pay. His solution started an argument that still continues, but it made welfare economics a major basis for public investment decisions. His proposed solution began with the familiar concept of demand (Figure 13.2). Given a competitive market, with a demand of *DD*, and a supply curve *SS* of an industry in equilibrium, price will be at *P* and the total amount paid for the good will be *QP*. The demand curve, however, shows that some individuals would be willing to pay prices higher than *P*; since they pay only *P*, they are in effect receiving an economic surplus. In fact, over the entire price range the total surplus for consumers will be equal to the area of the triangle *PAD* in the figure—the total *consumer's surplus*. So, what is the total willingness to pay *(WTP)*? Clearly, consumers pay the price *P* for the quantity *Q*, so they are willingly paying *P(Q)*—the area *OPAQ*. And they would also be willing to pay the total consumer's surplus—the area *PAD*. The total willingness to pay, therefore, should be the area *ODAQ*.

If it were necessary to value this total area, few would consider consumer's surplus and *WTP* as anything but theoretical concepts. To quantify this area would require knowing the shape and position of the entire demand curve above the price. This would be difficult, since data for the entire demand curve is almost never available.

Now refer to Figure 13.3. Here is the same situation, but with a new, large public project that will increase output from *Q* to *Q′*. At this new level of production, price will decrease to *P′*. *Total* consumer's surplus will increase substantially —to *P′A′D′*. But we are no longer interested in consumer's *total WTP*. What is wanted now is an estimate of the *change* in consumer's total *WTP* that results *from the project*. This is composed of two parts: (1) the new price times the project's output, which is $P'(Q' - Q)$, or in area terms, *QQ′A′B*, *plus* (2) the small triangle *AA′B*, which may be calculated as $\frac{1}{2}(P - P')(Q' - Q)$. These areas are shaded in

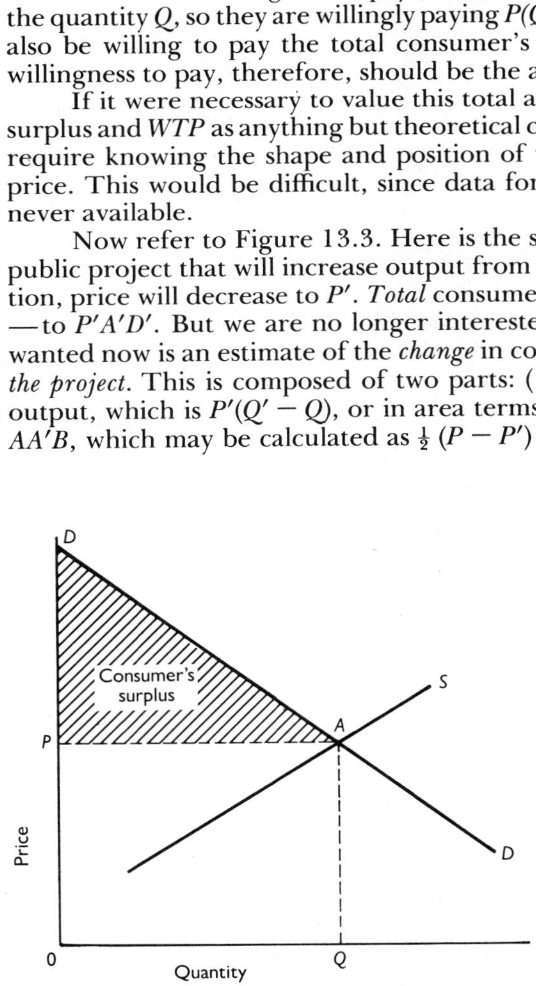

FIGURE 13.2. The Marshallian demand curve, with supply curve of a purely competitive industry in equilibrium. Consumer's surplus shaded.

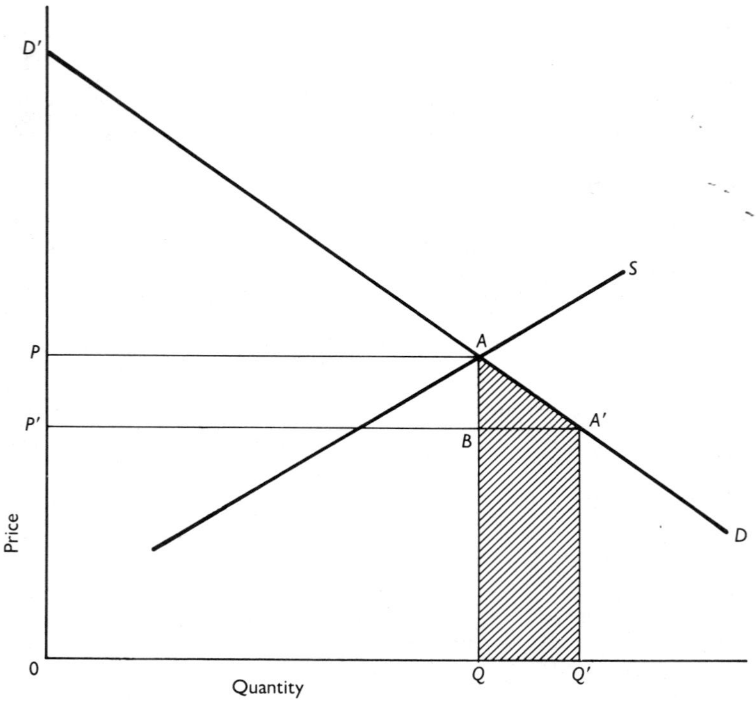

FIGURE 13.3. Demand curve of Figure 13.2, but with additional supply from new public firm. Total willingness to pay *for the project output* shown as shaded area.

Figure 13.3. (*Note*: this is the total *WTP* for the project output, not necessarily the total project benefit.)

The idea of consumer's surplus aroused a great deal of comment and study. One of the major sources of controversy stemmed from the partial equilibrium assumptions of Marshall's demand curve—particularly the assumption that income and all other prices were held constant. Surely (it was argued) if the price of a product were to drop significantly, this would have the same effect as a rise in real income. And, if this were so, Marshall's measurement of consumer's surplus and willingness to pay would no longer provide a true estimate of utility.

Income-Compensated Demand Curves

One obvious (but difficult) answer to this criticism was to develop an income-compensated demand curve. Unlike Marshall's demand relationship, which shows the quantities of a good demanded at various prices when income is held constant and utility varies, the compensated demand relationship shows quantities of a good demanded at various prices when utility is held constant and income varies (Hicks, 1943). Using this compensated demand curve, Hicks demonstrated that there were really four kinds of consumer's surplus. Only two afforded consumers true freedom of choice in response to price changes. These two, known as compensating and equivalent variation, have become the basis for most actual measures of welfare change. **Compensating variation** is defined as "the amount of money which, when taken away from an individual after an economic change, leaves the person just as well off as before." Compensating measures are therefore based on the pre-exchange or initial welfare level. **Equivalent variation** is defined as "the amount of money paid to an individual which—if an economic change does *not* happen—leaves the individual just as well off as if the change had occurred" (Just et al., 1982). Equivalent measures therefore reflect the post-exchange or subsequent welfare level (Mishan, 1972).

Note that neither equivalent nor compensating variation provide direct measures of utility! Their significance lies in the fact that they can be used to estimate the maximum amount someone is willing to pay, or the minimum they will accept, to change from one condition or situation to another.[2]

The practical aspect of all this is that under most conditions willingness to pay can be measured from Marshallian demand curves, just as would have been done had not Samuelson, Hicks, Willig, and others developed the new approach. When a project's output has little effect on price there is little or no income effect, and the Marshallian approach to consumer's surplus is OK. When these conditions do not hold, one must be careful, and there are many such conditions in public forestry. What would be the benefits and costs from a tariff on Canadian lumber imports? What benefits and costs would result from reducing (or increasing) the allowable cut in National Forests? What about extending wilderness areas or doubling entrance fees to National Parks or State Forest campgrounds? Any of these proposed changes could create significant effects on product prices or consumer demands, and in such instances the estimates provided using a Marshallian demand curve might well be inaccurate.

MEASURING COSTS AND BENEFITS

Measuring Costs

Costs are the goods and services a project takes from the economy. Federal government practice separates costs into two major types: associated costs and project economic costs.[3] *Benefits* are the additional goods and services a project will supply. In either case an appropriate guide is the "with and without" principle: a good or service that would be available to the economy *without* the project but will be unavailable *with* it is a cost; one that will be available *with* the project, but unavailable *without* it is a benefit.

From the standpoint of society, project costs are opportunity costs. They are things society gives up because of the project, so the cost of a public project is the alternative benefits foregone. Obviously, any cost incurred *before* the project would not be a project cost. One must be careful, however, in applying the with and without rule, for the alternative must be a realistic one—it must be politically and socially feasible as well as technically feasible. It would not be realistic, for example, to base National (or State) Forest land values on the sale price of nearby lakeshore recreational subdivision lots. With few exceptions, National Forest land is simply not for sale.[4]

---

[2]Recent empirical studies suggest that differences between willingness to pay and compensation demanded for loss are both frequent and substantial, with compensation demanded running from three to 20 times that of WTP (see Gregory, 1986).

[3]Associated cost is "the value of goods and services over and above those included in project costs needed to make the immediate products or services of the project available for use or sale." Project economic cost is "the sum of installation costs; operation, maintenance, and replacement costs; and induced costs." The induced cost is "uncompensated adverse effects caused by the construction and operation" of the project.

[4]The possibility of sale of public forests has been raised and discussed seriously. See *Selling the Federal Forests* (Gamacine, 1984).

As with benefits, the measuring device for costs is that of willingness to pay, with this difference. Most goods used by forestry projects will be producer goods—goods used in the production of some other good or service. In determining the social cost of such goods, one must determine if the amount of the good used will affect its price. If price will not be affected, then the market value of the good used will be its social cost, for this would be the value of the good if used elsewhere. If price will be increased, however, then the lost "producer's surplus" must be calculated and included in the social cost.

As an example, take the situation in Figure 13.4. The demand curve for the product in question is DD. Without the project the quantity marketed is Q at the price P. If the project requires the quantity $Q - Q'$ and if no new supply is forthcoming, price will rise to P'. Under these conditions the social cost will not be the quantity used by the project multiplied by either the beginning or ending price, but rather that indicated by the area $ACQQ'$. It is the quantity multiplied by the beginning price, plus the area of the small triangle ACW—the same procedure as that outlined for consumer's surplus.

If a factor of production used by the project is one produced under monopolistic conditions (the factor market is imperfect), then the true social cost is difficult to determine. The best method would be to estimate the profit realized by the producer—which, if accurately determined, would be the producer's surplus. This part, which is over and above any resource or other factor cost, is part of social *benefits*, not costs, and should be so treated.

In policy analysis, when the effect of regulation or legislation is being analyzed, the producer's surplus for an entire region, or over an entire industry, may be affected—for example, by a tariff, or a restriction or sale, or a subsidy for

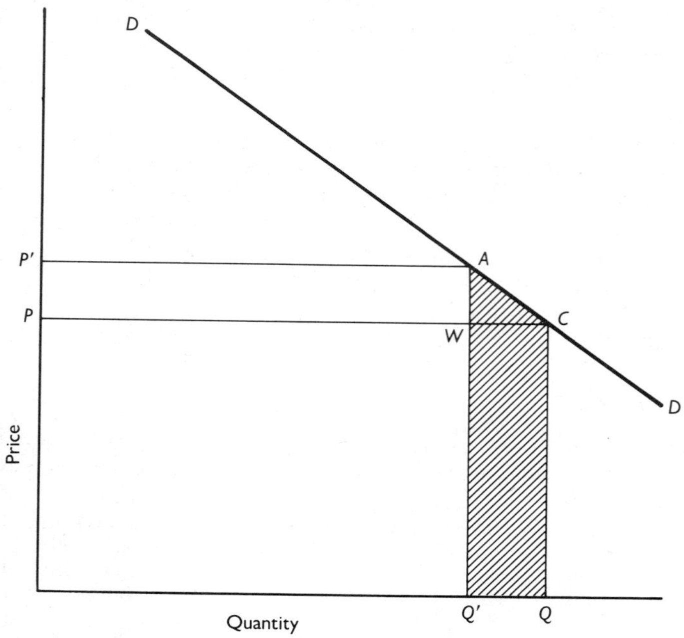

FIGURE 13.4. Social cost if purchase of productive factor for project use raises its price from P to P'.

planting, and so on. In such cases the effect on producer's surplus is calculated by determining the change in the area between the price line and the supply curve.

How often will this situation be encountered in the United States? The first would seem unlikely in any but the largest of forest projects, but could arise when dealing with certain policy questions. Imperfect markets for factors are not uncommon, but unless these profits are significant in the overall picture they would probably not affect project choice.

What about labor costs? This is a much more difficult question, and one more likely to arise. During the 1930s the national and state forest services assisted in alleviating regional unemployment. Should such situations arise, the proper approach would be to use a "shadow price" for labor. The shadow price would be the estimated social opportunity cost of the labor and might be considerably lower than the wage actually paid. Why? Because the opportunity cost to society of hiring otherwise unemployed people could be very low: if they are unemployed, their foregone production (their lost contribution to the economy) would be slight, and this is what the social cost to the nation would be—not the actual wage paid. In more usual cases, however, the social cost would be measured by the wage paid—as it would be for a private corporation.

As mentioned earlier, all taxes, custom duties, depreciation and depletion, and public subsidies are not costs to public projects, even when such a charge is levied. These charges are simply payments transferred from one part of the public to another—in the case of taxes, transferred from citizens to their government. Even subsidies are transfer payments, and if the subsidy happens to affect the price of the product, that effect has been included by weighting the quantity produced by its price.

There are also costs that are not valued in markets. These include negative externalities associated with the project—increased traffic on roads, increased silting of salmon or trout streams, destruction of scenic views, reduced wildlife, and so on. Determining nonmarket values is neither simple nor easy. Since this topic will be discussed at some length in a later chapter, we will not cover here other than to point out that all costs can be considered as negative benefits.

Measuring Benefits

Direct Benefits

Like costs, direct benefits are determined by willingness-to-pay criteria. If the market for the project product is purely competitive, then and only then will the quantity of the product produced, multiplied by its market price, yield the direct social benefit of that product. If the output of the project is sufficiently high to affect the product price, however, the problem is more complicated. Figure 13.5 provides an example: given the demand curve D and the before-project price of P, output of the product (in a competitive market) will be Q. If the project's production will be $Q' - Q$, then the new price will be P'. The total direct benefit from the project's production will be the value of the new output multiplied by its (new) price $(Q' - Q)P'$, *plus* $\frac{1}{2}(P - P')\ (Q' - Q)$. In terms of areas in the diagram, the direct benefit from the new output will be the area $QQ'BA'$, plus the small triangle of consumer's surplus, ABA'.

Although it does not affect the project's social benefit in any way, it is worth noting that a change such as that shown in Figure 13.5 also has a decided effect on producers. Producer's surplus is generally considered to be identical with profit, or with quasi-rent, and is measured by the area above the supply curve and below the

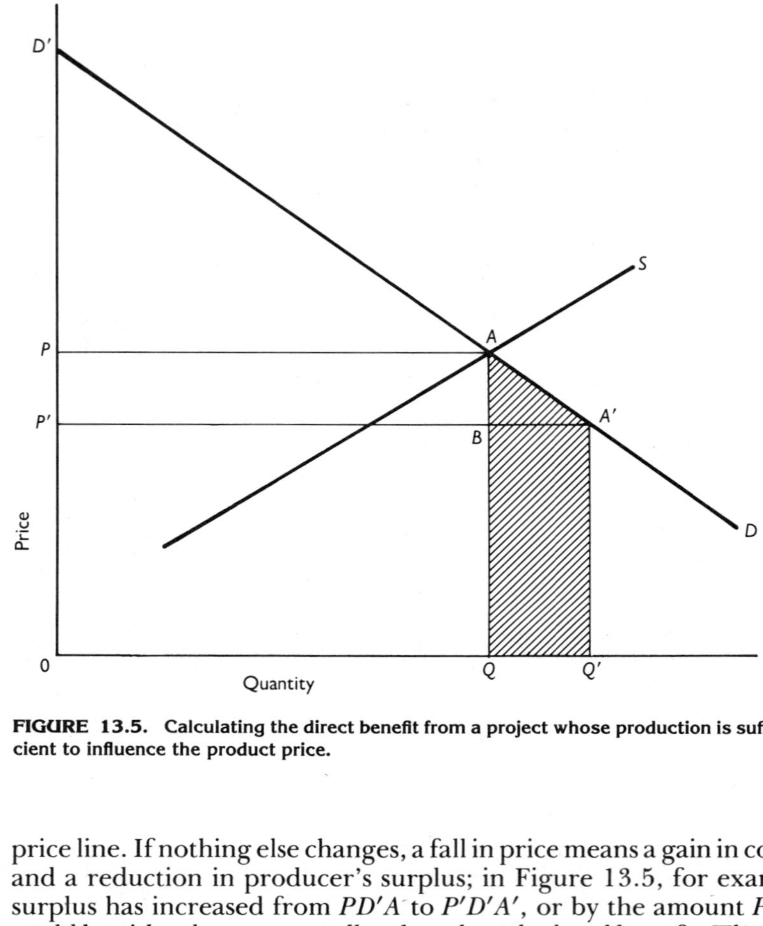

FIGURE 13.5. Calculating the direct benefit from a project whose production is sufficient to influence the product price.

price line. If nothing else changes, a fall in price means a gain in consumer's surplus and a reduction in producer's surplus; in Figure 13.5, for example, consumer's surplus has increased from $PD'A$ to $P'D'A'$, or by the amount $PAA'P'$. This area could be either larger or smaller than the calculated benefit. The point is that from a social perspective a gain to producers is as worthwhile as a gain to consumers. It can even be more valuable, for producers may be more likely to invest for future benefits.

If the market for project products is imperfect, the same analysis would apply, although determining the product supply curve might be difficult.

One more observation: many forest products are producer goods and their demand is derived from the final product. With a derived demand curve, the area above the supply curve for other factors is the producer's surplus. Moreover, this is a fully compensated demand curve, for it takes into account the change in other factor costs as output is changed.

Indirect Benefits and Costs

With public projects there are often net benefits to society additional to those gained by consumers of the project output. Roads built for access to timber, for example, often provide access for hunters, fishermen, and campers — social benefits that should be included in the project total. In fact, the amount and nature of almost any timber harvest can affect the production of other forest products — recreation, wildlife, forage — either positively or negatively. If the effect is positive, it constitutes a benefit and should be added to the project total. If negative, it is a project cost and should be so entered.

To obtain a measure of the total social benefits and costs of timber projects often necessitates an analysis going far beyond the returns from just the timber. Because a forest produces many of its products jointly, timber production usually results in additional products: recreation, wildlife, aesthetics, and so on. These are as much a product of the forest as timber is, and to society may have as much, or even more, benefit. But because they are not valued in the marketplace, these nontimber products pose a different set of problems to the analyst. Discussion of these products and of how their associated benefits might be measured will be delayed until Part 4, which deals directly with multiple use, joint production, and assessment of nontimber values.

UNCERTAINTY

Project analysis and the decisions such analyses are designed to aid are necessarily based on *estimated* information. But costs of materials and labor wages change frequently, as do interest rates and the prices of products produced by a project. Moreover, prices of substitutes change and new substitutes appear, technological improvements alter both methods and markets, and timber product yields will be affected by weather and are subject to losses through fire, insects, and disease. We live in an uncertain world: the question is "Can risk and/or uncertainty be brought directly into project analysis?"

Keep in mind that in economics, as in decision theory, risk is usually applied to events that have a known probability or probability distribution of outcome, while the term uncertainty is reserved for events whose outcome has no known probability distribution. The treatment of risk and uncertainty in project analysis is a major subject for debate among economists. The following discussion makes no pretense of being thorough, but will present the approaches often used in the mid-1980s.

In practice most forestry project analysis do not distinguish between risk and uncertainty. There is considerable justification for treating these the same, since the long time periods involved in timber production and the lack of precise information on timber yields or probable outputs of recreation, wildlife, and other forest products have discouraged most of those attempting to calculate production probabilities. The result is that with forestry projects much of the important data—future prices, future costs, future yields, future interest rates, and even future harvesting and marketing techniques, are "uncertain."

In most forestry projects certainty equivalents are used—that is, estimates of future costs, yields, and product prices are treated as if they were certain. Variations on the certainty equivalent approach include using prices calculated from regressions based on historic price data or those obtained by projecting current price trends, using yield increases adjusted for anticipated genetic or silvicultural developments, making trend estimates based on expected rates of technological development, or making (for example) estimates of future recreation visitor days or hunting days through any of several regression techniques.

When and if probabilities can be derived, then the *expected net benefits* of a project should be calculated, which means calculating both the expected costs and expected benefits. The *expected* value of a particular event is the sum of the weighted possible outcomes of that event—with each possible outcome weighted by the probability of its occurrence. To illustrate, if one knew that at the end of the selected rotation the yield per acre of a plantation would be

50 cords with a probability of $\frac{1}{4}$

60 cords with a probability of $\frac{1}{2}$

75 cords with a probability of $\frac{1}{4}$

then the expected yield would be $\frac{1}{4}(50) + \frac{1}{2}(60) + \frac{1}{4}(75)$ or 61.25 cords per acre. (The probabilities, of course, must add to 1.)

The same type of calculation might also give the expected price per cord of the output; the product of these two—the product of expected yield and expected price—would give the expected value per acre of the plantation output.

Unfortunately, our experience with forest plantations is such that information of this type is rarely available for even the physical yields, let alone for prices and costs. Still, those who deal with plantation projects usually have some idea of the way prices seem likely to go, and probably have considerably firmer ideas as to the likelihood of yield variability. In such cases it would seem sensible to *use* this information explicitly, rather than make all calculations based on current prices and standard yield tables.

To allow for both risk and uncertainty of future events, the usual practice is to adjust the discount rate. In effect, one starts with a risk-free interest rate—most often the rate obtained from U.S. Government bonds—and to this adds a risk premium. The premium is justified by the knowledge that future cash flows that are uncertain are less valuable than those that are certain. There are two problems with this: (1) there are few guidelines available on how large the premium should be, and (2) such an adjustment assumes that the risks or uncertainties involved are compounding (Clutter et al., 1983).

The other general method of handling uncertainty in project work is that of using a range of values for important items. One might, for example, carry out an analysis using high, low, and intermediate values for each item expected to be significant. In effect, this means carrying out a sensitivity analysis for each of the selected values. Such an analysis can be of substantial aid to a careful decision-maker. Suppose, for example, that an analysis of one project indicates that product prices will have to rise from current levels by 3% per year for the project to break even, while for another prices could drop by 20% from existing levels and the project would still be viable. This would surely be valuable information to the decision-maker—more valuable than an analysis presenting only a "most likely" set of prices and leaving the task of determining the validity of those prices to the decision-maker. Sensitivity analyses often reveal that a project is especially vulnerable (or sensitive) to changes in certain values—which is why they are named sensitivity analyses.

Because the interest rate selected is of such importance in most forestry projects, almost all analyses should be carried through using at least two, and preferably three, different rates. But there is considerably more to be said about interest rates for public projects.

DISCOUNT RATES FOR PUBLIC PROJECTS

From around 1950 until the 1970s there was much discussion of the discount rate appropriate to public projects. At first attention concentrated on suitability of the so-called "pure" rate of interest—one completely risk free—that was often defined as the average rate being paid on long-term U.S. Government bonds. During the 1950s this varied from around 2.5% to slightly over 3%, with the low rate justified by observing that this was the opportunity cost of capital to the U.S. federal government. After all, it was argued, the federal government could borrow all it needed or wanted at this rate. However, use of this same rate continued even when U.S. bonds were returning over 4%, and when long-term federal borrowing was halted due to a legal limit that fixed a $4\frac{1}{4}\%$ ceiling on debts maturing any more than five years from the time of sale (Grant, Ireson, and Leavenworth, 1976).

What is now considered the appropriate discount rate for public projects? The general rule for determining the social rate of discount is that it should be the opportunity cost of capital to society. Like most general rules, however, this one is much easier to quote than to follow. Government funds are obtained by taxing or by borrowing (selling securities). In either case, funds thus obtained are no longer available to the private sector. Consequently, the opportunity cost to society should be the projects foregone by the private sector—the projects in which the private sector would have invested if the money had not been taken by the government. By this reasoning, the discount rate appropriate for government projects would be the rate paid by private corporations on their marginal projects—adjusted, of course, for any differences in risk and project duration.

The social opportunity cost is also affected by the amount of investment. One cannot take the market rate in private markets as a guide to the rate for government funds unless the amount involved would not influence the total amount of money needed by government. So far as the United States is concerned, few forestry projects are likely to be of such a size, but this is a consideration in some developing nations.

Since 1969 federal agencies have been required by the OMB to use a discount rate of 10%—unless some other rate has been specified by law or by a legislated guideline. The Water Resources Development Act (1974) required the Water Resources Council to issue a discount rate as an annual basis for use in evaluating federal water projects. In 1981 this rate was 7.5%—well below the rate specified by OMB.

Both the 1974 Renewable Resource Planning Act and the 1976 National Forest Management Act directed that economic efficiency be a major criterion in resource planning and decision making, while the Management Act also requires the Forest Service to develop guidelines and to recommend an appropriate discount rate. The procedures followed in accordance with these directives may be summarized as follows (Row, Kaiser, and Sessions, 1981).

The starting point was taken to be the opportunity cost of capital in the private economy. If this is to be comparable to federal resource investments, it should be based on marginal (not average) investments, and should be adjusted to include corporate taxes ("Using a low after-tax rate of return to private capital for federal projects would bias capital allocation to the public sector") (Row et al., 1981). The discount rate should not include inflation or allowances for risk or environmental protection.

To determine the private opportunity cost of investment capital in forestry several statistical series were examined (Row et al., 1981). After considering these, however, it was "concluded that they had little overall advantage over a relatively straightforward measure of return to capital—interest rates on Aaa corporate bonds." Adjusted to real interest rates by using the GNP implicit price deflator, these interest rates appeared to have been "remarkably constant" during the 1960–1978 period examined, and were also relatively low when compared to average rates of return to capital in the private sector (the rates varied from 3.48% in 1961 to a low of −1.13 in 1974). Adjusted for an average corporate tax rate of 40%, the earning rates *before taxes* were around 3.6–3.9%. On this basis, the Forest Service recommended using a 4% discount rate for long-term land and resource planning.

The significance of the discount rate used in calculating benefits and costs stems from two quite different points. On the one hand this rate may be interpreted as the point at which funds should be taken out of the private sector and invested by the public sector. On the other hand, the discount rate also influences the scale of the project and whether or not the project will be justified.

The going market rate for private funds is not likely to be appropriate for public projects, since the social discount rate must balance the productivity of an investment against the reluctance of society to sacrifice current for future consumption. Social time preference tends to be considerably lower than individual time preferences, and properly so. Individuals take a myopic view of the future; their foresight is limited by their lifespan and often even more by their desire for present consumption. Society, however, must consider the needs of future generations. This strongly implies that social discount rates should be considerably lower than those for private investment.

How much lower? Some have argued as follows: imagine yourself returning to earth 60 years from now. What interest rate would you then say should have been used today? In answering, think again about how high discount rates influence resource allocation over time. High discount rates mean resources will be used rapidly, for future values are lowered. The present value of a tree worth $1000 sixty years from now is $3.28 at a 10% discount rate: at 18%, which was the going rate in 1982, the same tree would have been valued at somewhat less than 5¢.

The effect of high discount rates on resource utilization has led some economists to suggest that natural resource projects should be discounted at lower rates than other types of projects. Some natural resources are limited in quantity and future generations should not be handicapped by severe resource shortages. Others observe that resources constantly change—that technology always makes substitutes available when necessary and that if one is to judge from the past, then future generations will be a lot better off than we are today. So why make sacrifices now? In the previous chapter the impossibility of making valid interpersonal comparisons was seen to be a major problem to welfare economists, and therefore to benefit/cost analysis as well. The problem of social discount rates is even tougher, for its determination requires making *intergenerational* comparisons. Moreover, if we want to keep our decisions democratic, we should let the affected individuals express their own preferences—but how do we ask unborn generations how *they* would like resources allocated *to them*?

Fortunately, this problem does not stop society from making decisions between projects. We continue to use discount rates when evaluating public projects, for somehow or other future uses must be weighed against present uses. If discount rates are used consistently, allocation decisions can at least be made explicitly and on a comparable basis. The latter concern is particularly important, for even government funds are limited, and defensible choices must be made between projects.

QUESTIONS AND PROBLEMS

1. Distinguish between benefit/cost analysis and cost effective analysis.
2. Discuss the differences, and similarities, between objectives of public and private forestry, showing how these differences may lead to decidedly different management decisions.
3. What is meant when one says "A major objective of public investment is to increase *aggregate consumption*"?
4. What about objectives such as increasing employment or maintaining community

stability. Discuss the policy of making managerial adjustments in national forests to promote and protect the economic stability of local communities.

5. Public forest managers usually have more than one objective in mind when making forest management decisions and not all of these necessarily have an economic impact. How are such decisions made?

6. The 1983 report of the U.S. Water Resources Council described a four account system to be used in evaluating effects of alternative projects. Outline this system and give its major advantages. Do you think this system could, and should, be adopted by state and national forest administrators? Why?

7. Democratic principles normally require that *individuals* have a voice in public decisions. How is this handled in practice and how is the individual given a voice in the outcome of a benefit/cost analysis?

8. To what extent can, or should, benefit/cost analysis be used in making public investment decisions?

9. Why should a strict financial analysis be required for major public investment decisions? How and why does a social benefit/cost analysis depart from a financial analysis?

10. Assume that the elasticity of demand with respect to price for Douglas fir lumber is 0.8 and that by increasing the allowable cut in the Pacific Northwest the output of Douglas fir lumber can be increased 15%. Present lumber price averages $300 per thousand board feet, and present production 8 billion board feet. Assume a project is proposed to accomplish this increase.

    a. Assuming the demand is linear, draw the demand curve for Douglas fir lumber and indicate the consumer's surplus at the above price.

    b. Assuming all other things remain constant, what would be the yearly benefit from the project?

    c. If I argue that since stumpage price is a residual, a reduction in stumpage price of 15% on national forests would (since they are the largest producer and are therefore the price leader) soon produce a 15% drop in lumber price, and that therefore the same benefits could be attributed to a project designed only to lowering Douglas fir stumpage prices by 15%. Would you agree? If so, or if not, why?

    d. If, instead of the above, one assumes that the project would increase Douglas fir lumber output by 15% but would have no effect whatever on its price, what benefit should be ascribed to the project?

    e. In the situation of (c) above, it seems possible that lumber producers might realize some benefits. How would this alter the analysis?

    f. Would the result be any different if regional benefits were to be measured instead of national?

    g. These, of course, are not *net* benefits. What costs would one need to include in such a project?

11. When should one take into account the income effect on willingness to pay?

12. What sort of an adjustment should be made to labor costs when determining the benefits of a public project in areas of widespread unemployment?

13. How are taxes accounted for in public projects?

14. Since project analysis nearly always deals with future costs and benefits, how can one be sure that the data used in the analysis is (or will be) correct? If adjustments should be made, how are they quantified?

15. What is a social discount rate? How does it differ from the market (prime) rate? Is the difference really justifiable and, if so, on what basis?

chapter fourteen

CHOOSING AMONG INVESTMENT ALTERNATIVES

The economic basis for making private and social investment decisions, as well as some of the methodology of project evaluation, should now be reasonably well in hand. With corporate or other private investments a project will be judged acceptable as long as discounted project revenues exceed discounted project costs. Essentially the same guideline is used in public projects: a project is justified as long as the discounted benefits exceed the discounted costs. But what if more than one project meets this necessary condition? Or what if there is some constraint on investment funds at the time of decision? Are there practical economic guides for choosing among alternative investments?

In addition to these general questions, forestry poses a specific one. Because timber production often requires unusually long time periods and because the tree is at once product and factory, forest managers must determine how long a tree or stand should grow before it is harvested; in forestry terms, they must determine the best rotation. The question is most easily seen in plantation projects: when should the plantation be harvested? This chapter deals primarily with determining the rotation that will maximize *financial* returns to the owner from timber, although one should keep in mind that other factors and products may also have an influence. Before dealing with rotation determination, however, the more general problem of project choice should be explored.

PROJECT CHOICE

On the basis of the previous three chapters it would seem that choosing among alternative projects should be simple: choose the one with the largest net present value. Or should it be the one with the highest internal rate of return? Will the project with the highest NPV also yield the highest IRR? Or should the choice be made on some other basis? Like so many economic problems, the answers often depend on the conditions under which the decision is being made and on the nature of the alternatives.

Decision-makers need some way to rank projects in order of preference. Over the past few decades many rules of thumb have been devised, especially in the business world, but by 1980 most corporate executives and nearly all public decision-makers knew about and were using one of four basic approaches—net present value, the internal rate of return, the benefit/cost ratio, or the net benefit/investment ratio—as the major guide in project selection. Although they are closely related, each approach has important advantages and drawbacks.

GUIDES FOR PROJECT CHOICE

Net Present Value or Discounted Cash Flow

Three objections are often raised to using NPV as the criterion for ranking projects: (1) the NPV concept is not well understood by business executives, who have long measured investments by the internal rate of return; (2) NPV automatically favors large projects; and (3) calculation of NPV requires preselection of the discount rate.

The first objection may have been important a decade or two ago and could still have some relevance for those working in developing nations. In the United States, however, few in either private industry or public agencies are strangers to NPV calculations, although among nonindustrial private forest owners there would be a few who regard NPV as an unnecessarily complicated guide for forest investment.

The second objective is valid: total reliance on NPV rankings would lead one to favor large projects. Projects A and B in Table 14.1 would be ranked as listed, with A as the first choice. Yet project B is a substantially better investment: at a cost of only one-half that of A it has nearly the same return. This is a good reason for insisting that project analysts present both NPV and IRR in their reports.

The fact that calculation of a project's net present value requires preselection of the discount rate may cause occasional problems, particularly in public projects when there is no attempt to estimate the social opportunity cost of capital. In the United States, however, the analyst of public projects is usually given the discount rate, much the same as the situation in private industry. In either, the normal procedure would be to make the analysis using several discount rates—to create a sensitivity analysis with respect to discount rates (see page 302).

Internal Rate of Return

Many corporations make investment decisions on the basis of the particular project's IRR. If the IRR is above a preselected level—the alternative rate (hurdle rate, or cut-off rate) set by the company—the project is acceptable. If the investment budget will not accommodate all acceptable projects, they will then be taken in order as they are ranked by their IRR. Three objections are usually cited: (1) use of the IRR implies that all intermediate returns can be reinvested at the IRR rate, (2) in rare cases, calculation of the IRR may yield two or more answers, and (3) ranking by IRR can cause an error when choosing between mutually exclusive projects.

TABLE 14.1
Discounted Costs and Revenues, and NPVs, of Two Projects To Be Ranked

Project	Discounted Cost	Discounted Revenue	NPV
A	1,000,000	1,150,000	150,000
B	500,000	640,000	140,000

The first objection, at least in part, stems from problems of interpretation: if IRR is interpreted as "the rate of return on capital outstanding per period while it is invested in the project" (Merrett and Sykes, 1963), then no real problem exists. If, however, IRR is interpreted as the return on the total investment over the life of the project, then the reinvestment criticism holds. It is particularly troublesome when comparing projects requiring different capital outlays or with long investment periods. For example, if one must choose between earning 100% on $100 for one year, or earning only 50% on $5000 for a year, the 50% yield is obviously preferred ($2500 is better than $100).

Schallau and Wirth (1980) illustrate the difficulty with an example in which a Douglas-fir plantation, established by an "initial investment of $30,000 resulted in an annual net cash flow of $24,000 for 90 years, and an IRR (from the area) of about 80 percent." Schallau and Wirth point out that this rate of return can only be realized if all the receipts can be reinvested at the 80% rate, which seems highly unlikely. For example, if the project returns can only be reinvested at 10%, then the actual return from the area turns out to be around 13%, not 80%! Schallau and Wirth call this 13% the realizable rate of return (the RRR), which obviously depends on the actual rate that can be obtained on investment funds at the time the funds become available. Unfortunately, to use this RRR for comparative purposes one must know what the future interest rate will be when the income is received and, if one knew this, then all sorts of determinations could be made.

The second reason for avoiding the IRR as a guide to investment choice is because one may discover more than one solution. And if there are two or more IRR rates, which is the correct one? The reason is straightforward—the equation for IRR is a polynominal and so may have more than one root. If approximating the answer graphically, we would find that the two root situation might resemble that of Figure 14.1. It means that the net returns from the project, after becoming positive long enough to result in a positive NPV, become negative and remain so for a period sufficient to make the NPV negative. Each successive reversal in the sign of the calculated NPV will produce another IRR value.

Cases of two or more solutions are rare but possible in forestry. They are most likely to occur with long-term projects that, after a substantial series of positive

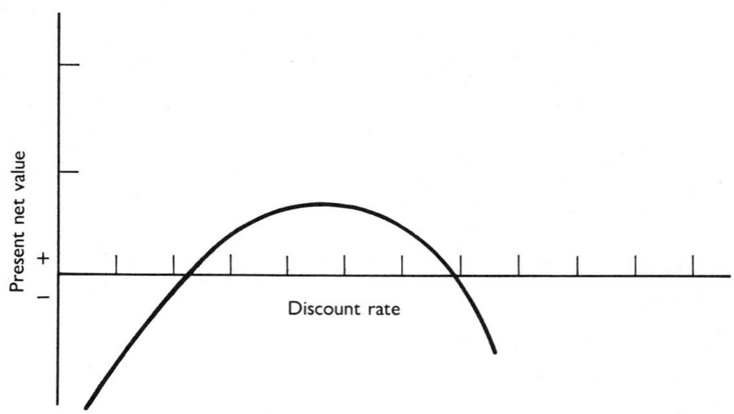

FIGURE 14.1. **When approximating the project IRR, one may discover there is more than one answer.**

earnings, are going to incur heavy costs—perhaps for new equipment, replanting, or some similar reason.[1]

The third problem arises only when dealing with mutually exclusive projects and is best discussed after dealing with the other two criteria.

The Benefit/Cost Ratio

Whether the benefit/cost ratio is calculated as the ratio of gross discounted benefits over gross discounted costs, or as net discounted benefits over total discounted costs, will make no difference as long as the same values for benefits and costs are used. According to U.S. government practice, however, the second procedure should be used by federal agencies (U.S. Govt., Water Res. Council, 1979). The ratio is calculated in a specific fashion—by subtracting the discounted associated costs from the NPV of gross benefits, then comparing this with the NPV of the "project economic cost". While it is true that a cost may be considered a negative benefit, it is also true that by subtracting from the benefits all the associated costs (the value of goods and services necessary to make the project benefits available) the ratio will be higher than if the associated cost had been included as part of the total project cost. For example, suppose the gross discounted benefit was 1,500,000, the gross cost was 1,000,000, and the associated cost 400,000. If associated costs are included as part of the gross discounted cost, the ratio is 1,500,000 over 1,400,000, or 1.071. If included as a negative benefit, the ratio becomes 1,100,000 over 1,000,000, or 1.1.

Net Benefit/Investment Ratio

This variety of the benefit/cost ratio, defined as the PV of the net benefits, divided by the PV of the investment, has some advantages over its better known parent. The major advantage of the net benefit/investment ratio is that it can be used to rank all but mutually exclusive projects (which we will see should be selected on the basis of NPV). The net benefit/investment ratio is seldom used at present, but because it is easily calculated from cash flow analyses it may achieve greater use in future years (Gittinger, 1982).

PRESENT PRACTICE IN PROJECT SELECTION

The four criteria for selecting projects do not exhaust the list of proposed methods, but they are the major ones used today. The first two, NPV and IRR, are generally used by individuals and industry, as well as by many agencies of state and federal governments, the World Bank, the Food and Agricultural Organization of the United Nations, UNIDO, and others. In practically all project work the usual procedure is to calculate NPV under several interest rates and also the IRR. The benefit/cost ratio and the net benefit/investment ratio are used primarily by public agencies.

Application of benefit/cost analysis by developing countries has resulted in major advances in the methodology, where subsidies of one kind or another, foreign exchange rates well below black market rates, rationing, imperfect capital markets, high unemployment, and underemployment are frequently encountered. A variety of techniques, many involving the determination and use of so-called shadow prices, have made project analysis in such nations far more revealing from a social standpoint. More will be said about this in Chapter 18.

[1]Analysis of cash flow series that have two or more reversals is treated at some length by Grant, Ireson, and Leavenworth (1976). See their Appendix B, pages 543–562.

Situations Influencing Project Selection

Four different situations are usually recognized as influencing the way project criteria are applied.

1. There are no budget constraints, and all projects under consideration are independent of each other.
2. The projects being considered have different lives—they cover different time spans. Any valid comparison between such projects requires correcting for the different investment periods.
3. Constraints: There are constraints on some factor—often available capital—so that not all the projects under consideration can be implemented.
4. Certain projects are mutually exclusive; if a particular project is implemented, some other project becomes impossible. (One can't have both a wilderness and intensive timber management on the same area; if you plant pine you rule out making a hardwood plantation on that area.)

In the first situation there is no problem. All projects with positive NPVs will have positive IRRs and B/C ratios, and all these will be profitable or beneficial to society. If there are no limiting factors and the projects are independent, the decision-maker can put the list up and throw darts to select projects, or choose them in alphabetical order, or simply say "let's take 'em all."

Comparing Projects with Different Lives

The second situation occurs when comparing projects that cover different time spans—projects with different lives. Consider the choice between two plantation projects, one of aspen on a rotation of 25 years, the other of sycamore on a rotation of 20 years. Present values of costs and returns, and the net present values of the two projects, are as follows:

	Sycamore (20 yrs)	Aspen (25 yrs)
PV of returns	\$30,000	\$31,000
PV of costs	\$20,000	\$20,000
NPV of project	\$10,000	\$11,000

Based on the above NPV values, the aspen project would be selected, but to choose correctly between these two projects we must put them on a basis of equal project life. Several ways to accomplish this can be used, one of which is to assume that the cash flows from the projects will be repeated to some common multiple of the project lives and determine the present value of this terminable periodic series. The lowest common multiple of the two projects above is 100 years. The formula for a terminating periodic series of payments (Clutter et al., 1983) is

$$A_t = PV \frac{(1 + .0i)^{nr} - 1}{(1.0i)r - 1\ (1 + .0i)nr}$$

where A_t = present value of a uniform terminating series of periodic payments
PV = present value of the first periodic payment
i = the selected discount rate, in decimal form
n = the number of periods
r = the length of the period (rotation)

At a discount rate of 6%, the sycamore project has a future value at 20 years of

$$\$10{,}000\ (1.06)^{20} = \$32{,}071$$

and the aspen project of

$$\$11{,}000\ (1.06)^{25} = \$47{,}211$$

The present value of five payments of $32,071 at 20-year intervals beginning 20 years from now is

$$\$32{,}071 \frac{(1.06)^{5\,(20)} - 1}{[(1.06)^{20} - 1]\ (1.06)^{5\,(20)}}$$

which equals $14,488. A similar calculation for the aspen project gives a value of $14,299 so the proper choice here is not the aspen, but the sycamore project: five payments of $32,071 at 20-year intervals is worth more than four payments of $47,211 at 25-year intervals.

One might also assume that each project will be duplicated as soon as it finishes, in effect giving both projects an infinite life (infinity is also a common multiple). In the above example this requires calculating the future value of both projects, but then calculating the present value of each of these as a series of perpetual periodic payments.

Using formula (4) from Chapter 7 yields the following present values of the two projects when duplicated an infinite number of times: For the sycamore project:

$$\$32{,}071 \frac{1}{(1.06)^{20} - 1} = \$14{,}531$$

Using the same procedure for the aspen project yields a value of $14,342. Again, the sycamore project would be selected.

Other methods of comparing projects of different lives include reducing both to equivalent uniform annual payments (this is the annual payment, or annuity, over the project life span that has a present value equal to that of the project—see Grant et al., 1976). Mishan (1971) developed a "normalizing" procedure that also permits comparing projects with different lives and is useful in certain other ways. In forestry, where projects of different species and rotations are frequently compared, the approach illustrated above seems most appropriate, for when one takes infinity as the common multiple for rotations, she is comparing projects on the basis of their land expectation value, or "Le," which will be developed later in this chapter.

Constraints on Projects

The third situation occurs when there is some constraint—some limitation—that inhibits choice of some projects. The most usual limitation is a budgetary one, although in forestry it is not rare to encounter land restrictions or environmental or political constraints.

In most cases NPV and IRR will result in the same project choice, but not always. One such instance is illustrated in Figure 14.2. Two projects, A and B, are shown on a graph similar to that used to approximate internal rates of return. Project A has an IRR of 13%, while that of Project B is 22.5%. At any discount rate between 10 and 22.5% Project B would be the clear choice, since both its IRR and NPV are above A's. But Project A is much more sensitive to changing interest rates than Project B. Suppose the appropriate discount rate is 8%? Now, using NPV as the guide, Project A would clearly be the choice, for at 8% its NPV is over $100,000 greater than that of B. At any discount rate below 10% Project A has the higher NPV, while above 10% Project B would be chosen. The two criteria, IRR and NPV, do not necessarily result in the same project choice.

There is one situation in which use of the NPV to guide project selection would lead to error. Suppose one were asked to rank the following three projects by their NPV as determined by discounting at 10% (Table 14.2).

Project A would be ranked first, C would be second, and B third. Now suppose, as is usual, there is a capital limit on expenditures for projects, and that the limit is set at 1200. The ranking by NPV would be unchanged: Project A would be number one, and only Project A would be implemented. But this would be wrong, for Projects B and C combined would not only also just exhaust the capital avail-

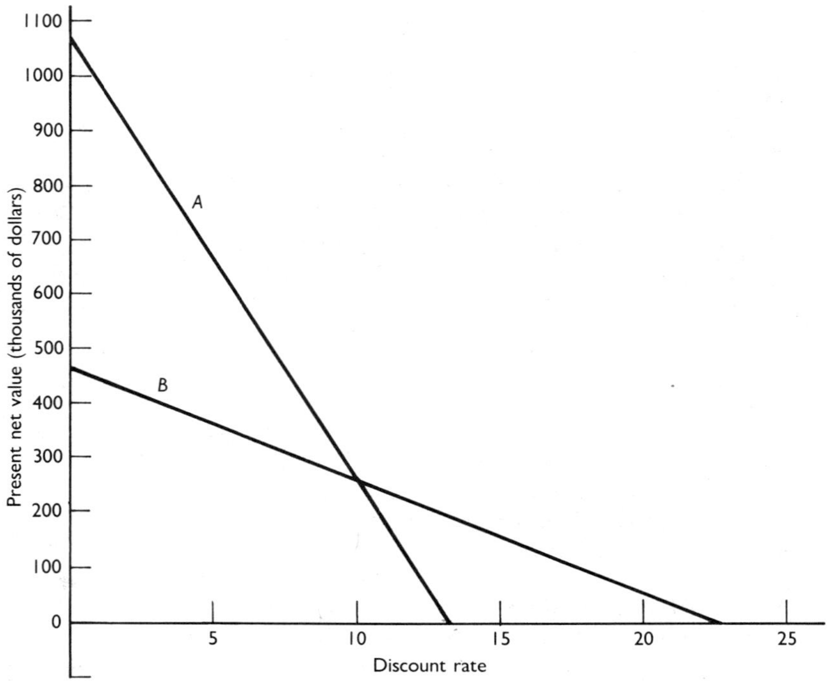

FIGURE 14.2. Present net value and internal rate of return do not always result in the same project choice.

TABLE 14.2
Ranking Projects by NPV and B/C

Project	Discounted Cost	Discounted Benefit	Net Present Value	B/C
A	1200	2200	1000	1.83
B	500	1100	600	2.2
C	700	1350	650	1.93

able, but would also give a decidedly higher yield: benefits would now total 2450 instead of 2200. The proper criterion in this case is to rank the projects by their benefit/cost ratio. In the above illustration, this would result in ranking Project B first, C second, and A third.

Mutually Exclusive Projects

Finally, there is the case of mutually exclusive projects. In these projects the easiest way is to use net present value as the decision guide, for this gives the correct answer. To use IRR correctly requires an additional step. Suppose, for example, A and B are two mutually exclusive projects, with A having the larger IRR. To ascertain which of the two is financially preferable, a fictitious third project, A-B, is created by subtracting the cash flow of B from that of A, each year, then calculating the IRR of this stream. If this IRR is larger than the previously determined cutoff rate, then Project A is preferred; if smaller, then Project B should be selected. This amounts to a two step rule:

$$IRR_A > IRR_B$$

and

$$IRR_{A\text{-}B} > IRR_B$$

then Project A should be selected. But if $IRR_{A\text{-}B} < IRR_B$, then Project B should be preferred. The IRR of A-B is the rate of return earned by the additional investment required by A as compared to B—its incremental cost. A direct comparison of the NPV of A with that of B would seem to be the easier method.

ROTATION DETERMINATION AS AN ECONOMIC PROBLEM[2]

Tree growth is a biological phenomenon, but one that can be altered through management. Growth rates depend on species, site, genetic characteristics, and other silvicultural variables such as spacing, thinning, fertilization, and so on. It is scarcely surprising that foresters, with training heavily weighted to biology, developed biologically based rotation guides. An economist, however, looks on managerial inputs as sources of costs, and views the incremental growth of the tree as growth of the forest capital investment. Financial rotations, consequently, are based on capital theory.

[2]The rotation for a forest is defined by the Society of American foresters as "the period of years required to establish and grow timber crops to a specified condition of maturity."

In earlier chapters reference was made to investment guides peculiar to forestry, particularly to those developed for determining the rotation maximizing the financial return from plantation timber production. The actual rotation length is often decided by factors unknown at the time of planting and is seldom critical from a practical standpoint. The planned rotation, however, is a significant input to the overall forest operation. As Leuschner puts it, "The rotation age determines the tree's age, hence, it largely determines the size, quantity, and quality of the timber" (Leuschner, 1984). It also influences the number of acres planned for annual cutting, the area to be planted each year, and many of the needed silvicultural inputs of forest management. Still, rotation ages not only can but always will be changed to suit the changing objectives of management. One paper company forester, when asked how his company determined their pulpwood rotation, replied "There's the mill manager's office. That's where our rotations are determined. If the mill needs wood, we cut!"

Forest growing stock is forest capital, and the question of how much capital should be invested for how long is as important to timber production as it is to every other enterprise. Hence rotation is an economic as well as a biological problem. We have generally assumed that a forest enterprise will be operated under the same economic pressures and motivations as nonforestry ones, so we have accepted the maximization of discounted cash flow as the appropriate objective for industrial industrial and private management. Public management objectives have also been discussed with the usual criterion for public projects accepted—maximization of aggregate consumption. Stated in these terms, the rotation problem is far from unique. It is certainly not a new problem, for the question "When should this stand be cut?" was inevitable as soon as conscious management for timber production began.

Treating Time as a Factor of Production

If time is treated directly as another factor necessary to the production of wood, both the biological and the economic approaches can be illustrated. As an example, suppose all physical inputs required for a one-acre plantation of *Pinus radiata* (Monterey pine) are assumed to be fixed, and time is regarded as the single productive factor. The resulting production function is shown in Figure 14.3, and is clearly of the same general family as most biological production functions.

It may surprise some that such curves have long been part of a forester's training. Only the terminology has been altered. Change output to yield, substitute age for time, replace total physical product with total yield, and the diagram is that of the traditional yield curve found in almost every forest management textbook published over the past century. And the parallel continues for the two curves derived from the total yield relationship. Divide the total physical product by the corresponding input, and one derives the average physical product (APP)—in forestry called mean annual increment or MAI. The *change* in total physical product that accompanies a *change* in input is called marginal physical product (MPP) by economists, but when time is the input and is kept in units of one year, foresters call this current annual increment, or CAI. The set of traditional forestry curves does not just resemble the physical product curves of production economics: the two are identical!

Maximization of *wood* yield from this single acre would mean holding the timber until the top of the yield curve was reached—a point well beyond the limits of Figure 14.3. Few foresters, however, have concerned themselves with the yield of a single acre. They prefer to strive for the highest possible *continuous* yield of timber from the entire forest under their jurisdiction. This means taking time into

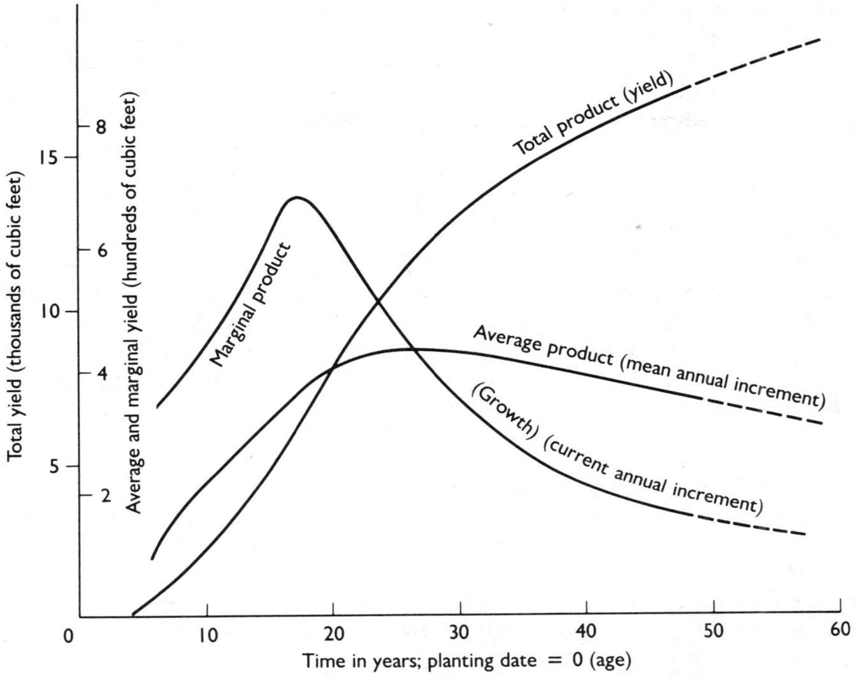

FIGURE 14.3. **Production function for *Pinus radiata*. Average and marginal product curves also shown (note different scales).**

consideration and requires managing on a rotation equal to the maximum current annual increment—harvesting at the culmination of mean annual increment. In the example used here, this would mean a rotation of either 26 or 27 years. We will return to this rotation shortly, for it is mandated for National Forests by the National Forest Management Act of 1976.[3] But first we should explore the various proposals for determining the rotation that would maximize the *financial* yield, often called the economic rotation.

GUIDES PROPOSED FOR DETERMINING OPTIMUM FINANCIAL ROTATIONS

To an economist aware of the identity between the yield curve and the production function, it would seem sensible to calculate costs and revenues, then select the time input that would maximize net returns from the operation—thus determining the optimum economic rotation. To do this, still using the physical data presented in Figure 14.3, one might proceed as follows.

Revenues and Costs

Revenues

Assuming for simplicity that only pulpwood will be produced, and that stumpage will be sold for 12¢ per cubic foot (about $10 per cord), then total revenue at any

[3]Although the act specifies that in National Forests harvest shall not be before culmination of mean annual increment, the specification is qualified in a manner that makes almost any rotation possible. See Dana and Fairfax (1980), p. 232.

time can be obtained by multiplying the yield at the selected time by the product price, 12¢. In using the Monterey pine example, this amounts to little more than relabeling the vertical axis in dollar units; the total product curve, although unchanged in shape or position (Figure 14.4), now shows plantation yields in dollars rather than in cubic feet. The calculations are given in Table 14.3. Now, what about costs?

Costs

Costs will also be kept as simple as reality allows. We will hold the question of land in abeyance and assume that planting costs including all costs of planting, soil preparation, seedling production, and so on, are \$70 per acre and are incurred at the end of the planting period (year 0). Two other costs are identified: a weeding cost of \$8 per acre incurred one year after planting, and the annual maintenance costs of \$5 per acre for taxes, fire protection, supervision, and so on. These costs are also shown in Figure 14.4. Of the three, only maintenance cost varies with time, increasing \$5 every year.

Maximizing Discounted Cash Flow

Following the procedure of Chapter 11, the discounted cash flows can now be calculated and the expected NPV of the plantation determined, using whatever happens to be the appropriate discount rate. In this example the NPV for each year, calculated at discount rates of 6, 8, and 10%, is given in Table 14.3

The procedure in forestry often has been to treat all costs as if they were met by borrowing the required funds at some specified interest rate.[4] The costs, together with the accumulating interest charges, are then carried to any desired date

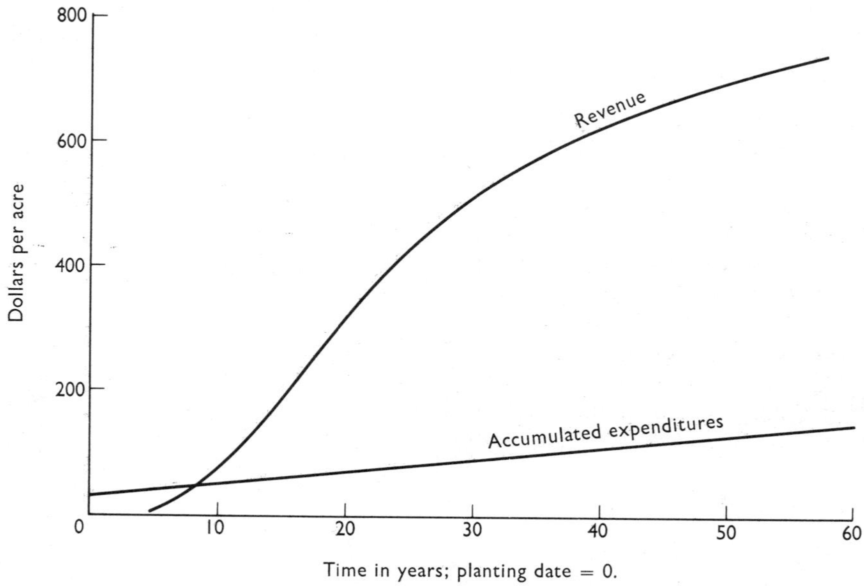

FIGURE 14.4. Revenues and accumulated expenditures per acre for *Pinus radiata*.

[4]Alternatively, one might argue that the interest charge is an opportunity cost: the funds could have been invested.

and can be compared directly with revenues at this same point in time. If a net revenue appropriate to this date is calculated and discounted to the present, the result will be identical to that obtained by subtracting discounted costs from discounted revenues.

Figure 14.5 depicts the same basic cost and return picture as that of Figure 14.4, but includes the interest charges (the shaded portion) as they accumulate at a compound rate of 8%. Note how the cost of interest increases as the time between planting and harvesting is increased.

Table 14.3 reveals that, according to the discounted cash flow criterion, the "optimum" rotation for this plantation changes with changes in the interest rate. If the discount rate is 6%, the maximum value for the stand will occur at a rotation of 22 years (NPV = $166.77). At a discount rate of 8% the maximum value of the stand will occur at a rotation of 20 years, although the value has declined to only $76.91. If the discount rate is further increased to 10%, the maximum of $22.69 is reached at just 18 years. And this is what one should expect. A higher discount rate means that future values are less important than present values—that the time value of money is higher. Future values decrease, and so does the time one can afford to wait before cashing in an investment.

An Alternative Guide to Maximization of NPV

Although there is no particular difficulty in calculating the rotation that maximizes discounted cash flow, there is an alternative of special significance to forestry. The method is often appropriate to uneven-aged management, and is especially suited to situations involving single tree selection systems. To explore this alternative we will temporarily leave the Monterey pine illustration.

Recall from Chapter 7 that two factors influence the slope and position of a curve tracing the growth of capital over time: the interest rate and the amount to which the rate is applied. In Figure 14.6 the curves originating at a and b show the growth of the two amounts, $0a$ and $0b$, at an 8% annual rate. Because discounting is the reverse of compounding, $0a$ and $0b$ are, respectively, the present values at year 0 of any point on aa'' or bb'' (see Figure 14.6).

Now suppose $0a$ were the total cost involved in establishing a forest and no other costs were involved. Under this special condition the annual interest charge would be the marginal cost of holding the stand for another year, for **there is no other factor causing any cost change**. Total revenue is increasing at exactly the same rate as total cost at point b'—where bb'' is tangent to the total revenue curve—for aa'' and bb'' are both increasing at 8%. The value $a'b'$ is the net revenue (total revenue − total cost), and its present net value is ab. Note that the *change in cost* does not equal the *change in revenue* at age t_1 (at t_1 the two curves have different slopes). It is the *rate* of change that is important here. The guide to maximization of present net value is that **the percent change in revenue must equal the percent change in costs**. And under the special condition that all costs are incurred at the beginning of the rotation, net present value will be maximized when total revenue is increasing at the preselected discount rate.

Financial Maturity

In a general sense the financial procedures discussed in this chapter are all designed to determine when timber is financially mature. The term financial maturity, however, has had a special connotation in forestry since Duerr and his co-workers introduced the guide to forest managers under the name *Financial Maturity* (Duerr et al., 1956). One significant change in the guide was suggested; that the rate of

TABLE 14.3
Calculation of Present Net Value, Land Expectation Values, and Forest Rent

Year	Cost	AC Cost	TC 6%	TC 8%	TC 10%	Yield	TR
0	75	75	75.00	75.00	75.00		
1	13	88	92.50	94.00	95.50		
2	5	93	103.05	106.52	110.05		
3	5	98	114.23	120.04	126.06		
4	5	103	126.09	134.64	143.66		
5	5	108	138.65	150.42	163.03	270	32.40
6	5	113	151.97	167.45	184.33	597	71.64
7	5	118	166.09	185.85	207.76	946	113.52
8	5	123	181.05	205.71	233.54	1320	158.40
9	5	128	196.92	227.17	261.89	1721	206.52
10	5	133	213.73	250.34	293.08	2150	258.00
11	5	138	231.56	275.37	327.39	2608	312.96
12	5	143	250.45	302.40	365.13	3096	371.52
13	5	148	270.48	331.59	406.64	3616	433.92
14	5	153	291.71	363.12	452.31	4171	500.52
15	5	158	314.21	397.17	502.54	4766	571.92
16	5	163	338.06	433.94	557.79	5403	648.36
17	5	168	363.35	473.66	618.57	6078	729.36
18	5	173	390.15	516.55	685.43	6763	811.56
19	5	178	418.55	562.88	758.97	7440	892.80
20	5	183	448.67	612.91	839.86	8095	971.40
21	5	188	480.59	666.94	928.85	8712	1045.44
22	5	193	514.42	725.29	1026.74	9295	1115.40
23	5	198	550.29	788.32	1134.41	9845	1181.40
24	5	203	588.31	856.38	1252.85	10363	1243.56
25	5	208	628.60	929.89	1383.14	10850	1302.00
26	5	213	671.32	1009.29	1526.45	11310	1357.20
27	5	218	716.60	1095.03	1684.09	11745	1409.40
28	5	223	764.60	1187.63	1857.50	12156	1458.72
29	5	228	815.47	1287.64	2048.25	12545	1505.40

return from the stand or tree should be equated to the available *alternative rate of return* on investment capital (instead of to the percent change in costs). The stand or tree was then said to be financially mature.

To calculate the rate of return from a stand or tree, the expected change in market value during the selected period is divided by the market value at the beginning of the period, then (normally) reduced to the corresponding one-year rate. Using the Monterey pine data again, at 20 years the expected rate of return is

$$\frac{(1045.44 - 971.40)}{-971.40} = 7.62\%$$

and at 22 years is

NPV 6%	NPV 8%	NPV 10%	Le 6%	Le 8%	Le 10%	FOR Rent
−75.00	−75.00	−75.00				
−87.26	−87.04	−86.82	−1541.67	−1175.00	−955.00	−88.00
−91.71	−91.32	−90.95	−833.74	−640.14	−524.05	−46.50
−95.91	−95.29	−94.71	−598.03	−462.21	−380.83	−32.67
−99.87	−98.97	−98.12	−480.37	−373.51	−309.55	−25.75
−79.40	−80.32	−81.11	−314.15	−251.46	−213.96	−15.12
−56.63	−60.38	−63.61	−191.94	−163.25	−146.05	−6.89
−34.96	−42.20	−48.36	−104.38	−101.32	−99.34	−.64
−14.21	−25.56	−35.05	−38.15	−55.60	−65.70	4.43
5.68	−10.33	−23.48	13.93	−20.67	−40.78	8.72
24.72	3.55	−13.53	55.97	6.61	−22.01	12.50
42.88	16.12	−5.06	90.62	28.23	−7.79	15.91
60.17	27.45	2.04	119.61	45.53	2.99	19.04
76.63	37.63	7.90	144.27	59.51	11.12	21.99
92.36	46.78	12.70	165.61	70.93	17.23	24.82
107.53	55.09	16.61	184.53	80.45	21.84	27.59
122.15	62.59	19.71	201.45	88.38	25.19	30.34
135.92	69.11	21.92	216.22	94.70	27.33	33.02
147.64	73.83	22.69	227.26	98.47	27.66	35.48
156.74	76.45	21.88	234.13	99.50	26.16	37.62
162.99	76.91	19.55	236.84	97.92	22.97	39.42
166.15	75.19	15.75	235.40	93.83	18.22	40.83
166.77	71.76	10.89	230.83	87.93	12.42	41.93
165.22	66.95	5.25	223.82	80.69	5.91	42.76
161.83	61.06	−.94	214.91	72.49	−1.05	43.36
156.90	54.33	−7.49	204.56	63.62	−8.25	43.76
150.76	47.04	−14.20	193.24	54.39	−15.50	44.01
143.66	39.36	−20.95	181.25	44.99	−22.68	44.13
135.79	31.42	−27.65	168.82	35.54	−29.71	44.13
127.33	23.37	−34.22	156.15	26.18	−36.52	44.05

$$\frac{(1181.40 - 1115.40)}{1115.40} = 5.92\%$$

The alternative rate of return is the opportunity cost of capital to the owner of a timber stand. If the best available alternative—bank deposits, savings bonds, municipal bonds, house mortgages, common stock, or investment in a different timber growing opportunity—should happen to pay 10%, and if the alternative is of equal risk, then 10% becomes the opportunity cost of investing in timber growth. When the return from timber production on the property in question drops below 10% of its market value, the stand should be harvested.

Actual costs of timber production are not specifically brought into the analy-

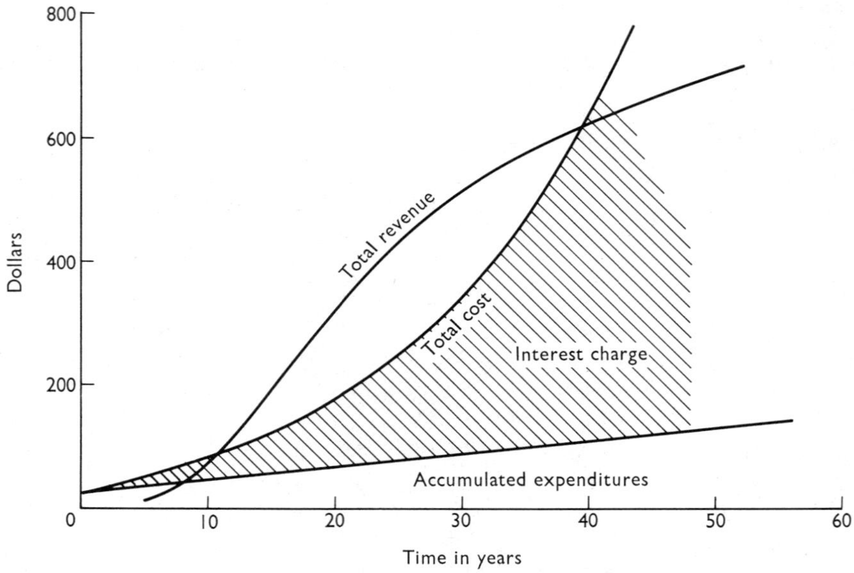

FIGURE 14.5. Costs and revenues per acre for growing *Pinus radiata*, including interest charge of 8%.

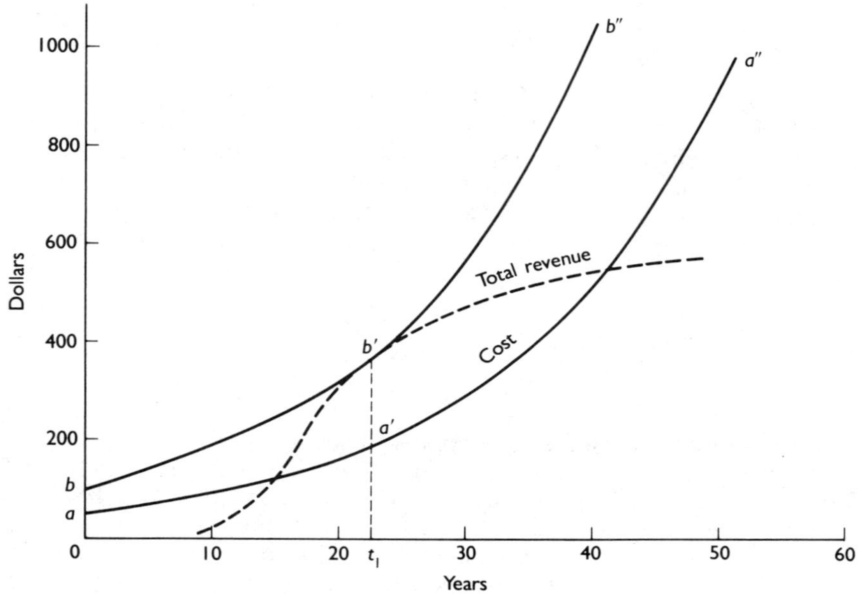

FIGURE 14.6. The guide to maximization of present net worth.

sis. In effect, the principle of sunk costs is employed: past costs of growing the stand are unimportant because, once made, they cannot be changed. As long as the timber stand value is growing at a satisfactory rate—satisfactory being defined as the fastest rate at which the capital can grow elsewhere—then it should be left to grow. If the rate of value growth falls below the best available alternative, capital should be taken out of timber and put to the alternative use. But bear in mind that the best alternative may be that of starting another rotation.

Land Expectation Value as a Guide to Rotation Determination[5]

Land, or soil, expectation values were first presented in 1849 by the German forester, Martin Faustmann, as a solution to the problem of evaluating forest land for tax purposes. These land expectation values (or Le values) have been a cornerstone of forest management since that time.

Herr Faustmann must have reasoned along lines somewhat as follows: If I were planning a forest from scratch, how much could I afford to pay for bare land? In calculating agricultural land values, one first determines the net income the particular field will produce, then decides how much must be invested to obtain an income stream of this same amount. If the going interest rate were 10%, one should be willing to pay $100 for land that will *net* 10 dollars or more every year—that is, after all costs including those of management have been deducted from the sale value of the crop. The value of agricultural land is equal to the cash amount that, when placed at interest, would return every year the same net income as the land. The land value is the "capitalized" value of the expected annual net income.

The transition from crop to forest land posed only minor difficulties to Faustmann, who might have had the following speculations:

> *Foresters say trees are a crop! OK! How much can I afford to pay for bare land if I intend to use it for* timber *production? To be consistent I'll have to treat it just the same as any other crop, but because of the time interval I'll have to include interest charges. And because my income won't be "incoming" until harvest, I'll have to figure its value today, which means I'll have to* discount *the net revenue. But that isn't all' there's one more step. Since I'll be operating on a sustained yield basis I'll be getting this income every so many years. Therefore, I'll have to figure the cash amount that would return this same net income once every rotation. It's obvious.* The value of bare forest land has to be the capitalized value of the infinitely long series of expected periodic incomes from the timber.

To determine Le values, net revenues are first calculated for each potential rotation by subtracting all costs and carrying charges from the total revenue. Each resulting net revenue is then capitalized as if it were the first payment in an

[5]We will follow Davis (1966) in this respect and use land expectation and Le. For a relatively recent review of the development of the Faustmann formula, see Samuelson (1976). See also Hyde (1980) who has a particularly good bibliography on treatment of the Faustmann formula. Also see Gaffney (1960).

infinitely long series of identical periodic payments.[6] The Le value represents the cash amount that would have to be invested today at the given interest rate to yield, at intervals equal to the indicated rotation, periodic incomes just equal to the net revenue expected from the forest. The answer to Faustmann's question of "How much could one afford to pay for bare land if it is to be used for timber production?" would be the highest of the values calculated at the appropriate interest rate, and could only be realized by adopting the specified rotation. In our example the maximum Le value at 8% will be $99.50, attained at a rotation of 19 years.

Forest Rent Rotations

Forest rent is the term given to the **average annual net revenue** when net revenue is defined as total (gross) revenue less total cash outlay (or total revenue minus all costs *except interest charges*). Figure 14.7*a* is simply a repetition of Figure 14.4, showing the estimated revenue and costs for *Pinus radiata* under the conditions stated earlier. The net revenue, defined as above, is shaded for identification.

Figure 14.7*b* illustrates the net revenue itself (the shaded area) plotted over the time axis. Dividing this net revenue by the corresponding age yields the average annual net revenue or forest rent, as shown in Figure 14.7*c*. In the example used here, the maximum forest rent is returned at either 27 or 28 years, when the average annual net revenue is $44.13.

Viewed geometrically, the average annual net revenue at any age is the tangent of the angle formed by connecting the origin (0) with the net revenue at that same age, since

$$\text{Average net revenue} = \frac{\text{Net revenue}}{\text{Age}}$$

The rotation of maximum forest rent, therefore, may be found by determining the point of tangency between a line drawn from the origin to the net revenue curve, Figure 14.7*b*. If forest production were entirely costless, and if timber prices were constant, then the rotation of maximum mean annual increment (the culmination of MAI) would coincide with that of maximum forest rent. But timber prices tend to rise with tree size (age) and forest production is seldom costless! In the United States one seldom hears of forest rent—it is an almost forgotten term. In Europe and in some of the Third World nations it may still be encountered.

SUMMARY—AND A SUGGESTION

This chapter began with a discussion of various guides for project selection. Net present value, the internal rate of return, benefit/cost ratio, and the net benefit/investment ratio were described and compared, with the general conclusion that NPV is the better overall guide, even though it favors large projects and requires one to select the discount rate in advance. Later, the special problem of rotation determination—how long to leave forest capital invested—was discussed, pri-

[6] There are many formulas for calculating Le values, varying primarily in the degree to which costs and revenues of different types and timings are explicitly recognized. A basic formula is

$$\text{Le} = \frac{\sum_{t=0}^{r} R_t(1+i)^{r-t} - \sum_{t=0}^{r} C_t(1+i)^{r-t}}{(1+i)^r - 1}$$

where R_t is a revenue received at time t, C_t is a cost incurred at time t, r is the rotation age, and i is the interest rate.

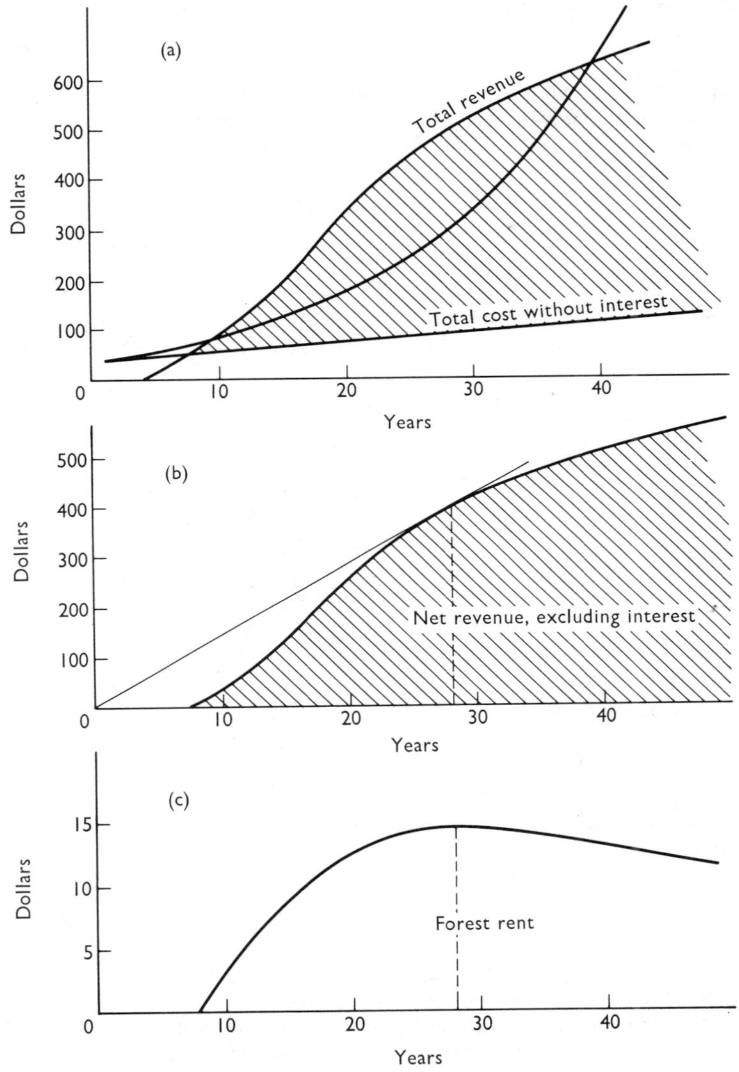

FIGURE 14.7. Derivation of forest rent.

marily as a means of introducing the special approaches that have been developed by foresters to deal with this problem. Determining logical objectives for today's managers requires a second look at the owners of forest land, for the goals of project selection, rotation determination, and management generally must be consistent with those of the forest land owners.

Many economists suggest maximization of Le values as the most rational financial objective, and with this there can be little argument from a theoretical standpoint. Practically, the choice is not so obvious, especially when genetic research promises future forests that will yield wood in far greater quantities and qualities, when transportation specialists promise wider access to recreation of all sorts, and when land values are generally increasing. Under these conditions it makes little sense to count on an infinite series of identical periodic payments. Why not take a simple approach, starting with land values, other costs, and prices as they actually are, and calculate the net present value of a forest using whatever discount rate is appropriate. The best estimates for costs of all future inputs and for salvage values at the end of the rotation can be used, along with the expected market value for land.

If land values, costs, and product prices do not change, and if the market value of land at the beginning and end of the rotation happens to be the same as that given by Le calculations, and if the same costs and interest rates are used as those assumed for the land expectation approach, then the calculated investment period (or rotation) will be identical to that obtained using the land expectation formula. This result, however, seems unlikely; we have not yet entered the "steady state."

Given the ease with which the necessary calculations can now be made, one need not be content with a single set of calculations. Several scenarios should be tested, embodying a range of assumptions regarding future growth rates, market prices, and interest rates. This approach not only provides the decision-maker with information regarding possible outcomes under a variety of conditions, but also permits one to test the sensitivity of the particular project to possible market or technological changes.

At present most corporations use either net present value or the internal rate of return as their guide to forest investments. Nonindustrial private owners, if interested in forestry as an investment, probably use net present value calculations and adjust these by "gut feel" for the nonmarket values they perceive.

Public forestry poses a different set of problems. The Forest Service is officially committed to using culmination of mean annual increment, somewhat constrained by its even-flow policy, to plan rotations and harvest cuts. From an efficiency standpoint it would seem preferable to move toward use of the net present value criterion, but incorporating both social values and a social discount rate; in other words, to use a social benefit/cost approach. To say that nonmarket values and externalities prohibit such an approach is valid only if one assumes the calculations are to provide rigid rules rather than guides to better decisions. It also implies that there is no way to estimate what these nonmarket values might be. After all, public forests exist primarily because we, the public, feel that some forest areas should not be managed by strict financial guidelines. We can, and should, expect efficient management of the national forests, but we also can demand that efficiency be defined in social terms. More will be said on this topic when multiple use is discussed.

QUESTIONS AND PROBLEMS

1. What are the four sets of conditions influencing the criteria selected for project choice, and which is the one most usually encountered?
2. Net present value (discounted cash flow) is generally regarded as the primary financial criterion used in selecting projects. What are the drawbacks to use of this guide? Can its use ever result in error? If so, under what conditions?

3. The internal rate of return is also often used as the criterion for selecting projects, especially by corporations. Again, what are the possible drawbacks, and under what conditions might use of the IRR lead to error?

4. Under what conditions can calculation of the IRR lead to two or more very different results? Are such conditions likely to arise in forestry? Why or why not?

5. Give an example from forestry of mutually exclusive projects. What guide should be used when selecting between projects that are mutually exclusive?

6. What is the result of following the recommended U.S. governmental practice of calculating a benefit/cost ratio by dividing net discounted benefits by total discounted costs? Under what conditions might this give a higher ratio than the more traditional B/C approach?

7. There are several guides to determining the optimum rotation of a forest or a stand for timber production. The major ones are:

   a. Culmination of mean annual increment (as used by the U.S. Forest Service).

   b. Maximizing the present net value of the current or planned stand.

   c. Determining the rotation that will maximize the internal rate of return.

   d. Maximizing the land expectation value.

   e. Maximizing the forest rent.

   You should be familiar with each of these—sufficiently familiar to know how each is calculated, the differences between them, how selection of certain variables such as the discount rate or taxes can change the rotation, whether they are particularly appropriate for certain private or public owners, and so on.

8. In the following problem on rotation determination, data will be supplied by the instructor, thus assuring that the data are relevant to the time and place.

   Assume you have just purchased 15,000 acres of loblolly pine land and intend to manage it as an even-aged forest after planting. Your task is to determine the optimum financial rotation under a particular (or perhaps all) criteria—NPV, IRR, B/C, and Le value.

Data

Yield Information

Stand Age (years)	Pre-thin Volume in Cunits
10	7.2
15	13.0
20	18.0
25	19.0
30	31.0
35	40.0
40	44.4
45	48.3
50	50.8

A thinning made at age 15 will remove 4 cunits per acre and another thinning at age 20 will remove 6 of the 18 cunits of standing timber volume.

Costs

Initial site preparation and planting costs $______ per acre.

Administrative costs average $______ per acre per year.

Property taxes average $______ per acre per year.

You plan an herbicide treatment at year 8 that will cost $16 per acre (this is a projected cost).

Revenues

Material from the initial thinning can be sold as stumpage for $______ per cunit.

Material from the 20 year thinning can be sold as stumpage for $______ per cunit.

Harvest aged material can be sold as stumpage based on the following schedule. (*Note*: expected price will be the present price, increased at a 1.2% compound rate.)

Stand Age	Present Price in $ per Cunit	Expected Price $ per Cunit
25		
30		
35		
40		
45		
50		

For a through d (below), all calculations should include depletion charges where appropriate, reflect the effect of a 25% capital gains tax, and use a discount rate of 10%. The ordinary income tax rate may be calculated at 35%.

a. Calculate the net present value of the proposed plantation for rotation ages 20,25, 30, and 35 years.

b. Plot these data.

c. Calculate the Le value and the IRR for the rotation that has the maximum NPV.

d. How might you build risk and (further) uncertainty into this set of calculations?

9. Now, using these same data, answer the above, but under the assumption that the plantation is on public forest land. As a further step, estimate the approximate culmination of mean annual increment.

PART 4

Nontimber Products and Multiple Use

After introducing the concept of investment projects, Part 3 began by presenting guidelines often used by corporations when making forest investment decisions (Chapter 11). Chapter 12 provided necessary background for making public investment decisions, while benefit/cost analysis—the principle guide to such decisions—was the subject of Chapter 13. Part 3's final chapter dealt with choosing between alternative investment opportunities, including that of rotation determination.

Part 4 takes up a major forestry problem—the necessity of dealing with the many nontimber products of the forest, including those whose values are not market determined. Chapter 15 presents the current production and importance of forest-produced mineral and energy (both with well-determined market values) then gives production estimates of forage, water, wildlife, recreation, and wilderness experiences. Specific methods of assessing the social benefits from such products using expressed or revealed preferences are described and discussed.

Chapter 16 deals with economic problems of multiple use management—their origin and the early development of the concept. An analytical model is developed through use of joint production theory. Current practice is discussed, including use of Clawson's compatibility matrix, linear programming and goal programming (and their required assumptions), and the special problem of below cost timber sales. The chapter ends with an example of multiple use involving changing values of timber, wildlife, and aesthetic values.

chapter fifteen

NONTIMBER PRODUCTS: PRODUCTION, MEASUREMENT, AND ASSESSMENT

To many foresters, reference to nontimber products implies forage, water, wildlife, and recreation — the products and services identified under the legal mandate of the Multiple-Use Sustained Yield Act of 1960 and since then publicized in Forest Service multiple-use plans and by its public relations work. The four product groups were well chosen, for they include the nontimber products of most concern to North American forest managers. If one takes a broader view, however, the product variety increases. In India, for example, the list of so-called minor forest products includes bamboo, fruit and nuts, flowers, cooking oils, honey, meat, medicinal herbs and leaves, animal skins, tussar silk, and many more. And they are far from minor; in several states these products are valued considerably above that of all timber, including fuelwood. A South American analyst once claimed that the sale value of nontimber products from that continent's forests was considerably in excess of that for all timber.

The easiest way to define nontimber products is by exclusion; in this book the term will include all the goods and services produced by or from a forest other than marketable wood products. So defined, nontimber production from North American forest areas includes (in addition to forage, wildlife, recreation, and water) products such as oil, natural gas, coal and minerals, naval stores, scenic beauty, wilderness experiences, protection of rare and endangered species, opportunities for scientific experimentation, educational experiences, and the option to participate in a forest-based experience even though the opportunity may never be used (option demand). There is growing public pressure to increase production of these products, and as forest land managers we have much to learn. Already the social value of these nontimber outputs has become sufficiently high to have an impact on forest management, particularly of public forests. And although one still hears some of these products referred to as intangibles, there has been substantial progress over the past decade in identifying them and estimating their value.

In this chapter we will first review briefly the status of this country's major nontimber products and, to some extent, the anticipated pattern of consumption through the remainder of the twentieth century. We will look at some of the ways to estimate benefits from their production, and examine the main factors believed to influence their demand.

PRESENT STATUS OF NONTIMBER PRODUCTION FROM FOREST LAND

The following review is in no way exhaustive. To permit a reasonable coverage of the many nontimber forest products, we will treat them in six product groups: subsurface materials, forage, water, wildlife, forest recreation, and wilderness experience.

SUBSURFACE MATERIALS

Subsurface materials are not forest products at all, but their presence, or even suspected presence, can influence production of both timber and nontimber products on the land surface. Petroleum, natural gas, and minerals are found beneath many forests. In fact, most of the mining in the United States is on forest or range land, and some of the larger private forest holdings were assembled for purposes of mining or well drilling—forests just happened to occupy the surface. Returns from mining or from oil and gas development can be very large. In 1975 the value of crude minerals produced was over $62 billion, up nearly 300% (in constant dollars) from 1950 (USDA For. Serv., 1981). In earlier years mining operations often ignored the forest cover, but legislation passed during the environmental movement of the 1960s and 1970s did much to prevent needless destruction of either timber or amenity values. The Surface Mining Control and Reclamation Act of 1977 (P.L. 95-87) has been particularly helpful in this regard, now requiring each state to regulate surface mining for coal.

Surface mining has the most impact on forests, although damage from underground mining also can be severe, especially from mine tailings, land subsidence, and access roads. In past years the strip mining of coal severely affected forests, particularly those in the Appalachian Mountain areas. Standing timber was destroyed, fish life in formerly clear-running streams was wiped out by acids from spoil banks and exposed coal seams, and former scenic beauty was replaced with ugliness. Today, state laws require restoration of surface soils and, often, replanting as well. Restoration of areas damaged in past years, however, still requires substantial expenditures of both time and money. In total, it is estimated that from the earliest settlement to mid-1977 over 5.7 million acres in the United States had been used for mining. Impressive though this figure is, it amounts to only around a quarter of 1% of the nation's land, and about 40% of this has been reclaimed (USDA For. Serv., 1981).

Oil and gas development interferes with timber far less than most mining does but can pose problems to other forest uses. Exploration and drilling may hinder recreational use and disturb wildlife, but once the well is producing or has been capped and the derrick removed, most uses of the forest can be reestablished. Royalties from producing wells, as from mineral production generally, can be very substantial. As one example, in 1983 the U.S. Forest Service received a total of 132.5 million dollars from rents, royalties, sales, and bonus bids for minerals (For. Serv. Report, 1984). Small wonder that subsurface rights are often reserved by the seller when forest land is sold.

In recent years the use of land for forest and range purposes has had a decided effect on mining—further proof that in nature there are usually two sides to any tradeoff. Establishment of wilderness and primitive areas, wildlife reserves, historic sites, and special uses such as recreation have resulted in restricting mineral exploration and development on as much as two-thirds of all federal lands (U.S. Dept. Int., 1977). There are some who believe this may be excessive, but little serious economic study of alternative management plans and their implications has been undertaken.

FORAGE

At one time in our development grazing lands would have provided a classic example of an open access resource, and in many of the developing nations they still do. *Open access resources* (often called *common property resources*) are those that are open to unregulated access by individuals (Hobley, 1985). The village commons of medieval times was the example used by Hardin in his classic article, "The Tragedy of the Commons" (Hardin, 1964). All such resources have the characteristic that while everyone has a right to use, no one has the responsibility to maintain or protect. In consequence, "everyone" finds it worthwhile to use the resource with no other thought than that of immediate consumption; each individual's motto becomes "What I don't get, the next person will!" As Hardin expressed it:

> *Therein is the tragedy. Each man is locked into a system that compels him to increase his herd without limit—in a world that is limited. Ruin is the destination toward which all men rush, each pursuing his own best interest in a society that believes in the freedom of the commons. Freedom in a common brings ruin to all. (Hardin, 1964)*

The problem is one of control. The use of many properties held by groups, tribes, or villages is carefully controlled through custom or tradition. Areas with unregulated access are often ruined by overuse. When the Forest Service began charging ranch owners for grazing privileges there was a great uproar, and until very recently the grazing fees on National Forest lands were, at best, nominal, although the number of animals grazed was controlled. In 1983 the Forest Service received a total of only 10.2 million dollars in grazing receipts from the more than 14,000 permittees—down from the high of 15.85 million reported for 1980, and somewhat less than one-third of the total spent on range management activities. Through 1985 Forest Service grazing fees were set according to a formula established by the Public Rangelands Improvement Act of 1978, a formula based on what livestock producers spend on production and receive from sales (For. Serv. Report, 1984).

The 1203 million acres in the contiguous 48 states classified as forest and rangeland in 1976 included 553 million acres of forest and 650 million acres of rangeland.[1] Fully 99% of the rangeland was in the 17 Western states, along with 37% of the forest land. Of the Eastern state *forest* land (63% of all U.S. forest land) only a small part was grazed. Slightly more than a third of the rangeland, and somewhat under one-fourth of the forest land, was under federal ownership.[2]

About two-thirds of the total forest and rangeland was actually grazed in 1976. Some 720 million acres (91%) of this grazed area was in the 17 states west of the Mississippi, with the remaining 69 million acres in the Eastern states made up primarily of forest land grazed by livestock (grazing in the East is generally confined to improved pasture lands not classified as forest or rangeland).

Federal lands, while making up 33% of the forest and rangeland grazed,

[1]Alaska has 232 million acres of rangeland, but it is virtually unused. Hawaii's 968 thousand acres of rangeland, though important locally, would add less than two-tenths of 1% to the total used above. In this discussion, forest land is defined as "land at least 10 percent stocked by forest trees of any size, including land that formerly had such tree cover and that will be naturally or artificially regenerated." Rangeland is defined as "land on which the potential natural vegetation is predominantly grasses, grasslike plants, forbs, or shrubs, including land revegetated naturally or artificially that is managed like native vegetation" (USDA For. Res. Rep., 1981).

[2]The first comprehensive study of forest, range, and inland water resources came as a response to the Resource Planning Act of 1974 (USDA For. Serv. Rep., 1977). This was updated by a second report (USDA For. Serv. Rep., 1981). See also USDA Forest Service (1984).

produced far less than one-third of the actual grazing as measured in animal unit months.[3] This seriously understates the importance of federal grazing lands, since livestock producers depend on these for seasonal forage to round out their year-long operations. This is true of Western beef cattle production — by far the largest use of U.S. grazing lands — and even more important to production of sheep, supplying some 42% of their annual forage requirement.

In addition to producing domestic livestock, grazing in forests and rangeland supports a growing number of wild horses and burros and furnishes food and cover for millions of other wild animals. In the 1930s wild horses and burros had increased to the point where they had become a serious problem on certain federal range lands. Action was taken to reduce the numbers of these animals during the late 1940s and early 1950s. Public concern over these actions resulted in the passage of the Wild Horses and Burros Protection Act of 1971, making management and protection of these animals a federal responsibility on both Forest Service and BLM lands. This resulted in a rapid increase in numbers of feral animals (horse numbers increased 25% in the 1974 – 1976 period), enough to cause considerable damage to some rangelands, and led to direct competition for grazing with other wildlife and, occasionally, with domestic stock. The Public Rangelands Improvement Act of 1978 provided some measures for disposing of surplus animals by the two agencies and may alleviate this problem.

While there are no reliable population data available for most wildlife species, an estimate by Wagner indicated big game populations in the 11 Western states (excluding the six Prairie states) of over 4 million animals (Wagner, 1978). To sustain these numbers would require some 11 million AUMs of forage. Available data indicated that as much as 154 million AUMs were available — far more than necessary to maintain existing populations.

Under most conditions, livestock grazing can be compatible with, and often complementary to, other resource uses. To evaluate range conditions under current grazing practices for its 1981 report the Forest Service classified areas by six management levels, with the levels depending in part on grazing intensity and in part on intensity of land management (USDA For. Serv., 1981). Table 15.1 gives the results of this survey.

TABLE 15.1
Production of Range Grazing, by Management Levels, in the 48 Contiguous States (in thousands of animal unit months)

Management Level	AUMs (thousands)	Percent of Total
1. No livestock	—	—
2. Some livestock	17,746	8.3
3. Extensive management	72,236	34.1
4. Intensive management	44,726	21.1
5. Livestock maximized	38,132	18.0
6. Exploitive grazing	39,145	18.5
Total grazing produced	211,986	

SOURCE: Forest Resource Report No. 22, pp. 169 and 172.

[3]An animal unit month, or AUM, is the amount of forage required by a 1000-pound cow or its equivalent in one month.

Since the first level allows no domestic livestock grazing whatever except for that by animals used for recreational purposes, none of the presently grazed lands fall into that category. The next four categories reflect increasing intensity of management, but the sixth, exploitive grazing, while it may be intensive, is one wherein soil and water resources are being sacrificed to obtain short term monetary returns. Note that of the 212 million animal unit months produced in 1976, 72 million came from lands *extensively managed*—lands managed to maintain full plant vigor and to achieve full livestock utilization of the available forage, but without seeding, fertilization, and so on. Exploitive grazing, aimed at short-run maximization of livestock production and definitely harmful to the resource, produced 18.5% of the total forage used.

The same study reported that under economically viable but much intensified management, grazing could be increased to 350 million AUMs, yet the area grazed would rise only 7%. On National Forest lands, grazing could be doubled *if* the potential areas were properly developed.

Unless otherwise specified, the above data apply to all forest and rangelands in the 48 contiguous states. Detailed estimates of grazing on *forest* land are not available, but the 1975 assessment gave some very broad indications (USDA For. Serv., 1977). The Eastern forest ecogroup produced 53.5 million AUMs and western forest ecogroup only 10.7 AUMs. These forested grazing lands, combined, also produced 20 billion cu. ft. of wood annually and 664 million acre feet of water, most of high quality (the sediment yield per acre from nonforested land averaged 5 to 6 times that from forested areas) (USDA For. Serv., 1977).

In an effort to include yields of other products in the same 1975 assessment, Western and Eastern forested and nonforested grazing areas were rated as bad, poor, fair, good, or excellent with respect to the production of 12 nontimber goods and services (USDA For. Serv., 1977). Figure 15.1 gives the results—a somewhat vague indication of production capabilities, and probably the first ever attempted for all our grazing lands. In examining the figure, one should keep in mind that, in addition to the products for which the areas were rated, these forested range lands also produced the wood, grazing, and water yields mentioned earlier.

Future demands for grazing will depend on the demand for red meat—primarily beef—in the United States (U.S. For. Serv., 1981). Population increases now seem likely to be less rapid than was assumed in the 1975 estimates, and recent medical discoveries might further reduce the assumed consumer preference for meat rather than poultry. With these facts in mind, and with the knowledge that large grazing increases could be permitted without a corresponding increase in grazing area if known and cost efficient management practices were applied, it seems unlikely that livestock production will take over additional forest land.

WATER[4]

Water uses can be divided into three major categories:

1. *Withdrawals*: Uses that withdraw water from some source, use it, then return it so it is available for another use.
2. *Consumptive Uses*: Withdrawals consumed through evaporation or transpiration, or by discharge to a location inaccessible for additional use.
3. *Instream Uses*: Boating, fishing, navigation, and so on. Uses that do not remove the water from its location in a lake or stream.

---

[4]Much of the information in this section comes from the For. Res. Rep. #22 (1981). The source for data on water use is cited as the U.S. Water Resources Council, the 1975 assessment of water and related land resources (in process).

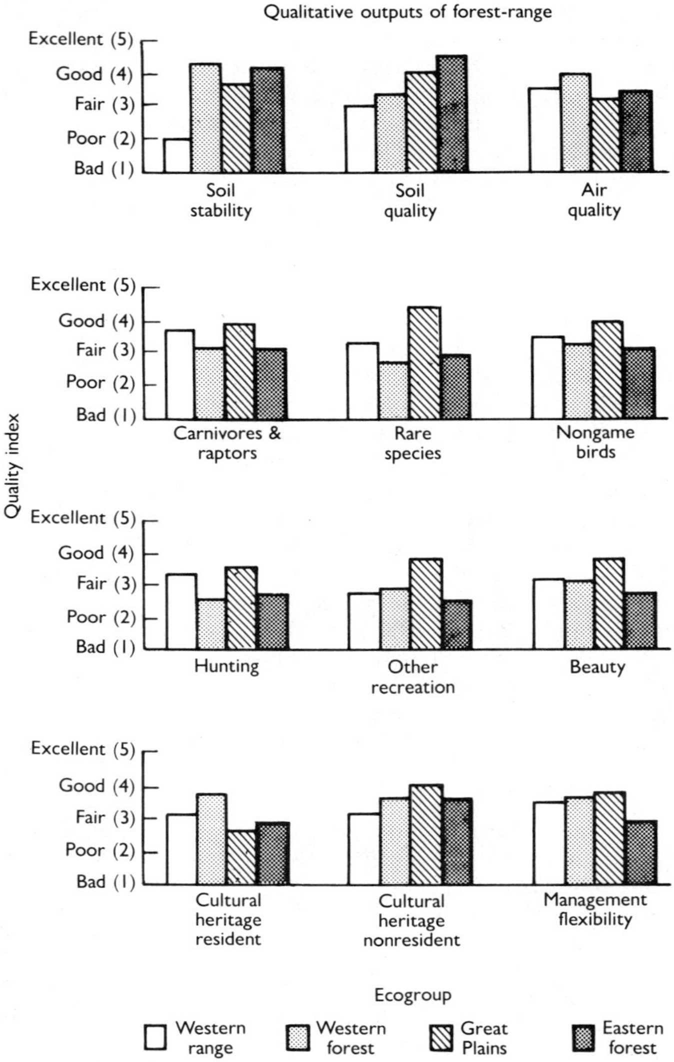

FIGURE 15.1. Quantitative estimates of nontimber outputs from forested rangelands. *Source*: USDA Forest Resource Report No. 21, p. 127, Fig. 27.

Water Withdrawals

Withdrawals account for a lot of water—in 1975, for an estimated 338.5 billion gallons per day. That's more than 1500 gallons for every person in the United States every day! Major uses for that year are given in Table 15.2 and in Figure 15.2. By far the largest use was for irrigation—47% of the total. Second was the 26% used for cooling purposes in steam electric generating plants, with another 15% taken by manufacturing. Most uses are expected to decline very substantially because of recycling procedures required by the Environmental Protection Agency.

TABLE 15.2
U.S. Freshwater Withdrawals in 1975, by Major Uses (millions of gallons per day)

Crop irrigation	158,743
Steam electric	88,916
Manufacturing	51,222
Domestic and commercial	28,786
Minerals	7,055
Livestock	1,912
Public lands	1,866

SOURCE: USDA Forest Service. No 22, Table 7.1, p. 275.

Consumptive Use of Water

Most of the water withdrawn for use is returned for reuse. Total consumptive use is estimated to be 107 bgd (billion gallons per day)—about 30% of all water withdrawn. Irrigation is by far the largest consumptive use, taking some 81% of all water consumed—in all, 86.4 bgd (Fig. 15.2). Domestic and commercial uses, which totaled only 7.4 bgd, includes water for drinking and sanitation and also for fire protection, street cleaning, and lawn and garden watering. About 1.4 bgd is consumed by the cooling operations in steam electric plants.

Instream Uses

These include some of the most important water uses, but unfortunately there is little or no data available. Navigational uses have increased rapidly since 1950 (there are now over 25,000 miles of navigable channels, canals, and reservoirs); swimming, boating, and other recreational uses have increased even more rapidly.

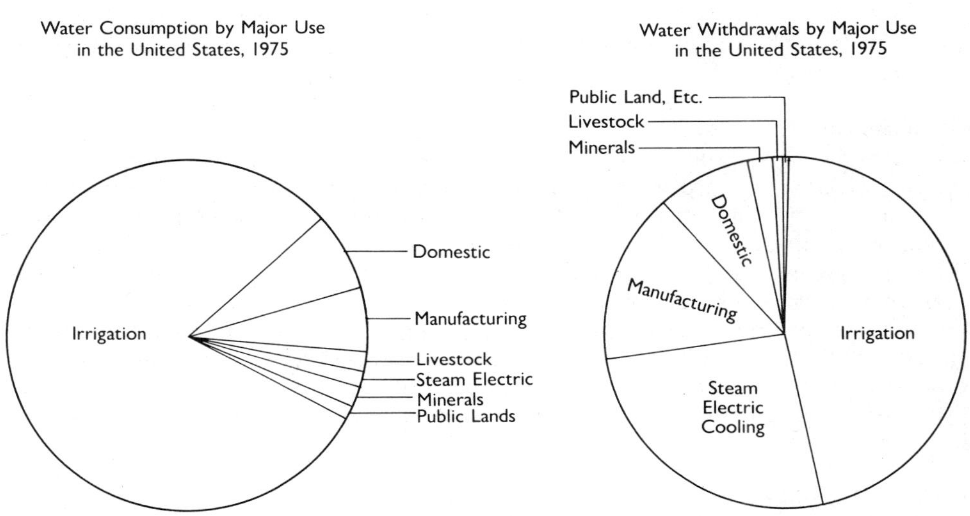

FIGURE 15.2. Water withdrawals and consumption by major uses in the United States, 1975. ***Source*:** **Forest Resource Report No. 22 (1981), p. 275–285.**

Water Demand/Supply

To evaluate water demand/supply relations, two measures are usually used: water quantity demanded and supplied and water quality. In general, the Eastern part of the country should experience few water shortage problems, although local difficulties may occur. In the South several critical areas exist. Four of the five subareas in the Texas Gulf region, for example, are expected to experience water shortages before the end of the century. Similarly, the entire Rio Grande Water Resource Region is expected to fall short of demands, as is virtually all of Southwestern United States. Almost all these shortfalls will result from the large quantities demanded by agricultural irrigation, and it is this use that will necessarily absorb the shortages. The high plains area and Southwestern United States presently draw much of their water supply from acquifers that are not being replenished.

Water quality goes far in determining the usability of water. Most quality problems result from using water to carry wastes—chemical wastes from manufacturing and mining, human wastes from municipalities, and so on. Regulations promulgated by the Environmental Protection Agency have had very salubrious effects; even Lake Erie may soon be (relatively) clean!

Forests affect water supplies in many ways: they intercept water and snow, they draw water from the ground and release it into the air through transpiration, and tree roots, leaves, and other residues aid the infiltration of precipitation into the ground and help prevent erosion. Whether forests can actually increase the amounts of available water cannot be answered categorically—it depends on the particular situation. Land management can, in some instances, be the best method for increasing supplies; in other situations it may prove ineffective.

WILDLIFE

Forest wildlife contributes to social well-being in so many different ways that to adequately cover even the most interesting details would take several volumes. Fortunately, the objective here is much less inclusive: to establish a perspective from which to assess forest wildlife, then fit it into the overall picture of nontimber products and values.

Like water, wildlife fills two distinctly different uses: *nonconsumptive*, wherein the wildlife itself is enjoyed and then left for others, and *consumptive*, in which wildlife is "captured" by an individual and is no longer available to others.

Nonconsumptive Uses

These uses range all the way from the elderly urban individual watching a migrating pine siskin at his birdfeeder to the wildlife photographer who packs 60 pounds of cameras, lenses, and tripods 12,000 feet up a mountain to photograph a mountain goat. Birdwatchers provide a good example of nonconsumptive wildlife use. The typical birdwatcher is not the middle-aged urban lady in tennis shoes of cartoon fame, but is male, white, married, considerably above average in education and income, and can be either retired or a young professional. In 1970 there were nearly 7 million of them! And the direct expenditures of U.S. citizens interested in enjoying nongame birds in 1974 amounted to some 500 million dollars—about one-sixth the total amount spent by all hunters, not including amounts spent for lodging, transportation, and food (Clawson and VanDoren, 1984)

Rare and Endangered Species

Under nonconsumptive uses one should also include the protection of rare and endangered species of birds, animals, or fish, since they may require very special (and, at times, costly) management practices. The Endangered Species Act of 1973 states:

> *Sec. 2 (c). Policy,—It is further declared to be the policy of Congress that all Federal departments and agencies shall seek to conserve endangered species and threatened species and shall utilize their authorities in furtherance of the purpose of this Act.*
>
> *Sec. 3. For the purpose of this Act . . . (2) The terms "conserve," "conserving," and "conservation" mean to use all methods and procedures which are necessary to bring any endangered species or threatened species to the point at which the measures provided pursuant to this Act are no longer necessary . . . (USDA For. Serv., 1977).*

In some cases this law will require preservation of all available habitat and full protection of the remaining species members. To assure nesting sites for the remaining pairs of Kirtland's warblers, for example, a good many acres of Michigan jack pine have been declared off limits to cutting, and in certain seasons of the year are off limits to everyone. Special controlled burning is a required part of this and of many other management schemes designed to assist endangered species.

The northern spotted owl provides an example of unusual interest to foresters in the Pacific Northwest.[5] Spotted owls are very particular about their nesting sites, preferring dense, multilayered coniferous old-growth forests. The problem is that if present rates of cutting continue, most old-growth forests on commercial timberland will be gone in another three or four decades. When their habitat is clear-cut, the owls either move to nearby old growth or simply disappear. On the basis of a study by a wildlife biologist, Oregon's Wildlife Commission recommended setting aside 1000 acres per pair of owls—which would amount to setting aside 400,000 acres of old-growth timber. "The timber folk, to put it mildly, were not pleased" (Heinrichs, 1983).[6]

The Assistant Secretary of Agriculture in charge of the Forest Service, after stating in a 1981 speech that the spotted owl management program would cost about 3.6 billion dollars in lost timber sales, asked "How many pairs of owls can we afford to protect?" To which a wildlife biologist quickly replied "How many owls can we afford to lose?" (Heinrichs, 1983, p. 214)[7]

The rapid increase in fuelwood use has raised a different kind of problem in some areas. The types of trees used by animals for den trees, along with the dead and dying, old or suppressed trees that supply both homes and food for woodpeckers and many other birds, are precisely those that are usually marked for fuelwood cutting. Forest managers have been urged to maintain a variety of species and age classes to protect these types of trees, but over time this could adversely affect available fuelwood supplies.

---

[5]The April 1983 *Journal of Forestry* featured a picture of this small owl, under which is the caption "The Next Snail Darter?"—a reference to the small and rare fish that became famous when, to prevent its extinction, the courts temporarily halted construction of the Tellico Dam—a multimillion dollar project.

[6]The study referred to was done by Eric Forsman—according to Heinrich's article cited here.

[7]Incidentally, the cost of lost timber sales is probably exaggerated. At the end of 1984 the Forest Service had 39 billion board feet of uncut timber sold and expected to sell only an additional 10.6 billion board feet during 1985.

Consumptive Uses

These uses will be broadly interpreted to include all hunting, fishing, and trapping. Together they cover a great range of outdoor activities that today are always considered to be recreational, though in earlier times these same pursuits were basic to survival. Most of the quantitative information presently available comes from the national surveys made by the U. S. Fish and Wildlife Service. Initiated in 1955, these surveys have been made every five years.

Hunting

As a minimum, most hunters would probably feel it necessary to distinguish between big game, small game, upland bird, waterfowl, and varmint shooting.[8] Others would probably separate white-tailed deer hunting from either moose hunting or hunting big-horn sheep, for the skills and stamina required as well as the nature of the terrain in which these animals are found are very different. The same might be said of hunting woodcock or snipe when compared to hunting Canada goose. Yet all these are often, but certainly not necessarily, forest-based activities.

In the United States most game is considered the property of the state, and state rather than federal law generally regulates hunting activities. Migratory waterfowl are an exception, since these birds visit many states—and nations—in their annual migrations. In this case, federal agencies determine the season and bag limits for the various species.

In 1955 a total of 11.8 million hunters—10% of all Americans over 12 years old—participated in hunting. By 1980 the number of such hunters had climbed to 16.78 million—an increase of 42%—although this was only 9% of the over-12 age group and indicated a slight decline from the 1975 Census. Over the same period the total number of days spent hunting rose from 169 million days in 1955 to 387 million in 1980, an increase of 228% (Clawson and VanDoren, 1984). Table 15.3 illustrates how the increase in hunting activity varied for three types of hunting.

Measured in terms of hunting *days* (not licenses sold), hunting seems to be increasing in popularity: the number of small game hunters increased at a compound rate of about 1.6%, while big game hunters increased much faster (4.3%), and numbers of those seeking wildfowl increased at an intermediate rate of slightly under 2%. The fact that most present-day hunters come from a rural background may be significant for the future popularity of hunting, for the U. S. population is becoming increasingly urban.

TABLE 15.3
Increase in Hunters: 1950–1980 (numbers in thousands)

	1955	1975	1980
Small game hunters	9,822	11,671	14,643
Big game hunters	4,414	7,774	12,632
Migratory waterfowl	1,986	2,894	3,190
Total days hunting	169,423	203,689	387,287

SOURCE: Clawson and VanDoren, 1984, Table 78, p. 248. Source cited as "Surveys by the U. S. Fish and Wildlife Service, various years."

[8]Shooting unprotected, nongame birds such as crows, starlings, and sparrows, or animals such as porcupines, rats, and prairie dogs.

Hunting is predominantly a male sport: 20% of all males over the age of 12 years old participated in hunting in 1955, but only 18.8% hunted in 1980. The 1.6 million females who hunted in 1980 were only 1.5% of the total female population over 12 years old—a percentage that had gone up from 0.8% in 1955. In 1980 about 35% of all hunters were under 25 years old, and 40% were between 25 and 45. As a group, these hunters spent $8.938 billion—an important source of income to many rural areas (Clawson and VanDoren, 1984).

Income from the sale of licenses provides the bulk of the funds used by the State Fish and Game agencies—82% of these funds in 1973. Viewed nationally, funds for fish and wildlife management came from a variety of sources (Figure 15.3), but licenses still provided the major amounts. Revenues generated by sale of hunting licenses increased almost every year from 1956 through 1978—from $46.6 million in 1956 to $185.1 million in 1978.

Habitat improvement would seem to be the primary means of increasing game populations. For big game (including deer, moose, elk, antelope, big-horn sheep, mountain goats and—when present—turkeys) this generally implies some alteration of timber-harvesting practices. White-tailed deer, for example, much prefer second-growth timber and do particularly well in cutover, re-sprouting aspen areas. Elk must have wooded migration routes to and from their winter forage areas. Some animals, particularly the large carnivores, require large areas maintained relatively free from intrusion by humans. Grizzly bears and wolves provide specific examples.

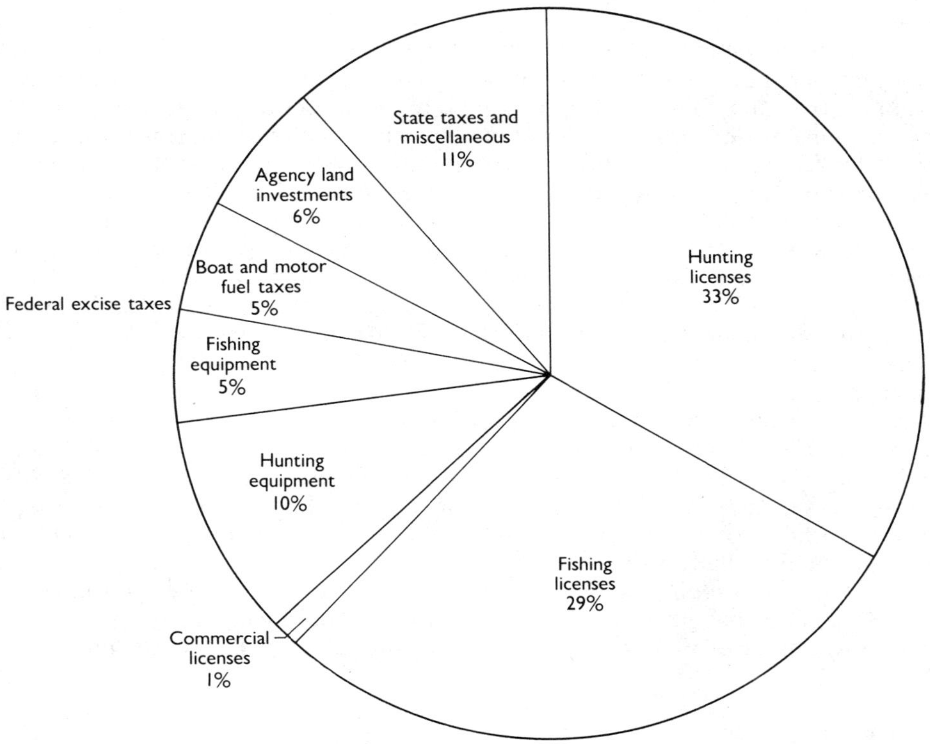

FIGURE 15.3. Sources of funds for fish and wildlife management, 1971. ***Source*****: Forest Resource Report No. 22, op. cit., p. 134.**

Much of the big game hunting is on public forest lands, in part for reasons of open public access, but also because public forests provide more areas of suitable habitat. Small game and upland bird hunting probably occurs to at least some extent on almost every area where game might reasonably be found and hunting is not specifically prohibited. The major problem facing small game hunters is access to hunting areas. Fee hunting on private lands may be a partial answer, and shooting preserves—areas stocked with game raised for the purpose, where one can hunt with whatever degree of privacy one is willing and able to pay for—are becoming quite common. Management practices for improving habitat for small game are well established, but since most small game hunting is on private lands it is probably unreasonable to expect action unless owners are compensated. Eventually, suitable compensation methods will be worked out—they have been in other nations, and some examples can be found in this country.

Forest management aimed at enhancing wildlife habitat is part of multiple use management—the usual practice on public lands, but rare on industrial forests or on private areas managed for timber. A. S. Leopold has summarized the private land situation:

> *The management procedures that enhance wildlife habitats are nearly all of a sort that cut profits to the timber operator: leaving strips or corners of mature trees uncut, leaving snags and potential snags in the forest, keeping clearcut blocks small, desisting from excessive use of herbicides and pesticides, maintaining some uneven-aged stands when even-aged stands are simpler to manage mechanically. (Leopold, 1978)*

Fishing

Fishing is one of the most popular forms of outdoor recreation in the nation. In 1980 there were nearly 47 million fishermen—nearly three times as many as there were hunters—and they spent approximately 18 billion dollars for their sport (Clawson and VanDoren, 1984). Water quality in lakes and streams is an important concern for forest managers and the provision of access to streams and lakes and occasionally to camping space might require special attention of forest managers. In addition, many fish species are particularly susceptible to poisoning by pesticides or herbicides. Special measures are often required to prevent pollution from aerial applications and these can substantially raise spraying costs. When damage does occur, whether to birds, animals, or fish, the social costs involved must be subtracted from any benefits attributed to the spraying program.

Trapping

Trapping has been declining since the depression years. Between the 1970–1971 and 1975–1976 seasons, however, the national harvest increased from 7 million to 13 million pelts, while the value of the harvest increased by six times (to $123 million). Muskrat accounted for about half the total number of pelts, while another third were either raccoon or nutria. Somewhat surprising was the fact that raccoon accounted for fully half of the total value.

Much of the trapping takes place on wetland areas, or along river and lake shores. So far as management is concerned, trapping would seem to offer no major problems to timber production, although local difficulties can arise (with beaver, for example).

Endangered Species

As observed earlier, the most difficult problems arise from endangered wildlife or aquatic species. In general, these are species that now inhabit "islands" of a particular ecological environment—mature timber that includes dead and dying trees, riparian or wetland areas (about half the wetland areas have been lost over the past two centuries and probably 80% of the riparian habitat found in wetlands), or species that require large, undisturbed areas (grizzly, elk, and wolf). One of the more difficult problems of endangered species is that the whole idea is new; we have virtually no experience in preventing the extinction of a habitat. In fact, we are still very much in the process of identifying essential habitat requirements for many of these endangered species. How successful our efforts will be remains unknown; to date, only 10 species have been delisted since the federal endangered species list was first prepared in 1981 (U.S.D.A. For. Serv., 1981).

FOREST RECREATION

Since World War II more and more people in the United States have participated each year in outdoor recreation of some kind. This has been partly because of population growth, but is probably more the result of rising incomes, changing age distributions, more available leisure time, and a combination of increased mobility and easier access to recreation sites. Forest recreation is only part of a broad group of activities. For our purposes, forest recreation will be defined as leisure-time activities voluntarily undertaken in a forest environment for the primary purpose of enjoyment. Even this encompasses a wide range of activities: hunting, fishing, birdwatching (all three of which involve wildlife and fish), hiking, horseback riding, camping, mountain climbing, snowmobiling, skiing, swimming, canoeing, automobile riding, and many others.

While forest recreation is usually associated with some kind of activity, *recreation* and *activity* are not identical. Many forest activities are undertaken by people for work as well as for fun. A timber cruiser ploughs 15 miles through the timber, eats his sandwich by a brook, and calls it hard work. A New Yorker hikes 15 miles through the forest along the Appalachian Trail, eats a sandwich by a brook, and enthusiastically calls it recreation. There is a very complex interaction between the forest environment and the individual, his or her objectives and usual environment, cultural background, companions (if any), and many other things as well. All work together to determine the degree to which a forest "experience" will be enjoyed, merely tolerated, or strongly disliked.

One distinction often sufficient to determine when an activity is recreation and when it is not is the freedom of choice to participate. A recreational activity is discretionary—the individual or group must be free to choose whether or not to join in. Social pressures, the status gained through participation, peer pressures, family pressures—all may play a role in the recreational activity an individual chooses and may constrain the discretionary nature of recreation, but they do not eliminate it.

Economic Peculiarities

There are some economic peculiarities of forest recreation. From a production standpoint one of the most important is that, like timber, **most forest recreation is produced on site**. It is resource oriented, rather than user oriented, and the consumer must nearly always travel to the area.

Like almost all outdoor recreation, **forest recreation is necessarily a kind of package deal**, involving anticipation, travel to the forest site, the activity (or lack of it) on the site, the return travel, and, finally, recollection (Clawson and Knetch, 1966). The fact that it is convenient to refer to the activity by the place where the action occurs does not mean the other parts of the package can be ignored. In thinking of forest recreation, one must bear in mind that the forest part of the experience represents only part of this package of goods. In some ways the demand for forest recreation resembles the demand for houses: both are packages and our primary concern is with a component of that package. The demand for the forest site, like the demand for lumber or sawlogs, is derived from the demand for the total package.

Forest recreation is often produced jointly. Private timberlands, and some areas of national forests and BLM lands, for example, are primarily managed for timber production but also supply forest recreation. On such areas the recreation is often complementary to the production of timber and other products, but examples of competitive production are common (as Clawson's diagram, discussed in Chapter 16 well illustrates). At the opposite end of the scale one can cite the National Parks and many State Parks, where management is primarily directed toward recreation. Here again other products are often produced, with cases of competitiveness and complementary easily identified. Both timber cutting and hunting are prohibited in National Parks, yet excellent watershed protection, fishing, and scenic viewing opportunities are provided.

Users and Use of Forest Recreation

Forest recreation data are far from uniform and are frequently of questionable accuracy. Visits to National Parks, for example, can be reasonably well monitored; campgrounds are well developed, entry fees must be paid, and so on. In contrast, visitors to National Forests go through no manned entryway, often pay no fees, may or may not encounter daily checks on the campgrounds, and find that much of the recreation—particularly hunting and fishing—is widely dispersed. Yet National Forests make up only 20% or so of the nation's forest land, and statistics covering recreation on privately owned forests is, practically speaking, nonexistent. Some indication of the importance of private areas might be gained from the fact that in 1977 more than one-half (52%) of the nation's campgrounds were privately owned and operated.

National Forest statistics probably provide the best indications of the nature of forest recreation; Table 15.4 gives the estimated outdoor recreational use on National Forests, by type of activity, for three years, 1968, 1975, and 1982. Note that the rate of increase varies from activity to activity and that the data is in terms of "visitor days" (recreational use aggregating 12 person hours, which may entail 1 person for 12 hours, or 12 persons for one hour, or any equivalent).

Camping is still the most popular activity, but the rate of increase is apparently leveling off. Winter sports are now showing the most rapid increase, although less than one-third of all ski lifts were on National Forests in 1978. As would be expected, young people have a higher rate of participation than those in older age groups, especially in the more strenuous activities such as skiing, backpacking, mountain climbing. Participation rates drop quickly after age 65. Income also plays an important role, with those in middle income groups having the highest participation rates, especially in activities such as camping, boating, or using off-road vehicles.

Nearly two-thirds of the camping activities were in developed campsites. Usually this is a family activity—studies in the Northeast indicate that 90% of camping parties were families (USDA For. Serv., 1977). Tents are still the most popular shelter and represent the largest single investment for most campers. Recreational vehicles, and in recent years, the smaller vans especially, are increasing in importance and tend to create a demand for more expensive facilities.

TABLE 15.4
Estimated Outdoor Recreation Use (in thousands of visitor days) on National Forest Land in the United States, by Type of Activity, 1968, 1975, and 1982

	Year		
Activity	**1968**	**1975**	**1982**
Camping	39,211	53,092	57,089
Picknicking	6,682	7,481	9,160
Recreational travel	36,519	44,836	53,789
Boating	3,832	6,341	8,063
Water skiing, etc.	605	1,040	986
Swimming and diving	2,841	3,930	5,209
Fishing	14,520	16,151	16,461
Winter sports	5,679	9,412	14,173
Games and team sports	542	915	1,396
Organized camping	4,495	4,263	4,157
Recreational residences	8,153	7,428	6,376
Hunting	14,043	14,472	16,235
Hiking and mountain climbing	4,548	9,059	12,734
Horseback riding	2,053	2,617	3,671
Resort use	3,794	3,927	4,364
Gathering forest products	1,284	2,274	5,388
Nature study	854	1,029	1,961
Viewing scenery, sports, etc.	5,064	7,634	8,520
Visitor information	1,763	3,927	3,603
Totals	149,847	199,201	233,438

SOURCE: Clawson and VanDoren (1984), pp. 198–199.

To meet the rising demand will require locating and constructing more developed campsites. This should pose no problem, for there appears to be no significant limit to the number that might be developed other than one imposed by agency budgets. Private developers have also entered the market and can (but do not always) offer more privacy, along with swimming pools, hot showers, and other frequently desired facilities. At present it appears that in this country the less developed recreational sites will be on public areas and the more elaborate ones developed privately. The shared "hut" system, so important in Europe, New Zealand, and elsewhere, has yet to make any significant impact on forest recreation in this country.

There are people who prefer to camp under much more primitive conditions—people who feel the desire, or even necessity, to "get away from it all." These are the dispersed campers—perhaps as many as one-third of all campers, although (almost by definition) accurate data on this kind of activity is much harder to obtain. The over 200,000 miles of Forest Service roads would seem to offer ample opportunity for expanding this type of camping, although it might increase fire hazards, health and sanitation problems, and could create difficult maintenance situations unless carefully planned.

The use of data from public forests, and especially from National Forests, should not be interpreted as an indication that privately owned forest land does not contribute to the supply side, or that forest recreation poses no problems for private forest management. In many instances, corporate owners have developed campgrounds of much the same caliber as those on public forests, some free of charge, others available on payment of a modest fee. Corporate and other private forest owners have been particularly active in expanding the market for recreational homesites. Public forest land is seldom available for such use now, although "special use permits" granted in former years are still held in some places.

Forest recreation, especially activities such as camping, can create real management problems. All too often, foresters have lacked experience with people problems, for these are of recent vintage. Guidelines for the physical layout of developed campsites, for access roads, buffer strips around the site, and so on, are being established, as are economic guidelines aimed at cost effective development; acceptable guidelines for people are difficult to formulate and even more difficult to apply. We are gradually beginning to realize that "Managing for recreation requires different kinds of data and management concepts than does most other activities. While recreation must have a physical base of land or water, the product —recreation experience—is a personal or social phenomenon. Although the management is resource based, the actual recreational activities are a result of people, their perceptions, wants, and behavior" (Committee of Scientists Report, 1979).

WILDERNESS EXPERIENCE

The wilderness concept is sufficiently unique, and new, to warrant a reasonably full description. The following definition provides a start:

> *Wilderness: An area of undeveloped Federal land retaining its primeval character and influence, without permanent improvements or human habitation, which is protected and managed so as to preserve its natural conditions and which (1) generally appears to have been affected primarily by the forces of nature, with the imprint of man's work substantially unnoticeable; (2) has outstanding opportunities for solitude or a primitive and unconfined type of recreation; (3) has at least 5,000 acres of land or is of sufficient size as to make practicable its preservation and use in an unimpaired condition; and (4) may also contain ecological, geological, or other features of scientific, educational, scenic, or historical value. (For. Res. Rep. #21, 1977, p. 235)*

The wilderness idea was officially recognized in 1939, when the Secretary of Agriculture promulgated regulations defining Forest Service wilderness policy—the "U-regulations."[9] Directly attributable to Robert Marshall, these caused much debate within the agency.[10] Most of the areas established by these regulations later became designated as wilderness areas by the Wilderness Act of 1964.

---

[9]Regulation U-1 pertained to areas of 100,000 or more acres and allowed no roads, timber harvesting, motorized transportation, or occupancy under special use permits. Grazing, water storage projects, and others were allowed. U-2 applied to smaller areas—5000 to 100,000 acres. Called wild areas, these were managed under the same constraints as the U-1 areas. U-3 areas were recreation areas that were to be managed for use "substantially in their natural condition," but allowed timber harvest, roads, and so on at the Chief's discretion (Dana and Fairfax, 1980, p. 157).

[10]Marshall, a founder of the Wilderness Society, was "energetic, wealthy, and totally dedicated to wilderness preservation." Transferred in 1937 from the Bureau of Indian Affairs, he became Chief of the Recreation and Lands Division in the Forest Service without fulfilling the usual requirement of working up through the ranks. He died two years later, 38 years old, two months after the U-regulations had been declared in force. See Dana and Fairfax (1980), pp. 157ff.

In addition to guides for establishing such areas, this Act also set out guidelines for management *precluding the use of motorized vehicles and of harvesting timber*—a restriction opposed by many forest service officers and even more by the timber industry.

By the mid-1980s, following two Roadless Area Reviews and Evaluations (RARE I and II) and the substantial additions made by the 98th Congress (162 new wildernesses and additions to 46 others) the situation seemed to be moving toward a more settled condition with around 32.1 million acres in the wilderness system (West. Tim. Assoc., 1985).

By definition, use of a wilderness area must be low to fulfill the intentions behind the Wilderness Act, which includes maintaining "outstanding opportunities for solitude." Intensity of use varies from area to area. As one would anticipate, those wilderness areas nearest population centers are the most visited, and primarily by people living relatively nearby. The Boundary Waters Canoe Area of Minnesota and the John Muir Wilderness in California were the more heavily used in 1978, with both reporting over a million visitors. Others, however, were very lightly used: visitor days per acre per year varied in 1978 from nearly 7 down to 0.04. In several wilderness areas limitations on visitor numbers have been imposed; licensing has been suggested, as have lotteries, or permits for a limited number. Fee systems have been suggested for some areas with fee levels set to hold visitors to some preestablished limit. New methods of management may also be needed, since use is often concentrated along particular trail segments of camping locations.

In 1978 Forest Service wilderness (the Forest Service administered well over 80% of area designated as wilderness) had over 8 million visitor days of use—about 4% of the total recreational visitor days spent on National Forests. Most users were young urban adults, frequently white collar workers of high educational levels and with above average incomes.

ASSESSING SOCIAL BENEFITS

Social benefits are derived from producing nontimber products on either public or private forest lands, often without any conscious managerial effort. The most obvious problems arise in public production, because public forest land allocation is often influenced by nontimber production. If private forest land allocation is affected, one can assume either that the owners have weighed the costs and benefits involved before making their decision, or have decided that the costs are minor.

As with other products, the social benefits from public production of nontimber products should equal the consumer's surplus plus (in some cases) the producer's surplus, generated by production of the product. Both are measured by willingness to pay (see Chapter 13). The only problem is that many of the nontimber products are not marketed. This fact raises two dilemmas: (1) consumer's and producer's surplus require determination of the relevant portion of the product's demand curve, and (2) somehow one must come up with a price if willingness to pay is involved. The discussion of several ways to solve these two problems will take up the remainder of this chapter. We will concentrate on two basic approaches for determining preferences: (1) by asking people what they prefer *(their expressed preferences)* and (2) by observing what people do *(their revealed preferences).*

Expressed Preferences

This seems a very direct approach: if you want to know what people prefer, ask them! Questions can be asked in many ways, ranging from formal public opinion polls through detailed personal interviews. But in any case, the questions must be the right ones.

A major advantage of this approach is that questions can be very precisely directed. "How much will you pay to spend one night in a forest campground with 50 campsites, with every camping space at least two acres in size and fronting on a sand beach offering good swimming but no fishing, 10 miles off an all-weather highway, with water faucets for every two campsites, pit toilets, and located 50 miles from the nearest store?" Or they might be so general as to yield almost no information ("Do you like public campgrounds?"). But if the question is overly complex, it can lead to meaningless answers, for people can deal with only so much detail; any more and the question becomes confusing. Furthermore, the questions must be framed in terms that mean the intended thing to the respondent. With campgrounds, for example, one should know if the respondent has had experience as a camper, something about the duration of his stay, has she ever camped overnight in a Forest Service developed campground, and so on.

There are often sources for error. People may give strategic rather than truthful answers in order to influence the result or enhance their status. Hypothetical questions ("How much would you be willing to pay as an entry fee to this campground?") run the danger of producing only hypothetical replies from respondents if they feel that their answers will not influence user fees or other payment schedules. Other biases must be guarded against but, fortunately, there are ways of avoiding most of these.

Revealed Preferences

Revealed preferences are derived from the way people act. The results somehow seem to be more realistic than those from surveys—perhaps because one feels people are reacting in the real world instead of one they imagine when responding to questions. Probably the most common use of revealed preference analysis is that employed in economic studies that take market values as true indicators of consumer preference. But market choice is influenced heavily not only by income (which greatly limits choice) but also by the availability of alternatives, by market imperfections, by peer pressure, by desire for status, by family situations, and so on. And one must keep in mind that revealed preferences, when used to plan for future needs, imply a continuation of the social status quo.

At least three techniques have been used to assess benefits from nonmarket products and all rely on revealed preferences. The first, generally referred to as the travel cost method, is the approach most often selected by those dealing with forest-based nonmarketed products. Substantially less use has been made of the second method, which yields estimates of shadow prices for the particular good or activity. A third technique (the hedonic approach) derives a demand curve for the concerned product by observing how values of observed marketed goods vary with changes in specified characteristics that are also possessed by the nonmarketed good. Although seldom used by forest economists up to the mid-1980s, it can provide highly useful insights. The remainder of this chapter will discuss these three methods.

Travel Cost Method

The travel cost method stems from the seminal paper by Clawson (1959) and has been elaborated by several others.[11] The principles of the travel cost method can be illustrated by the following example.

---

[11]The initial approach was first described by Clawson (1959). A considerably refined version (sometimes referred to as the CK travel cost method was presented in Clawson and Knetsch (1966). One of the best descriptions of the method may be found in Freeman (1979).

Assume a developed forest campground in which campers must check in and out through a single entry. Each group spending a night is given a card to fill out. The card asks:

Please circle the figure that best represents the distance in miles from your home to this campground

50 100 150 200 250

Each of these figures represents the average distance from the camp to the visitor's home. At the end of the season the information is tabulated, providing the data of Table 15.5.

Two more bits of information are needed: (1) the number of people within each of the five travel zones, and (2) the average cost of making a round trip to the campground from each zone. To determine the first, a series of concentric rings could be drawn about the campground, with one ring at the midpoint of each distance zone. Then, using population density maps, county census reports, city population figures, the total population within each of the five distance zones could be determined. Average cost data could be obtained from automobile clubs, from national studies, or from any other suitable source. These data could finally be recorded in a table such as that of Table 15.6 and the visitation rates calculated for each zone.

The usual problems of demand and supply shifts over time have been eliminated since the data pertain to a single camping season. Two key assumptions, however, must now be made, one with respect to demand, the other to supply. On the demand side *it is assumed that the demand curve for the entire recreation experience is constant throughout the market area.* People respond to travel cost changes in the same manner regardless of the travel zone in which they reside. No attention is given urban concentrations, education, age levels, income differences, or other factors. On the supply side *it is assumed that all variation in cost of the total forest recreation experience is included in the travel expenditure.* Here there is no recognition of highway patterns, road types, public transportation facilities, cost of equipment, and so on.

Now, since all costs are assumed to be included in the travel expenditure, it follows that within each of the five zones the supply curve will be perfectly horizontal. Each zone will have a different (horizontal) supply curve, and it will be located at the level of average travel costs from that zone. The five supply curves derived from the data of Table 15.6 are shown in Figure 15.4. The *consumption* points on the five supply curves necessarily trace out the demand curve we have been seeking, for *demand was assumed constant* throughout the entire market area. By connecting these points, we may draw the demand curve for the total forest recreation

TABLE 15.5
Number of Visitors from Each Mileage Zone

Number	Miles	Zone
250	50	1
400	100	2
4,500	150	3
1,600	200	4
3,000	250	5

TABLE 15.6
Summary Table, Forest Campground Study

Zone	Cost (dollars)	Number of Visitors	Total Population	Visits per 1000 Population
1	2	250	5,000	50
2	3	400	10,000	40
3	4	4,500	150,000	30
4	5	1,600	80,000	20
5	6	3,000	300,000	10

experience, as shown in Figure 15.4. The demand curve for the total recreation experience for each one of the five zones can now be drawn by multiplying the per 1000 population demand value (i.e., the visits per 1000 population values from the horizontal axis of Figure 15.4) by the respective zonal populations in 1000s. To illustrate, the demand curve for zones 1, 3, and 5 are reproduced in Figure 15.5.

After obtaining the demand curve for each of the five zones, it should now be possible to derive the demand curve for the site itself. Derived demand, we know, requires knowledge of the supply conditions of all factors of production other than

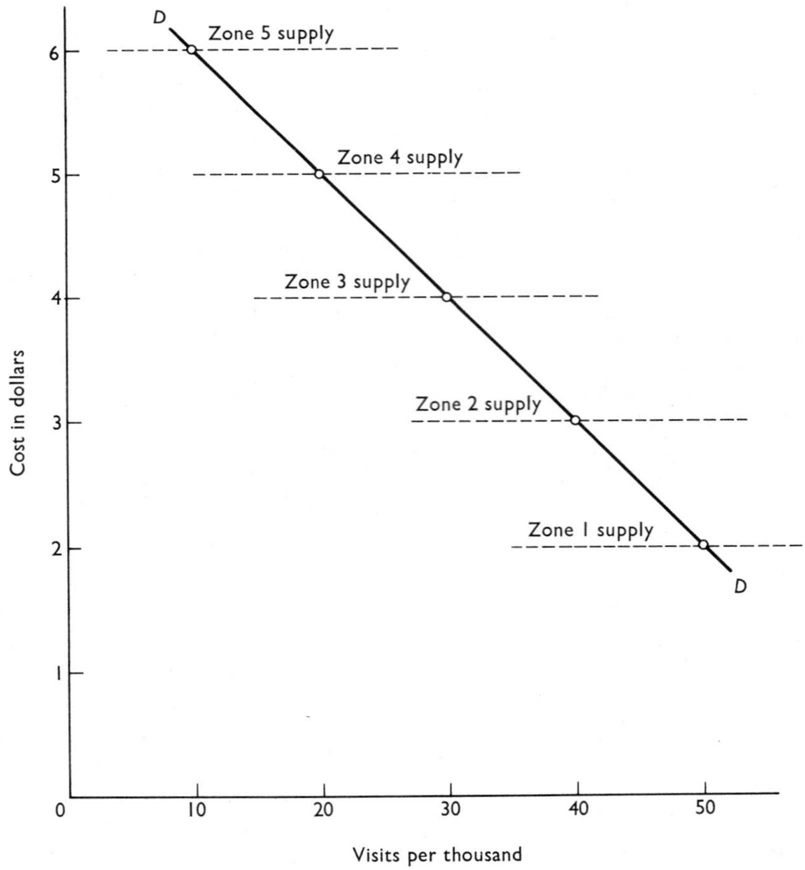

FIGURE 15.4. The five supply curves serve to identify the demand curve.

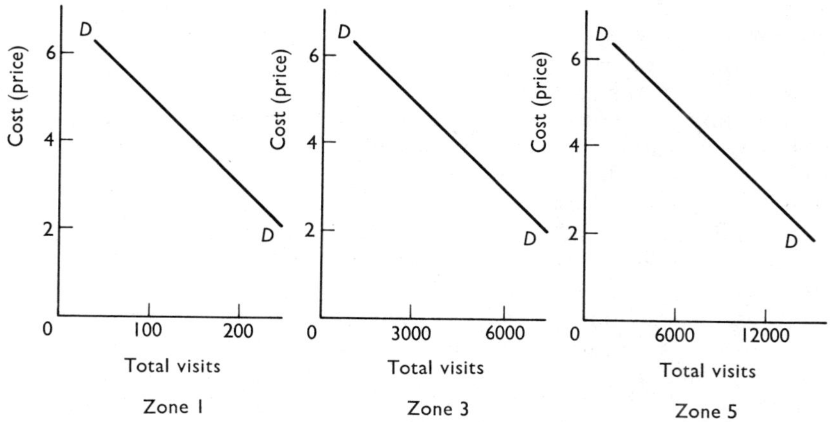

FIGURE 15.5. The demand curves for zones 1, 3, and 5 (note change in horizontal scale).

the one being considered, and *by assumption there are no production factors other than travel*. Consequently, for each zone, the travel cost must be subtracted from the appropriate supply curve, leaving the demand curve for the forest recreation site in each of the five zones. The five demand curves, shown in Figure 15.6, differ from each other in two respects. The visitor scale has been changed by multiplying the per 1000 scale by the population of the zone in 1000s, and in each case the demand curve has been shifted down by the amount of the travel cost (the supply curve of other productive factors).

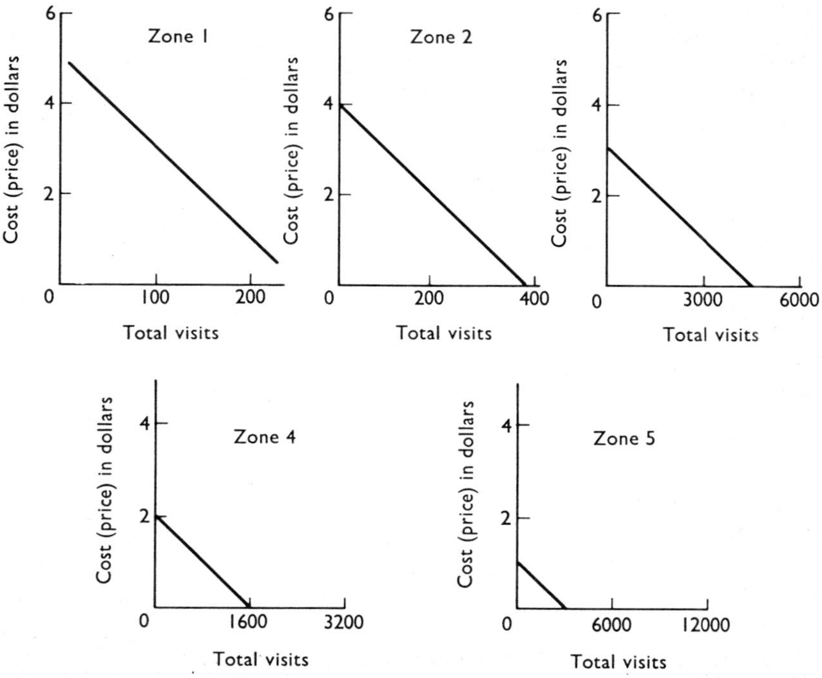

FIGURE 15.6. The demand for the forest recreation site (the campground) in each of the five zones (given the assumptions stated in the text).

The final step can now be taken. To construct the *total demand* curve for the recreation site we need only aggregate the five demand curves—add up the number of visitors that will come from each travel zone at any specified price level. The result is shown in Figure 15.7, from which one can read the effect on number of visitors of any change in price, such as might result from imposing an entrance fee to the campground. Provided the set of assumptions is valid, this curve is the demand curve for the forest campground.

Given the demand for the campground, the problem of determining its social value is no different from that for any marketed good—with one significant exception: *there is still no market determined price*. In fact, for most public campgrounds the admission price is nominal, and may even be zero. In the latter case the total area under the demand curve is consumer's surplus and should be used to calculate the campground's value. One still must estimate the number of years over which this demand curve will be appropriate. (Is demand increasing or decreasing, and at what rate? Will admission always be free?) And, of course, the discount rate will be important in the value determination.

To obtain the *net* social value, it is necessary to determine the costs involved over whatever time period has been selected for the analysis. Here one must include any opportunity costs stemming from the expected use. Foregone timber values are one example, but these must be revenues that are truly foregone. To illustrate, the fact that establishing a specific campground means that sawtimber worth $50,000 will not be cut does not necessarily mean society has lost $50,000. One must ask "Would the timber have been cut this year, or 10, or 15 years hence? Would failing to cut this quantity of timber mean an actual reduction in the national timber supply of this same amount, or did the campground establishment only mean a larger cut in some other area? As always, the "with and without" test should be applied.

There are a good many weaknesses in the travel cost approach, only some of which can be overcome by more careful design. More data might be gathered from the visitor interview, enabling one to weed out those campers taking multiple stop trips or who use the campground only as an overnight hotel. Other important information might include the actual travel time, cost of meals, total miles traveled,

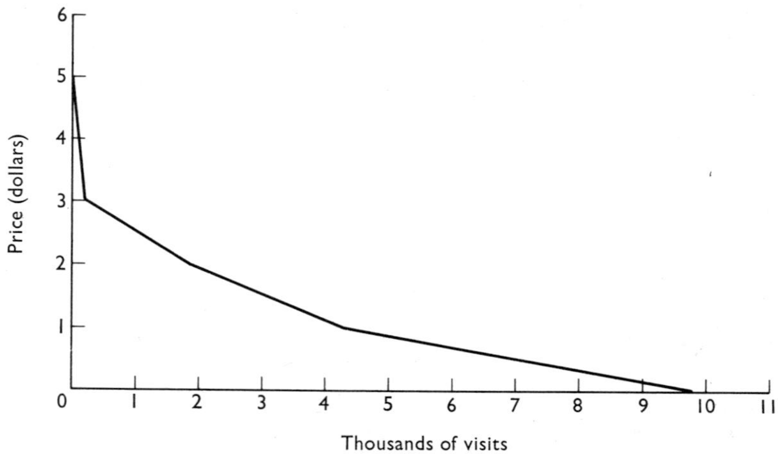

FIGURE 15.7. The demand for the forest campground (given the assumptions stated in the text).

number in party, and income, all of which might be used to improve or extend the applicability of findings.

Travel time is of particular significance, since time cost can have a very direct effect on both the shape and position of the demand curve. For example, take two travel zones of 25 miles, a per mile travel cost of 20¢, and trip times for two zones of 1 and 2 hours.

Zone	Distance	Travel Cost	Trip Time
1	50	$10	1 hr
2	100	$20	2 hr

If there are no other costs to be considered, an individual from zone 1 has a total cost of $10 plus 1 hour, and one from zone 2 has a total cost of $20 plus 2 hours. If the time cost is ignored, the cost difference between visitors from the two time zones is $10—the extra hour of travel time does not enter the calculation. This omission makes the computed demand curve more elastic (flatter) than it should be, so the benefits from the campground (the total area under the demand curve) is underestimated. Figure 15.8 illustrates the effect of ignoring time costs by assuming a time cost of $5 per hour. Clearly, the higher the value of travel time, the greater the underestimation of benefits. To correct for this bias Freeman suggests calculating a shadow price for the time cost, then adding this to the other travel costs (Freeman, 1979, pp. 204–209).

Multipurpose trips and the availability or nonavailability of alternative sites

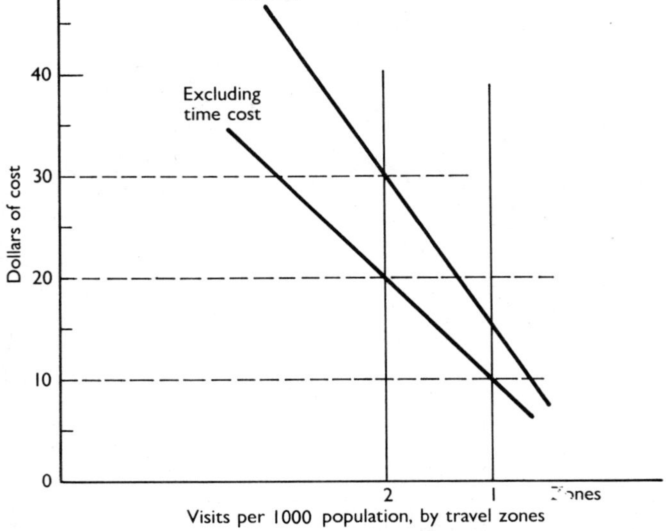

FIGURE 15.8. Effect of ignoring time cost when using the travel cost approach. Benefits will be underestimated.

pose particularly difficult problems, especially if one hopes to generalize results from one site to another (Freeman, 1979). Pertinent to this and other problems, Clawson has suggested that a new index, or measure, be developed to permit converting different outdoor recreation sites to a common denominator. Called effective acreage equivalents, the measure would "convert surface acres *apparently* available for outdoor recreation into a meaningful number of acres that are *effectively* available for that purpose" (Clawson, 1984).

Shadow Price Method

This approach has several variants that lead to the calculation of shadow prices for nonmarket forest products. Suppose a particular State Forest is being managed as a multiple-use unit, with emphasis on the joint production of deer and timber. Discussions with both timber and wildlife specialists familiar with the forest conditions permit construction of the production possibility curve showing annual yields for each 100 acres (Figure 15.9). The State Forestry Department has held public hearings during the last two winters and feels that both the timber and hunting interests are reasonably satisfied. Records show that average yields have been at point *a* on the possibility curve—about 8 deer and 15,000 board feet from each 100 acres. A line just tangent to the production possibility curve at this point intersects the two axes—at 25,000 board feet and at 20 deer.

If stumpage is selling for \$60 per thousand board feet, then 25,000 board feet is worth \$1500. And if the area is satisfying public demands for timber and

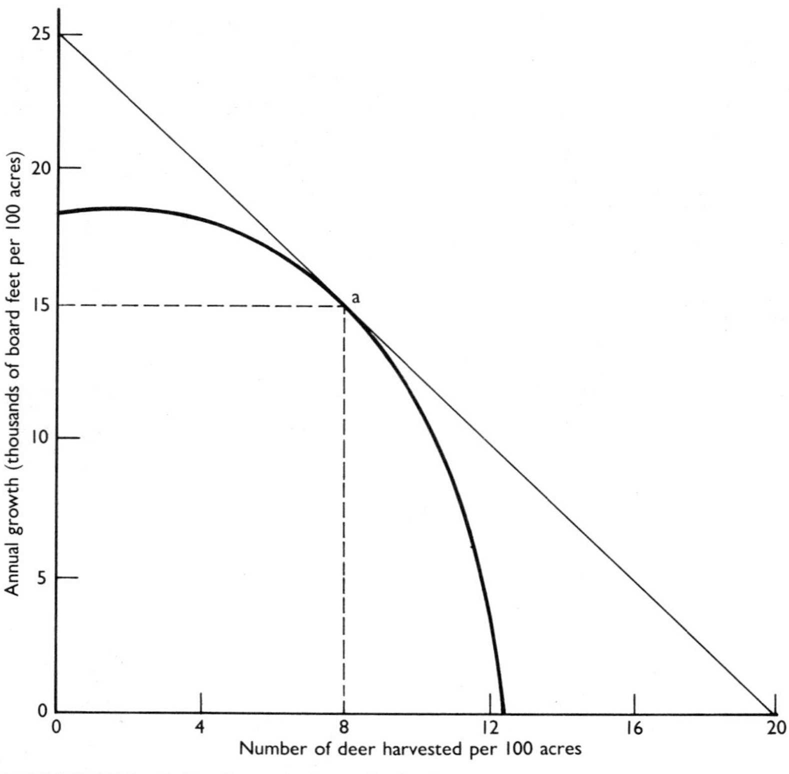

FIGURE 15.9. Estimating a shadow price for deer.

deer, then 20 deer would have the same value, which implies deer are worth 1500/20, or $75.00 each. This may be more or less than either the forest manager or the department chief believes likely, but the area has been managed *as if* this were the market value of harvestable deer. By accepting present practices, society apparently agrees with this implied value—a shadow price under which management can continue to operate. No one has been asked to value deer, but social preference has been revealed by action, not words.

The example is overly simplistic, but in fact it is more sophisticated, and might even lead to a more accurate indication of value, than some of the methods used in the mid-1980s.

The Hedonic Approach[12]

The premise on which this approach is based is that the value of a nonmarketed good can be measured by observing how consumers change their selection of certain goods that are market priced as specified characteristics change. Because the specified characteristics are also possessed by the nonmarketed good, the analyst reasons that demand for the latter will respond in the same way as did demand for the marketed good. The approach can be applied only when certain conditions are met: (1) individuals (or families) must have freedom and a sufficient knowledge of alternatives to decide their level of consumption of the good or activity concerned; and (2) a choice of the specific good or activity is made by selecting a group of goods that are market priced. Big game hunting, white-water rafting or canoeing, trout fishing, skiing, and camping are examples of forest-based products to which the method might be applied. Among the hunting public, for example, the *quality of certain big game hunting areas* might be measured by three characteristics—abundance of game (likelihood of success), potential for trophy specimens, and the accommodations available. Hunters who participated during the season would be sampled, with each one asked to provide information on the area selected, the number of trips, the days spent on each trip, and the accommodations and equipment chosen. If the information were not already at hand, each hunting area visited by hunters in the selected sample would be examined and the characteristics of the area measured or estimated.

Two steps are now necessary. The first consists of estimating the *implicit* prices of these characteristics, usually by regression analysis. In the imaginary example of hunting, probably two sets of prices would be required—a dollar price, and a time price. Both would be estimated by regressing the cost (time and dollar) of the hunting trips as a function of the three characteristics (game abundance, potential of trophies, and accommodation quality). These two cost types would probably be combined into a single cost at a later point by placing a monetary value on time (as suggested in the travel cost discussion). By this means a price equation would be determined in which the price paid (the cost) for each trip is a function of the identified characteristics. In our case this equation might be

$$P_x = F(c_1A,\ c_2T,\ c_3H)$$

where P_x = price (cost of the area x), and F is the hedonic, or implicit price function in which

A = game abundance
H = accommodation quality
T = potential of trophies
c = regression coefficients

[12]For an excellent description of this method, see Freeman (1979), Chapter 8, pp. 195–229.

From this hedonic price equation the implicit price of each characteristic can be determined by differentiating the implicit price function with respect to each of the specified characteristics. The resulting equations give the increase in cost (price) required to obtain one more unit of that characteristic, other things being held constant. If these relationships are linear in a particular characteristic, then that implicit price will be constant per unit. But because individuals vary in many ways, a regression of this type is not likely to be linear; any change in the implicit price will probably depend on the quantity of the particular characteristic present. And there may be any (or all) the other familiar problems encountered in regression analysis—problems such as multicolinearity, and so on. It also seems likely that considerable experimentation may be needed before a defendable transformation of time cost to monetary cost could be devised.

The results, however, could be very useful. Under some conditions these implicit price functions for the characteristics are, in fact, inverse demand functions, showing the change in demand price for a particular characteristic as more or less of that characteristic is selected. For this to be true, either supply conditions must be perfectly elastic or both supply and demand functions must be determined simultaneously. Unlike the other approaches to valuation of nonmarket forest products, the hedonic approach provides direct assistance to resource managers. How much is an increase in game abundance worth? How much should be devoted to producing trophy-sized animals? Should better accommodations be furnished, and so on.[13]

Linear programming models also permit an estimation of shadow prices. By making small changes in the constraint levels one can estimate the tradeoffs between products. Every tradeoff involves some opportunity cost, and if a market value can be determined for that opportunity cost then values for the alternative product may be implied (Connaughton and Fight, 1984).

Some Remaining Problems

One point about any kind of revealed preference study cannot be overemphasized. Tradeoffs, especially when coupled with market values, do not necessarily reflect true social preferences or social values. They represent a certain state of society and technology and are dependent on the state of knowledge as well. In the deer example, did the stumpage sale value reflect social appraisal or that of a single large buyer? Did the people attending the hearings know all the choices that were available? Did they really *have* any choice? To what extent should public hearings held in a given locality influence decisions on national or state forest land? There are many such questions!

Perhaps the most difficult questions in terms of assessing nonmarket benefits come from wilderness areas. One might think, especially after reading the earlier definition of a wilderness area, that these would raise few management problems. After all, how can one "manage" an area that is to be maintained in a primitive state? The problems arise because these areas produce a product, that product being a wilderness experience. But some areas near population centers have attracted so much use that the wilderness experience per se is no longer available.

One approach to this problem of overuse would limit the number of visitors allowed into a wilderness area at any one time. Some studies indicate that most

---

[13] An example applicable to forestry is that reported by Brown and Mendelsohn (1984), who applied the technique to value steelhead fishing density in Washington state streams. See also Brookshire, Thayer, Schultz, and d'Arge (1982).

potential users would agree to this type of control (Stankey, 1979), while others suggest that licenses be issued—which would raise the problem of selecting the proper fee or other allocation mechanism. There has also been much discussion about the degree of freedom wilderness users should be granted: should they be free to take any risk and bear the consequences, must public rescue units be kept in readiness, or should special "no risk wildernesses" be established (McAvoy and Dustin, 1981; Wagar and Allen, 1981)?

Equally interesting problems are posed by the substantial forest areas withdrawn from timber production by the Wilderness Act. With over 32 million acres declared out of bounds to timber harvesting, some opportunity costs must surely be involved. Measuring these costs, however, has not proved as simple as one might think. Analysts have not always remembered that local worker displacement may not be a social cost to the nation, or that removing even a substantial forest area from commercial timber production might not immediately, or even eventually, affect lumber supplies or consumer prices. How much, for example, has the national harvest of timber been reduced by passage of the Wilderness Act? By how much have lumber prices been increased? Do wilderness area set-asides really have any effect on timber supply?

Option values also enter the assessment of many nontimber benefits. Option values are demands by individuals who are not yet actual consumers of a good, or who anticipate that they might want to increase their consumption of a good or activity sometime in the future. In such cases, people may be willing to pay to perpetuate the future availability of some good or service. Wilderness provides a good example: an individual presently living in a New York apartment may have absolutely no plans for backpacking into any wilderness area, yet he or she may be willing to pay to maintain areas in a primitive, wilderness condition. In other words, people will pay to keep open the *option* to some day have a wilderness experience, not just for themselves but also for their children, or even their grandchildren! The theory of option demand has been well developed by Weisbrod (1964), Krutilla (1967), Krutilla and Fisher (1975), and others.

The significance of option demand lies in the knowledge that, although presently difficult to quantify, in some cases it undoubtedly exists and would surely increase consumer's surplus for such services. How much option demand would shift outward the demand curve for (say) wilderness experiences or for grizzly bears, thus increasing the consumer's surplus and hence the social value of these forest products, is a moot point.

Clearly, multiple-use and nontimber products have had, and will continue to have, far-reaching effects on forest management and on the social well-being of consumers in the United States. Measuring the costs and the benefits that result from multiple use of a nation's forests is not a simple exercise, and there is much room for the development of new methods that would help in the task.

QUESTIONS AND PROBLEMS

1. Nontimber products were treated in six groups: subsurface minerals, forage, water, wildlife, forest recreation, and wilderness experience. Taking each group in turn, what would you state as the major problem that particular group of products poses to (a) private forest managers, and (b) public forest managers.

2. How important are the National Forests to Western cattle production?
3. Discuss, briefly, the problem created for Western national forests by wild horses.
4. "In the near future the United States will face real difficulties in providing its citizens with the beef cattle they need simply because they are running out of suitable grazing land." Discuss the validity of this statement.
5. What are the three major categories of water use and what is their relative significance to water supplies in the United States?
6. Distinguish between consumptive and nonconsumptive uses of forest wildlife, giving two examples of each and discuss both uses in relation to forest management activities.
7. The average age of U.S. citizens is increasing, as is the number and proportion of people over 60 years old. What implications for forest recreation might this change hold, if any?
8. Wilderness set-asides have come under much criticism by some sources, but are vigorously advocated by others. In your view, do we need more, or fewer, wilderness areas? Support your stated viewpoint as strongly as you can.
9. Distinguish between the two basic approaches to determining preferences (expressed and revealed).
10. The travel cost method has most often been used in studies designed to assess the value of forest parks, campgrounds, (and to a lesser extent) hunting and fishing opportunities. Describe this method, and give the major advantages and disadvantages of this approach.
11. Two other approaches were discussed—that of using shadow prices and the hedonic method. Outline the conceptual basis for these two approaches.
12. How well do you believe market prices reveal individual preferences?
13. What is an option value. Should option values influence management practice on national forests? Support your answer.
14. A forest manager prepared three multiple-use management alternatives for review by the supervisor. All presented potential ways of managing the particular forest for varying quantities of timber, water, summer recreation visits, and wildlife. The three plans are outlined in the table below.

Kind of Annual Yield	Plan A	Plan B	Plan C
Timber (dollars)	450,000	420,000	400,000
Water (acre feet)	30,000	30,000	30,000
Recreation (visitor days)	20,000	20,000	20,000
Wildlife (deer harvested)	3,000	4,000	4,500

a. What can you say about the value the supervisor implicitly places on deer if he selects Plan A?
b. What can you say about the value she places on deer if she selects Plan B?

chapter sixteen
MULTIPLE USE: THEORY AND APPLICATION

Since the turn of the century forest managers have been unable to say whether the timber they produced would be used to build a house or furniture, to make grocery bags or books, to stoke a fireplace, or for any of hundreds of other products. Now the complexity of wood products is only part of the story. Today's forester may be managing a watershed, a game refuge, a public or private hunting facility, or a ski area. Timber production may not be even a secondary consideration. He or she may need to devise a cutting method to provide scenic overviews from a high speed throughway, or to save old growth on a national park or roadside strip, or to manage a wilderness area. But whether the objective is panorama or pulpwood, sawlogs or songbirds, water or wilderness, the forester is manipulating a land-based ecological system to produce products desired by society.

This fundamental concern with forest land management and ecological relations rather than with the product per se had been a characteristic of forestry in earlier days. During the first two decades after World War II, however, lumber, plywood, and pulp were the principle objectives of forestry efforts in the United States. Perhaps most important was the fact that the wood-using industries became the primary source of employment for foresters.

The Environmental Crisis

As the U.S. population grew, the number of special interest groups pressing for recognition of their own claim to forest land use also increased. The demand for new housing had soared immediately after the war and the lumber and plywood industries pressed the Forest Service to increase timber sale volumes from federal forests—a reversal of their prewar attitude toward public stumpage. At the same time Western stockowners were revolting against reductions in grazing allotments on public lands and the increased fees charged for the grazing permits issued.

Other groups were equally vocal. The Wilderness Society, founded in the 1930s by Bob Marshall, attracted many new members and began pushing for recognition of the wilderness concept—a concept which, if accepted, would require large areas of primitive, unspoiled natural areas to be set aside in perpetuity, areas that sometimes included large amounts of merchantable timber. More and more people were using National Parks and forests for outdoor recreation. The Park Service, whose parks had deteriorated severely during the war years, was aggressively seeking support for increased budgets and, apparently, hoped to become *the* federal recreation agency. The Forest Service saw this as a threat to its turf and recognized recreation as a forest product, equal (at least on paper) to timber and grazing. Relations between the two services were, at best, somewhat less than cordial.

Other groups joined the fun. The Sierra Club, Conservation Clubs, hunting and fishing clubs, garden clubs—all became highly concerned with how public forests, and particularly the National Forests, were to be used. By the late 1950s the timber interests, recreationists, and wilderness enthusiasts seemed to be on a collision course in their efforts to increase *their* individual use of public forest lands.

And other important factors were at work. Both population and incomes were expanding. Paid vacations, retirement programs, and a greatly increased mobility were combining to bring outdoor recreation of all kinds within the range of vastly increased numbers of people. The result was what Fairfax has called an era of mass recreation and a widespread interest in National and State Forests and Parks (Fairfax and Dana, 1980).

Then, in 1962, Rachel Carson's *Silent Spring* awakened the public to the dangers a short-sighted science might create when coupled with a highly efficient production and distribution system (Carson, 1962). This awareness, coupled with the ignition of an activism fueled by obvious problems of civil rights, integration, and questionable military involvement in southeast Asia, had a profound effect on the way Americans viewed their forests.

Much of the change can be traced to the realization that increased *quantities* of material things would not necessarily raise the *quality* of living—and, in fact, there seemed considerable evidence that it might have precisely the opposite effect. Questions were raised about the diminishing availability of wild space, the reduction in aesthetic variety, the stresses of urbanization stemming from our increasing separation from nature, and the serious results that might come if this separation could not somehow be bridged. A second and related outcome was the rapidly growing demand for outdoor recreation, much of it based on the belief that through outdoor activities one might find relief from both physical and mental stress.

The energy crisis of the early 1970s gave emphasis to political and economic factors, illustrating all too well that control of a key natural resource could effectively nullify vast military power and demonstrating the industrial world's vulnerability to resource shortages. Then came indications that rising emission of CO_2 might induce global climatic changes, followed in the early 1980s by what many believed to be a greater threat—that posed by acid deposition, one form of which is acid rain. Widespread tree mortality in West Germany's world-famous Black Forest, in Scandinavia, in Canada, and even in our own Northeastern timberlands seemed to be an obvious consequence.

One major result of the new appreciation of the environment was a sudden upsurge of interest in ecological problems by members of the so-called hard science disciplines; a major sign of this interest was the elevation of ecology to a position of academic respectability. For a short time in the mid-1970s, many people were optimistic that "good" science could solve the major ecological problems but by the 1980s the pendulum had reached full swing: we were beginning to understand that many environmental problems could not be solved by hard science alone. Instead, attention began to turn to the way in which our political, social, legal, and economic institutions function—or fail to function!

The Challenge to Forestry

By the mid-1980s many new laws had been added in an attempt to assure that our public forests, and much of our environment, would be managed so as to serve both present and future generations. And although no one knows exactly when or how these new problems will be met, it seems reasonably certain that few foresters in this country will be able to confine their attention to growing timber for wood products. Forest managers will necessarily be more attuned to people and forestry will become a truly interdisciplinary profession—no less scientific, but far more socially oriented, and with much greater appreciation of the fact that *all* the products of the forest are important to the maintenance and growth of our society.

A good many of the forester's difficulties, or challenges, arise because forests may be regarded as, and are also a source of, *public* goods—goods that, once produced, are available to all. Yet, as stressed in an earlier discussion, many forest products produced in the public sector are also produced privately. Being an ecological system, a forest almost inevitably produces a variety of products. The Timber Report of the Public Land Law Review Commission (PLLRC, 1969) put it this way:

> *Rain falls, and water comes from the forests as an almost inevitable and completely natural sequence of events. Timber is grown, wildlife reproduces, and beauty is created without any conscious effort by man. But through either positive management and protection, or through negative neglect and carelessness, man can and does alter the amounts of the various products in both relative and absolute terms.*

The term **multiple use** has been used by forest managers for years to indicate their recognition of this product multiplicity. Now more than recognition is demanded. Methods must be devised, and used, to assure that when humans alter the mix of forest products the change will benefit society.

MULTIPLE USE IN FORESTRY

Early European foresters were often more concerned with producing good hunting than with timber production, and some of the oldest European town forests were established as much for protecting municipal water sources as for furnishing fuel and other wood products to the local citizenry. In this country the multiple use idea was clearly embodied in then Secretary of Agriculture Wilson's famous 1905 letter to the Chief Forester (written by Gifford Pinchot, then himself the Chief Forester) officially transferring the "Forest Reserves" from the Department of Interior to the Department of Agriculture (Pinchot, 1947).

> *In the administration of the forest reserves it must be clearly borne in mind that all land is to be devoted to its most productive use for the permanent good of the whole people and not for the temporary benefit of individuals or companies. All the resources of forest reserves are for use. . . .*
>
> *You will see to it that the water, wood, and forage of the reserves are conserved and wisely used. . . . The continued prosperity of the agricultural, lumbering, mining, and livestock interests is directly dependent upon a permanent and accessible supply of water, wood, and forage. . . . In the management of each reserve local questions will be decided upon local grounds; the dominant industry will be considered first, but with as little restriction to minor industries as may be possible; sudden changes in industrial conditions will be avoided by gradual adjustment after due notice; and where conflicting interests must be reconciled the*

> *question will always be decided from the standpoint of the greatest good of the greatest number in the long run.*

For many years multiple use was a major rallying point for those favoring public control of all forest land. Today, for a variety of reasons, the larger timber-owning corporations are moving toward a multiple-use view. Forests can and do produce many goods and services. If these products are in demand, one should expect to find private operators willing to provide them wherever and whenever they can be supplied at a profit. As a management principle, however, multiple use has been largely a development of the public sector, a fact no doubt reflecting the large area in public ownership and the variety of forest goods and services these owners demand.

Key Legislation[1]

More than a half-century after Wilson's famous letter, multiple use was given statutory basis by the United States Congress when it passed, on June 12, 1960, the "Multiple Use-Sustained Yield Act" (United States Code 16, 1960). The act was largely a restatement of existing statutes and procedures the Forest Service had followed for many years. Perhaps its principle objective was that of giving support to officials who were under increasing pressure from special interests to use the national forests for particular purposes rather than for multiple purposes (United States Congressional Record, 1960). Four years later, legislation similar to the Multiple-Use Sustained Yield Act officially extended (on an interim basis) multiple use to lands supervised by the Bureau of Land Management.

Multiple use was defined (in Section 4 of the Multiple Use Sustained Yield Act of 1960) as follows.

> *"Multiple Use" means the management of all the various renewable surface resources of the national forests so that they are utilized in the combination that will best meet the needs of the American people, making the most judicious use of the land for some or all of these resources or related services over areas large enough to provide sufficient latitude for periodic adjustments in use to conform to changing needs and conditions; that some land will be used for less than all of the resources; and harmonious and coordinated management of the various resources, each with the other, without impairment of the productivity of the land, with consideration being given to the relative values of the various resources, and not necessarily the combination of uses that will give the greatest dollar return or the greatest unit output.*

One could scarcely be surprised that this definition has fallen somewhat short of completely clearing up all questions regarding the multiple use of forest lands.

In 1964 came the Wilderness Act, establishing a National Wilderness Preservation System—and making the Secretary of Agriculture responsible for reviewing all the Forest Service primitive areas (established during the 1930s) as to their suitability for inclusion in the new system. Other legislation followed: The Wild and Scenic Rivers Act and the National Trail System Act in 1968 and the highly important National Environmental Policy Act of 1969 (which required environmental impact statements on federal actions affecting quality of the environment). Also, in the late 1960s, a group of West Virginia citizens became irate when the

[1]For reasonable detail about these events (which reshaped public forest management in less than two decades) see Fairfax and Dana (1980).

Forest Service clearcut certain areas in the Monongahela National Forest and demanded that their representatives take action. Within a short time, virtually any clearcut scheduled by the Forest Service was drawing bitter criticism (Fairfax and Dana, 1980). The Senate Subcommittee on Public Lands (under Senator Church) investigated national forest harvesting techniques and issued the Church Guidelines, which were immediately adopted by the Forest Service as their clearcutting policy.

RARE I (Roadless Area Review and Evaluation) was started by the Forest Service to identify potential wilderness areas in 1971, and in 1974 the Forest and Rangeland Renewable Resources Planning Act (RPA) required the Forest Service to prepare long-range plans, to conduct a decennial assessment of the renewable resource supply-demand situation, and to update management plans at least every five years. Then came the National Forest Management Act of 1976, which not only repealed that part of the language of the 1897 legislation that had been the basis of the Monongahela litigation, but also substantially amended the planning process of the Resource Planning Act (RPA). This "landmark in the history of the nation's natural resource legislation" (McGuire, 1976) plays a highly significant role in the management of National Forests,[2] primarily through the many changes it made in the Resource Planning Act—changes that establish the land management plan required for each unit of each national forest as the "cornerstone for management" of the National Forest System.[3]

Clearly, the two decades of legislation that began with the Multiple-Use Sustained Yield Act established that management of public forest lands would not be left entirely to trained professionals. Congress had served notice that from now on the general public would insist on being involved, even though day-to-day management decisions would be left to foresters, and that the decisions that were made would be under constant scrutiny. The immediate question posed to foresters was "Just how do we make plans that will satisfy these different directives?"

The essential economic framework for multiple use management has already been developed in Chapter 5 under the general topic of joint production. After placing multiple use within this framework, we will review two traditional interpretations as well as a more recent modification of the concept. An introduction to linear programming, its application, and limitations will follow. Finally, we will look at some of the management implications and problems that necessarily follow acceptance of nontimber products and services as true jointly produced outputs from a forest under multiple use management.

Technical aspects of production will necessarily be very quickly covered—in part because the production framework has already been discussed, but also because quantitative information on most nontimber products is sparse. In all these areas there is a growing body of literature, some of which is noted in subsequent chapters.

---

[2]For a presentation and discussion of the revised RPA, see Le Master (1976).

[3]Section 6 of the RPA states, among other things, that "the Secretary of Agriculture shall develop, maintain, and, as appropriate, revise land and resource management plans for units of the National Forest System. Coordinated with the land and resource management planning processes of State land local governments and other Federal agencies" and "shall use a systematic interdisciplinary approach to achieve integrated consideration of physical, biological, economic, and other sciences." [The Secretary] "shall assure that such plans—provide for multiple use and sustained yield of the products and services obtained therefrom in accordance with the Multiple-Use Sustained-Yield Act of 1960 and, in particular, include coordination of outdoor recreation, range, timber, watershed, wildlife and fish, and wilderness." The plans shall "form one integrated plan for each unit of the National Forest System." See Le Master (1976).

MULTIPLE USE AND JOINT PRODUCTION

From a conceptual standpoint it makes little difference whether the products being jointly produced are as similar as sawlogs and pulpwood or are as dissimilar as poplar pulpwood and white-tailed deer or scenic beauty and Douglas-fir peelers. The same production model can be used; it is in its application that differences (and difficulties) arise.

An Economic Model

Joint production, it will be recalled, occurs whenever the same production facility is used to produce two or more products. In the multiple-use model, two products may be produced in "technically variable" proportions (Chapter 5). The production of timber and forage provides a straightforward example.

With per-acre production of timber measured in cubic feet and of forage in animal unit months (AUMs), Figure 16.1 permits any possible combination of the two products to be shown. A set of production possibility curves can be constructed by connecting all timber and forage combinations that can be produced from the area at a given cost. Since cost is constant on such curves they may also be termed isocost curves.

Assuming that the amounts being produced are not large enough to influ-

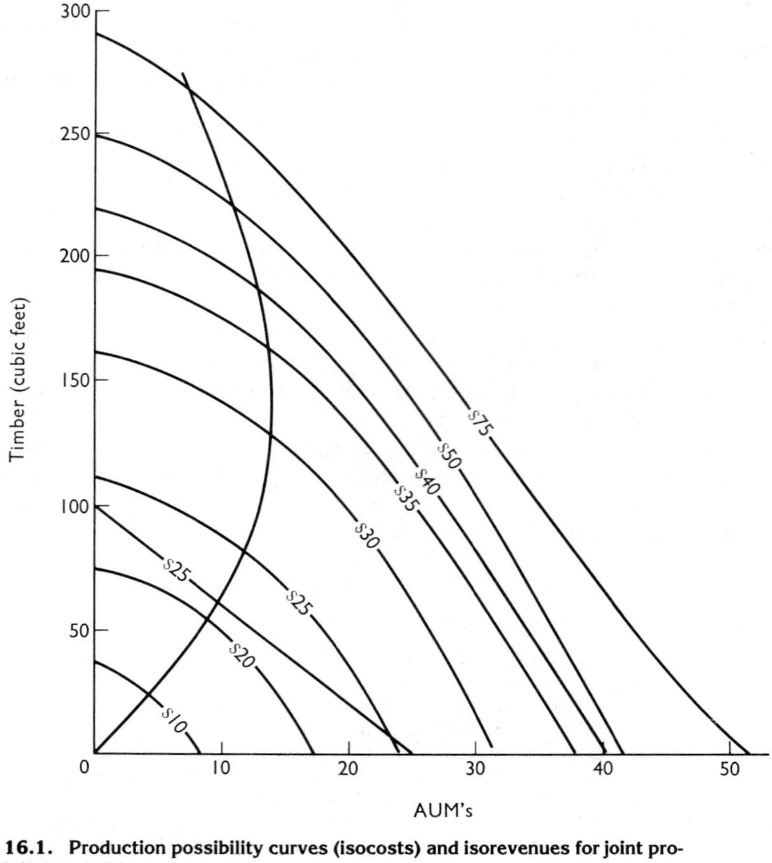

FIGURE 16.1. Production possibility curves (isocosts) and isorevenues for joint production of timber and forage.

ence the price of either product, the isorevenue curves (curves connecting all points yielding equal total revenue) will be straight lines, with the slope of the lines inversely proportional to the price ratio of the two products. In the example of Figure 16.1, with timber at 25¢ per cubic foot, and AUMs at $1 each, either 100 cubic feet of timber of 25 AUMs will yield a total revenue of $25. And every combination of timber and forage indicated on the line connecting these two points will yield this same return. As long as prices remain unchanged, all isorevenue curves will be parallel.

The points of tangency between the two families of curves (isocosts and isorevenues) identify the various combinations of products that will produce the indicated revenue for the lowest possible cost (economic efficiency). Connecting these points of tangency yields the "expansion path," or scale line, the path of product combinations along which production would logically be expanded. The problem now becomes one of determining which point on the expansion path maximizes net revenue, a problem readily solved by standard graphic or mathematical methods. The entire process can be combined into a single diagram such as that of Figure 16.2, where the product combinations and expansion path are shown below and the usual total cost and revenue curves directly above. Maximum net revenue is attained (in the example) at an output of 178 cubic feet of timber and 13.2 AUMs. With this as a basis we can examine alternative approaches to multiple-use forest management.

Traditional Interpretations

Two approaches emerged from early discussions about multiple use. One called for applying the term only to large areas—a ranger district, a working circle, or even an entire forest holding. The area mapped in Figure 16.3, for example, produces several products, but each specific area is devoted to a single (or at least a primary) use. Under this interpretation, the manager would settle on a *primary* use for each specific area. *Secondary* uses would be permitted on the area only if they would not interfere with the primary objective. Timber production, for example, might not be prohibited on a designated recreation area, but would not be permitted to interfere with recreational use.

The alternative approach makes no area subdivision and declares no primary uses. Instead, the objective of management is assumed to be the maximization of social returns, measured in whatever units are deemed appropriate. Following this approach, management should produce the combination of products that would maximize net returns to the owners.[4]

Clawson's "Compatibility Matrix"

These two traditional interpretations of multiple use have been criticized by several authors, most notably by Clawson who wrote (1974):

> *If by multiple use one means that every use should occur on every acre of forest every year (or every day), then this is absurd—it has never occurred and never can; if by multiple use one means that all uses of the forest are proper and appropriate, that one use should be reconciled with another wherever possible, and that some tracts of land may have to be given over*

[4]The list of authors writing on multiple use is far too long to attempt complete citation. Probably the first direct application of joint production theory to forestry problems was that of Ciriacy-Wantrup (1938). It was applied to multiple use by Gregory (1955), and is now being more extensively and rigorously applied by Bowes and Krutilla (1985).

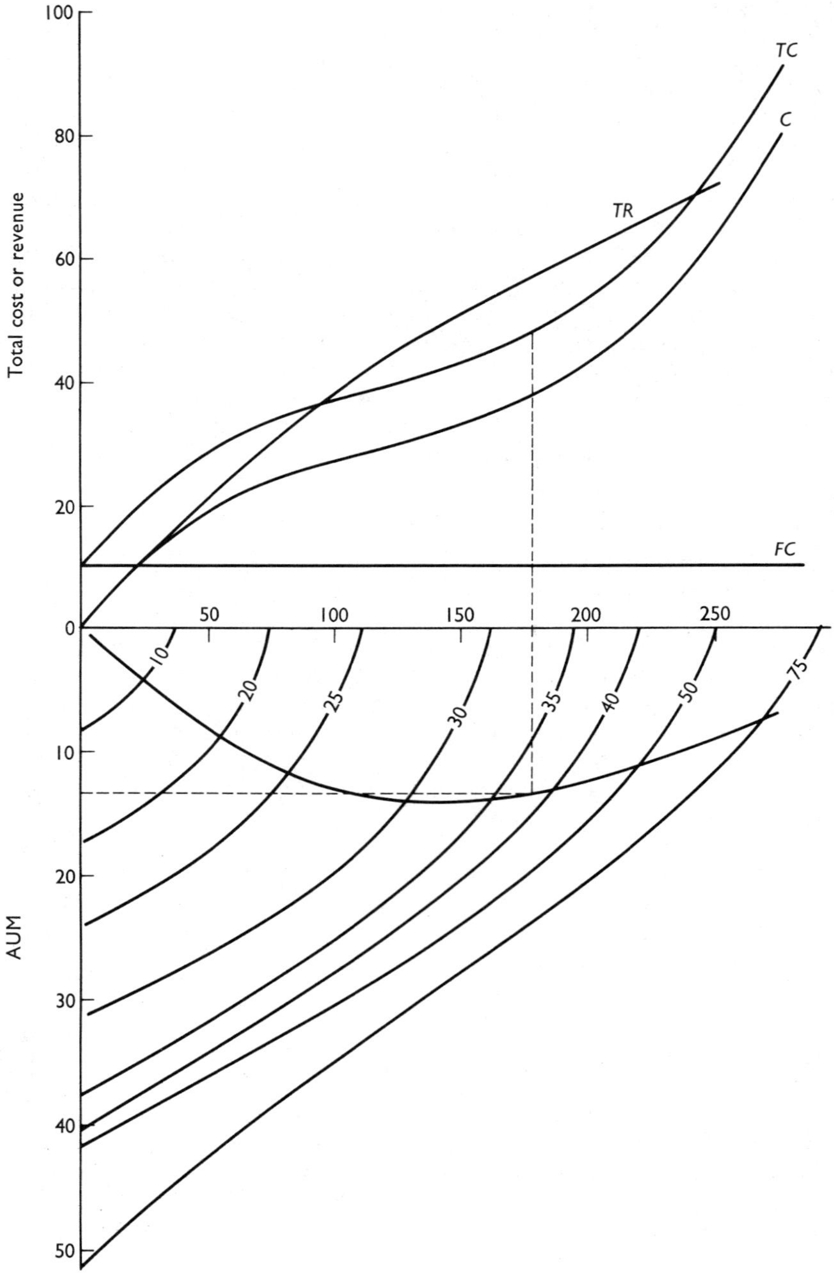

FIGURE 16.2. The expansion path and the point of maximum net revenue, indicating the optimum combination of timber and forage under the assumed conditions.

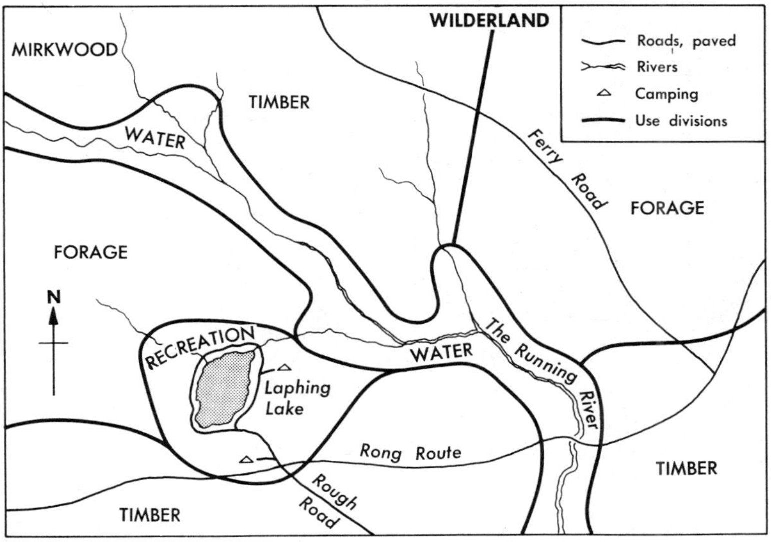

FIGURE 16.3. Map of forest area designating primary uses: multiple use on total area, but single use on any specific area.

> *largely to one use while other adjacent or intermingled tracts of forest are largely given over to other uses, then multiple use is highly sensible. . . . To the extent that dominant use means one use of the forest is exclusive of all other uses, it is impossible. . . . (From "Conflicts, Strategies, and Possibilities for Consensus in Forest Land Use and Management" by Marion Clawson, in* Forest Policy for the Future, *RFF Working Paper LW1, copyright 1974 by Resources for the Future, Washington D.C., p. 119. Reprinted by permission.)*

Clawson felt that terms such as multiple use, dominant use, and single use were of little use to management since they are far too imprecise. Instead, he suggested an approach based on compatibilities and incompatibilities. Three types of forest uses were identified (Clawson, 1974, p. 111):

1. Those that use forest land but do not really use the forest. Mining, road construction, and grazing are examples.
2. Those largely or wholly intolerant of each other (timber *harvest*, wilderness, and intensive recreation).
3. Those that occur to some extent regardless of human action, but are influenced by that action (watershed protection, wildlife, aesthetics).

Using these concepts, Clawson (1975) built a compatibility matrix of uses that displayed his judgment of the compatibility (or lack of it) between seven uses of forests (Table 16.1). The degree of compatibility ranges from "fully compatible" to "incompatible," with other qualifying adjectives indicating intermediate stages. Of the 42 combinations, only two are "completely incompatible," and those pair wilderness with wood harvesting and wilderness with recreation. Some other uses are of "limited compatibility," or "moderately compatible," or may be "incompatible for some kinds" (such as recreation and wildlife).

TABLE 16.1
The Degree of Physical Compatibility of Secondary and Primary Forest Uses

	Secondary Use						
Primary Use	Attractive Environment	Recreation Opportunity	Wilderness	Wildlife	Natural Watershed	General Conservation	Wood Production and Harvest
Maintain attractive environment		Moderately compatible; may limit intensity of use	Not inimical to wilderness but does not insure	Compatible to most wildlife, less so to a few	Fully compatible	Fully compatible	Limited compatibility; often affects amount of harvest
Provide recreation opportunity	Moderately compatible unless use intensity excessive		Incompatible; would destroy wilderness character	Incompatible for some kinds; others can tolerate	Moderately compatible; depends on intensity of recreation use	Moderately compatible; incompatible if use too heavy	Limited compatibility depends on harvest timing and intensity; roads provide access
Wilderness	Fully compatible	Completely incompatible, can't tolerate heavy use		Highly compatible to much wildlife, less so to others	Fully compatible	Fully compatible	Completely incompatible, precludes all harvest
Wildlife	Generally compatible	Limited compatibility; use intensity must be limited	Mostly compatible though some wildlife require vegetative manipulation		Generally fully compatible	Generally fully compatible	Generally compatible but may require limiting volume or conditions of harvest

Natural watershed	Fully compatible	Moderate compatibility; may require limitation on intensity	Not inimical to wilderness but does not insure	Generally compatible		Fully compatible	Moderate compatibility; restricts harvest methods but does not prevent timber harvest
General Conservation	Fully compatible	Moderately compatible; if use not excessive	Not inimical to wilderness but does not insure	Generally compatible	Fully compatible		Compatible but requires modifications in methods of timber harvest
Wood production and harvest	Compatible if harvest methods strictly controlled	Moderately compatible	Completely incompatible; would destroy wilderness	Compatible if harvest methods fully controlled	Compatible if harvest methods fully controlled	Compatible if harvest methods fully controlled	

SOURCE: Marion Clawson, *Forests for Whom and for What?*, copyright 1975 by Resources for the Future, Washington D.C., p. 40. Used by permission.

In a later paper Clawson took the two uses most often completely incompatible—wilderness and wood harvesting—and combined them with a scale of forest productivity for the two products (Clawson, 1979). Instead of scaling the diagram in cardinal units, he developed five managerial "situations" (Figure 16.4), and used these to identify the type of problem that might result from each situation. Situation II, where much National Forest land would be classified, is characterized by low wood values, low scenic values, and so on. Here serious conflicts are unlikely (although clearcutting might promote objections).

Situation III would include some public forests, much of the private, nonindustrial forests, and most of the industrial forest land. Situation I would probably contain much of the designated wilderness area: high and rocky mountain areas, and so forth. Situation IV is where the real battleground is—a value battle between giants for areas like the Olympic Peninsula, the Redwood Park addition—while V contains a little of everything, and use would probably be determined by location.

The high values of situation IV poses a somewhat different problem. Here the potential for intensive management of both products is high and, at the extremes, very high. But intensive management for either one of the products precludes the other.

While none of the approaches reviewed above provides direct guides to the manager who must decide how much of each product should be produced from a particular forest area, Clawson's classification system aids one in determining the

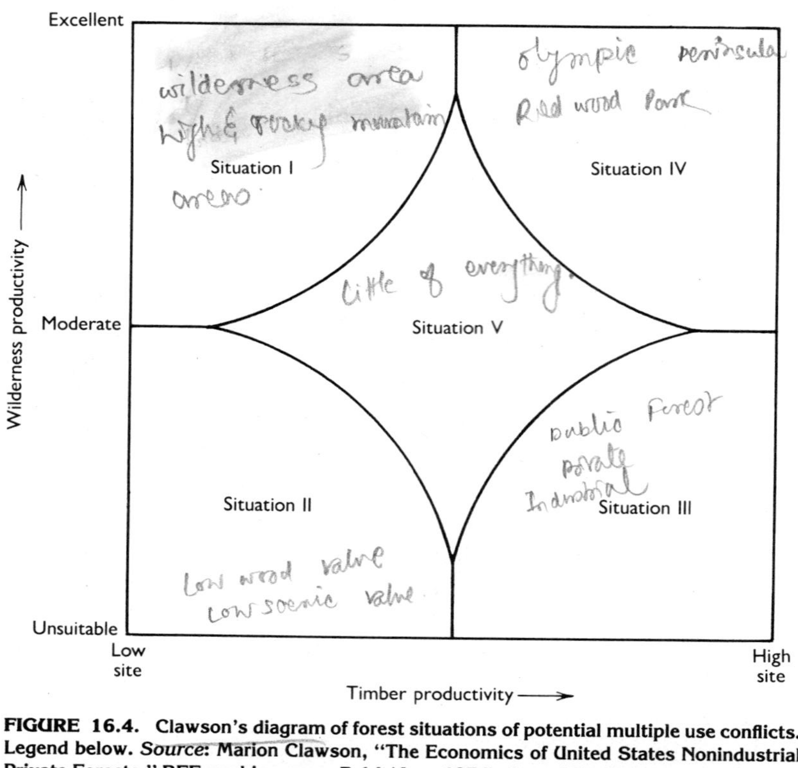

FIGURE 16.4. **Clawson's diagram of forest situations of potential multiple use conflicts. Legend below. *Source:* Marion Clawson, "The Economics of United States Nonindustrial Private Forests." RFF working paper R-14 (June 1974). Copyright 1979 by Resources for the Future, Washington, D.C. Fig. 7-3, p. 96. Used by permission.**

areas where these decisions will be significant. Study of Table 16.1 and Figure 16.4 leads to two simple but important conclusions: naturalness (or maintaining an attractive environment), intensive recreation, and intensive vegetation management (whether for wildlife or timber production) are uses that will eventually conflict if attempted on the same small areas. If forest resources are to be successfully *managed* for these products, the areas should be separated.

Isocost Patterns

Isocost diagrams may also throw some light on such decisions.[5] Four possible sets of joint production isocost and revenue curves of recreation and timber production are shown in Figure 16.5. Additionally, each resembles to some degree one or

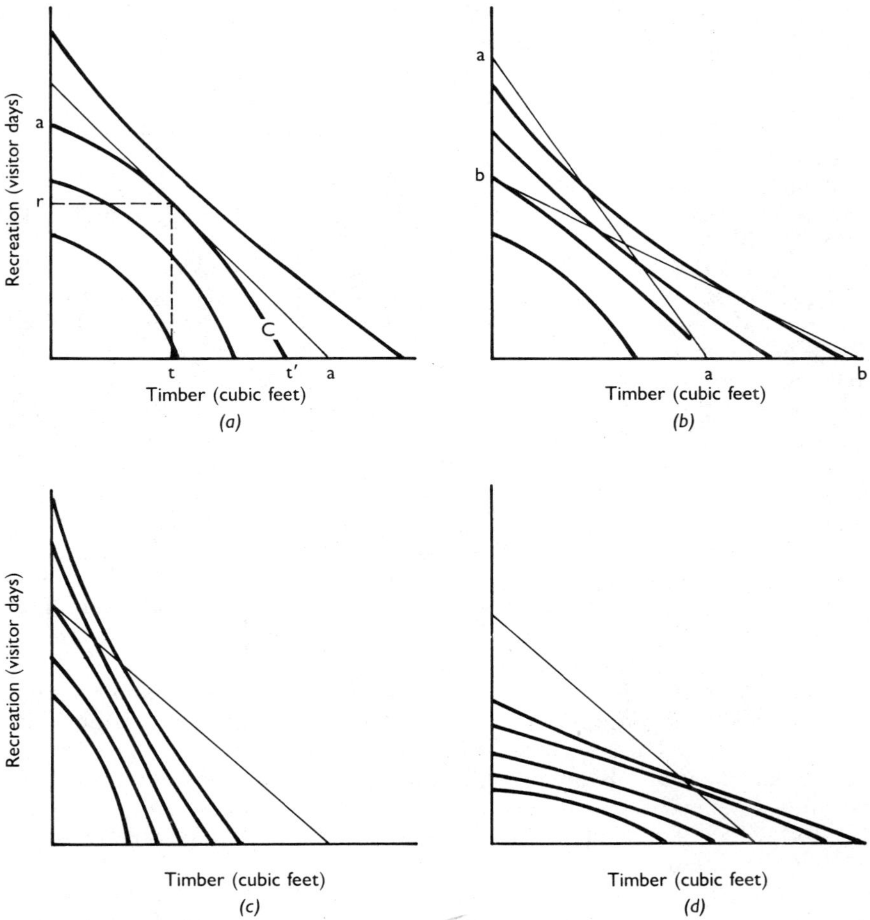

FIGURE 16.5. Four of many different possible sets of isocost and isorevenue curves when timber and recreation are produced jointly. Note that as intensity of management increases, joint production ceases to be economic. Single use becomes the better management approach.

[5]Other approaches have been suggested, particularly those of a purely biological nature. See, for example, Biesterfeldt and Boyce (1978). Their approach attempts to put into usable form the presently existing ecological knowledge about the effects of managerial input on any selected number of outputs, however measured. As such it is purely biological, leaving the choice of product mix to judgment alone.

more of the situations of Figure 16.4. Of particular interest here is the shape of the isocost curves (which may be viewed as production possibility curves under specified cost levels). Note also that as the intensity of management increases (as costs increase) the isocosts tend to become concave—*intensive* uses are generally incompatible.

In Figure 16.5*a* the isocost curves indicate that the two products are competitive throughout their entire output range, but competitive in a somewhat special sense. Output of either timber, or recreation, or of both timber *and* recreation, could be increased by spending more money on production. But if costs are held constant, timber (recreation) output can only be increased if recreation (timber) output is decreased—that is, by reallocating the production budget.

Now suppose the relative prices of the two outputs were such that the isorevenues curve was that of $a-a$ (in Figure 16.5*a*). Given these relative values, if an expenditure of C dollars is to be made for managing the area, the optimum combination of the two products would be t cubic feet of timber and r visitor days of recreation. Here, if timber were to be made the primary (or key) product the decision would cost the equivalent of $a-t'$ cubic feet of timber, which is the cost of the foregone revenue that could have been obtained if the timber and recreation combination had been selected. This would be quite similar to many of the production combinations in situations II and V toward the central part of Figure 16.4 (Clawson's diagram).

In Figure 16.5*b* the cost and revenue pattern is quite different. Here, with $a-a$ depicting the possible revenues from the two products, only timber should be produced. Under these particular conditions a strong case might be made for making this a key area for timber production. But suppose the relative values change, that recreation values increase relative to those for timber until the isorevenue curves take on the slope of $b-b$. Now the area should be devoted solely to outdoor recreation. Note that at low intensities of management it might well be desirable to produce both products. Under intensive management levels, however, one or the other product would be produced—production of one product rules out the other.

This set of curves depicts to a considerable degree the changing production cost and revenue relationships that would hold in Figure 16.4 as one moves out diagonally through situation I, then through V, and into situation IV. Each new position on the diagonal would require rescaling the diagram, since productivity increases for both products as one moves out along the diagonal.[6]

The two sets of isocosts in Figure 16.5*c* and *d* would, under the same relative revenue conditions, permit production of only one product—in *c*, recreation, and timber in *d*. Areas such as that described by Figure 16.5*c* would be near the vertical (wilderness) axis of Clawson's diagram, while *d* would resemble those lying close to the horizontal (timber) axis. At the extremes, the isocost curves would become concave—good wilderness management is incompatible with timber harvesting, and the reverse is also true.

Now suppose, for example, the situation were that illustrated by Figure 16.6—a condition somewhat similar to the one in Clawson's situation V, but with cardinal measures for both timber and wildlife. Assume also that this is an area being managed on a production budget of $10 per acre, and that someone has

[6]Because each point on Clawson's diagram indicates a specific combination of site and topographic qualities and characteristics, a specific set of isocost and revenue curves would be required for every point on Figure 16.4.

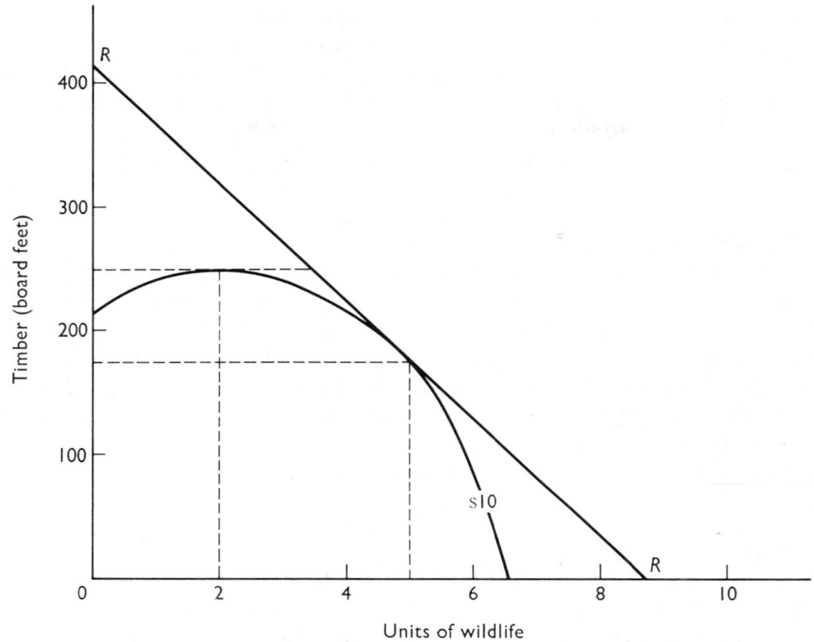

FIGURE 16.6. Single product management can be expensive. Here, wildlife production would be increased 150% at a cost of 75 board feet of timber.

decided the money should be devoted exclusively to timber production—it is a timber key area. Under these conditions 250 board feet of timber would be produced, along with two units of wildlife (a unit of wildlife might be a specific combination of birds and animals). As pointed out earlier, under these conditions the implication is that wildlife has no value. *If* we had information that would allow us to plot the \$10 isocost curve, we could use the diagram as a sort of sensitivity analysis diagram by testing the changes in output combinations that would optimize returns at various relative values for wildlife. Suppose, for example, we assume that $R-R$ indicates the true relative values of timber and wildlife (roughly, a unit of wildlife is assumed to be worth 50 board feet of timber). With these relative values, the optimum product mix would be 175 board feet of timber and 5 wildlife units. If the area is presently under the timber key designation, one could point out that three additional units of wildlife could be gained at a cost of just 75 board feet of timber (we can get a 150% increase in wildlife production for just 75 board feet—a reduction of only 30%).

MULTIPLE-USE PRACTICE

For over a decade before passage of the Forest Management Act of 1976 both public and private forest management groups had been using linear programming for scheduling harvest cuts, planning log transportation systems, calculating allowable cuts under various restrictions, and so on. So when the 1976 legislation required the Forest Service to use a "systematic interdisciplinary approach" when making plans for each forest unit that would "in particular, include coordination of outdoor recreation, range, timber, watershed, wildlife and fish, and wilderness" (Le Master, 1976), linear programming was quickly pressed into service. In fact,

Timber RAM, an early linear programming model developed by the Forest Service, was originally presented in 1967 as a tool for multiple-use management (Navon, 1967). And as one would expect, additional examples of linear programming use for this purpose quickly followed the Management Act.[7] While omitting a detailed presentation of linear programming techniques, some discussion of the application of the procedure to multiple-use planning and management is certainly in order.

Linear Programming and Multiple Use

Linear programming has been defined as a mathematical computing technique dealing with "maximizing or minimizing a linear expression of variables, called the 'objective function,' while the variables satisfy given linear equations or inequalities, referred to as constraints" (Vajda, 1962). As applied to multiple-use problems of forestry, this has usually resulted in an objective function specifying maximization of timber yield or timber value, subject to constraints that include the production of given nonzero amounts of other products such as water, forage, wildlife, recreation, and so on. The best way to introduce this subject is to work through a simple example.

An Example

Assume we are managing a hunting club of 40,000 acres. We have worked out two alternative management systems, X and Y, each producing different combinations of deer, timber, and recreation. If we were to manage the entire 40,000 acres under system X we could produce, annually, 6800 visitor days of camping and related outdoor recreation, 3200 deer for harvest, and 5,000,000 board feet of timber. Under system Y the annual production from the 40,000 acres would be 11,333 visitor days of camping and related outdoor recreation, 4800 deer, but only 2,400,000 board feet of timber. Expressed on a per-acre basis, this means system X would provide 0.17 camping days, 125 board feet of timber, and 0.08 deer annually, and system Y would yield 0.283 camping days, 60 board feet of timber, and 0.12 deer per year.

As a hunting club, the objective is to maximize the number of deer produced from the property each year. Our objective function, then, can be expressed as

$$0.08X + 0.12Y = \text{harvestable deer}$$

where X = number of acres managed under regime X

y = number of acres under regime Y

It is this function that we wish to maximize.

We have also guaranteed to supply a local sawmill with 3.5 million board feet of timber per year and have an agreement with the County Commission to host no less than 8500 visitor days of camping and related outdoor recreation on the club's property. (In return we are given a substantial break on our property tax, and some costs are thereby made tax deductible as contributions to nonprofit organizations.) These constraints can be expressed as the following inequalities:

$$\text{For timber: } 125X + 60Y >= 3{,}5000{,}000 \text{ bd. ft.}$$

$$\text{For camping: } 0.17X + 0.28333Y >= 8500 \text{ visitor days}$$

[7] A complete listing of such publications is impractical, but a few examples are: Biesterfeldt and Boyce (1978), Benninghoff and Ohlander (1978), Calish, Fight, and Teeguarden (1978), Kemper and Davis (1976), Fight and Randall (1980), and Kent (1980).

and, since we have only 40,000 acres:

$$\text{For area: } X + Y <= 40{,}000$$

The coefficients for timber and camping are the per-acre production yields we can achieve through the two management schemes. When multiplied by the areas devoted to X and Y, the results must total not less than the promised output of the two products. The area inequality states that the managed area cannot exceed 40,000 acres.

For more complex problems we would probably use any one of several computer programs, put our equations into a format appropriate to the program at hand, feed the data into the computer, and print out the answers. This problem, however, is simple enough to permit a graphic solution—a considerable aid to understanding how linear programming (L.P.) works. Figure 16.7 illustrates the problem and the way it is solved.

The space between the two axes of Figure 16.7 includes all possible area combinations of X and Y, the two management systems. An L.P. program works, essentially, by (1) finding the region that meets all the stated constraints (called the feasible region); then (2) examining this feasible region to discover the best—the optimal—choice among all the feasible possibilities.

Our first constraint, $X + Y <= 40{,}000$, is represented by the line $F'F$ in the figure; we can use no more than 40,000 acres to produce our required products.

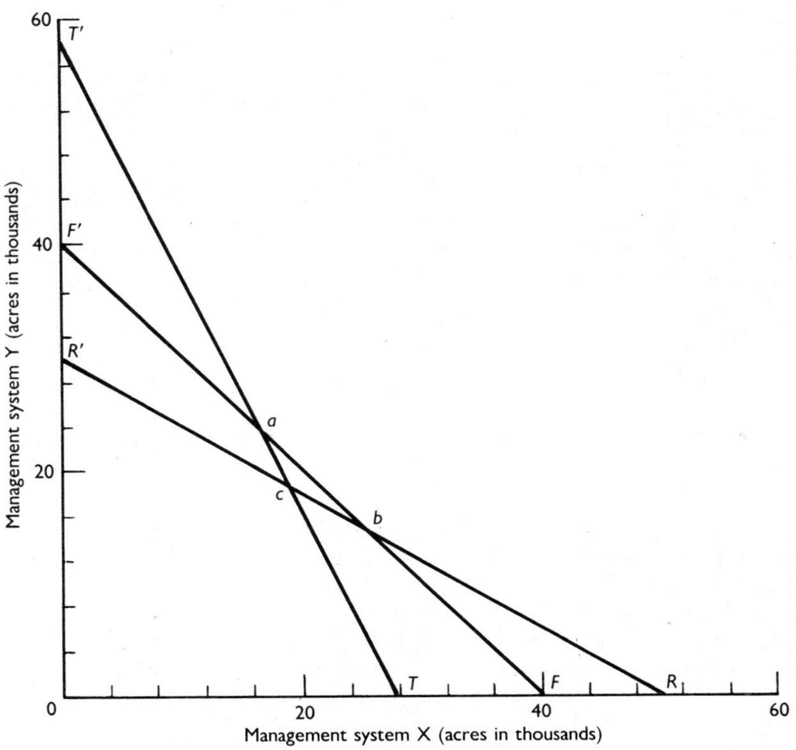

FIGURE 16.7. Linear programming model for multiple use. A simple example of timber, camping, and deer hunting, with timber and camping as constraints, and maximizing number of deer produced.

This means that only those area combinations bounded by $X = 40{,}000$ and $Y = 40{,}000$ are feasible; all points outside (to the right of) $F'F$ are infeasible.

The timber and camping constraints can also be represented graphically. At least 3,500,000 bd. ft. of timber must be produced. Under the Y management system this would require 3,500,000/60 or 58,333 acres, and under the X system no less than 28,000. The line connecting these two points, $T'T$, thus represents the timber constraint. All combinations of X and Y to the right of $T'T$ are feasible so far as the timber constraint is concerned; they will all produce at least 3,500,000 bd. ft. Combinations to the left of $T'T$ are impractible.

Similarly, since 8500 visitor days of camping must be produced, combinations of X and Y to the right of $R'R$ are feasible (so far as camping is concerned) while those to the left are not.

Clearly, the combined effect of the three constraints is to eliminate from consideration all combinations except those in the small triangle *abc*. This is the feasible region; it contains all the combinations that can produce at least 3,500,000 bd. ft. of timber and 8500 camping days from 40,000 acres or less. In some problems no feasible region may exist—there may be no combination that will permit meeting all the constraints. In such situations it may be appropriate to make at least one of the constraints less severe—or to change the objective! Now the combination of X and Y within *abc* that will maximize deer production must be found. Fortunately, we do not have to consider all the combinations, but only the corners: *a*, *b*, and *c*. In moving from *b* to *a* we are keeping the acreage constant and substituting system Y for system X. Since system Y produces more deer, *a* would be preferred. And in moving from *c* to *a* we are keeping timber production constant while both increasing the acres under management and substituting Y for X. Of course, we can also read the X and Y values at *a*, *b*, and *c* from the graph, substitute them in the objective function, and thus identify the combination with the largest deer yield.

We might also determine the equations for the three constraints and solve for the X and Y values of *a*, *b*, and *c* algebraicly. The equation for $T'T$, for example, can be determined by calculating its slope (the tangent of the angle $T'T$ makes with the X axis) or $58{,}333/28{,}000 = -2.08333$. The equation of $T'T$ is, therefore

$$Y = 58{,}333 - 2.08333X$$

Similarly, $R'R$ can be calculated to be

$$Y = 30{,}000 - 0.6X$$

and the equation for $F'F$ is simply

$$Y = 40{,}000 - X$$

The points of intersection can now be found by simultaneously solving equation pairs. At point *a*, 16,923 acres would be under X and 23,077 under Y. At *b*, 25,000 acres would be under X and 15,000 under Y, and at both *a* and *b* the entire 40,000 acres would be under management. At *c*, 19,101 acres would be under X, and 18,539 under Y, leaving 2,360 acres unmanaged.

Substituting these values in the initial equations, we find that we would have

$$2{,}115{,}375 + 1{,}384{,}625 = 3500{,}000 \text{ bd. ft. of timber}$$

$$2{,}876.9 + 6{,}538.1 = 9415 \text{ visitor days of camping}$$

$$1{,}353 + 2{,}769 = 4122 \text{ deer for harvesting}$$

At b we would have

$$3{,}125{,}000 + 900{,}000 = 4{,}025{,}000 \text{ bd. ft. of timber,}$$
$$4{,}250 + 4{,}250 = 8500 \text{ camping days}$$
$$2{,}000 + 1{,}800 = 3800 \text{ deer for harvesting}$$

and at c we would have

$$2{,}387{,}640 + 1{,}112{,}360 = 3{,}500{,}000 \text{ bd. ft. of timber}$$
$$3{,}247.2 + 5{,}252.8 = 8500 \text{ camping days}$$
$$1{,}528 + 2{,}225 = 3753 \text{ deer}$$

Note that option c would exactly satisfy the camping and timber requirements, but produce only 3753 deer and leave part of the area unmanaged. Combination a, with the maximum deer output, would, of course, be selected.

Admittedly, the example is a simple one, though it resembles many of the decisions forest managers make. Linear programming doesn't make the decision—it only indicates how one might maximize one output level while meeting other production targets. But note again the way L.P. programs operate. By considering the constraints imposed on the problem, the program eliminates all the solutions that are not feasible—that are ruled out by the assumed conditions. Then, starting at some corner solution, it systematically moves around, testing each of the possible corners (like those of a, b, and c), finally determining the one with the maximum output of the single selected product. More generally, we should say it determines the value that maximizes (or minimizes, as the case may be) the variable selected in the objective function.

Using computer capabilities one can compare many programs under long lists of constraints and using large numbers of variables.[8] In fact, some feel the speed and simplicity of computation make it far too easy to make decisions entirely by the numbers spewed out by the printer—one can test so many options that the meanings underlying the figures are forgotten. Still, L.P. and its more advanced offspring such as integer programming, nonlinear programming, dynamic programming, and so forth, can be tremendously useful because they permit the systematic comparison of multiple-use options.[9]

A Second Look at the Assumptions

Use of L.P. in multiple-use planning and management is restricted at times because the objective function can specify only one product or one criterion for optimization at any one time. But multiple use means producing many products simultaneously. To get around this problem the usual practice has been to rerun the program, with the objective function changed each time to specify optimization with respect to a different product. The results obtained are of considerable aid to

[8]As an illustration, consider this quotation from the operation manual for "Optimization Programs at the University of Michigan (Hall, McWhorter, and Spivey, 1980). Describing SIMPLEX II, for implementation on the MTS/360 computer system, the manual states "The program is designed to handle relatively small linear programming problems in which the initial tableau contains 10,000 elements or less. To solve problems larger than this, one should use. . . . "

[9]For a variety of models and applications, see Meadows, Bare, Ware, and Row (1975). See also Johnson and Scheurman (1977). One model much used by the Forest Service is FORPLAN, described in Johnson, Jones, and Kent (1980).

management, but do not necessarily indicate the results that might be obtained if multiple objectives had been specified. Unfortunately, no generally accepted way to do this is at hand.

Goal Programming

Goal programming has been suggested as one way to handle this difficulty (Schuler, Webster, and Meadows, 1977; Dane, Meador, and White, 1977). The procedure — really a form of linear programming — is not a maximizing exercise; instead it minimizes the sum of weighted deviations from a set of goals selected by the manager. The goals must be fully stated and quantified, and the variables defined so they contribute to reaching the goals. One of the major problems in application stems from the necessity of defining the relative importance of the goals (i.e., determining that meeting the deer production target is twice as important as achieving the timber production target).

Perhaps the most serious problem with goal programming, however, has to do with its information requirements. Our knowledge of relations between inputs and outputs for timber production are scarcely what one would term precise, but we know these much better than we do the relationship between, say, clearcutting aspen and changing numbers of white-tailed deer, or even that between a harvest cut and the number of visitor days that will be spent in a nearby campground. Yet any goal programming or linear programming analysis requires knowledge not only of the variables to be entered in the objective function and the constraint inequalities, but also knowledge that enables reasonably accurate coefficients to be attached to these variables.

Linearity, Additivity, and Risk

The assumption of linearity in both objective and constraint equations has been mentioned before. One of the saving features of some important forestry relationships (yield over time is a good example) is that while the relations are certainly not linear over the entire period, if one is careful it is sometimes possible to approximate given relations with a linear equation over critical segments of the curve without causing serious error. However, one must constantly keep this linearity problem in mind when using L.P. models.

A related assumption is that of additivity. This means that the various activities are assumed to be completely independent, so that one can add the results of any two of the activities being considered and have this sum equal the result of the two being carried on together in the same area. In multiple-use problems this assumption is violated whenever the outputs are truly joint products that can be produced in variable proportions.

Problems of risk and uncertainty are also difficult to incorporate as part of L.P. models. Generally, risk is handled by using certainty equivalents — treating unsure estimates as if they were certain. Other procedures are also available, but L.P. models quickly become very complex when attempts are made to explicitly incorporate risk and uncertainty; for one reason, linear relationships seldom remain that way after an element of uncertainty is added.

TIMBER PRODUCTION AND MULTIPLE-USE MANAGEMENT

As one would expect, the emphasis placed on recreation, wildlife, soil protection, and water by the Multiple Use-Sustained Yield Act of 1960 was received somewhat less than warmly by the wood-using industries, especially those dependent on public timber. The RPA of 1974 and its revision by the Forest Management Act of

1976 increased fears that timber outputs might be severely reduced under multiple-use management. Others, fully aware that almost all monetary returns to forestry are realized from the sale of timber, worried that reductions in timber receipts might impact adversely on agency appropriations and thereby affect nontimber production as well. One major result was that several studies were initiated to determine what the probable effects of increasing the emphasis on nontimber uses might be.

A relatively early basic analyses was that of Hartman (1976). To the traditional Faustman rotation (maximizing present net value of a forest property under continuous management) Hartman added an amenity service, the value of which increased with stand age. Using this model, Hartman showed that with such a service (which could be wildlife, beauty, water yields, forage, or all of these) one would be less likely to harvest a forested area and then abandon it when the present value became negative. (This situation could arise with an area that has valuable standing timber on a site not worth holding after harvest — like much of the Lake States pine forest in the 1880s, or perhaps much of the tropical forest today). Note that this result would not come about if the amenity service and the timber were independent, as assumed when using an L.P. model. If they were truly independent, one could cut the stand and still have the amenity service.

If the amenity value is assumed to rise as the stand age increases, the optimum rotation age of the stand will also increase. In fact, if the amenity value is large and increasing with stand age, it is entirely possible that the stand should be left forever. On the other hand, if the amenity value is decreasing with stand age (as it might if the amenity value comes from forage, from water yields, or some kinds of wildlife), then the optimum rotation age will be shorter than it would for timber alone. In general, the Hartman rotation will lie between that determined by the usual Faustman calculation and the age that would maximize only the value of the amenity service.[10]

Like all management models, Hartman's builds on a set of assumptions, including those inherent to the Faustman approach. Additionally, every stand is considered independent of all others, an assumption of questionable validity for multiple-use forest management since there can be little doubt that the various stands constituting a forest are not independent. Moreover, any immediately adjacent stands will surely play an important part in determining the production of, for example, wildlife and recreation.

A later study (Calish, Fight, and Teeguarden, 1978) examined how nontimber values might affect Douglas-fir rotations by calculating Faustman (land expectation) values for timber and for seven nontimber products. The author's conclusion was that optimum rotations could be either lengthened (if nontimber values increased continuously, though perhaps at a declining rate) or shortened (if these values decreased after reaching some maximum). They concluded that economic rotations "may do a good job of providing for the nontimber benefits under extensive timber management."[11]

Multiple stand management has also been approached by assuming the managed unit consists of several stands, some or all of which may already be stocked,

[10]An excellent review of Hartman's work is presented together with a more general model by Bowes and Krutilla, 1985.

[11]This analysis also treats single stands. In addition, it regards the various products as being independent in their production.

and all of which are jointly related (Bowes and Krutilla, 1985).[12] Using a dynamic programming model, Bowes and Krutilla present an example that, while simple when compared to real-life situations, yields significant results. Although concentrating entirely on rotation determination via the Faustman formula, they demonstrate that substantial departures from the timber only approach will be necessary. They show, for example, that a multiple-use forest with many stands, under socially optimum management, might never converge to an even flow state. Moreover, if it did so converge, it might bear very little resemblance to the usual timber management solution: some areas might be maintained as clearings, others as old growth reserves, some on a short rotation, others on long rotations.

Their final conclusions are very thought-provoking:

> *The general multiple use harvesting policy is seen to be complex. No simple rule of thumb is likely to describe the harvest. We see that sometimes younger stands are harvested, leaving older ones uncut. We may choose to briefly delay regeneration. We rarely cut a particular stand at the same age twice in succession during the initial periods. The forest areas may be managed with some areas set aside for specialized purposes—old growth or clearing for wildlife. We may choose to specialize over time with the land producing high timber yields in some periods and high recreation benefits in others. Even flow policies are not inherently desirable long-run goals. The optimal harvest age is unlikely to be at the age of maximum sustained yield. Indeed, long even flow rotations, far from being the desirable compromise policy for multiple use management, may simply provide both uneconomic timber and poor balance of age classes for nontimber uses. Perhaps most importantly, we see from these examples, that the harvesting decision can be extremely sensitive to factors about which we have little empirical knowledge. (From Michael Bowes and John V. Krutilla, "The Economics of Multiple Use Public Forest Land Management." Draft. Washington D.C.: Resources for the Future, 1985. Used by permission.)*

Clearly, multiple-use management of forests, either public or private, is likely to call for decisions very different from those foresters have made in past years—or are making in the late 1980s. The tools for estimating nontimber product values discussed in Chapter 15 are much improved over those used 10 years ago, but we still have much to learn. Particularly needed is empirical knowledge of the biological relationships between forest products that are jointly produced.

Below Cost Sales

During discussions of the 1985 Forest Service budget the issue of below cost timber sales received much attention. Studies were developed by the General Accounting Office, the House Appropriations Committee survey and investigations staff, the Wilderness Society, and the Congressional Research Service. All of these argued that "the Forest Service was failing to recover the costs of preparing timber sales in the receipts for the timber sold" (Journal of Forestry, Jan. 1986; Emerson et al.,

[12]As a specific example, they treat 8 stands of equal area and productivity. Stand age increments are in 20-year units, with a maximum age of 100 years after which there is no increment. Amenity values are developed in index form, then treated much as in the Hartman model. Various stumpage values are tested.

1985). A Forest Service report to Congress[13] proposed a multiple account system somewhat similar to that of the U.S. Water Resources Council in 1983 (see Chapter 13). Three different accounts were suggested (1) a financial account, (2) an economic account that includes long-term costs and benefits and reflects values of both timber and nontimber products, and (3) an employment, income, and program account (Journal of Forestry, 1986).

The problem is mentioned here because it highlights one of the inherent difficulties of multiple-use management. Whether certain sales should or should not have been made may be worth exploring, but whenever budgets are prepared on a line item basis (this much for timber sales, that much for recreation, this much for wildlife, that much for wilderness areas, and so on), problems are certain to arise when products are jointly produced. Costs incurred for products produced jointly cannot be allocated in any other than an arbitrary fashion. Moreover, any allocation to individual products is likely to lead to a misallocation — to mistakes in output levels of products that are economically and socially desirable.

Jointly produced products are, by definition, products whose marginal costs are influenced by the output level of another product or products being produced. A road built into a formerly unaccessed area may first be used for timber extraction, then for hiking, hunting, and fishing, and later for sightseeing. How should the cost be allocated? It is possible that any one of the three uses might justify the road building in that the costs of the road would be exceeded by the benefits achieved. It is also possible that while benefits from no one of the three would, by itself, cover the road costs, the benefits from the three uses would far exceed the costs.

True, there may well be some multiple-use costs that can be specifically assigned to a particular use. A well drilled for supplying drinking water to a campground would seem a good example — but it might also supply water for fire suppression in a timber stand. The fact is not just that a forest produces many products valued by society, but that forest activities affecting one product generally influence the production of some, or many, other products that are being jointly produced.

TIMBER PRODUCTION AND MULTIPLE USE: AN EXAMPLE

The following example has two objectives. One is to illustrate application of benefit/cost analysis to multiple-use forest management. The second is to show that if we had (1) reasonably accurate knowledge of the interrelationships involved, (2) an accepted means of quantifying some important products, and (3) a way of measuring significant social nonmarket determined values, public forest management might undergo substantial changes.

To facilitate comparison this example will be based on essentially the same biological data as that used for the discounted cash flow analysis of the pulpwood plantation in Chapter 11. But, as will be seen, substantial differences exist. Two will be apparent immediately: the major *wood* product will be sawtimber (as in most public forests today) so a longer rotation will be necessary — sawtimber requires larger, and therefore older, timber. More important, however, is the effect of actively managing the forest for certain nontimber products. In this regard the specific data used, although reasonable and appropriate for illustrative purposes,

[13]See also the comments on below-cost timber sales and on the much broader impact on forestry the questions raised, discussed by James R. Lyons (1985) in his article "Reflections on the Image of the Forest Service." For further comment, see Reid (1985) and Beyerhelm (1985).

are not the result of any specific study. The example is intended to illustrate concepts and approaches; the numerical results, per se, should not be applied unless the specified conditions are matched and the numbers checked through independent research.

Assume that the run-down 2000-acre cattle ranch of Chapter 11, instead of being picked up by an investment corporation, somehow came into public ownership—perhaps the former owner willed it or presented it as a gift to the Forest Service or State Department of Natural Resources (the DNR). The decision to plant the area in loblolly pine, along with the other decisions implied, followed discussions with game management specialists and the recreation and tourist development divisions.

While recognizing that thinning schedules vary considerably with the selected rotation, the analysis assumes planting in year 0 and an initial thinning at age 20. Since the major wood product will be sawtimber, yields are given in board feet, but pulpwood will probably be sold from the initial thinning and all material remaining after the sawlog harvest will also go for pulp. The estimated yields of both products are given in Table 16.2 and the yield curve used is that of Figure 16.8.

The game division points out that both small game and deer could be greatly benefited by having a few areas open, and suggests leaving a two-acre plot with irregular boundaries in each 100 acres of plantation. These areas will require fertilizing before planting with game food species and should be burned off and replanted each four years. Both forestry and game management agree that beginning at age 30 the forest should be burned over about every five years. Including these and other expenses, the anticipated costs are given in Table 16.3.

The value of sawlog stumpage is assumed to rise with the age of timber for two reasons: (1) as the trees become larger their value per thousand board feet rises because the quality of the lumber these logs will yield is increased; (2) lumber values, from which stumpage values are derived, have been rising for the past 150 years. The present stumpage value for sawtimber 30 years old is assumed to be \$110 per M bd. ft., and increasing at 1.5% per year. Pulpwood stumpage price is assumed to be \$14 per cord at present, and rising at only 0.5% each year.[14]

[14]These prices are somewhat lower than the average price for Southern Pine (1980–1985) reported in *U. S. Timber Production, Trade, Consumption, and Price Statistics, 1950–1985*, USDA Forest Service Misc. Publ. No. 1453.

TABLE 16.2
Volume per Acre for (Fictitious) Public Multiple Use Forest

Year	Pulpwood (cds)	Sawtimber (bd. ft)
20	15.0	
25	harvest	
30	16.3	9.0
35	16.7	12.4
40	17.1	16.2
45	17.5	20.2
50	18.0	24.0
55	18.6	27.7
60	18.9	30.0

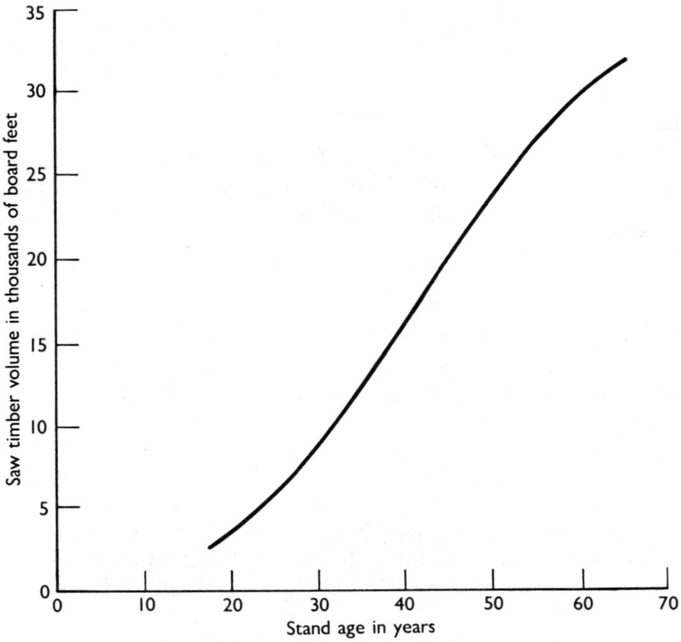

FIGURE 16.8. Sawtimber volume for illustrative problem.

TABLE 16.3
Anticipated Costs for Proposed 2000-Acre Public Forest Plantation

Cost in $/Acre	Item
20	Site preparation for timber planting
100	Site preparation for game food areas
40	Planting timber
60	Planting game areas
5	Fertilizing game areas in year 0
4	Fertilizing game areas in year 4
3	Fertilizing game areas every 4 years, beginning in year 8
2	Burning off forest every 5 years, beginning in year 30
3	Burning off game areas every 4 years, beginning in year 4
3	Administration costs

Harvest preparation costs are assumed to be \$4 per acre for thinning in year 20, and \$8 per acre for harvesting from age 30 onward. These are subtracted from the stumpage sale receipts, yielding the net returns from the timber sales.

Amenity Values

Amenity products, including wildlife, recreation, water, most grazing, and aesthetics, are produced jointly with timber. One should begin by estimating a joint production function, the general form of which might be

$$TP_f = F(L,M,E:Q_t,Q_{w1},Q_g, \ . \ . \ . \ Q_n)$$

where TP_f is the total forest product determined by a joint function of the various inputs used (labor, machinery, and energy), *and* the outputs of all the different products of the forest—Q_t, the quantity of timber produced, Q_{w1}, the quantity of wildlife, and so on down through the *n*th product. Having this, one might proceed to derive joint cost and revenue functions and thus determine the optimum combinations of products to be produced—given that values could be attached to each product.

Unfortunately, the production relations have yet to be determined. The three curves in Figure 16.9 use a sort of shortcut approach. Recognizing that many (though not all) of the nontimber products are heavily influenced by the *size* of the timber, and that size is a function of age, the curves were drawn after reviewing appropriate literature, and after many discussions with game management specialists, forest and resource economists, and forest managers. They seem useful for exposition, and do permit illustrating certain effects—they are employed as one might use shadow prices for the specified amenities. The prices, however, should not be regarded as scientifically derived estimates of what is really a very complicated joint relationship.

Hunting values are given per day of hunting; in both cases they are expected to rise until year 10, then gradually decline as the timber becomes larger. The

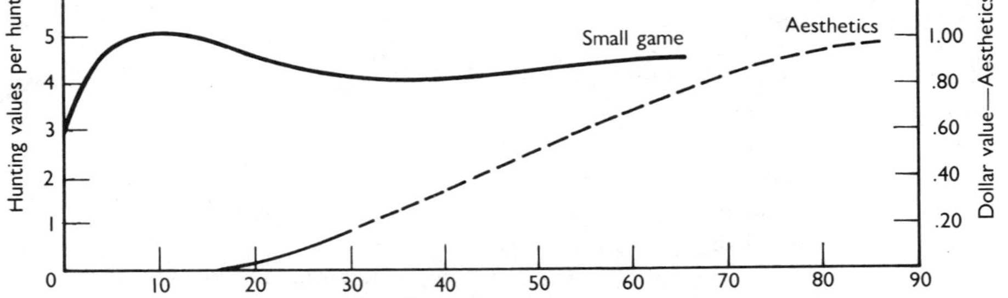

FIGURE 16.9. Values of deer hunting and small game hunting (per day) and aesthetic value per viewer experience. Values of use only for illustrative purposes.

constant maintenance of the game food plots is expected to hold hunting values at substantial levels. Numbers of hunters are also expected to increase in the early years as the abundance of game increases, but should level off around year 6. Total values of hunting are determined by multiplying the value per day by the number of hunter days during the season.

Aesthetic values are probably the most controversial. One might argue that during the early years the aesthetic value of a plantation is probably negative. Since this project began as a worn-out cattle ranch it was decided to give it no aesthetic value until the 18th year. The remaining aesthetic values were derived as follows: on average, 65 passenger cars per day pass by or through the forest area. In 18 years, if traffic increases at 1.2% annually, there will be about 80 cars each day, and at 1.5 passengers per car, around 120 people per day, or 44,000 per year, will view the forest. Because forests of large trees are unusual, the (imaginary) DNR estimates aesthetic values will increase at 1.5% annually. If three-fourths of the people are regular travelers, only one-quarter of those travelling by may receive aesthetic enjoyment from the forest, and that may be very small—today only 1¢ per person, but increasing at 1.5% annually. In year 18 this will still amount to only $143.81. It is also estimated that from year 18 on, the trees increase in size and beauty—and value—so traffic will increase at 2.5% annually. This gives rise to the values used in the analysis.

Several sets of calculations are presented in Table 16.4. First, the optimum rotation from a strictly financial standpoint—the rotation maximizing discounted cash flow—is calculated at a 4% discount rate—probably acceptable from a public viewpoint, although certainly debatable. The stumpage value alone is maximized at a rotation of 54 years—at $1,265,510. If the discount rate is increased to 7%, the optimum rotation drops to 38 years with a value of $232,203. The discount rate for this project is obviously very important! But when the amenity values are included, the values change significantly. Using a 4% discount rate, the forest values are maximized at the same rotation (at 54 years), but the discounted value rises to $1,782,152. With 7% discount, the forest value is maximum at 40 years, with a (social) value of $495,788.

But suppose that aesthetic values rise faster—that when population increases, income increases, and with increasing size of the timber these values rise at, say, 7% annually—still beginning only after 18 years and with hunting values unchanged. Now, with a discount rate of either 4 or 7%, *the forest would never be cut*—or, at least, even with a 7% discount rate, the present value will continue to increase until sawtimber values decrease faster than the combined nontimber values increase, and the combination of sawtimber and nontimber values are increasing at a rate greater than the discount rate. This is hardly surprising if one understands the behavior of discounted values under differing discount rates, but it illustrates an outcome quite possible when one includes nontimber products in the total social value mix.

More important, the example illustrates the basic difficulty of multiple-use management. Surely, there is some cutting regime that would permit harvesting the old growth timber without destroying the other values. The project is treated here as a traditional timber plantation: clear-cut when the NPV is maximized. But with the other products increasing in value, no *maximum* would be reached until the stand begins to deteriorate—and our plantation yield curves aren't carried out that far. We know, intuitively, that some of the high quality trees could be removed, and that their removal might even increase some of the nontimber values. The real problem is that we don't know all the biological, or economic, relations to optimize the social values of a multiple product forest.

self evident (without reasoning)

TABLE 16.4
Calculations for Example of Timber Production under Multiple-Use Management

	A	B
1	YEAR	0
2	Joint Costs	
3	Site prep, timber, $20 per acre	39200
4	Site prep, game, $100 per acre	4000
5	Planting, timber, $40 per acre	78400
6	Planting, game, $60 per acre	2400
7	Fertilizer, game	200
8	Burning, timber	
9	game	
10	Administration	6000
11	TOTAL JOINT COSTS	130200
12	Timber yields & costs/acre if cut	
13	Sawlog vol per acre in M. bd. ft.	
14	Sawlog stumpage price per M	
15	Total sawlog stumpage value	
16	Pulpwood available per acre, in cds	
17	Pulpwood stumpage price per cd.	
18	Total forest pulpwd st. value	
19	Harvest prep cost	
20	Stumpage value net of prep cost	0
21		
22	Amenity values	
23	Small game, per hunting day	3
24	" " hunting days	1200
25	Large game, per hunting day	6.5
26	" " hunting days	700
27	Aesthetics, per person	
28	" no. of persons	
29	AMENITY BENEFITS	
30	Small game (incr at 1%/yr)	3600
31	Large game (incr at 1%/yr)	4550
32	Aesthetics (incr at 1.5%/yr)	
33	TOTAL AMENITY VALUES	8150
34	TOTAL JOINT COSTS	130200
35	STUMPAGE VALUE NET OF PREP COST	0
36		
37	PNV at r'th year, 4%	-122050
38	PNV of stumpage, r'th year, 4%	-130200
39		
40	PNV at r'th year, 7%	-122050
41	PNV of stumpage, r'th year, 7%	-130200
42		
43	Aesthetics (incr at 7%/yr)	
44	New total amenity values	8150
45	PNV of forest,amen val incr 7%	-122050

	C II	D II	E II	F II	G II	H II
	1	2	3	4	5	6
				1600 at $40/acre		
	at $5/acre			160 at $4/acre		
				120 at $3/acre		
	6000	6000	6000	6000	6000	6000
	6000	6000	6000	7880	6000	6000
	Price in year 0 for young sawtimber is $110/M. Increase at 751.5%/year					
	Price in year 0 is $14.00/cord. Increase at 0.5% per year					
	$4/acre for thinning, $8/acre from age 30 on					
	0	0	0	0	0	0
	3.5	3.9	4.15	4.5	4.65	4.8
	1300	1400	1500	1600	1700	1800
	7.1	7.6	7.9	8.2	8.5	8.65
	800	900	1000	1050	1050	1050
	4595.500	5569.746	6413.624	7492.349	8308.234	9171.534
	5736.800	6977.484	8139.378	8959.601	9380.265	9641.257
	10332.30	12547.23	14553.00	16451.95	17688.50	18812.79
	6000	6000	6000	7880	6000	6000
	0	0	0	0	0	0
	-117884.	-111831.	-104227.	-96900.1	-87293.0	-77166.9
	-135969.	-141517.	-146851.	-153586.	-158518.	-163260.
	-118001.	-112283.	-105301.	-98761.2	-90427.5	-81889.8
	-135807.	-141048.	-145946.	-151958.	-156235.	-160233.
	10332.30	12547.23	14553.00	16451.95	17688.50	18812.79
	-118001.	-112283.	-105301.	-98761.2	-90427.5	-81889.8

TABLE 16.4 *(continued)*
Calculations for Example of Timber Production under Multiple-Use Management

I	A	II	I	II
1	YEAR			7
2	Joint Costs			
3	Site prep, timber, $20 per acre			
4	Site prep, game, $100 per acre			
5	Planting, timber, $40 per acre			
6	Planting, game, $60 per acre			
7	Fertilizer, game			
8	Burning, timber			
9	game			
10	Administration			6000
11	TOTAL JOINT COSTS			6000
12	Timber yields & costs/acre if cut			
13	Sawlog vol per acre in M. bd. ft.			
14	Sawlog stumpage price per M			
15	Total sawlog stumpage value			
16	Pulpwood available per acre, in cds			
17	Pulpwood stumpage price per cd.			
18	Total forest pulpwd st. value			
19	Harvest prep cost			
20	Stumpage value net of prep cost			0
21				
22	Amenity values			
23	Small game, per hunting day			4.9
24	" " hunting days			1000
25	Large game, per hunting day			8.8
26	" " hunting days			1050
27	Aesthetics, per person			
28	" no. of persons			
29	AMENITY BENEFITS			
30	Small game (incr at 1%/yr)			9456.234
31	Large game (incr at 1%/yr)			9906.531
32	Aesthetics (incr at 1.5%/yr)			
33	TOTAL AMENITY VALUES			19362.76
34	TOTAL JOINT COSTS			6000
35	STUMPAGE VALUE NET OF PREP COST			0
36				
37	PNV at r'th year, 4%			-67012.3
38	PNV of stumpage, r'th year, 4%			-167819.
39				
40	PNV at r'th year, 7%			-73568.1
41	PNV of stumpage, r'th year, 7%			-163970.
42				
43	Aesthetics (incr at 7%/yr)			
44	New total amenity values			19362.76
45	PNV of forest,amen val incr 7%			-73568.1

J II	K II	L II	M II	N II	O II
8	9	10	11	12	13
1600				1600	
120 at $3/acre				20	
120				120	
6000	6000	6000	6000	6000	6000
7840	6000	6000	6000	7840	6000
0	0	0	0	0	0
4.95	5	5	5	5	4.95
1800	1800	1800	1800	1800	1800
8.9	8.95	9	9	8.98	8.95
1050	1050	1050	1050	1050	1050
9648.253	9843.167	9941.599	10041.02	10141.43	10140.41
10119.30	10277.91	10438.68	10543.07	10624.83	10695.23
19767.55	20121.07	20380.28	20584.08	20766.26	20835.64
7840	6000	6000	6000	7840	6000
0	0	0	0	0	0
-58296.9	-48375.7	-38660.9	-29187.3	-21113.6	-12203.7
-173548.	-177763.	-181817.	-185714.	-190611.	-194215.
-66626.2	-58945.2	-51635.0	-44706.3	-38966.8	-32810.6
-168533.	-171797.	-174847.	-177697.	-181178.	-183668.
19767.55	20121.07	20380.28	20584.08	20766.26	20835.64
-66626.2	-58945.2	-51635.0	-44706.3	-38966.8	-32810.6

TABLE 16.4 *(continued)*
Calculations for Example of Timber Production under Multiple-Use Management

I	A		P
1	YEAR		14
2	Joint Costs		
3	Site prep, timber, $20 per acre		
4	Site prep, game, $100 per acre		
5	Planting, timber, $40 per acre		
6	Planting, game, $60 per acre		
7	Fertilizer, game		
8	Burning, timber		
9	game		
10	Administration		6000
11	TOTAL JOINT COSTS		6000
12	Timber yields & costs/acre if cut		
13	Sawlog vol per acre in M. bd. ft.		
14	Sawlog stumpage price per M		
15	Total sawlog stumpage value		
16	Pulpwood available per acre, in cds		
17	Pulpwood stumpage price per cd.		
18	Total forest pulpwd st. value		
19	Harvest prep cost		
20	Stumpage value net of prep cost		0
21			
22	Amenity values		
23	Small game, per hunting day		4.9
24	" " hunting days		1800
25	Large game, per hunting day		8.93
26	" " hunting days		1050
27	Aesthetics, per person		
28	" no. of persons		
29	AMENITY BENEFITS		
30	Small game (incr at 1%/yr)		10138.36
31	Large game (incr at 1%/yr)		10778.04
32	Aesthetics (incr at 1.5%/yr)		
33	TOTAL AMENITY VALUES		20916.41
34	TOTAL JOINT COSTS		6000
35	STUMPAGE VALUE NET OF PREP COST		0
36			
37	PNV at r'th year, 4%		-3589.86
38	PNV of stumpage, r'th year, 4%		-197679.
39			
40	PNV at r'th year, 7%		-27025.7
41	PNV of stumpage, r'th year, 7%		-185995.
42			
43	Aesthetics (incr at 7%/yr)		
44	New total amenity values		20916.41
45	PNV of forest,amen val incr 7%		-27025.7

Q II	R II	S II	T II	U II	V II
15	16	17	18	19	20
	1600				1600
	120				120
	120				120
6000	6000	6000	6000	6000	6000
6000	7840	6000	6000	6000	7840
					8
					15.46854
					242546.7
					7840
0	0	0	0	0	234706.7
4.85	4.8	4.72	4.65	4.6	4.55
1800	1800	1800	1800	1800	1800
8.88	8.8	8.75	8.7	8.65	8.65
1050	1050	1050	1050	1050	1050
			.01	.02	.03
			11000	11275	11556.88
10135.26	10131.08	10061.85	10011.75	10003.14	9993.356
10824.87	10834.63	10880.80	10926.81	10972.65	11018.32
			143.8075	299.2274	466.9630
20960.13	20965.71	20942.65	21082.37	21275.02	21478.64
6000	7840	6000	6000	6000	7840
0	0	0	0	0	234706.7
4716.968	11724.89	19396.04	26841.13	34091.30	147432.9
-201011.	-205197.	-208277.	-211239.	-214087.	-110548.
-21603.5	-17157.4	-12426.9	-7964.58	-3740.91	127563.6
-188170.	-190825.	-192725.	-194500.	-196159.	-137532.
			371.7926	815.5270	1341.644
20960.13	20965.71	20942.65	21310.35	21791.32	22353.32
-21603.5	-17157.4	-12426.9	-7897.12	-3530.69	127999.9

TABLE 16.4 *(continued)*
Calculations for Example of Timber Production under Multiple-Use Management

I	A	W
1	YEAR	21
2	Joint Costs	
3	Site prep, timber, $20 per acre	
4	Site prep, game, $100 per acre	
5	Planting, timber, $40 per acre	
6	Planting, game, $60 per acre	
7	Fertilizer, game	
8	Burning, timber	
9	game	
10	Administration	6000
11	TOTAL JOINT COSTS	6000
12	Timber yields & costs/acre if cut	
13	Sawlog vol per acre in M bd. ft.	
14	Sawlog stumpage price per M	
15	Total sawlog stumpage value	
16	Pulpwood available per acre, in cds	
17	Pulpwood stumpage price per cd.	
18	Total forest pulpwd st. value	0
19	Harvest prep cost	
20	Stumpage value net of prep cost	0
21		
22	Amenity values	
23	Small game, per hunting day	4.47
24	" " hunting days	1800
25	Large game, per hunting day	8.53
26	" " hunting days	1050
27	Aesthetics, per person	.04
28	" no. of persons	11845.00
29	AMENITY BENEFITS	
30	Small game (incr at 1%/yr)	9915.826
31	Large game (incr at 1%/yr)	11037.92
32	Aesthetics (incr at 1.5%/yr)	647.7556
33	TOTAL AMENITY VALUES	21601.50
34	TOTAL JOINT COSTS	6000
35	STUMPAGE VALUE NET OF PREP COST	0
36		
37	PNV at r'th year, 4%	154279.3
38	PNV of stumpage, r'th year, 4%	-110785.
39		
40	PNV at r'th year, 7%	132780.7
41	PNV of stumpage, r'th year, 7%	-138981.
42		
43	Aesthetics (incr at 7%/yr)	1961.930
44	New total amenity values	22915.67
45	PNV of forest,amen val incr 7%	133534.3

X II	Y II	Z II	AA II	AB II	AC II
22	23	24	25	26	27
		1600			
		120			
		120			
6000	6000	6000	6000	6000	6000
6000	6000	7840	6000	6000	6000
0	0	0	0	0	0
0	0	0	0	0	0
4.4	4.35	4.3	4.25	4.21	4.18
1800	1800	1800	1800	1800	1800
8.46	8.4	8.32	8.25	8.2	8.15
1050	1050	1050	1050	1050	1050
.05	.06	.08	.09	.1	.12
12141.94	12445.49	12756.63	13075.54	13402.43	13737.49
9858.150	9843.586	9827.746	9810.605	9815.452	9842.964
11056.81	11088.18	11092.40	11109.07	11152.16	11195.00
842.3859	1051.677	1458.851	1707.471	1973.789	2464.177
21757.35	21983.44	22379.00	22627.14	22941.40	23502.14
6000	6000	7840	6000	6000	6000
0	0	0	0	0	0
160928.2	167413.1	173085.1	179322.2	185432.8	191502.8
-113418.	-115949.	-119130.	-121471.	-123722.	-125886.
137691.6	142328.9	146740.9	150909.9	154860.3	158642.5
-140336.	-141601.	-143147.	-144252.	-145286.	-146251.
2689.684	3539.893	5176.504	6386.997	7783.265	10243.56
23604.64	24471.66	26096.65	27306.67	28750.87	31281.52
138862.2	144024.4	149169.3	154200.5	159151.3	164185.4

TABLE 16.4 *(continued)*
Calculations for Example of Timber Production under Multiple-Use Management

I	A	II	AD	II
1	YEAR			28
2	Joint Costs			
3	Site prep, timber, $20 per acre			
4	Site prep, game, $100 per acre			
5	Planting, timber, $40 per acre			
6	Planting, game, $60 per acre			1600
7	Fertilizer, game			120
8	Burning, timber			
9	game			120
10	Administration			6000
11	TOTAL JOINT COSTS			7840
12	Timber yields & costs/acre if cut			
13	Sawlog vol per acre in M bd. ft.			
14	Sawlog stumpage price per M			
15	Total sawlog stumpage value			
16	Pulpwood available per acre, in cds			
17	Pulpwood stumpage price per cd.			
18	Total forest pulpwd st. value			0
19	Harvest prep cost			
20	Stumpage value net of prep cost			0
21				
22	Amenity values			
23	Small game, per hunting day			4.15
24	" " hunting days			1800
25	Large game, per hunting day			8.1
26	" " hunting days			1050
27	Aesthetics, per person			.13
28	" no. of persons			14080.93
29	AMENITY BENEFITS			
30	Small game (incr at 1%/yr)			9870.044
31	Large game (incr at 1%/yr)			11237.58
32	Aesthetics (incr at 1.5%/yr)			2777.307
33	TOTAL AMENITY VALUES			23884.93
34	TOTAL JOINT COSTS			7840
35	STUMPAGE VALUE NET OF PREP COST			0
36				
37	PNV at r'th year, 4%			196853.5
38	PNV of stumpage, r'th year, 4%			-128605.
39				
40	PNV at r'th year, 7%			162234.9
41	PNV of stumpage, r'th year, 7%			-147430.
42				
43	Aesthetics (incr at 7%/yr)			12170.84
44	New total amenity values			33278.46
45	PNV of forest,amen val incr 7%			169190.6

AE II	AF II	AG II	AH II	AI II	AJ II
29	30	31	32	33	34
			1600		
			120		
	3920 at $2/acre				
		120			
6000	6000	6000	6000	6000	6000
6000	9920	6000	7840	6000	6000
	9	9.6	10.3	11	11.7
	148.2634	150.4839	152.7412	155.0323	157.3578
	2615366.	2831505.	3083539.	3342496.	3608528.
	5.5	6	6.5	7	7.5
	16.25960	16.34090	16.42260	16.50472	16.58724
0	175278.5	192169.0	209224.0	226444.7	243832.4
	15680	15680	15680	15680	15680
0	2774965.	3007994.	3277082.	3553261.	3836681.
4.12	4.1	4.09	4.08	4.07	4.06
1800	1800	1800	1800	1800	1800
8.05	8	7.95	7.9	7.85	7.8
1050	1050	1050	1050	1050	1050
.14	.16	.18	.2	.21	.23
14432.95	14793.78	15163.62	15542.71	15931.28	16329.56
9896.681	9947.125	10022.09	10097.56	10173.54	10250.03
11279.89	11321.93	11363.68	11405.13	11446.28	11487.11
3111.705	3699.818	4330.348	5005.761	5468.263	6230.857
24288.28	24968.87	25716.12	26508.46	27088.08	27968.00
6000	9920	6000	7840	6000	6000
0	2774965.	3007994.	3277082.	3553261.	3836681.
202717.6	1062931.	1104953.	1152682.	1198232.	1241259.
-130606.	721786.7	756114.1	796197.5	834256.8	869849.4
165648.9	454296.1	461478.1	470335.8	477613.8	483292.2
-148274.	214962.2	218986.9	224803.0	229176.3	232051.8
14375.16	18018.24	22231.69	27091.79	31198.57	37475.79
35551.74	39287.30	43617.47	48594.49	52818.39	59212.93
174187.8	464716.0	474095.8	485487.7	495524.9	504334.7

TABLE 16.4 *(continued)*
Calculations for Example of Timber Production under Multiple-Use Management

I	A II	AK II
1	YEAR	35
2	Joint Costs	
3	Site prep, timber, $20 per acre	
4	Site prep, game, $100 per acre	
5	Planting, timber, $40 per acre	
6	Planting, game, $60 per acre	
7	Fertilizer, game	
8	Burning, timber	3920
9	game	
10	Administration	6000
11	TOTAL JOINT COSTS	9920
12	Timber yields & costs/acre if cut	
13	Sawlog vol per acre in M bd. ft.	12.4
14	Sawlog stumpage price per M	159.7181
15	Total sawlog stumpage value	3881789.
16	Pulpwood available per acre, in cds	8
17	Pulpwood stumpage price per cd.	16.67018
18	Total forest pulpwd st. value	261388.4
19	Harvest prep cost	15680
20	Stumpage value net of prep cost	4127498.
21		
22	Amenity values	
23	Small game, per hunting day	4.05
24	" " hunting days	1800
25	Large game, per hunting day	7.75
26	" " hunting days	1050
27	Aesthetics, per person	.24
28	" no. of persons	16737.80
29	AMENITY BENEFITS	
30	Small game (incr at 1%/yr)	10327.03
31	Large game (incr at 1%/yr)	11527.60
32	Aesthetics (incr at 1.5%/yr)	6764.273
33	TOTAL AMENITY VALUES	28618.91
34	TOTAL JOINT COSTS	9920
35	STUMPAGE VALUE NET OF PREP COST	4127498.
36		
37	PNV at r'th year, 4%	1280804.
38	PNV of stumpage, r'th year, 4%	902041.6
39		
40	PNV at r'th year, 7%	487127.6
41	PNV of stumpage, r'th year, 7%	233206.6
42		
43	Aesthetics (incr at 7%/yr)	42888.60
44	New total amenity values	64743.24
45	PNV of forest,amen val incr 7%	511553.5

AL II	AM II	AN II	AO II	AP II	AQ II
36	37	38	39	40	41
1600				1600	
120				120	
				3920	
120				120	
6000	6000	6000	6000	6000	6000
7840	6000	6000	6000	11760	6000
13.1	13.8	14.6	15.4	16.2	17
162.1139	164.5456	167.0138	169.5190	172.0618	174.6427
4162436.	4450630.	4779267.	5116761.	5463306.	5819095.
8.344	8.702792	9.077012	9.467324	9.874418	10.29902
16.75353	16.83729	16.92148	17.00609	17.09112	17.17657
273991.2	287201.7	301049.1	315564.2	330779.1	346727.7
15680	15680	15680	15680	15680	15680
4420748.	4722151.	5064636.	5416646.	5778405.	6150143.
4.05	4.05	4.06	4.07	4.08	4.09
1800	1800	1800	1800	1800	1800
7.73	7.71	7.69	7.67	7.65	7.64
1050	1050	1050	1050	1050	1050
.26	.28	.3	.32	.33	.35
17156.25	17585.15	18024.78	18475.40	18937.29	19410.72
10430.30	10534.61	10666.23	10799.42	10934.22	11070.62
11612.83	11698.62	11784.95	11871.85	11959.30	12063.10
7623.829	8541.767	9521.401	10566.22	11336.35	12508.85
29666.97	30774.99	31972.58	33237.48	34229.87	35642.58
7840	6000	6000	6000	11760	6000
4420748.	4722151.	5064636.	5416646.	5778405.	6150143.
1317349.	1352341.	1392796.	1431064.	1465966.	1500062.
931281.0	959006.2	992203.9	1023221.	1050895.	1077804.
489416.8	490785.2	493683.4	495450.0	495788.1	495594.2
232898.9	231749.6	232203.3	231594.9	229647.2	227228.7
50957.91	60187.17	70725.30	82739.17	93579.94	108854.0
73001.05	82420.40	93176.48	105410.4	116473.5	131987.8
517636.0	523229.4	530807.1	537730.8	543561.2	549380.3

TABLE 16.4 *(continued)*
Calculations for Example of Timber Production under Multiple-Use Management

I	A	II	AR II
1	YEAR		42
2	Joint Costs		
3	Site prep, timber, $20 per acre		
4	Site prep, game, $100 per acre		
5	Planting, timber, $40 per acre		
6	Planting, game, $60 per acre		
7	Fertilizer, game		
8	Burning, timber		
9	game		
10	Administration		6000
11	TOTAL JOINT COSTS		6000
12	Timber yields & costs/acre if cut		
13	Sawlog vol per acre in M bd. ft.		17.8
14	Sawlog stumpage price per M		177.2624
15	Total sawlog stumpage value		6184329.
16	Pulpwood available per acre, in cds		10.74188
17	Pulpwood stumpage price per cd.		17.26246
18	Total forest pulpwd st. value		363445.1
19	Harvest prep cost		15680
20	Stumpage value net of prep cost		6532094.
21			
22	Amenity values		
23	Small game, per hunting day		4.11
24	" " hunting days		1800
25	Large game, per hunting day		7.63
26	" " hunting days		1050
27	Aesthetics, per person		.37
28	" no. of persons		19895.99
29	AMENITY BENEFITS		
30	Small game (incr at 1%/yr)		11236.01
31	Large game (incr at 1%/yr)		12167.79
32	Aesthetics (incr at 1.5%/yr)		13757.55
33	TOTAL AMENITY VALUES		37161.34
34	TOTAL JOINT COSTS		6000
35	STUMPAGE VALUE NET OF PREP COST		6532094.
36			
37	PNV at r'th year, 4%		1532243.
38	PNV of stumpage, r'th year, 4%		1102782.
39			
40	PNV at r'th year, 7%		494579.4
41	PNV of stumpage, r'th year, 7%		224046.3
42			
43	Aesthetics (incr at 7%/yr)		126207.7
44	New total amenity values		149611.5
45	PNV of forest,amen val incr 7%		554924.6

AS II	AT II	AU II	AV II	AW II	AX II
43	44	45	46	47	48
	1600		1600		
	120		120		
		3920			
	120				
6000	6000	6000	6000	6000	6000
6000	7840	9920	7720	6000	6000
18.6	19.4	20.2	21	21.8	22.6
179.9213	182.6201	185.3594	188.1398	190.9619	193.8263
6559210.	6943947.	7338750.	7743834.	8159420.	8585731.
11.20378	11.68554	12.18802	12.71210	13.25872	13.82885
17.34877	17.43551	17.52269	17.61030	17.69836	17.78685
380968.6	399337.0	418591.1	438773.4	459928.9	482104.4
15680	15680	15680	15680	15680	15680
6924499.	7327604.	7741661.	8166927.	8603669.	9052155.
4.13	4.15	4.17	4.19	4.2	4.21
1800	1800	1800	1800	1800	1800
7.62	7.61	7.6	7.6	7.6	7.6
1050	1050	1050	1050	1050	1050
.38	.4	.42	.44	.45	.47
20393.39	20903.22	21425.80	21961.45	22510.48	23073.24
11403.59	11573.40	11745.47	11919.82	12067.75	12217.45
12273.36	12379.82	12487.19	12612.06	12738.18	12865.56
14699.84	16098.26	17585.64	19166.88	20393.95	22160.35
38376.79	40051.49	41818.30	43698.77	45199.88	47243.36
6000	7840	9920	7720	6000	6000
6924499.	7327604.	7741661.	8166927.	8603669.	9052155.
1562517.	1590709.	1616876.	1641828.	1665454.	1687611.
1125906.	1146911.	1165851.	1183559.	1199993.	1214923.
492809.7	490293.0	487104.0	483514.2	479532.5	475158.9
220184.6	215627.4	210447.3	204913.0	199051.5	192841.8
142159.3	164119.2	188997.6	217153.8	243576.2	279015.2
165836.3	188072.4	213230.3	241685.6	268382.1	304098.2
560103.0	565127.5	570100.0	575320.4	580620.2	586229.8

TABLE 16.4 *(continued)*
Calculations for Example of Timber Production under Multiple-Use Management

I	A	II	AY II
1	YEAR		49
2	Joint Costs		
3	Site prep, timber, $20 per acre		
4	Site prep, game, $100 per acre		
5	Planting, timber, $40 per acre		
6	Planting, game, $60 per acre		
7	Fertilizer, game		
8	Burning, timber		
9	game		
10	Administration		6000
11	TOTAL JOINT COSTS		6000
12	Timber yields & costs/acre if cut		
13	Sawlog vol per acre in M bd. ft.		23.3
14	Sawlog stumpage price per M		196.7337
15	Total sawlog stumpage value		8984435.
16	Pulpwood available per acre, in cds		14.42349
17	Pulpwood stumpage price per cd.		17.87578
18	Total forest pulpwd st. value		505349.0
19	Harvest prep cost		15680
20	Stumpage value net of prep cost		9474104.
21			
22	Amenity values		
23	Small game, per hunting day		4.22
24	" " hunting days		1800
25	Large game, per hunting day		7.6
26	" " hunting days		1050
27	Aesthetics, per person		.49
28	" no. of persons		23650.07
29	AMENITY BENEFITS		
30	Small game (incr at 1%/yr)		12368.93
31	Large game (incr at 1%/yr)		12994.22
32	Aesthetics (incr at 1.5%/yr)		24036.14
33	TOTAL AMENITY VALUES		49399.29
34	TOTAL JOINT COSTS		6000
35	STUMPAGE VALUE NET OF PREP COST		9474104.
36			
37	PNV at r'th year, 4%		1702723.
38	PNV of stumpage, r'th year, 4%		1222770.
39			
40	PNV at r'th year, 7%		469045.5
41	PNV of stumpage, r'th year, 7%		184934.0
42			
43	Aesthetics (incr at 7%/yr)		319031.6
44	New total amenity values		344394.8
45	PNV of forest,amen val incr 7%		590831.8

AZ II	BA II	BB II	BC II	BD II	BE II
50	51	52	53	54	55
		1600			
		120			
3920					3920
		120			
6000	6000	6000	6000	6000	6000
9920	6000	7840	6000	6000	9920
24	25.25	25.9	26.5	27.2	27.7
199.6847	202.6800	205.7202	208.8060	211.9381	215.1172
9393169.	10030633	10443180	10845383	11298843	11679141
15.04370	15.69058	16.36527	17.06898	17.80294	18.56847
17.96516	18.05499	18.14526	18.23599	18.32717	18.41880
529714.4	555254.6	582026.2	610088.6	639504.0	670337.7
15680	15680	15680	15680	15680	15680
9907204.	10570208	11009526	11439792	11922667	12333798
4.23	4.25	4.27	4.28	4.29	4.3
1800	1800	1800	1800	1800	1800
7.6	7.6	7.6	7.6	7.6	7.6
1050	1050	1050	1050	1050	1050
.51	.52	.54	.56	.57	.58
24241.33	24847.36	25468.54	26105.26	26757.89	27426.84
12522.23	12707.25	12894.72	13054.16	13215.51	13378.78
13124.16	13255.40	13387.96	13521.84	13657.06	13793.63
26027.27	27609.07	29828.54	32182.24	34079.48	36077.47
51673.66	53571.72	56111.22	58758.24	60952.05	63249.87
9920	6000	7840	6000	6000	9920
9907204.	10570208	11009526	11439792	11922667	12333798
1716216.	1758739.	1767167.	1772501.	1782152.	1780712.
1228936.	1264178.	1265266.	1263220.	1265510.	1256710.
462652.0	463193.6	455711.8	447740.8	440930.6	431972.0
176786.2	175628.2	166482.7	156883.4	148494.6	138005.0
364179.5	407245.5	463825.2	527541.0	588912.4	657221.1
389825.8	433208.1	490107.8	554117.0	615785.0	684393.5
595917.8	608504.1	613890.9	619647.0	627206.1	633281.8

TABLE 16.4 *(continued)*
Calculations for Example of Timber Production under Multiple-Use Management

I	A	II	BF II
1	YEAR		56
2	Joint Costs		
3	Site prep, timber, $20 per acre		
4	Site prep, game, $100 per acre		
5	Planting, timber, $40 per acre		
6	Planting, game, $60 per acre		1600
7	Fertilizer, game		120
8	Burning, timber		
9	game		120
10	Administration		6000
11	TOTAL JOINT COSTS		7840
12	Timber yields & costs/acre if cut		
13	Sawlog vol per acre in M bd. ft.		28.3
14	Sawlog stumpage price per M		218.3439
15	Total sawlog stumpage value		12111100
16	Pulpwood available per acre, in cds		19.36692
17	Pulpwood stumpage price per cd.		18.51090
18	Total forest pulpwd st. value		702658.1
19	Harvest prep cost		15680
20	Stumpage value net of prep cost		12798078
21			
22	Amenity values		
23	Small game, per hunting day		4.32
24	" " hunting days		1800
25	Large game, per hunting day		7.6
26	" " hunting days		1050
27	Aesthetics, per person		.6
28	" no. of persons		28112.51
29	AMENITY BENEFITS		
30	Small game (incr at 1%/yr)		13575.42
31	Large game (incr at 1%/yr)		13931.56
32	Aesthetics (incr at 1.5%/yr)		38828.37
33	TOTAL AMENITY VALUES		66335.35
34	TOTAL JOINT COSTS		7840
35	STUMPAGE VALUE NET OF PREP COST		12798078
36			
37	PNV at r'th year, 4%		1783985.
38	PNV of stumpage, r'th year, 4%		1252570.
39			
40	PNV at r'th year, 7%		424267.5
41	PNV of stumpage, r'th year, 7%		128800.0
42			
43	Aesthetics (incr at 7%/yr)		745662.6
44	New total amenity values		773169.6
45	PNV of forest,amen val incr 7%		641566.6

BG II	BH II	BI II	BJ I
57	58	59	60
6000	6000	6000	6000
6000	6000	6000	6000
28.3	28.8	29.3	30
221.6191	224.9434	228.3175	231.7423
12292767	12697603	13111818	13626446
20.19969	21.06828	21.97422	22.91911
18.60345	18.69647	18.78995	18.88390
736536.7	772048.8	809273.2	848292.3
15680	15680	15680	15680
13013624	13453972	13905411	14459058
4.34	4.36	4.38	4.4
1800	1800	1800	1800
7.6	7.6	7.6	7.6
1050	1050	1050	1050
.62	.64	.65	.67
28815.32	29535.70	30274.09	31030.95
13774.65	13976.51	14181.03	14388.24
14070.88	14211.59	14353.70	14497.24
41742.60	44828.86	47367.56	50796.33
69588.13	73016.96	75902.29	79681.81
6000	6000	6000	6000
13013624	13453972	13905411	14459058
1759092.	1757737.	1756075.	1762835.
1220211.	1211324.	1202134.	1201296.
411229.2	403255.1	395491.6	389519.7
114290.5	104873.8	95708.76	88361.76
845065.7	956723.4	1065681.	1204749.
872911.2	984911.5	1094216.	1233634.
645511.3	655554.3	666594.4	680536.6

QUESTIONS AND PROBLEMS

1. In many ways one might say that multiple use as a practicing managerial concept came of age during the 1960–1980 period. Explain.
2. Sketch the joint production model for multiple use. What factors make this particular model somewhat difficult to apply in practice?
3. What is the compatibility matrix, and in what ways does it assist a forest manager?
4. Draw an isocost diagram that, if quantified, would permit one to determine the optimum output of two forest products produced from the same forest area.
5. With reference to linear programming, define, and illustrate with an example from forestry:
   a. An objective function.
   b. A constraint, expressed in words and as in inequality or equation.
   c. A "corner" solution.
6. What are the advantages and disadvantages of a linear programming approach to multiple use?
7. Why is nonlinearity a problem? Discuss the assumption of additivity and its suitability to multiple-use forest management.
8. Suppose one were to add some other product or service to be jointly produced with timber, and that the product were known to increase in value as the stand became older (aesthetic values, for example, might increase as a redwood, or Douglas-fir stand grows older). If the stand is to be managed so that the total value of the two products will be maximized, what would be the likely effect on the optimum timber rotation?

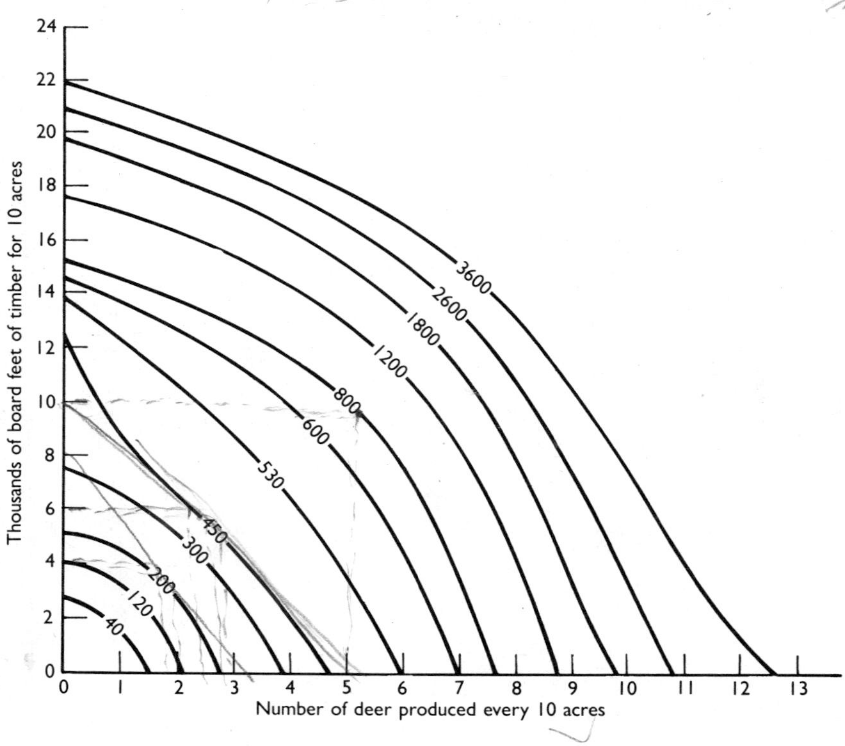

9. What is the general conclusion reached by Krutilla and Bowes in their study of multiple use?

10. Suppose, as an ardent student of multiple use, you have worked out production cost relationships for producing timber and white-tailed deer on a 3000-acre tree farm not far from Hollywood, California. Naturally you'll take a small part of the area to raise most of the deer efficiently—with corn and hay, etc.—and use to keep the tree farm stocked. The costs, figured over the total area involved and *put on a 10-acre* basis for convenience, then plotted as isocost curves, are reproduced in the diagram. Assume that:

   a. You can sell shooting privileges to Hollywood "hunters" at a rate of $200 per deer.

   b. Timber can be sold as stumpage for $80 per thousand board feet.

   Answer the following, and show on a copy of the diagram how your results were derived.

   A. If you had 20 dollars *per acre* to spend on production, what would be the most profitable combination of timber and deer to produce?

   B. If you had 450 dollars per 10 acres to spend on production, what would be the most profitable combination of timber and deer to produce?

   C. If deer remain at a price of $200 each, and assuming that profits will be maximized at a combination of 10,000 board feet of timber and 5 deer from each 10 acres, how much will timber be worth?

   D. With the combination you decided on in (C) above (5 deer and 10,000 bd. ft. of timber from each 10 acres) and under the assumption that (a) deer can be sold for $200 each and timber for $100 per M ft., and (b) *all* costs are included in the isocost diagram, how much profit will the tree/deer farm yield each year?

PART 5

Some International Aspects of Forest Economics

This final part treats some of the more important international aspects of forest economics. Beginning with a review of the current world forest situation, Chapter 17 proceeds to a discussion of world trade in timber products—the basis for such trade, the quantities and values involved, and barriers that either already exist or may be imposed. The chapter concludes with a brief presentation of current U.S. trade in forest products.

The final chapter, Chapter 18, is devoted primarily to forestry and its potential contribution to economic development, especially as it pertains to developing nations. This is an emerging area with which U.S. foresters must become familiar, for in some nations forests have a key role to play—one that has often been ignored or overlooked by the nations and agencies offering assistance and probably even more so by the very nations seeking development.

chapter seventeen
THE WORLD'S FORESTS AND INTERNATIONAL TIMBER TRADE

THE FORESTS

Most of the information on forest areas, volumes, productivity, and trade in forest products comes from the work of the Forestry Department of the Food and Agriculture Organization of the United Nations, more commonly known as FAO. This information is compiled from data sent by ministries of member nations and is generally the most accurate available. The problem is that in many cases the available figures are not very accurate! In large and important areas of South America, Africa, and Southeast Asia, for example, almost no forest surveys have been made, and only about a third of the world's forests have been inventoried. New satellite information is providing better estimates of current forest areas and, to a limited extent, their condition, but at present, much of this new data has not been analyzed.

The lack of data is immediately apparent when one looks at estimates of the world's total forest area. Figure 17.1 and Table 17.1 give the figures compiled by FAO and used by the Forest Service in their December 1981 "Analysis" (USDA For. Serv., 1983).[1]

Of the nearly one-third of the world's surface covered with forests (very close to the corresponding figure for the United States), somewhat less than two-thirds is closed forest — land with at least 20% or more crown cover. The remaining third is typically open brush land or scrub forest, often valuable to local inhabitants for fuel and fodder, but offering little opportunity for industrial wood production. About half the world's forest land is in the temperate climate regions, including the forests of North America, Europe, and the USSR.

In terms of growing stock, about a third of the world total is made up of

[1]In contrast, the "Global 2000 Report to the President" (Tech. Rep. Vol. 2, p. 118) gives a total forest land area of 4130 million hectares (10,206 million acres), and 2655 million hectares (6561 million acres) of closed forest. In 1985, using FAO and other data sources, a Forest Resource Assessment Committee of the Economic Commission for Europe gave the total forest land area for 1980 as 4,320.5 million hectares (10,676 million acres), and 2948 million hectares (7285 million acres) of closed forests. This last is slightly less than 3% under that used in Forest Resource Report No. 23.

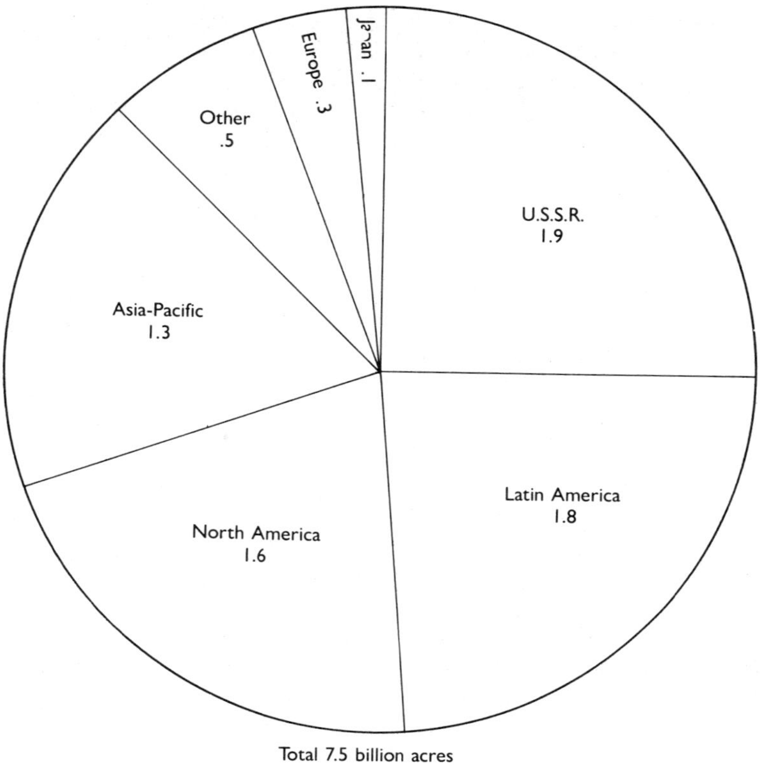

FIGURE 17.1. Closed forests of the world, by country or region, 1973. ***Source:*** **Forest Resource Report No. 23, p. 89.**

softwoods, over half of which is in the Soviet Union, and an additional quarter in North America. Nearly half the hardwood volume is in Latin America's tropical and semitropical forests, with much of the remainder being African and Asian tropical hardwoods. Europe and North America have about 18% of all hardwood growing stock. Timber growing stock in each of the world's regions is given in Table 17.2.

Temperate Forests

Within countries or regions, temperate forest areas have varied considerably over the 30 years between 1950 and 1980, but in total have increased by about 5%. North American forest areas (Canada and U.S.) declined about 5%, European forest areas increased substantially, while the USSR showed identical figures for 1950 and 1980. Some of the variation can be attributed to changes in the accuracy of reported data, but the indicated trends are probably in the right direction.

Temperate regions produced about half the wood used in the world, with North America, Europe, and the USSR together producing nearly 75% of the world's *industrial* wood. South America and Africa, together producing nearly 25% of all wood, account for only slightly over 8% of the industrial wood. For the most part, forest areas in temperate regions are quite stable; they not only produce most of the industrial wood, but their forest areas are increasing and they are very active in plantation establishment. They are, however, experiencing some difficult ecological problems, particularly the loss of trees which may or not be caused by acid deposition, the lack of acceptance by many of monoculture, especially of conifers, the reluctance to accept even-aged stands and their harvest by clearcutting, and other difficulties.

Tropical Forests

Tropical forests are of particular interest for several reasons, not the least of which is the frequently heard claim that they are disappearing at an alarming rate. A major project, initiated under the UNEP's Global Environmental Monitoring System, is beginning to develop the first solid data on which comparisons, and judgments, might be based. This study divides tropical forests into four types: closed tree, open tree, fallows (both open and closed), and shrub formations (Steinlin, 1982). Data from 76 countries were included, along with information from Landsat Satellite images wherever data were inadequate.

Closed tree forests (those with a tree canopy such that no continuous grass cover can develop) cover some 1200 million hectares in the tropics. Of these, 96.6% are broadleaved, and only 2.9% coniferous, with 0.5% bamboo. Some 57% of all tropical forests are in South or Central America, with 25% in Asia and 18% in Africa. Forests of industrial quality **(productive forests)** make up slightly less than 75% of all broadleaved tropical forests—about 860 million hectares. Most of the remaining broadleaved areas are not exploitable because of either stand or accessibility conditions.

TABLE 17.1
Land and Closed Forest Areas of the World, by Regions or Country, 1972

Region or Country	Total Land Area	Closed Forest Land[1]: Area	Closed Forest Land[1]: Proportion of Total Land Area	Closed Forest Land[1]: Per Capita Area
	Million acres	*Million acres*	*Percent*	*Acres*
North America	4,549	1,559	34	6.7
Latin America	5,088	1,791	35	5.9
Nordic[2]	278	127	46	7.4
Europe (except Nordic)	1,078	229	21	.5
Near East	1,476	8	1	.1
Africa	7,487	469	6	1.2
Asia-Pacific (except Japan)	7,104	1,265	18	.7
Japan	93	62	67	.5
U.S.S.R.	5,289	1,940	37	7.7
Total	32,442	7,450	23	2.0

[1]Forest land with 20 percent or more tree crown cover.

[2]Finland, Iceland, Norway, and Sweden.

SOURCE: Forest Resource Report No. 23, p. 88. United Nations, Food and Agriculture Organization and Economic Commission for Europe, *European timber trends and prospects 1950 to 2000.* Timber Bull. for Europe. Supp. 3, vol. XXIX, Geneva, 1976 Reidar Persson. *World forest resources.* Rapporter och Uppsatser, Nr. 17, Royal College of Forestry, Stockholm, 1974.

TABLE 17.2
Timber Growing Stock of the World, by Softwoods and Hardwoods, and Region or Country, 1973 (billion cubic feet)

	Closed Forests[1] and Open Woodlands				Closed Forests[1]			
			Hardwoods				Hardwoods	
Region or Country	All species	Softwoods	Temperate	Tropical	All species	Softwoods	Temperate	Tropical
North America	1,288	953	335	(2)	1,288	953	335	(2)
Latin America	3,260	92	166	3,002	2,981	92	134	2,755
Europe	526	335	191	(2)	526	335	191	(2)
Africa[3]	2,134	4	11	2,119	1,494	4	7	1,483
Asia-Pacific (except Japan)	1,330	201	176	953	1,208	201	124	883
Japan	71	39	32	(2)	71	39	32	(2)
U.S.S.R.	2,790	2,366	424	(2)	2,578	2,154	424	(2)
Total	11,399	3,990	1,335	6,074	10,146	3,778	1,247	5,121

[1]Forest land with 20% or more tree crown cover.

[2]Not represented in region.

[3]Includes the Near East.

SOURCE: Adapted from S. L. Pringle, *Tropical moist forests in world demand, supply and trade.* Unasylva 28 (112–113): 106–118. Rome, 1976; Forest Resource Report No. 23, p. 90.

Moist tropical forests, more than half of which are in Latin American countries, have very high volumes of biomass and a tremendous diversity of vegetative cover. Soils are often relatively infertile. The nutrients, locked in the vegetation, are recycled as plants die and decompose, but little is stored for long in the soil. The variety of life forms in these forests is only now beginning to be recognized, but far exceeds that found in any other part of the world.

The heterogeneity of tree species in tropical moist forests has long posed a serious problem to their industrial use. There are exceptions, especially in Indonesia and Southeast Asia, where nearly pure stands of valuable dipterocarp timber occur. A major part of moist tropical forest species, however, have no commercial value: some because they are unsuitable for construction or decorative veneers, but more often because they are unavailable in the required sizes and quantities or because their wood structure is unknown. Logging intensity varies significantly in tropical forests—from 8 m^3/hectare in tropical America and 13 m^3/hectare in tropical Africa to 31 m^3/hectare in tropical Asia. Undisturbed diptocarp forests in Asia are being logged at 40 to 100 m^3/hectare (Steinlin, 1982). As one consequence, most of the clearing of tropical forests has been for opening new areas to agriculture.

Shifting agriculture (also called swidden, or slash and burn agriculture) has been a major problem. In earlier times, when population pressure was less severe, the interval between farming a particular area was long enough to permit nearly full restoration of soil nutrients and the system was probably reasonably efficient. Properly carried out, it can be ecologically sound, but requires that the land be used for agriculture only one or two years, then left to fallow for 7 to 15 years until a secondary forest is established. Now, however, the cycle is repeated so frequently that crop yields are already low and decreasing rapidly.

One alternative is to shift to perennials such as coffee, rubber, and oil palm, but while such practices can result in a steady cash income, food production is lowered. Conversion to permanent agricultural crop use requires very special soil management practices, and to institute these over the large areas affected will require a huge educational effort on the part of governments ill equipped for the task (Wehnstock, 1984; Pottier, 1984).

Open forests are characteristic of drier tropical areas. They cover 734 million hectares, compared to the 1200 million of closed forest. Africa has two-thirds of these areas, with tropical America having 30% and Asia only 4%. In more severe climatic conditions, the "shrub" formation takes over. This covers 624 million hectares—only 15% less area than that occupied by open forest. Again Africa leads the list, with 71% of the total. Tropical America follows, with 146 million hectares (23%), and Asia with 35 million (6%) (Steinlin, 1982). These areas are often overlooked by forestry officials, since they offer no opportunity for industrial timber production. Yet millions of people depend on these types of forest for fuel, much food, and shelter.

Plantation forestry is at long last beginning to have an impact on timber production, and the areas planted are increasing rapidly. FAO has estimated that by 1975 a total of around 90 million hectares (222 million acres) were in forest plantations of one type or other (Table 17.3). Although only 9% of this total (about 8 million hectares or 20 million acres) were in the developing nations of Africa, Latin America, and Asia, industrial plantations in tropical and subtropical areas were projected to triple by the end of the century.

Because plantations can be so much more productive than natural forests, these areas have the potential of making a major impact on world production in the coming years. Annual growth of 10 to 20 cubic meters per hectare (143 to 285 cu. ft. per acre, roughly similar to annual growth of loblolly pine on an average U.S.

TABLE 17.3
Distribution of World's Forest Plantations

Region and Economic Class	Millions of Acres
Developed market economies	
North America	27
Western Europe	32
Other	27
Developing market economies	
Africa	6
Latin America	8
Asia	6
Centrally planned economies	
Europe and USSR	42
Asia	74
Total	222

SOURCE: U.S.D.A. Forest Service, Forest Resource Report No. 23, p. 98.

plantation site) is common, and examples of growth rates of 35 cubic meters per hectare (500 cu. ft. per acre) are not unusual.

One other element of tropical forestry is often overlooked—the fact that wood is not the only important product. From the standpoint of those who live in tropical forest areas, wood per se may be far less important than the leaves, medicinal herbs, honey, nuts, flowers, gums, oils, fibers, birds, animals, and so on that make up much of their life support system (Poulsen, 1982).

Tropical Forest Devastation

Are tropical forests in danger of being completely devastated? On this question there is wide disagreement, but the regular reports of FAO throw some light on the matter.

During each of the first five years of the 1980s about 7.5 million hectares of the closed forest and 3.8 million hectares of open forest will probably disappear due to population growth and subsistence agriculture (Steinlin, 1982). This means an annual reduction of about 0.62% of the closed and 0.52% of the open forests (an overall rate of about 0.6%) and is about the same in each of the three major tropical forest regions (America, Africa, and Asia).

There are substantial differences within the three regions. In West Africa, as much as 4% of the total closed forest and 6% of the *productive* closed forest area is disappearing each year. This contrasts with central Africa, where no more than about 0.2% is being destroyed. Most of the African loss is caused by shifting agriculture (small areas are cleared or partially cleared, cropped for 2 or 3 years, then abandoned to restart the cycle). Government planned deforestation is more important in Asia, particularly in Malaysia, Indonesia, and Sri Lanka, where large scale resettlement projects are underway. In all, some 11.3 million hectares are being lost to some form of agriculture each year.

Another 4.4 million hectares of the tropical forests are affected by logging activities, making a total "disappearance" of around 15.7 million hectares each year. Much of this logging leaves a substantial tree cover, although one with a much altered species distribution. This means we are losing tropical forests at an 0.8% annual rate (Steinlin, 1982).

Somewhat surprisingly, the study indicates that the undisturbed ("virgin") forests have a much lower rate of deforestation than the total—only about 0.27% annually. The pressure on the more genetically diverse and untouched areas is apparently quite low; at this rate it would take 92 years to reduce the area 25%!

This, however, is probably an overly optimistic view. The FAO report in Unasylva pointed out that "new logging roads open up hitherto undisturbed closed forests to even more unplanned, uncontrolled subsistence farming. . . . Only a small part of forest loss is due to planned projects" (Steinlin, 1982). Moreover, only a small part of the annual loss in the tropics is offset by plantations. But although these planted areas would produce far more wood than a natural forest of equal size, the genetic diversity and uniqueness of the undisturbed tropical moist forest have no substitute. It has been estimated that 25 to 50% of the world's species live in tropical forests, and their open access property nature makes them extremely vulnerable. These "genetic resources are like an unopened treasure chest in that their long-term value to humanity is quite uncertain" (Sedjo and Clawson, 1983).

Added to this is the concern raised by some about the possible effects arising from increased atmospheric CO_2 that could significantly raise (or some say lower) temperatures throughout the world. How important a "sink" the tropical forests may be is highly uncertain, as is the role played by the oceans in the absorbing of CO_2.

To summarize: the area of tropical forest is being reduced—but at a far slower rate than many believe and have stated. We are not likely to run out of tropical forests—even untouched tropical forests—in this century or (to be very safe) in the first half of the next century. But there are risks; unique areas, once destroyed, cannot be replaced. We most certainly should play it safe.

It must be recognized, however, that many of the necessary decisions are not within the province of those living in developed nations. The common property (open access) nature of most tropical forests requires major institutional changes in land tenure. In some developed countries this has been resolved through the institution called private property, in others by adequate public control and management, and in a few (including the United States) by a combination of these two. Until this problem is settled satisfactorily, destruction of tropical forests will continue. And there are other major problems not of a forest nature but that make the future outlook for tropical forests uncertain. Chief among these is the rapid population growth in the developing nations and the slow growth rate in per capita incomes.

INTERNATIONAL TRADE

In 1982 Western sawmill operators were crying about the depressed housing and lumber markets. Mills were closing for lack of sales and U.S. production of softwood lumber was cut to 24.9 billion board feet, down from 33.3 billion in 1973.[2] Given this situation, why did we, in this same year, import 9.15 billion board feet of softwood lumber from Canada—nearly 37% of our reduced production. And in 1985, although domestic production had climbed to 30.3 billion board feet, imports from Canada had risen to over 14.6 billion—fully 48% of domestic production. Also in 1985 we produced only 824 million square feet of hardwood plywood but imported 1729 million square feet (U.S. Dept. Commerce, 1984). Why were we buying hardwood plywood from Korea, Taiwan, and Japan? We have plenty of hardwoods; for the past 30 years we have been growing more hardwood sawtimber than we harvest, and in 1985 almost surely grew at least twice as much as we cut!

To understand why this happens it is necessary to go back to fundamentals. In part, international trade takes place for exactly the same reason almost any trade takes place: because *both* parties gain. But two reasons for international trade differ from those that motivate one to trade stamps or bicycles—the principle of specialization and the principle of comparative advantage.

Principle of Specialization

The increased ability and superior performance an individual gains through specialization is easily seen; if you specialize in some productive effort you eventually produce more. In industrial production there are other gains: bringing together a team of research people who are specialists in a particular field, developing a group of management executives who together can solve corporate problems, or creating a pool of skilled or semiskilled laborers having the needed set of techniques. These, and similar internal and external economies of scale, are generally obtained only if a corporation, an area, or sometimes a nation, specialize in producing particular products.

---

[2]According to reports from the U.S. Census. Production of lumber reported by the Western Wood Products Association in 1979 and 1982 was lower by nearly 1.8 and 1.1 billion board feet in the corresponding years.

Principle of Comparative Advantage

The second principle, that of comparative advantage, is less obvious but equally important. Probably the most used example is that of the town's leading lawyer who also happens to be the best at a word processing keyboard. If the town is of reasonable size, it will surely pay this lawyer to concentrate on the legal problems of her clients and hire a secretary to do her paperwork. When measured in earning power, her *comparative* advantage lies in legal affairs, not in secretarial duties.

As a very simple example of how this applies to trade, consider the following. Two individuals, Walton and Crockett, have retired to a remote wilderness and become hermits. One day they happen to meet and discover each has certain talents. In terms of food production per day, Walton can catch four units of fish but can capture only two units of game, while Crockett can get either three units of fish or four of game during the same time. Then Walton gets a brilliant idea. "Suppose I do the fishing while you do the hunting. We'll meet at this same place after two days and swap."

When they meet, Crockett has produced the expected 8 units of game and Walton 8 of fish. Together they now have 16 units of food instead of the 13 they would have had if each spent a day producing each product. Thanks to their *specialization*, both men can be better off.

What about the terms of trade? Walton will go no higher than a two fish for one game exchange, since he can do this well on his own. For the same reason Crockett will figure he must get at least a three fish to four game exchange rate. This means the exchange rate of fish for meat will lie someplace between 2/1 and 3/4 (or between 2 and 0.75), but we can't tell exactly where. The final result will depend on the bargaining strength of the two people—and remember that Crockett has a squirrel rifle and Walton a fly rod! Nevertheless, trade *could* be mutually beneficial, and if it is going to continue it will have to be.

Now suppose the production possibilities are slightly different—that the possible production of the two men (over the same time period) is changed to that of Table 17.4. Crockett is now clearly the better provider, for he can outfish and outhunt Walton. He has an *absolute* advantage in producing both products. But suppose they decide (in spite of Crockett's superiority) to trade after each has concentrated on what he can do best, *comparatively*. Walton still has a comparative advantage in fish production (his fish-to-game ratio is 2.0 as compared to Crockett's ratio of 1.25) while Crockett's game-to-fish ratio is .80 (to be compared to Walton's .50). With each man working 2 days, their total production will be 16 units of food (compared to only 15 if each spends one day on each kind of production). The exchange limits within which the final exchange will be made can easily be determined, but not the exact ratio.

Note that two things had to exist before Walton and Crockett could benefit. Each had to specialize in the thing he did best comparatively, and the two had to be willing to exchange the two commodities.

TABLE 17.4
Production Possibilities of Walton and Crockett

Man	Fish	Game
Walton	4	2
Crockett	5	4

Extension to an International Basis

To put this on a nation-to-nation basis, imagine two countries with differing capabilities for production. With one hundred units of resources, country A can produce either 16,000 square feet of plywood or 4 computers, while country B can produce either 15,000 square feet of plywood or 3 computers (Figure 17.2). The production possibilities for the two countries are indicated by the lines connecting the production outputs *(PP)*. The trade possibilities are indicated by *TT*, which show the exchange that could be made if nothing else was involved.

Given this situation, suppose each country concentrated on what it can do best. A's ratio is 4/16, or 1/4 (1 computer for 4 thousand sq. ft. of plywood), but A can get as much as 20,000 sq. ft. of plywood for 4 computers if trade could take place, for this is B's opportunity cost (in terms of plywood) for making 4 computers. By the same reasoning, B's ratio is 3/15, or 1/5 (1 computer for 5 thousand sq. ft. of plywood), and B could get as many as 4 computers for 16,000 sq. ft. of plywood (which the latter is A's opportunity cost for plywood in terms of computers). Note that in both cases the trade opportunity curve lies *above* the production possibility curve. In effect, trade would expand the possibilities for both A and B, for both can get *more* from their 100 resource units by using them in what they produce best.

What would be the actual terms of trade? This can't be answered from the given data, for the terms of trade would be determined by (a) the nature of demand and supply for the products in each of the two countries, and (b) the *world* demand and supply for the products. To see how the terms of trade would work out, and how they might be affected by factors other than the production costs per se, another example will be used.

In Figure 17.3, *D* and *S* are demand and supply curves for paper in each of two countries, Alpha and Beta. Alpha's currency is the dollar and Beta's the kudzu, the present exchange rate being 25 kudzus for 1 dollar. Paper prices are given in the domestic currencies, but the two scales are equal at the present exchange rate.

The demand and supply curves show that paper prices are at equilibrium levels in both countries, at \$290 per ton in Alpha and K12,000 in Beta (the latter equal to \$480 at the current rate of exchange). Now suppose that whatever restrictions exist are removed—that completely free trade is permitted. Alpha would

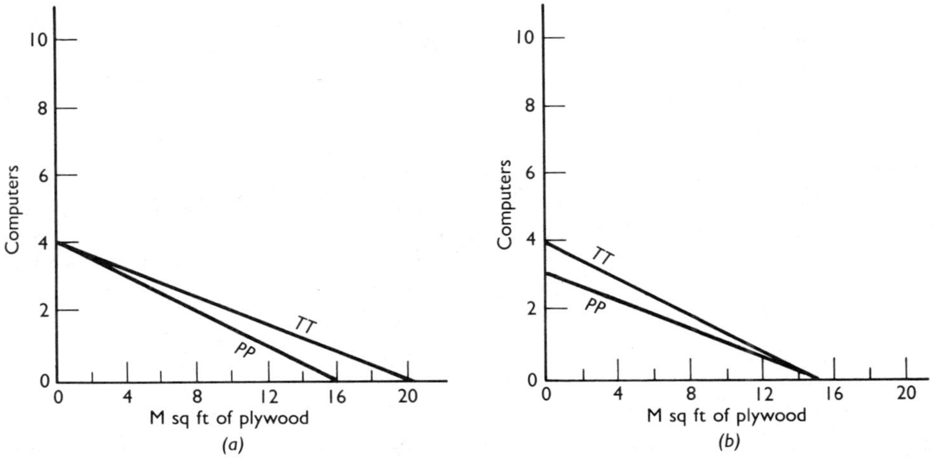

FIGURE 17.2. Production possibility curves and trade curves for two countries producing computers and plywood, using 100 units of resources.

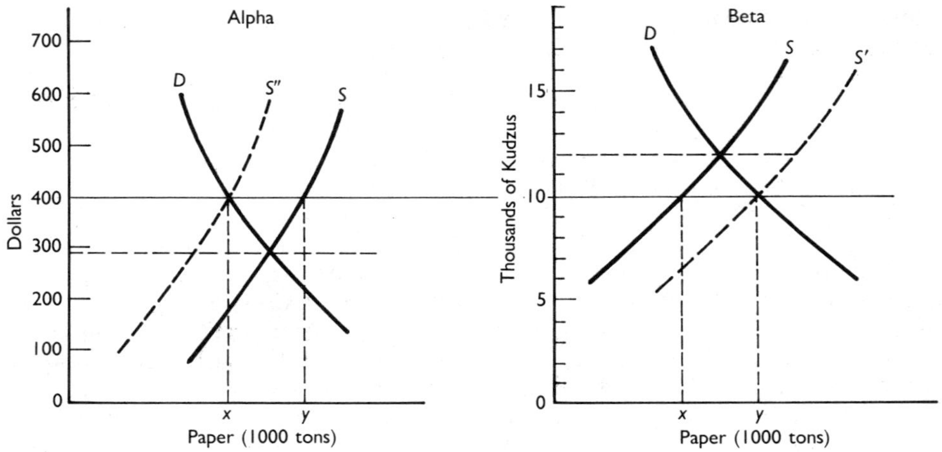

FIGURE 17.3. Determination of quantities of paper to be imported and exported—providing trade is completely free, transport costs are negligible, and all else remains constant.

soon be shipping paper to Beta, where it would find a ready market. Paper prices in Beta would start to fall and, as supplies were diverted from Alpha to Beta, prices in Alpha would begin to rise. Eventually, a new equilibrium would be established. If transportation were costless and all else remained as before, Beta's supply curve, composed of its own supply *plus* the quantities imported from Alpha, would shift out to S'. Beta would then be importing x-y tons from Alpha, and the Beta price would drop to K10,000 per ton. In Alpha the supply curve would shift to S'' and paper price would rise to $400 per ton. Both countries would continue to manufacture paper, but because Alpha has a comparative advantage in paper production it will trade paper for Kudzus, which it can use to buy pottery from Poland, or beans from Boston—or possibly bacon from Beta (Beta's best bacon is better!).

In the real world things are seldom this simple. One would expect the price in Beta to be higher than in Alpha by the cost of shipping, handling, and insurance. Yet even in the simple form used here the principle of comparative advantage throws considerable light on world timber production and trade today, and even more on the probable trends in timber production over the next few decades.

In the world, as in our own country, timber comes from three major sources:

1. Old growth forests. In the United States from the West Coast and from British Columbia, and tropical hardwoods from several areas.
2. Second growth forests, essentially unmanaged, volunteer stands. Our hardwood stands of the Eastern states north, south, and central areas.
3. Increasingly from man-made forests. In the United States those in the Southern states, and for the world, primarily from Latin America and Africa, from New Zealand, and (to a lesser extent) Australia.

When timber was primarily from old growth forests, the United States held a decided comparative advantage in high quality structural timbers: white and red pines from the Northeast and the Lake states, then the Southern pines, and for many years, the Douglas fir from the Pacific Coast forests. In turn, each of these areas enjoyed a comparative advantage within the United States. In Europe, the

Nordic nations played a similar role—with some help from Canada and the United States.

Both nationally and internationally, we seem now to be in a transition period. As old growth forests are exhausted, those areas best capable of *growing* good timber begin to take over the supply side (look again at Figure 10.5). In our own country we see this happening on a large scale as the wood-using industries increasingly concentrate in the Southern states and some of the better areas of the West. And, in Europe, the biggest suppliers are the Nordic nations, and these supplies are coming from intensively managed forests, often from plantations. But we also see new sources of timber coming to the forefront on world markets, and most of these are in semitropical nations, Chile and Brazil in South America, New Zealand (and to a lesser extent) Australia, and South Africa. Production in these areas was still quite small in the late 1980s, but is growing rapidly. By the end of this century these countries could have a substantial impact on world timber trade, for they can produce large quantities of wood at relatively low cost. Sedjo and Clawson (1983) estimated that "only about five percent of the world's total forestland could economically produce on a sustained yield basis the (world) wood requirements forecast for the year 2000." The point being made here is a simple one: a region's comparative advantage in wood-based industries will increasingly depend on the economics of timber growing.

International trade, be it timber, wood products, or any or all of thousands of goods, is not just a flow back and forth along a two-way street. There are many streets and they run to many places, and traffic is often heavy. More important, road blocks often impede and sometimes completely stop the flow of trade traffic. Four groups of trade barriers are particularly significant: (1) tariffs, (2) quotas, dumping, and other barriers, (3) discriminatory shipping charges, and (4) foreign exchange rates. Before examining these, however, some comments on the negative side of comparative advantage may be ventured.

First, when taking a world view, one should keep in mind that the comparative advantage possessed in (for example) manufacturing by the developed nations is to a considerable part due to the fact that they *are* developed. To developing nations it may well seem that the principle of comparative advantage works in the direction of keeping, or even enlarging, the technological gap between developed and developing nations. To be sure, we now have a comparative advantage in high technology manufacturing—making items such as computers, robots, electronic gear, and so forth. It is also true that this ability came, at least in part, because of our relatively advanced educational system and our store of capital and trained labor. But these are surely part of the development being sought by the developing world. To at least some degree the developing nations can argue "Give us a chance to develop our own skills and our own resources—*then* let's talk comparative advantage."

Second, there are matters of security that might make domestic production of some items in a nation desirable even if manufacturing costs are considerably higher domestically than in another country. The United States, for example, would be highly unlikely to rely on any foreign nation for most of its military hardware!

Third, economic power often brings at least some political power, and developing nations—even though recognizing the validity of the principles of specialization and comparative advantage—may feel that economic domination by developed nations endangers their own political freedom.

With these caveats in mind, we can proceed to examine trade barriers.

Tariffs

Tariffs can have two objectives: to protect the domestic industry from foreign competition and to provide government revenue. The effect of a tariff on trade is easily seen from the previous illustration, Figure 17.3. Suppose Beta decides that the paper industry must be protected from the "cheap labor in Alpha" (which might also happen to be more efficient). Clearly, if a tariff of $190 per ton or higher is imposed, there will be no paper shipped from Alpha to Beta, for this would make the price of Alpha paper as high or higher than that in Beta. But suppose a tariff of K1,000 per ton is levied. In that case the amount of paper coming from Alpha would be considerably reduced (the supply curve wouldn't shift as far) but imported paper would still be sold in Beta, now at a price closer to K11,000.

What would be the result if, after Beta's imposition of the new tariff, Alpha were to retaliate by placing an equivalent one on all paper imports from Beta? From what we know here, this would have no effect whatever. Beta's production costs are so much higher than Alpha's that any sales Beta might make there would be at a decided loss. Alpha would be much more likely, though, to put a tariff on something that *is* imported from Beta—maybe butter—in which case Beta's butter producers would be hurt, but *so would Alpha's butter consumers.* Competitive tariff "wars," like any war, hurt both participants.

Quotas, Dumping, and Other Trade Barriers

When a quota is placed on some product, only a stated amount is allowed to be shipped into the importing nation. In 1981, for example, the U.S. government negotiated a "voluntary" quota on Japanese-built automobiles to be admitted to this country, thereby providing (it was hoped) time for U.S. automobile manufacturers to retool for making smaller, more energy-efficient models. The quotas were not requested in 1985. It seemed to most consumers, and apparently to the President, that the objective had not been accomplished. The Japanese, taking advantage of the higher prices (and profits) for U.S.-produced cars, raised their own prices for the limited quantities, and were apparently able to substantially increase the profit margin on the smaller number of cars they were allowed to export.

Quotas are generally regarded by economists as being even less desirable than tariffs. They provide no revenue, and they protect the domestic industry from competition more completely than a tariff. If a foreign producer is efficient enough, it can overcome the effect of the tariff, but there is no way it can exceed a quota except through illegal means. However, as apparently happened with Japanese automobiles, it is not always necessary to increase the quantity shipped to increase profits.

"Dumping" is another ploy—in some ways the opposite of quotas. Modern paper manufacture (for example) has large economies of scale—so large that world trade in paper and wood pulp tends to be dominated by firms with very large manufacturing plants. Moreover, the paper industry is ologopolistic in structure; prices tend to be quite sticky. If prices become "soft" in a firm's domestic market, it may prove desirable for the firm to continue manufacturing at its optimum rate. With the lower per ton costs it may sell only enough on the domestic market to maintain its market share at the going price, then ship the surplus abroad at really bargain prices. This is dumping. In the exporting country this is called good business, and it may be a good deal for the consumers in the receiving nation. By the producers in the countries receiving the bargain-priced goods, however, it is called unfair competition.

There are other barriers to trade, some of them much more subtle than tariffs or quotas. If, for example, the specifications for a product are carefully written, they may cause foreign manufacturing costs to rise considerably, thereby reducing their competitive edge. Inspection at the customs office may be so rigid as to require special handling. Restrictive grading rules or size specifications can be imposed that may raise foreign costs. Quarantine laws, or restrictions on shipping, can also prove bothersome. As a general rule, free trade proves to be beneficial to producers and consumers alike. Anything that makes trade more difficult tends to reduce welfare in both importing and exporting nations.

If trade restrictions are so clearly harmful, why do some industrialists and labor representatives, and most politicians, spend so much effort to secure them? Two explanations can be offered.

First, people involved in producing something that can be better or less expensively made elsewhere are definitely hurt when these goods are imported. Sales are lost, profits are reduced, workers lose their jobs, whole plants may be made useless, and the capital they represent effectively destroyed. When something as large as the automobile or steel industry is affected, a lot of jobs and investments may be at stake. Some forest product industries are also large; the largest single employer in Washington and Oregon and in several California counties is the lumber industry. To the extent that U.S. output is replaced by Canadian lumber, for example, these areas are hurt. But we should also note that Canadian supplies tend to hold down lumber prices for *all* consumers.

Second, ignorance of their real effect probably plays some role in tariff support, especially among politicians. As Professor Mansfield has put it, "There is no law that prevents people with little understanding of economics from holding public office" (Mansfield, 1983). And there are other reasons for political support. Special interest groups may have much to gain from tariffs or quotas and such groups occasionally have quite a bit to do with the success or failure of political campaigns. When some people will be hurt and some will benefit, the problem becomes difficult. When a small number will be hurt a lot while a large number will benefit, though each individual gains relatively little, that small number may become very vocal—and thereby important—in political circles. They might even be contributors to party campaigns!

Shipping Rate Differences

Shipping rates for international trade "are set by shipping conferences, each of which controls a specific route and acts independently of other conferences in setting rates" (Radcliffe, 1978). These conferences can, and often do, act much as international cartels and tend to be unresponsive to actions by an individual government.

If shipping rates are set so that exports (for example) from the United States to Japan are considerably higher than from Japan to the United States, U.S. exporters can be at a decided disadvantage. Shipping rates depend on value per ton, on cubic feet per ton (bulkiness), and on whether or not the commodity can be shipped by tanker (Radcliffe, 1978). Most forest products typically are low valued per ton, bulky, and cannot be shipped by tanker (although pulpwood chips are being shipped in a very similar fashion), and their low value per ton makes shipping costs a higher proportion of delivered price than most products. In addition, most tariffs are changed on a c.i.f. (cost + insurance + freight) basis, and foreign consumption taxes are charged on a landed duty paid basis (the taxable base includes the duty charges)—which multiples the effect of the shipping charge.

Shipping rates, tariffs, and taxes change from year to year, from country to country, and are specific to commodities; any figures given for wood based products could be misleading. For timber products, however, the protection afforded by shipping charges is often substantially higher than that granted by tariffs or quotas.

Foreign Exchange Rates

Changes in the foreign exchange rate can, and generally do, effect trade patterns, as is seen in Figure 17.4*a* and *b*. Figure 17.4*a* is identical to that of Figure 17.3 after free trade has been established. Alpha is shipping $x-y$ tons of paper to Beta, causing Beta's supply curve to shift from S to S'', and her paper prices to drop from K12,000 to K10,000 (equivalent to $400 in Alpha's currency). At the same time, paper prices in Alpha rise from $290 to $400 as Alpha's supply curve shifts from S to S'' The exchange rate is $1 = K25.

Now suppose the dollar rises to where it takes K50 to buy a dollar. Relative to Alpha, Beta's price scale has doubled, and on the new scale Beta's demand and supply curves for paper are D'' and S''. Now it would take a price of K20,000 to buy $x-y$ tons from Alpha! But at K20,000, Beta will not import paper. Alpha's market has disappeared. In fact, at the domestic equilibrium price in Beta (K12,000), Alpha will start buying from Beta, for this will now cost Alpha only $240 per ton (12,000/50 = 240). If exchange rates remain at this new level, a new equilibrium will eventually be reached, *this time with Alpha buying paper from Beta*. (You figure out how much Alpha will import and at what price.)

From the end of World War II until 1973 most countries kept their exchange rates essentially constant. In that year the principle trading nations adopted a policy of letting their exchange rates "float." Since then the U.S. dollar has gone through a period of weakness (in the late 1970s), then into a sustained period of unparalleled strength. American tourists "never had it so good," French and Italian wines were flooding the American market, German goods were at all time lows, and the United States was importing far more goods than they were exporting. Our trade balance was running at a negative 12 *billion* per month, and many of our exports were becoming too expensive to compete abroad.

The outcome of this has yet to be learned. It does show, however, one effect

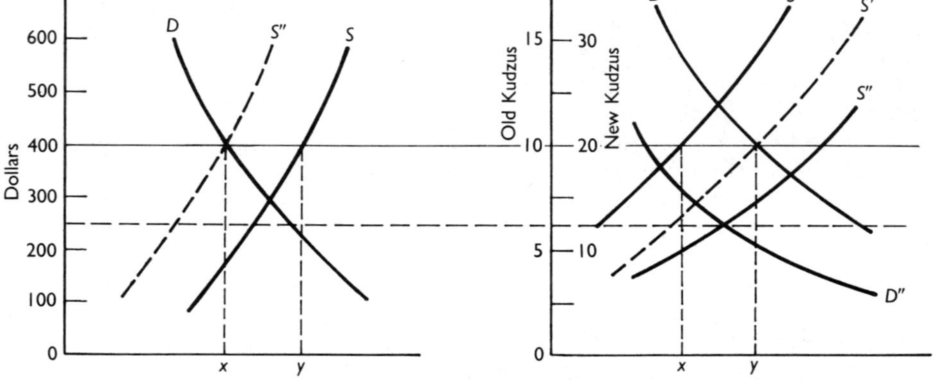

FIGURE 17.4. ***(a)* and *(b)*. Effect on trade of a change in foreign exchange rates. Direction of trade can be reversed.**

very expensive dollars can have on foreign trade. One other effect, often overlooked, was the very positive impact this had on European production and employment. One economist observed that, although this was hurting our exports, it quite possibly may have saved the financial structure of the industrial world from collapse, since the principle European nations were at long last making a recovery from the general recession (McCracken, 1985).[3]

U.S. AND WORLD TRADE IN FOREST PRODUCTS

World Trade

Fuelwood, charcoal, pit-props (mine timbers), sleepers (cross ties), and certain other primary forest products are seldom important items in world trade. They are too bulky and too heavy to make shipping long distances economically feasible. Other forest products, particularly pulp, paper and paper products, plywood, particleboard, and veneer are important elements of trade. Lumber is also heavy and bulky, but trade in some species is very important; cross-border shipments between neighboring nations (for example, Canada and the United States) can be highly significant.

Total world production of wood in 1982 was given by FAO as 2,958,470 thousand cubic meters (104,477,481 thousand cubic feet) (FAO, 1984). Asia and North and Central America together accounted for 51.2% of the world's production, followed by Africa, USSR, Europe, and South America in that order. Oceania (Australia, New Zealand, and the islands) produced the remaining 1.2% (Figure 17.5).

Of the total production, a little over 54% was used as fuel, either burned as wood or as charcoal. About 44% of this was produced in Asia, another 25% in Africa, and almost 13% in South America. Less than one-tenth of 1% of the world's fuelwood production was exported.

Total world trade in wood and wood products is not insignificant, from 1979 through 1984 averaging 56 billion dollars per year. From 1971 through 1980 the total value of forest product imports, measured in current dollars, increased by over 400% to 62.3 billion, but lost some of the increase during the next three years, declining to 55.8 billion in 1984 (FAO, 1986). Most of the importing was by developed nations, although the developing nations' share was rising (from 13.3 to nearly 19%).

Imports

Viewed globally, by far the major market for sawlogs and wood chips is Japan; in 1982 Japan's share of major wood product imports included 43% of the total volume of industrial roundwood traded. This included 56% of the coniferous and 47% of all traded nonconiferous sawlog and veneer logs, 34% of traded pulpwood, 73% of wood chips and particles, but only 9.6% of wood pulp, and 7% of the sawnwood. Europe was second to Japan in industrial roundwood (36%) and in wood chips and particles (20%), but was the major market for wood pulp (55%), paper and paper board (50%), wood-based panels (60%), and sawnwood (44%). Major details of forest product imports are given in Table 17.5.

---

[3]By the end of January 1987 the dollar had dropped substantially, and was still dropping. The pound was at $1.52, the franc at 16.45¢, and the West German mark at 54.87¢. The Japanese yen, which had been around 250 yen per dollar in May 1985, had dropped to 152 yen per dollar.

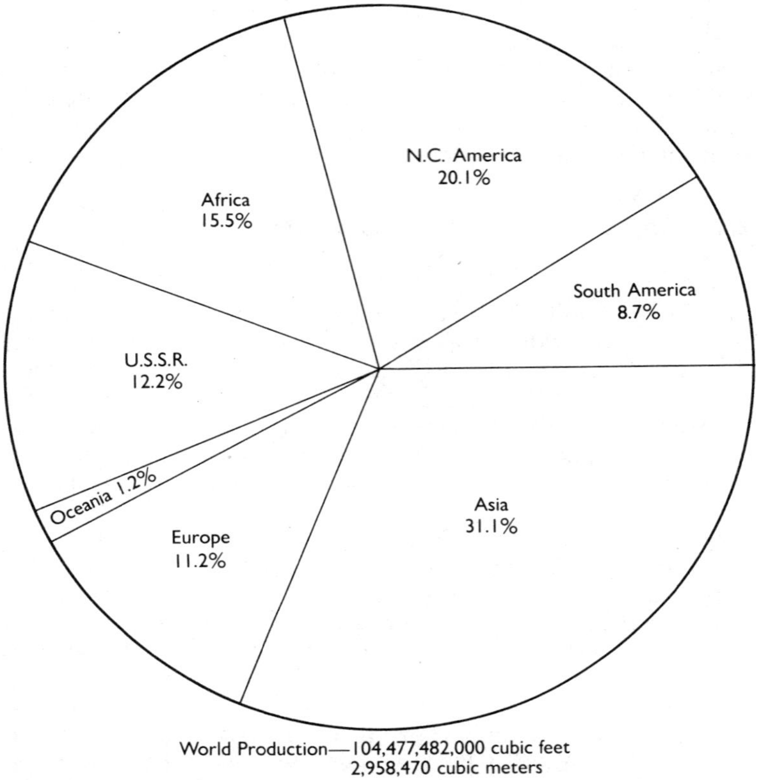

FIGURE 17.5. Distribution of world population of roundwood (total harvest of wood) in 1982. *Source*: FAO, 1984.

TABLE 17.5
Percent of World Value of Major Wood Products Imported by Europe, Japan, and United States in 1982

Item	Europe	Japan	United States
Industrial round wood	35.7	42.9	2.6
Wood chips and particles	20.0	73.0	5.3
Sawlogs and veneer logs, conif.	22.0	56.0	—
Sawlogs and veneer logs, nonconif.	20.0	47.0	1.7
Wood pulp	55.0	9.6	18.0
Sawnwood, conif.	43.0	7.1	35.6
Sawnwood, nonconif.	44.0	6.4	4.4
Total sawnwood	44.0	7.0	30.3
Wood-based panels	60.0	1.8	14.0
Paper and paperboard	50.0	2.2	21.0
World	48.1	12.4	15.0

SOURCE: 1982 Yearbook of forest products. FAO Forestry Series Food and Agriculture Organization of the United Nations, Rome, 1984.

Exports

The other side of the coin—exports—shows somewhat less concentration. In 1982 Australia was the largest exporter of chips and particles (40%) and the United States second (30%). The United States was the largest exporter of coniferous sawlogs (53.5%), and the second was the USSR (23.4%), while the major exporter of nonconiferous sawlogs, Malaysia, shipped nearly 55% of the total, Indonesia about 9%, and Africa 13%, most coming from the Ivory Coast. The United States and Canada combined to ship 50% of the wood pulp exported, while Europe exported 33% (Finland and Sweden accounting for nearly two-thirds of Europe's total). In export paper and paperboard, Canada shipped 25.6% of the total, and Europe 55% (Finland and Sweden accounting for half of the European total).

U.S. Trade[4]

The value of both U.S. forest product imports and exports has been growing. Imports increased in all but two of the 12-year 1971–1982 period, exports in all but four. In 1982 imports of forest products were valued over 8 billion dollars, and exports at 5.7 billion. Table 17.6 gives quantities and values of the major forest products traded by the United States.

The United States has relatively few barriers to international trade in forest products. In 1985 the highest tariff was on plywood and, compared to many other

TABLE 17.6
U.S. Trade in Forest Products—1982

	Import			Export		
	Quantity			Quantity		
	(cu. m)	(cu. ft)	Value $	(cu. m)	(cu. ft)	Value $
Sawlogs, conif.	447	15,786	22,930	14,111	498,326	1,174,180
Sawlogs, nonconif.	83	2,931	3,500	422	14,903	83,905
Pulpwood	1,443	50,959	52,148	4,389	154,996	238,614
Chips & particles	814	28,746	36,700	4,151	146,591	231,139
Sawnwood, conif.	2,118	748,248	1,569,285	3,818	134,832	575,576
Sawnwood, nonconif.	479	16,916	98,443	819	28,923	220,954
Wood-based panels	2114	74,655	504,468	875	30,900	251,198
	mt tons			mt tons		
Wood pulp	3,296		1,484,974	3,060		1,407,655
Newsprint	5,942		3,748,652	279		142,964
Print & writing	676		388,023	186		190,363
Other paper & paperboard	374		201,807	2,821		1,380,758
			5,897,306			8,110,930

SOURCE: 1982 Yearbook of forest products. FAO Forestry Series, Food and Agriculture Organization of the United Nations, Rome, 1984.

[4]This section draws heavily from the work of Roger Sedjo of Resources for the Future, especially from his paper prepared for the World Forestry Congress of 1985 "The role of forestry trade in economic development" (Sedjo, 1985).

countries, that was relatively small: 8% advalorem (8% of the declared value) on hardwood plywood and 20% on softwood plywood. Certain paper and paperboard products carried a 5% tariff, some had lower rates.

In 1950 Canada supplied 8.7% of the U.S. softwood lumber market and no one was particularly upset. In 1960 Canadian imports reached 12.1%, and by 1970 were 17.8%. A few Western producers raised mild objections, but since much of the imported lumber was from U.S. subsidiaries in British Columbia the objections were not taken very seriously. The increase continued, however, and in 1980 Canada supplied 26.7% of the U.S. market. Pressure for enactment of tariffs or quotas on Canadian softwood lumber exports increased enough to make Canadian producers uneasy (*Agricultural Law Letter,* 1985; *Wall Street Journal,* 1983).

In 1982 the U.S. domestic lumber industry filed a countervailing duty suit with the International Trade Administrations (the ITA), claiming that the low Canadian stumpage prices constituted an unfair subsidy to Canadian producers—a claim that was disallowed by the ITA (*Journal of Forestry,* December 1986).

In 1985 Canadian imports of softwoods were 33.5% of our total softwood lumber consumption and Pacific Northwest producers were very upset. In May 1986 the United States raised tariffs on Canadian cedar shakes and shingles by 35%, and in October imposed a countervailing duty of 15% on all Canadian softwood lumber exports. Canada objected strenously and retaliated with duties against U.S. products "ranging from computer parts to oatmeal" (Tarnapol, 1986). Several conferences were held, and in January 1987 Canada imposed a 15% export duty on softwood lumber, while the U.S. removed its countervailing duty. This agreement was expected to maintain higher lumber prices in the United States, but permitted Canada instead of the United States to collect the 15% duty on Canadian shipments (*Christian Science Monitor,* 1987).

Softwood lumber prices in the United States rose almost immediately, but the play has not finished: many dealers expect Eastern and Southern home builders to continue buying Canadian lumber because of its perceived higher quality—it is softer, straighter, and lighter than Southern pine. Consumers appear willing to pay higher prices for the Canadian lumber rather than buying the harder, though stronger, southern pine (*Wall Street Journal,* 1987). Canadians are considering the substitution of a stumpage tax for the current export duty, but this may not be enough to satisfy U.S. producers. Either way, U.S. consumers will pay higher prices for lumber, fewer homes will be built, and the market for lumber substitutes will improve.

There are restrictions on some exports. Softwood logs from federal lands west of the 100th meridian cannot be exported.[5] Log exports are also prohibited from National Forests in Alaska, as determined by the Alaskan Regional Forester. These restrictions were enacted to prevent export of materials that might otherwise be lost to U.S. manufacturers. There are, however, no restrictions on export of *processed* materials.

Restrictions against importation of U.S. forest products by other nations change from year to year. In 1985, for example, the European Economic Community (EEC) had a duty-free quota of 600,000 cubic meters for importation of softwood plywood from nonEEC nations. Any import over this limit incurred a duty of 11.5%. European importers, as a consequence, tended to purchase the

[5]The 100th meridian runs through the approximate center of North and South Dakota and Nebraska, cuts of the western third of Kansas, and the Oklahoma Panhandle, and probably a little less than the western half of Texas.

plywood up to the quota early in the year, then stop purchasing altogether. Canada had (in 1985) import duties of 4 to 17% on U.S. exports of moulding and millwork and hardboard. Japan imposed a duty of 15% or less on U.S. exports of veneer, softwood plywood, fiberboards, and spruce-pine-fir lumber (Darr, u.d.). Paper products of various kinds are also subject to duties in both Canada and Japan.

Most forest products exported from the United States are at least partially processed. Since many are used in construction, any restrictions on wood use for that purpose have an effect on trade. Perhaps one of the major barriers to exports of U.S. forest products is not specific to the product per se—the fact that the United States has not adopted the metric system. Metrification has been discussed for many years, but progress has been painfully slow. In most parts of the world, lumber, wood-based panels of all kinds, and even paper sizes are based on the metric system of measure: 2 by 4 studs of standard 8 foot length or 4 by 8 plywood sheets just don't fit the construction needs of Japan or Europe. Additionally, our grading standards are different from theirs, which raises marketing problems as well as problems in actual use.

Finally, a major handicap during the early 1980s was the almost phenomenal strength of the U.S. dollar. As observed earlier, as our dollar grows stronger our products become more expensive to foreign buyers. In the early 1980s our forest products *were* expensive in overseas markets and Canadian softwood lumber *was* cheaper, although plywood and paper exports had climbed in 1984 (McCarl and Haynes, 1985).

There seems little doubt that the United States has a comparative advantage in timber production over most countries. Additionally, our manufacturing facilities are probably at least as efficient as those anywhere. Intensive management, particularly from plantations in the South and West, holds a promise of providing an abundance of high quality, reasonably priced raw material for almost any kind of forest product. Whether we can, or will, take advantage of our opportunity will depend on many things, not the least of which is the way foresters respond to the challenge.

QUESTIONS AND PROBLEMS

1. Can you fill in the blanks of the following paragraph correctly?

   Forests cover nearly ________ of the earth's land surface, with about ________ in the temperate climatic zones. Close to ________ of these forests are softwoods, with (country?) having by far the largest amount—over ________% of all softwoods. About ________% of the hardwoods are found in South America. North America, Europe, and the USSR together produce nearly ________% of the world's industrial wood, while most of the wood produced in Asia and Africa is used as ________.

2. World authorities appear to hold decidedly different opinions on the speed with which tropical forests are being harvested. How severe is the tropical forest situation? Why are tropical forests threatened—if they are? What are the problems that tropical forests pose to industrial use?

3. What is shifting agriculture?

4. Japan and the United States both produce hardwood plywood. They also both pro-

duce wheat. Suppose (for the purposes of this problem) that when measured in terms of U.S. dollars, the costs of production in the two countries are as follows.

Country	Hardwood Plywood (cost per M sq. ft.)	Wheat (cost per bushel)
Japan	$100	$3.33
United States	$200	$4.00

Ignoring all other commodities and countries, and assuming constant production costs, full employment, etc., answer the following questions. (Show all calculations.)

a. Will trade take place?

b. If so, who will trade what?

c. What can you say about the price that will be established for the two products?

d. Assume that the price of wheat settles at $5 per bushel, including shipping costs, insurance, and so on. What can you now say about the price of Japanese plywood?

e. Suppose the U.S. plywood industry succeeds in getting Congress to impose a tariff of 10¢ per square foot of plywood (to equalize the costs of production). What will happen to the trade picture?

5. Why do developing nations often maintain that "comparative advantage is a good argument for the industrialized world, but it really hurts a nation that is trying to get development underway"?

6. Admittedly, tariffs and other trade barriers may, at least for a time, protect domestic producers against competition from foreign producers of similar products. Who pays for the protection?

7. Economists, some politicians, and many business executives have long supported "free trade" policies. So why do other industrialists, labor unions, and many politicians support quotas, tariffs, and other trade barriers designed to protect producers of products that can be produced at lower cost in other countries?

8. What nation provides the major import market for sawlogs and wood chips? Why isn't this same nation a large importer of manufactured finished wood products—of paper, lumber, and plywood?

9. What is the current position of the United States as an importer and exporter of logs? Of lumber? Of plywood? Of paper?

10. If the exchange rate for Canadian dollars were to rise from (for example) 80¢ to $1.10, what effect would this probably have on

a. Lumber prices in the United States.

b. Lumber imports from Canada to the United States.

11. Under the conditions stated in Question 10, would you expect Canada to impose a duty on U.S. softwood lumber? Explain.

chapter eighteen
FORESTRY AND ECONOMIC DEVELOPMENT

Many forestry problems in industrialized nations, perhaps especially in the United States, stem less from biological reasons or destructive cutting practices than from the growing affluence of our society. The demand for recreation, for wilderness experience, the option demand expressed by urban dwellers, all became forestry problems only after development of an industrialized society capable of supplying more than basic needs and a distribution system permitting at least a major part of its members to express and satisfy these demands.

Most of the world's people, however, face quite a different set of problems. At the individual level the set reduces to a daily routine of getting enough food, water, and shelter to maintain the family. For governments it becomes one of reshaping an entire national economy—its production system, often much of its political and legal system, and inevitably its citizen's work habits and lifestyles—so that enough can be produced to satisfy a growing population not only with enough food, water, and shelter to maintain life, but to satisfy other demands that are, or will be, expressed by their people. Since World War II much has been said and written about the struggle of African, Asian, and Latin American nations to "catch up" to the standard of living enjoyed by the industrially advanced countries. The gap is widening; yet catching up is not the problem, nor even a rational goal, for much of the world. It has become crucial that *all* people be able to enjoy at least the minimum of health and leisure necessary for physical and mental peace and comfort—crucial to them *and to us*, because people have learned that these goals are realizable within the framework set by modern technology and they have high expectations.

The Developing Nations

Using 1981 income data, the World Bank divided the developing nations into two groups. The low income countries include 42 nations with incomes averaging less than $405 per capita, while the 62 middle income nations are those with incomes averaging above $405 and up to somewhat over $5000 (IBRD, 1984).[1] In contrast are the 21 industrial market economies (the United States, with an average income of $12,820 per capita, was twelfth), and the 8 oil-exporting economies (headed by Qutar, where incomes averaged $27,719 per capita). The fifth group of nations is the 8 East European Non-market Economies, including USSR.

For present purposes, the low income group poses the major problem. The 10 poorest nations had per capita incomes averaging between $82 (Bhutan) and $200 (Malawi). Above this, but still below $405 in average per capita income, are 32 nations. This group includes both India and China, which together make up nearly half the world's population. It is painfully evident that much of the world is pitifully poor.

In addition to low incomes, developing nations tend to share certain of the following features. Developing nations have low rates of capital accumulation: low incomes mean low savings and savings are the main source of investment capital. Almost all these nations have high population growth rates, considerably higher than those of the industrial, market-oriented countries. Most of the people living in developing nations gain their living from agriculture—nearly always more than 50% and sometimes as much as 80%. Characteristically, there is a high rate of unemployment, and an even higher rate of underemployment in both rural and urban areas.[2] Developing nations are also characterized by high illiteracy rates and low levels of technology. Many, but not all, are located in the tropics.

A lack of natural resource endowments is *not* a characteristic of developing nations. In fact, some of these countries are undoubtedly among the most richly endowed of all countries in terms of natural resources, with Indonesia and Brazil providing outstanding examples. This richness is often contrasted with that of Switzerland, a highly industrialized nation that has virtually no stock of many important natural resources but an average per capita income of $17,434—the highest of any nation not in the oil-exporting group.

Requirements for Development

While it is fairly easy to describe the common characteristics of developing nations, there is far less agreement on the basic requirements for development. **Capital** is usually accorded a high place in the needs for development; without capital there is little hope for building a modern industrial society. **Managerial ability** and **entrepreneurship** are also recognized as important factors. Partly because **technical knowledge** is always needed, **education** is usually ranked high on the list of necessities. Some reasonable degree of **political stability** is believed a prerequisite, for without this there will be little investment from either internal or external sources. A **will to develop** is often included; people, not all, but enough to serve as a catalyst, must want to see their country develop and want it badly enough to work at it so hard that they inspire others by their example.

---

[1]The average incomes quoted here were calculated by converting domestic currencies to U.S. dollars through use of foreign exchange rates. Purchasing power can be decidedly different from that indicated by foreign exchange rates. Based on 1975 dollars (for example), the U.S. GDP (gross domestic product) per capita in 1981 was $7176, India's was $146, and Malawi's $138.

[2]Underemployment exists when people work at tasks that do not fully occupy them. In some rural areas of Asia it has been estimated that over 25% of the workers could cease work entirely and national income would remain unchanged; the remaining 75% could perform all the needed work. It may also occur to the reader that many characteristics ascribed to developing nations also apply to the pockets of poverty found in industrial nations—in Michigan's Upper Peninsula, or in Appalachia, for example. The conditions in these areas, however, are far less severe than those common in many of the developing nations.

Market or Command Economy?

Do nations with a market-based economy develop more rapidly than those with a command economy? This is partly a matter of opinion, partly a matter of the measuring system employed, and at least partly a matter of what is measured. The fact seems to be that both systems *can* work, but examples of failure are far too frequent in both types.[3] It appears probable that a nation's history and culture have something to do with whether a command or a market system will best promote development. It also seems that the *stage* of development is a factor. Certainly, there are times when it appears highly advantageous for a country to have central control of natural resources and investment funds and to have authority and ability to act promptly. But having centralized planning does not assure *good* planning, and ease of establishing priorities does not assure optimum selection of projects. Moreover, the ability to act quickly is apparently accompanied by an ability to make drastic mistakes. China had an enviable record of progress until the "great leap forward" was applied to agriculture during the late 1960s. Industrial progress was virtually halted during the 1970s. Now, after a traumatic interlude (and a pronounced move toward free markets), industrial development seems to be spurting ahead again. Like many economic phenomena, such aberrations seem easier to explain after their occurrence than to forecast. One thing does seem clear: one should avoid drawing conclusions about advantages gained through a particular system, especially when the concerned nation is a few thousand miles from the analyst.

Development Theory

Current theories of development are primarily the product of academicians in industrialized nations. Quite naturally, since economic development implies economic growth, development theory began by building on growth theories. Economic growth had been a topic of study for years, and since the 1930s had been pointed toward establishing conditions of stability. Prominent among the findings were those of Harrod and Domar, who in 1946 had independently derived relations between a nation's economic growth rate, the savings rate, investment, and the "capital/output ratio" (Harrod, 1948; Domar, 1946).

With Y = aggregate income, I = investment, s = the percent of income saved during a year, and k = the incremental capital/output ratio, (or amount of new capital invested/new production), it was shown that, since $sY = I$,

$$\Delta Y = \frac{I}{k'} \quad \text{or} \quad \Delta Y = \frac{sY}{k,} \quad \text{so} \quad \frac{\Delta Y}{Y} = \frac{s}{k}$$

which implies that if all the new investment (capital) is fully utilized, and if market demand is sufficient to assure purchase of all new output, then $\Delta Y/Y$ (the growth rate of the economy) equals the percent of income saved (and invested) divided by the capital/output ratio.

To see the impact of this relationship, suppose we were assisting a developing country with an annual population growth rate of 2.5% and were striving for a 3.5% growth in per capita income. To achieve this objective, our economy must grow at a 3.5% + 2.5%, or 6% annually. If the nation's new investments have a capital/output ratio of 4 (which means that each year's new investments will add about $\frac{1}{4}$ the amount invested to the national product), then

[3]Comparison between market and command economies is difficult, since the measures used (gross domestic product and gross material product) are not directly comparable.

$$6\% = \frac{s}{4'} \qquad \text{or} \qquad s = 24\%$$

This implies that the countries must *save* 24% of their income each year!!

It was widely recognized that these relationships were derived while searching for stability conditions in developed economies. Nevertheless, the theory focused attention on investment and on the capital/output ratio as important factors of economic growth and became a cornerstone of development economics for several years. It was quickly modified to bring in employment, the contribution to capital that might be made by foreign trade, by outside assistance, and similar factors. Dual economies, in which there are two widely different economic sectors (the industrial and the agricultural, for example) were brought into the picture. Technological change was introduced as a factor. Much was learned and progress began in understanding the process of economic development.

Along with these insights came a historic perspective: five stages of growth, including an analogy to an aircraft's takeoff, were identified in the late 1950s and widely accepted as a basis for study. From (1) a traditional society structured to maintaining existing production methods and institutions, nations eventually achieve (2) the preconditions for takeoff. This is followed by (3) the takeoff, then (4) the drive to maturity, and finally, (5) the age of high mass-consumption (Rostow, 1960).

Other developments soon followed. Debates in learned journals took place on whether growth should be balanced relative to economic sectors or proceed from growing points (leading sectors), whether funds might be borrowed from developed nations and repaid by the developments that would follow, and whether possibly a managed inflation in the currency of developing nations might not serve to repay the borrowed funds.

Central to all the development theories presented in Western economies were two assumptions: (1) steady, incremental adjustments would accompany the steady increments of growth in the developing world. Development meant change, but change at the margin, and (2) change was almost always defined as industrialization. The developed economies had once been agricultural and had developed through industrialization. This was the obvious path. Development would spread among nations and trickle down among classes so everyone would benefit from the process (Yotopoulos and Nugent, 1976).

In contrast to the gradual adjustment and change presumed by Western economists is the assumption underlying the theories supporting development efforts of the planned, but developed, economies—particularly that of the USSR. Marx was not a believer in marginalism. He preached a different sermon, one aimed at changing existing systems, but through revolution, not evolution and marginal change. Development, he claimed, is not a gradual process but one of disequilibrium which, unless corrected, will often lead to increasing differences in welfare.

Meanwhile, the developing nations have been generating ideas and questions of their own—such questions as "Do our people really want to copy the lifestyle of the Western materialistic societies? No doubt an increase in living standards is desired, and new ways of doing things would ease the hard lives of our rural people, but to what extent is it necessary to give up our culture? Your wonderful machines let each factory worker produce as much as 10 men did last year, so what do the other 9 do?"

Their workers, and scientists, are answering some of these questions, and in several developing nations new ways of solving old problems — ways that fit their own needs — are being created. Some of the most exciting of these are in forestry!

FORESTRY AND DEVELOPMENT

It would scarcely be surprising if the preceding comments on development theory left some forestry students unclear about the process and objectives of economic development. The intention, however, was not one of making development specialists, but to provide enough background for the reader to fit the discussion of forestry in developing nations into the general scheme of economic development. The literature of forestry's role, while not as formidable as that of development theory, is becoming quite extensive. Some of the best writings are still those of Westoby, whose work laid the foundation for most of the international efforts in this field.[4]

Forestry in Developing Nations

American forests and American culture, though differing in many aspects from that of Europe, appeared sufficiently similar to permit an easy transfer of methodology, so forestry education in the United States followed the European pattern without serious difficulties. But this does not hold for developing nations, where cultures are very different from ours, and whose forests are tropical rather than temperate. Management techniques common in Europe and America often require major modifications if they are to be applied successfully in developing regions, and some simply won't work! Now, after three decades or more, new approaches are being devised — approaches that hold promise of permitting forestry to contribute much more directly to solving some of the development problems. The general picture of world forestry was presented in Chapter 17; some of the significant details pertaining to developing nations are highlighted below.

Supply conditions are extremely variable. Over 50% of South America is covered with some type of forest, mostly tropical hardwoods of virtually unknown quality. Central and West Africa, Central America, and Southeast Asia have some magnificent stands of tropical hardwoods. Central America, particularly Nicaragua and Honduras, have substantial areas of native conifers, as does Mexico (although Mexico is scarcely a developing country in the sense used here). Southeast Asian forests, especially those of Indonesia and Malaysia, are particularly rich, with as much as 75% of the volume reported to be of commercial species.

Fast-growing plantations offer a very different kind of opportunity, providing conditions are right. Growth rates of 2, 3, and even 4 times the average obtained in most developed nations (including our Southern pine region) are often reported. Because land is frequently low-priced and available, some LDCs offer good opportunities for investment in pulp and paper manufacture. At the same time it should be pointed out that when large-scale forest plantation programs are instituted there is typically a displacement of an impoverished population that had been dependent on grazing or on shifting agriculture. The support of such local people during the early stages of plantation development, when labor requirements are relatively low, can be a major cost in such schemes.

---

[4]Jack C. Westoby provided continuing inspiration and guidance to FAO's Forestry and Forestry Industries Department from the early 1950s until his retirement in the early 1980s. The author gratefully acknowledges his own particular debt to this leader.

Markets

The part forestry and forest products will play in development depends considerably on markets. The traditional market relationship, established during the colonial epoch, was one in which raw materials were shipped from developing nations to developed, and finished materials from the developed to the developing. That the pattern still exists is evident from Chapter 17 (see Table 17.4). Japan, for example, is the major importer of forest raw materials, but imports a very small percentage of its finished products.

To counter this established flow, some developing nations restricted export of raw materials, insisting that manufacturing take place before export, but change in trade has been slow. In 1971 (for example) developing nations exported 95% of the world's total export of nonconiferous sawlogs, and in 1982 this had dropped to 92%. Export of nonconiferous sawnwood by developing nations over the same period climbed from 53% of the world total to 67%. Plywood exports from developing nations somewhat more than doubled between 1971 and 1982, while those from developed nations declined 15%.

Developing nations have two major reasons for wanting to increase exports: to earn foreign exchange and to increase employment opportunities. In past years the quickest way to increase foreign exchange from forest products was to export logs to developed nations. In effect, this was simply exchanging forest capital (which some tropical nations had in abundance) for foreign exchange that could be used to build schools, hospitals, and factories (and, in too many cases, supply television sets and Mercedes automobiles for a select few). Now new employment opportunities are becoming much more important.

A New Approach

Gradually, as more people in developing countries become professionally educated, the need for a change in development strategy became apparent. In part, this was due to the slow rate at which the fruits of industrialization were trickling down to the general population, and in part to a realization that perhaps the technologies they were importing from developed nations were not the most appropriate for their needs. Capital-intensive factories often seemed inappropriate for economies that were, and always had been, labor intensive in their production. Even more important to forestry was the realization that although exporting logs might bring in foreign exchange, it sometimes left forests in a condition that handicapped local inhabitants. And far more important was the gradual realization that a forest managed for industrial wood production would not supply local inhabitants with the forest materials essential to their welfare.

Probably the biggest factors in changing attitudes were the energy crisis and the realization that most rural people are uninterested in industrial forestry. Yet much of the nonindustrial forest—forest with trees of species, sizes, and qualities that would almost never make sawlogs or even decent pulpwood—serves two very important functions: (1) protecting land from both wind and water erosion, and (2) providing local inhabitants with a source of fuel, fruits, and oil seeds, along with agricultural implements, fencing, and fodder and shelter for livestock.

The result of this was the eventual reclassification of forests into three major types: production, protection, and "social" forests.[5] Production and protection

---

[5]Who first used the last term may never be determined. India's Forest Service began discussing Social Forestry around 1970, and the plans developed by the National Agricultural Commission made Production, Protection, and Social Forestry the three major divisions of the agency.

forests had long been recognized, although little thought had been given to managing protection forests. Social forestry, however, was a new concept, and between 1970 and 1985 it became the leading edge of forestry in developing nations.

Production forests are managed for commercial production of wood—sawtimber, pulpwood, or very occasionally, fuelwood. The management tools and techniques are essentially those used throughout the world in industrial forests, and the principle objectives are financial.

Protection forests are managed primarily for their protective capacity—as cover for watersheds, to prevent erosion on steep slopes or on easily erodible land, or as protection against the spread of deserts, and so on. But these forests must also satisfy the needs of rural inhabitants for wood products, fuelwood, building poles, fruit and nuts wherever possible, and also fodder for their livestock. Overgrazing is endemic and the demand for fuelwood had been increased by rising populations and the increased cost of fuel oil. Multiple-use management is an absolute necessity on most of these protection forests. Almost all are publicly owned, with vast areas in Africa, South America, and Asia held as common property either by nomadic herdspeople or by rural villagers. Many of these lands are encumbered with vested rights to grazing, the passage of herds, hunting, or fuelwood gathering.

The problems of managing such lands are many and difficult. Economic analysis is complicated by the necessity of treating nonmarket values, by the difficulties of identifying those who benefit, measuring the benefits, and (if costs are to be recovered) allocating the costs and devising an administratively feasible payment procedure. Almost every management activity is complicated by the area involved, the ingrained habits of the people living in the area, and most of all by the absence of alternatives. Regulations against fuelwood cutting or against grazing are easy to write and proclaim; they are virtually impossible to enforce unless reasonable alternatives are provided. An old political saying "You can't beat somethin with nothin!" emphasizes that a measure or candidate can't be defeated by pointing to defects or inadequacies—it is equally appropriate to protection forest management.

What makes this problem unique is its critical nature. The hill region of the Himalayas provides a dramatic example, and equally critical areas may be found in Africa, China, or South America. A few decades ago these Himalayan foothills (some over 12,000 feet high) were tree covered and the ecological stability achieved over a few thousand years was being maintained. Then, as agriculturists were crowded out of the plains, hill dwellers were pushed further and further up the mountainsides. In the late 1880s, as populations began growing rapidly, the situation deteriorated. By 1960 floods in India and Pakistan were becoming more common and reservoirs vital to irrigation began filling with silt. These were reservoirs providing water to the huge Indo-Gangetic Plain, the world's most productive agricultural area, which supplied food for half the subcontinent's population—around 500 million people. Most hill farmers are doubly dependent on forests: for food and for the fodder necessary to feed the only nonhuman power available to them, their bullock.

This area has turned out to be one of the world's most fragile environments. After 20 years of research the Swiss Association for Technical Assistance in Nepal reported that none of the approaches they had tested successfully reversed the ecological destruction—this from people who know mountains (Schmid, 1976). World Bank officials concluded, after reviewing five World Bank watershed protection projects, that in many areas adequate protection could be achieved only if authorities were willing to force people to move out of endangered areas or to spend the money necessary to carry out large-scale rehabilitation projects. The first would be politically difficult, and the second highly unlikely, since short-term development goals are usually given higher priority (Spears, 1982).

These are substantial technical problems, but the most formidable ones are social and economic: what does one substitute for fuelwood in a country with an average income of little more than $100 per year? Can one prevent local inhabitants from grazing their cattle or from defoliating the trees for fodder if this is their way of life? Reforestation would probably stop the erosion and hence the flooding in the plains below, but what sort of a social arrangement will transfer the needed funds from the agriculturists on the plains below (the ones who really benefit) to the hill people (who must change their traditional lifestyle and become forest dwellers)? How will the displaced rural inhabitants survive until the trees are large enough to provide some income? New species such as *Leucaena glauca, Albezia,* and *Morus* spp. hold much promise, since they (particularly *Leucaena*) are genuine dual purpose trees, providing excellent fodder and good fuelwood in addition to being exceptionally fast growers. But there are no universal answers and no miracle trees. Each site has its own potential, each species its own site requirements, and each project necessitates not only that a suitable species/site combination be discovered, but also that the needs of the rural inhabitants be satisfied.

Protection forests are critically important, but we have much to learn about their management. Integrated projects will be needed in many situations. The five World Bank projects referred to previously were in different parts of the world and involved a total expenditure of over $150 million. Review of these projects revealed that expenditures on forestry and soil conservation averaged only a third of the total cost. Agriculture, irrigation, farm improvements, and extension and training inputs often accounted for a significantly higher proportion of cost than did forestry and soil conservation work. Grass and tree combinations appear to be more effective than reforestation alone; grass results in lower water losses through evapotranspiration and provide grazing areas as well (Spears, 1982).

Social Forestry

Often defined as forestry for people in the developing nations, social forestry represents a dramatic new approach to the problem of getting communities and individuals involved in solving their fuelwood and fodder problems. It is clearly related also to the protection forest problem. China began local involvement with its "four square" program—plant trees around villages, around houses, along roads, and along canals. India's Forest Service had long planted trees along roads, railroads, and canals. Social forestry, however, meant taking forestry into the grade schools and into India's villages. Progress has been most rapid in areas where foresters first took time to discover what the villagers felt they needed from the forest: fuelwood, grass and fodder leaves, fruit, shade, building poles—but not sawtimber and pulpwood. They also discovered that to involve the villagers it was first necessary to prove to them that a reforested area could be productive—more productive than any of their current pasture land—quickly. And villagers meant next year, not in 10 or 15 years!

To almost everyone's amazement, India's foresters are succeeding in many places. Where they fail it is generally through lack of good understanding and cooperation between foresters and the local inhabitants. To achieve the needed cooperation, foresters had to have very different training from what most had normally received.[6]

[6]S. A. Shah, Chief Conservator of Gujarat, where social forestry has made great progress, doubts that regularly trained foresters can carry out this job; they are "too indoctrinated with timber growing" (from personal conversations with the author). See Schmithusen (1983) and Gregersen (1979).

Leaders of a major Nepal project point out that agricultural extension—often the adopted model—can succeed as long as a few farmers are convinced that a new crop or new method is worth trying. If their trials prove successful, neighbors will follow their lead, and in a few years substantial changes can occur. But community forestry is different: success can be achieved only if a consensus has first been established among the villagers and a concerted effort is made by the entire community. "It is not enough for a handful of dedicated villagers to plant tree seedlings in a communal forest if these are going to be trampled the next day by cattle being grazed by other villagers" (Pelinck, Manandhar, and Gecolea, 1984).

Agroforestry is forestry *integrated with* farming. It is not tree farming, for the objective is to get farmers to view forestry and trees as a *complement* to agriculture, not as an alternative land use. Agroforestry, like social forestry (of which it is usually considered a part) also had its origin in developing nations. The term is used most where the pressure for food production is so great that virtually any land that can produce some sort of food crop is converted to farmland. The result can be seen most vividly in Africa south of the Sahara, where (according to the FAO survey of tropical forests) in 1980 some 50 million people no longer have enough fuelwood to cook their food or heat their homes. Moreover, an additional 130 million rural people were getting their fuel only by overexploiting the surrounding vegetation. These 180 million will become 500 million by the end of the century. In these areas it is essential to integrate forestry with agriculture, for the solution to the genuinely critical problem cannot come from governments: it must be accomplished by the people themselves.

An initial beginning can be made by planting multipurpose trees—usually fruit or nut trees or fodder trees—around houses and in villages and along paths and roads. The fact that much of the land is open access property raises major protection difficulties, for fuelwood has traditionally been available to anyone who would take it, and virtually all land is open to grazing.

Taungya plantations techniques may also be of help, but are only applicable in areas where permanent agriculture is being practiced and require dedication of at least some area to tree production.[7] Shelterbelts have also proved worthwhile, although few studies of their economic value have been made.

One of the more promising developments has been the emergence of nongovernment organizations, the NGOs, that can sometimes produce results when official efforts fail. Farm cooperatives that are already established and are well accepted provide a ready-made channel for extension efforts, and in some instances have started "greening" movements that have been highly successful. The NGOs are often viewed with some suspicion by government agencies, but with problems as intractable and as serious as those posed in Africa, Asia, and in other parts of the tropics all possible avenues should be explored.

PROJECT ANALYSIS IN DEVELOPING COUNTRIES

In general, forestry projects for developing countries are formulated and analyzed using the same benefit/cost format and much of the same methodology as that presented in Chapters 11 and 14. Costs, revenues, and the appropriate technical aspects of a project can be very different in developing nations (i.e., one seldom uses elephants when logging in the United States), but if the project is in the private sector the analytic procedures are virtually identical to those in Chapter 11. How-

[7]The Taungya method apparently originated in Burma and India, when plantations to replace the original forest after harvesting were established. A farmer is permitted to use a specific area for crop production, but must also plant trees on the same land. Trees are planted in rows that are spaced 3 or 4 meters apart, and the crop is planted between the rows. After three or four years the tree shade prevents further cropping, and the farmer is moved to another area.

ever, since forests in most developing nations are publicly owned, *economic* analyses are often required whether the project is financed by a private corporation (either local or foreign), by the national government, an international agency such as FAO or the World Bank, or through some bilateral aid program such as AID, SIDA, or CIDA.[8] Like projects everywhere, development projects require *financial* analysis, often in substantial detail, for the needed money must be budgeted and approved somewhere. Because of the public nature of most forestry projects, the cost and benefits must also be analyzed from the standpoint of public values. Hence the second type of analysis required for most development projects is an economic analysis in which costs and benefits are determined from a public rather than a strictly financial standpoint. The general framework needed for this economic analysis was that presented in Chapter 14, and the general procedure is usually termed social benefit/cost analysis.

The primary objective of social benefit/cost analysis is usually taken to be maximization of aggregate consumption, already defined in Chapter 13. Because unemployment and underemployment are widespread in developing nations, employment per se is often an important objective. Moreover, projects may be sited in particular localities so regional development and regional incomes will be enhanced, and at times the most important objective may be that of introducing a new methodology, a new product, or a new manufacturing process. Any of these additional objectives can necessitate some modification in the analysis.

The Food and Agriculture Organization of the United Nations (FAO) has been the principal implementing agent for projects funded by the World Bank, and the two have cooperated in pioneering application of benefit/cost analysis to forestry projects in developing nations.[9] If these multilateral organizations are to fulfill their role in assisting developing countries, the countries must prepare and submit projects in a way permitting project evaluation and comparison, for development funds are very limited. Bilateral development agencies (who generally seek country-to-country projects) have shown less interest in applied economic analysis, but if one reviews past efforts of agencies such as USAID, SIDA, CIDA, and ODA, their need for such analyses appears to be at least as great.

The work for FAO by Gregersen and Contreras on project analysis for forestry projects (1979), by Busby on financial analysis of tree growing (1985), by Gray on revenue systems (1983), and the many works of Arnold, Wardle, Pringle, and other regular staff members has been of great help. Several developing nations now have forestry units within their planning commissions that are able to carry out project formulation and evaluation; others work almost independently, although it is gradually being recognized that forest projects must fit into a nation's economic program for development if it is to be implemented.

Two quite different approaches to these problems have been developed; the primary difference between them is the basis for measuring costs and benefits. One, the Little-Mirrlees approach, uses border prices (or world prices) for determining both costs and benefits (Little and Mirrlees, 1969; 1974). Border prices are

---

[8]AID is the U.S. Agency for International Development, SIDA is the Swedish International Development Agency, and CIDA the Canadian International Development Agency. Most of the developed nations have such agencies. Sweden, and to a lesser extent Canada, have both tended to concentrate their development assistance programs on forestry projects. Until the late 1970s AID virtually ignored forestry.

[9]FAO has also been the implementing agency for many bilateral aid forestry programs, particularly those sponsored by Sweden and Canada.

readily determined for traded goods, but determining costs and benefits for untraded goods becomes more difficult. These require one to estimate accounting prices, defined as the marginal social cost of the particular good, and may require estimating both the demand and supply curves for the goods concerned. Whether a good is to be treated as one traded or one untraded is sometimes considerably more difficult to determine in practice than in theory. The accounting price for labor and for investment funds pose more difficult problems.

A different procedure is used by UNIDO (the United Nations Industrial Development Organization) and is the basis for project analysis by FAO, the World Bank, and many of the governments of developing nations (Dasgupta, Sen, and Marglin, 1972). It differs from the Little-Mirrlees method primarily in the basis for calculating social costs and benefits, and ties back to a different way of looking at the principle objective. Little and Mirrlees use "uncommitted social income measured at border prices," while the UNIDO Guidelines use "aggregate consumption measured at domestic market prices." They both identify foreign exchange, savings, and unskilled labor as requiring correction, and both employ essentially similar techniques to carry out the corrections. Both use a discounted cash flow analysis and a social discount rate, and recommend making adjustments for inequality in income distribution (Little and Mirrlees, 1974). So far as this author can determine, a decision to accept or reject a project by one method would be supported by the other.

The UNIDO approach will be followed here, although details of specific calculations will not be presented. Three particular shadow prices almost always must be determined: those for labor, for goods requiring foreign exchange, and for investment funds. Formulas for their calculation are presented in the Guidelines, along with some explanation as to why they are necessary. The social discount rate will be assumed given by the planning agency. A full derivation of the formulas and discussion of alternative methods, however, is not believed appropriate here. And in keeping with the objectives of this text, the presentation of benefit/cost analysis in developing countries will be through an example—ideally, a fitting way to end this chapter and book.

DONDELA EUCALYPTUS PULP AND SAWLOG PLANTATION PROJECT

Dondela is an imaginary developing nation in the tropical-semitropical part of the world. Granted independence following World War II, Dondela has had less than average success in its development efforts. It is currently classed by the World Bank as a low income country, has an average per capita gross domestic product of around $350 per year and a population of 15 million, increasing about 2.5% annually. The official exchange rate of its currency, the Kruzero (Kz), is .085 dollars and has been reasonably stable for the past three years. Eighty percent of its population is rural, living by traditional subsistence agriculture. Cash crops—particularly bananas, coffee, and cashew nuts—provide some foreign exchange earnings, but the country has a growing negative balance of trade.

Recently, a World Bank loan enabled the government to construct a large earth-fill dam in the foothills area. The reservoir permits irrigating a major part of the lower plain area: agricultural experts say crop production in the irrigated area can be tripled and Dondela can become a food exporter. Additionally, the power produced can be used to initiate industrialization. The National Planning Commission (NPC) had just started to consider the kinds of industries that might be appropriate and where they might locate, when a new problem surfaced: erosion from the hill region is producing heavy silt deposits in the reservoir; unless something can be done, and soon, the reservoir will become useless in 25 years.

The Dondela Forest Service (DFS), which controls all forested land in the nation, suggests a major plantation project as an ideal solution, and lists the following reasons:

> The reservoir watershed area is well suited to growing eucalyptus, a species with which the DFS has been experimenting for some 20 years.
>
> A forest cover will effectively stop existing erosion, and eucalyptus will provide the needed cover quickly.
>
> Rural inhabitants are suffering from an acute fuelwood and building material shortage.
>
> The country presently imports all its paper at high foreign exchange costs.
>
> Lumber prices have tripled during the past 15 years and sawtimber is scarce.
>
> The watershed region is one of the most severely depressed areas in all Dondela with high unemployment and underemployment. Forest plantations can use substantial amounts of unskilled labor.
>
> The plantation project would provide the needed raw material for a pulp and paper mill to be located in the area: wood, power, water and labor would be available, creating a potential industrial development center for the region.

The Planning Commission is enthusiastic about the possibilities and requests the DFS to prepare a project for their study and evaluation.

DFS surveys establish the availability of at least 100,000 hectares (247,000 acres). Almost all of this is now heavily grazed, and fuelwood gatherers have left only stubs of trees and brush. Grazing and cutting are completely uncontrolled. No cities exist in the area, but there are many small villages. Unemployment and underemployment is indeed high—official figures show a minimum unemployment level of over 30%, with most of the workers classed as landless laborers.

DFS foresters decide on a 15-year rotation. Fifteen hundred hectares will be planted each year for 15 years, with the anticipated per hectare yields shown in Figure 18.1.[10] Fuelwood will be available in five years, and by the end of the fifteenth year's growing season the plantation will support a sawmill cutting 60 million bd. ft. per year, a pulp and paper plant producing 100,000 tons per year, over 70,000 m^3 of fuelwood, and 50,000 m^3 of building poles. All in all, 22,500 hectares of plantation will be established, located with two objectives in mind—ease of transport to the proposed mill site and erosion control. The biggest problems will probably be persuading the local inhabitants that eucalyptus—a new species to most—will not make an ecological disaster area of their surroundings, and getting them to control their wandering cattle and goat herds. For these reasons the plantation will not be established in a single block, but scattered over much of the 100,000 hectares. Additionally, a separate but companion project is being proposed for planting Leucaena in badly gullied areas. This would provide rapid erosion control, and at the same time, supply fodder for farm animals and supplement fuelwood needs in the early years of the eucalyptus project.

[10]Yield data is based on experience gained from seeing and working with eucalyptus in several areas of the world. The yields used are above average, but well below those obtained in many countries.

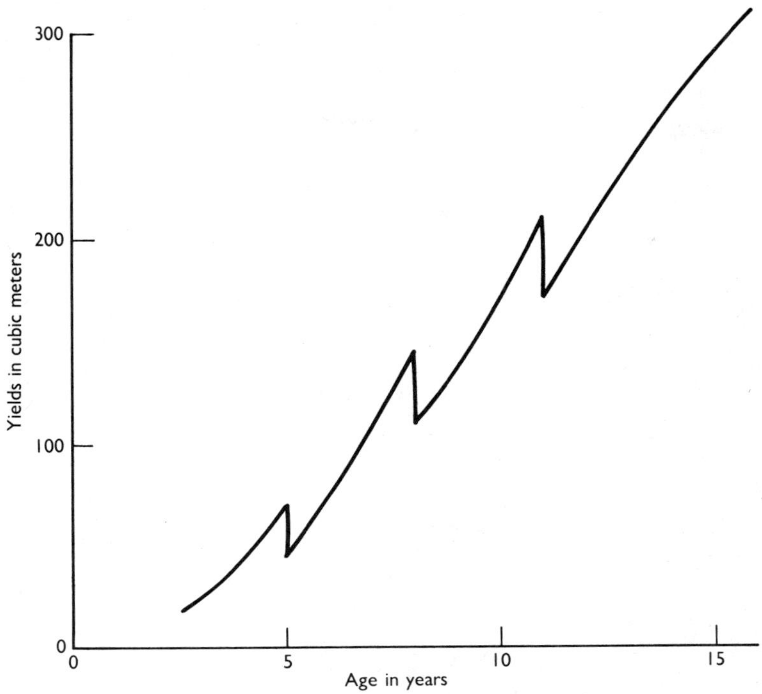

FIGURE 18.1 Yields, in m³, for Dondela plantation project.

The Financial Analysis

The financial analysis of the eucalyptus project has been prepared with the product yield and price assumptions, skilled and unskilled labor requirements, wage rates, and domestic and foreign material and equipment costs, based in part on those used by FAO (Busby, 1985). Table 18.1 summarizes these by year. Note that all costs are on a per hectare basis—for total expenditures these must be multiplied, each year, by the area involved—1500 hectares planted each year over the 15-year period.

In somewhat similar fashion, after the costs have been totaled for each year of the project's life (Table 18.2), the total costs (line 20) and total revenues (line 39) are for a 15-hectare sample, where each hectare goes through every stage of the 45-year sequence.

Past experience in Dondela and nearby countries shows that eucalyptus produces at least two coppice rotations after the first (fifteenth year) harvest, and that yields the second and third rotations will be at least 15% higher than from the seedlings. Revenues from these latter rotations will therefore be increased accordingly, and no planting costs will be involved after the fifteenth year. The nursery will be abandoned after year 15, with no salvage value. Thinnings will be made at age 5, 8, and 11 on each rotation. The final harvest will come at the 59th year (one year is gained the first coppice rotation, since no planting period is needed). No land cost is incurred since by law the area must be devoted to forestry.

The complete cost and revenue picture for the 60-year period is given in the computer printout of Table 18.2 Note that the nursery establishment cost is not shown separately in this table, but is included as a cost in year −1 and carried ahead

TABLE 18.1
Labor, Materials, and Equipment Needs, and Yields in m^3 and Kz

Year	-1	0	1	2	3	4	5	6	7
Labor, skilled	90	473	128	96	96	96	96	128	96
Labor, unskilled	840	2848.4	770.5	456.5	456.5	456.5	456.5	566.5	456.5
Materials	500	460	110	160	160	160	160	170	160
Domestic equipment	50	334	48	26	26	26	26	34	26
Foreign equipment		485	35	35	35	35	35	35	35
Harvest cuts									
Labor, skilled							30		
Labor, unskilled							144		
Materials							30		
Domestic equipment							40		
Foreign equipment							50		
Total costs	1480	4600.4	1091.5	773.5	773.5	773.5	1067.5	933.5	773.5
Percent unskilled	.5675676	.6191636	.7059093	.5901745	.5901745	.5901745	.5625293	.6068559	.5901745
Average percent of total cost due to unskilled labor					.5838204				
Yields in cu. m.									
Fuelwood							12		
Pulpwood							10		
Building poles							2		
Sawlogs							0		
							24		
Revenues	Product Price								
Fuelwood	25						300		
Pulpwood	40						400		
Building poles	75						150		
Sawlogs	125						0		
Total revenues							850		

TABLE 18.1 *(continued)*
Labor, Materials, and Equipment Needs, and Yields in m³ and Kz

	Year	8	9	10	11	12	13	14	15
2									
3	Labor, skilled	96	96	96	96	96	96	96	96
4	Labor, unskilled	456.5	456.5	456.5	456.5	456.5	456.5	456.5	456.5
5	Materials	160	160	160	160	160	160	160	160
6	Domestic equipment	26	26	26	26	26	26	26	26
7	Foreign equipment	35	35	35	35	35	35	35	35
8									
9	Harvest cuts								
10	Labor, skilled	30			30				60
11	Labor, unskilled	144			156				240
12	Materials	30			40				80
13	Domestic equipment	40			40				50
14	Foreign equipment	30			30				60
15									
16	Total costs	1047.5	773.5	773.5	1069.5	773.5	773.5	773.5	1263.5
17									
18	Percent unskilled	.5732697	.5901745	.5901745	.4268350	.5901745	.5901745	.5901745	.5512465
19	Average percent of total cost due to unskilled labor								
20									
21									
22									
23									
24	Yields in cu m.								
25									
26	Fuelwood	12			12				12
27	Pulpwood	13			13				164
28	Building poles	10			10				15
29	Sawlogs	0			5				100
30									
31	Revenues	35							291
32	Fuelwood	300			300				300
33	Pulpwood	520			520				6560
34	Building poles	750			750				1125
35	Sawlogs	0			625				12500
36									
37	Total revenues	1570			2195				20485
38									

TABLE 18.2
Financial Analysis of Dondela Plantation Project

		-1	0	1	2	3	4	5	6	7	8	9	10	11
1	Year	-1	0	1	2	3	4	5	6	7	8	9	10	11
2														
3	Nursery	25	27	29	31	34	37	40	43	46	50	54	58	63
4	First Planting		4600.4	1091.5	773.5	773.5	773.5	1067.5	933.5	773.5	1047.5	773.5	773.5	1069.5
5	Second Planting			4600.4	1091.5	773.5	773.5	773.5	1067.5	933.5	773.5	1047.5	773.5	773.5
6	Third Planting				4600.4	1091.5	773.5	773.5	773.5	1067.5	933.5	773.5	1047.5	773.5
7	Fourth Planting					4600.4	1091.5	773.5	773.5	773.5	1067.5	933.5	773.5	1047.5
8	Fifth Planting						4600.4	1091.5	773.5	773.5	773.5	1067.5	933.5	773.5
9	Sixth Planting							4600.4	1091.5	773.5	773.5	773.5	1067.5	933.5
10	Seventh Planting								4600.4	1091.5	773.5	773.5	773.5	1067.5
11	Eighth Planting									4600.4	1091.5	773.5	773.5	773.5
12	Ninth Planting										4600.4	1091.5	773.5	773.5
13	Tenth Planting											4660.4	1091.5	773.5
14	Eleventh Planting												4600.4	1091.5
15	Twelfth Planting													4600.4
16	Thirteenth Planting													
17	Fourteenth Planting													
18	Fifteenth Planting													
19														
20	Total Costs		4627.4	5720.9	6496.4	7272.9	8049.4	9119.9	10056.4	10832.9	11884.4	12661.9	13439.4	14513.9
21														
22	Revenues													
23	First Planting							850			1570			2195
24	Second Planting								850			1570		
25	Third Planting									850			1570	
26	Fourth Planting										850			1570
27	Fifth Planting											850		
28	Sixth Planting												850	
29	Seventh Planting													850
30	Eighth Planting													
31	Ninth Planting													
32	Tenth Planting													
33	Eleventh Planting													
34	Twelfth Planting													
35	Thirteenth Planting													
36	Fourteenth Planting													
37	Fifteenth Planting													
38														
39	Total revenues		0	0	0	0	0	850	850	850	2420	2420	2420	4615
40	Revenues up 15%							978	978	978	2783	2783	2783	5307
41														
42	NPV total rev., 6%	198450.3		NPV total costs 6%		189956.5		Net benefits at 6%=			8493.857			
43	NPV total rev., 8%	121443.7		NPV total costs 8%		142885.9		Net benefits at 8%=			-21442.2			
44		154166.1				159082.2		Net benefits at 7%=			-4916.06		IRR=6.7	
46	Increased rev. by 15%							With prices up 15% net ben at 8%=-3225.62						
47	NPV of new rev, 8%	139660.3									at 7%=18208.86		IRR=7.5	
48	NPV of new rev, 7%	177291.1												
49														

TABLE 18.2 *(continued)*
Financial Analysis of Dondela Plantation Project

Year	12	13	14	15	16	17	18	19	20	21	22	23
Nursery	68	73	79									
First Planting	773.5	773.5	773.5	1263.5	1091.5	773.5	773.5	773.5	1067.5	933.5	773.5	1047.5
Second Planting	1069.5	773.5	773.5	773.5	1263.5	1091.5	773.5	773.5	773.5	1067.5	933.5	773.5
Third Planting	773.5	1069.5	773.5	773.5	773.5	1263.5	1091.5	773.5	773.5	773.5	1067.5	933.5
Fourth Planting	773.5	773.5	1069.5	773.5	773.5	773.5	1263.5	1091.5	773.5	773.5	773.5	1067.5
Fifth Planting	1047.5	773.5	773.5	1069.5	773.5	773.5	773.5	1263.5	1091.5	773.5	773.5	773.5
Sixth Planting	773.5	1047.5	773.5	773.5	1069.5	773.5	773.5	773.5	1263.5	1091.5	773.5	773.5
Seventh Planting	933.5	773.5	1047.5	773.5	773.5	1069.5	773.5	773.5	773.5	1263.5	1091.5	773.5
Eighth Planting	1067.5	933.5	773.5	1047.5	773.5	773.5	1069.5	773.5	773.5	773.5	1263.5	1091.5
Ninth Planting	773.5	1067.5	933.5	773.5	1047.5	773.5	773.5	1069.5	773.5	773.5	773.5	1263.5
Tenth Planting	773.5	773.5	1067.5	933.5	773.5	1047.5	773.5	773.5	1069.5	773.5	773.5	773.5
Eleventh Planting	773.5	773.5	773.5	1067.5	933.5	773.5	1047.5	773.5	773.5	1069.5	773.5	773.5
Twelfth Planting	1091.5	773.5	773.5	773.5	1067.5	933.5	773.5	1047.5	773.5	773.5	1069.5	773.5
Thirteenth Planting	4600.4	1091.5	773.5	773.5	773.5	1067.5	933.5	773.5	1047.5	773.5	773.5	1069.5
Fourteenth Planting		4600.4	1091.5	773.5	773.5	773.5	1067.5	933.5	773.5	1047.5	773.5	773.5
Fifteenth Planting			4600.4	1091.5	773.5	773.5	773.5	1067.5	933.5	773.5	1047.5	773.5
Total Costs	15292.4	16070.9	16850.4	13434.5	13434.5	13434.5	13434.5	13434.5	13434.5	13434.5	13434.5	13434.5
Revenues												
First Planting				20485					978			1805
Second Planting	2195				20485					978		
Third Planting		2195				20485					978	
Fourth Planting			2195				20485					978
Fifth Planting	1570			2195				20485				
Sixth Planting		1570			2195				20485			
Seventh Planting			1570			2195				20485		
Eighth Planting	850			1570			2195				20485	
Ninth Planting		850			1570			2195				20485
Tenth Planting			850			1570			2195			
Eleventh Planting				850			1570			2195		
Twelfth Planting					850			1570			2195	
Thirteenth Planting						850			1570			2195
Fourteenth Planting							850			1570		
Fifteenth Planting								850			1570	
Total revenues	4615	4615	4615	25100	25100	25100	25100	25100	25228	25228	25228	25463
Revenues up 15%	5307	5307	5307	28865	28865	28865	28865	28865	29012	29012	29012	29282

TABLE 18.2 *(continued)*
Financial Analysis of Dondela Plantation Project

1	Year	24	25	26	27	28	29	30	31	32	33	34	35
2													
3	Nursery												
4	First Planting	773.5	773.5	1069.5	773.5	773.5	773.5	1263.5	1091.5	773.5	773.5	773.5	1067.5
5	Second Planting	1047.5	773.5	773.5	1069.5	773.5	773.5	773.5	1263.5	1091.5	773.5	773.5	773.5
6	Third Planting	773.5	1047.5	773.5	773.5	1069.5	773.5	773.5	773.5	1263.5	1091.5	773.5	773.5
7	Fourth Planting	933.5	773.5	1047.5	773.5	773.5	1069.5	773.5	773.5	773.5	1263.5	1091.5	773.5
8	Fifth Planting	1067.5	933.5	773.5	1047.5	773.5	773.5	1069.5	773.5	773.5	773.5	1263.5	1091.5
9	Sixth Planting	773.5	1067.5	933.5	773.5	1047.5	773.5	773.5	1069.5	773.5	773.5	773.5	1263.5
10	Seventh Planting	773.5	773.5	1067.5	933.5	773.5	1047.5	773.5	773.5	1069.5	773.5	773.5	773.5
11	Eighth Planting	773.5	773.5	773.5	1067.5	933.5	773.5	1047.5	773.5	773.5	1069.5	773.5	773.5
12	Ninth Planting	1091.5	773.5	773.5	773.5	1067.5	933.5	773.5	1047.5	773.5	773.5	1069.5	773.5
13	Tenth Planting	1263.5	1091.5	773.5	773.5	773.5	1067.5	933.5	773.5	1047.5	773.5	773.5	1069.5
14	Eleventh Planting	773.5	1263.5	1091.5	773.5	773.5	773.5	1067.5	933.5	773.5	1047.5	773.5	773.5
15	Twelfth Planting	773.5	773.5	1263.5	1091.5	773.5	773.5	773.5	1067.5	933.5	773.5	1047.5	773.5
16	Thirteenth Planting	773.5	773.5	773.5	1263.5	1091.5	773.5	773.5	773.5	1067.5	933.5	773.5	1047.5
17	Fourteenth Planting	1069.5	773.5	773.5	773.5	1263.5	1091.5	773.5	773.5	773.5	1067.5	933.5	773.5
18	Fifteenth Planting	773.5	1069.5	773.5	773.5	773.5	1263.5	1091.5	773.5	773.5	773.5	1067.5	933.5
19													
20	Total Costs	13434.5	13434.5	13434.5	13434.5	13434.5	13434.5	13434.5	13434.5	13434.5	13434.5	13434.5	13434.5
21													
22	Revenues												
23	First Planting			2524				23558					978
24	Second Planting	1805			2524				23558				
25	Third Planting		1805			2524				23558			
26	Fourth Planting			1805			2524				23558		
27	Fifth Planting	978			1805			2524				23558	
28	Sixth Planting		978			1805			2524				23558
29	Seventh Planting			978			1805			2524			
30	Eighth Planting				978			1805			2524		
31	Ninth Planting					978			1805			2524	
32	Tenth Planting	20485					978			1805			2524
33	Eleventh Planting		20485					978			1805		
34	Twelfth Planting			20485					978			1805	
35	Thirteenth Planting				20485					978			1805
36	Fourteenth Planting	2195				20485					978		
37	Fifteenth Planting		2195				20485					978	
38													
39	Total revenues	25463	25463	25792	25792	25792	25792	28865	28865	28865	28865	28865	28865
40	Revenues up 15%	29282	29282	29661	29661	29661	29661	33195	33195	33195	33195	33195	33195
41													
42													
43													
44													
45													
46													
47													
48													
49													

TABLE 18.2 *(continued)*
Financial Analysis of Dondela Plantation Project

1	Year	36	37	38	39	40	41	42	43	44	45	46	47
2													
3	Nursery												
4	First Planting	933.5	773.5	1047.5	773.5	773.5	1069.5	773.5	773.5	773.5	1263.5		
5	Second Planting	1067.5	933.5	773.5	1047.5	773.5	773.5	1069.5	773.5	773.5	773.5	1263.5	
6	Third Planting	773.5	1067.5	933.5	773.5	1047.5	773.5	773.5	1069.5	773.5	773.5	773.5	1263.5
7	Fourth Planting	773.5	773.5	1067.5	933.5	773.5	1047.5	773.5	773.5	1069.5	773.5	773.5	773.5
8	Fifth Planting	773.5	773.5	773.5	1067.5	933.5	773.5	1047.5	773.5	773.5	1069.5	773.5	773.5
9	Sixth Planting	1091.5	773.5	773.5	773.5	1067.5	933.5	773.5	1047.5	773.5	773.5	1069.5	773.5
10	Seventh Planting	1263.5	1091.5	773.5	773.5	773.5	1067.5	933.5	773.5	1047.5	773.5	773.5	1069.5
11	Eighth Planting	773.5	1263.5	1091.5	773.5	773.5	773.5	1067.5	933.5	773.5	1047.5	773.5	773.5
12	Ninth Planting	773.5	773.5	1263.5	1091.5	773.5	773.5	773.5	1067.5	933.5	773.5	1047.5	773.5
13	Tenth Planting	773.5	773.5	773.5	1263.5	1091.5	773.5	773.5	773.5	1067.5	933.5	773.5	1047.5
14	Eleventh Planting	1069.5	773.5	773.5	773.5	1263.5	1091.5	773.5	773.5	773.5	1067.5	933.5	773.5
15	Twelfth Planting	773.5	1069.5	773.5	773.5	773.5	1263.5	1091.5	773.5	773.5	773.5	1067.5	933.5
16	Thirteenth Planting	773.5	733.5	1069.5	773.5	773.5	773.5	1263.5	1091.5	773.5	773.5	773.5	1067.5
17	Fourteenth Planting	1047.5	773.5	773.5	1069.5	773.5	773.5	773.5	1263.5	1091.5	773.5	773.5	773.5
18	Fifteenth Planting	773.5	1047.5	773.5	773.5	1069.5	773.5	773.5	773.5	1263.5	1091.5	773.5	773.5
19													
20	Total Costs	13434.5	13434.5	13434.5	13434.5	13434.5	13434.5	13434.5	13434.5	13434.5	13434.5	12343	11569.5
21													
22	Revenues												
23	First Planting			1805			2524				23558		
24	Second Planting	978			1805			2524				23558	
25	Third Planting		978			1805			2524				23558
26	Fourth Planting			978			1805			2524			
27	Fifth Planting				978			1805			2524		
28	Sixth Planting					978			1805			2524	
29	Seventh Planting	23558					978			1805			2524
30	Eighth Planting		23558					978			1805		
31	Ninth Planting			23558					978			1805	
32	Tenth Planting				23558					978			1805
33	Eleventh Planting	2524				23558					978		
34	Twelfth Planting		2524				23558					978	
35	Thirteenth Planting		978	2524		1805		23558	2524				978
36	Fourteenth Planting	1805			2524				23558				
37	Fifteenth Planting		1805			2524				23558			
38													
39	Total revenues	28865	29843	28865	28865	30670	28865	28865	31389	28865	28865	28865	28865
40	Revenues up 15%	33195	34319	33195	33195	35271	33195	33195	36097	33195	33195	33195	33195
41													
42													
43													
44													
45													
46													
47													
48													
49													

TABLE 18.2 *(continued)*
Financial Analysis of Dondela Plantation Project

1 Year	48	49	50	51	52	53	54	55	56	57	58	59
2												
3 Nursery												
4 First Planting												
5 Second Planting												
6 Third Planting												
7 Fourth Planting	12 63.5											
8 Fifth Planting	773.5	1263.5										
9 Sixth Planting	773.5	773.5	1263.5									
10 Seventh Planting	773.5	773.5	773.5	1263.5								
11 Eighth Planting	1069.5	773.5	773.5	773.5	1263.5							
12 Ninth Planting	773.5	1069.5	773.5	773.5	773.5	1263.5						
13 Tenth Planting	773.5	773.5	1069.5	773.5	773.5	773.5	1263.5					
14 Eleventh Planting	1047.5	773.5	773.5	1069.5	773.5	773.5	773.5	1263.5				
15 Twelfth Planting	773.5	1047.5	773.5	773.5	1069.5	773.5	773.5	773.5	1263.5			
16 Thirteenth Planting	933.5	733.5	1047.5	773.5	773.5	1069.5	773.5	773.5	773.5	1263.5		
17 Fourteenth Planting	1067.5	933.5	773.5	1047.5	773.5	773.5	1069.5	773.5	773.5	773.5	1263.5	
18 Fifteenth Planting	773.5	1067.5	933.5	773.5	1047.5	773.5	773.5	1069.5	773.5	773.5	773.5	1263.5
19												
20 Total Costs	10796	10022.5	8955	8021.5	7248	6200.5	5427	4653.5	3584	2810.5	2037	1263.5
21												
22 Revenues												
23 First Planting						5000						4000
24 Second Planting							5000					3000
25 Third Planting								5000				2000
26 Fourth Planting	23558								5000			500
27 Fifth Planting		23558								5000		
28 Sixth Planting			23558								5000	
29 Seventh Planting				23558								5000
30 Eighth Planting	2524				23558							4500
31 Ninth Planting		2524				23558						3000
32 Tenth Planting			2524				23558					2500
33 Eleventh Planting	1805			2524				23558				2000
34 Twelfth Planting		1805			2524				23558			1000
35 Thirteenth Planting			1805			2524				23558		500
36 Fourteenth Planting	978			1805			2524				23558	
37 Fifteenth Planting		978			1805			2524				23558
38												
39 Total revenues	28865	28865	27887	27887	27887	31082	31082	31082	28558	28558	28558	51558
40 Revenues up 15%	33195	33195	32070	32070	32070	35744	35744	35744	32842	32842	32842	59292
41												
42												
43												
44												
45												
46												
47												
48												
49												

at the selected interest rate, then added to the NPV calculation for the 15-hectare sample.

The pulp and paper plant is to be an entirely separate project — it will require very different funding and management and comes under the Agency for Industrial Development, which is enthusiastic at the prospect. The DFS emphasizes, however, that construction of the pulp and paper plant should begin no later than year seven of the plantation project, and notes that there is ample market for the timber if the paper mill is not feasible, and that no planning has been done for pulp supplies after the second coppice rotation since the proposed pulp and paper plant will then be at least 45 years old. When an official comments that 100,000 tons of paper seems a huge amount, the DFS economist observes that this is only slightly more than Dondela presently imports, and that it would provide little more than 2% of the average U.S. per capita paper consumption.

The Planning Commission is pleased with the outcome of the financial analysis, although a little disappointed that the project fails to show a positive return at an 8% discount; the IRR of just 6.6% might be a handicap if they seek outside financial assistance. They agree that the prices used are ultraconservative: the 15% price increase and resulting IRR of almost 8% makes the project more attractive. However, they also want an economic analysis. If that is favorable, they are sure they can fund a major part of the project internally, and external assistance can then be obtained easily. The planning commission agrees to supply the needed shadow prices.

The Economic Analysis

Planning commission analysts say that the 12 Kz wage rate was correct for the financial analysis, but unemployment is so high in the hill region that a shadow wage rate should be used in any economic analysis. Landless laborers there currently get about 4 Kz per day — when they can find work. But because their savings will be small even at a wage of 12 Kz, and because as the project progresses some workers will undoubtedly move into the area (thus incurring moving costs), the social cost of a worker is more than his usual 4 Kz per day. The direct cost to society of hiring an unskilled laborer is the marginal product of the worker in his or her private employment; this is what society gives up by putting the worker on the public payroll. But there are indirect costs as well — particularly the loss of investment funds that might otherwise be used to improve consumption at a later period. The planning commission recommends using 8 Kz per day as the rate for unskilled project workers.[11]

Importation of materials or equipment from foreign sources causes no problems. In Dondela the exchange rate has been stable for three years, there is no black market for foreign currencies, and the DFS has already calculated any needed foreign goods at their landed purchase price; no tariff charges have been included. Hence, no foreign exchange shadow price will be used.

Benefits are more difficult. There are three areas of real concern: (1) to what extent will the wages paid unskilled workers add to aggregate consumption? (2) will the market price of the products be affected by the project output? (3) what are the indirect benefits and how may they be valued?

The planning commission settles the first point. Previous studies have shown

[11]Shadow wage rates are not easy to calculate. They require (for example) estimation of the shadow rate of investment for both workers and investors, along with other data. For further information, see DasGupta, Sen, and Marglin (1972), Chapter 15.

that consumption in the hill region increases by one-third when public wage levels are paid. Most of the increase results from the increased consumption by the workers themselves, but there is a small component due to increased consumption by those with whom the unemployed workers had been living. (Extended families are the rule in Dondela, and especially so in the hill region. If a family member is out of work, the others somehow take care of them.) For these reasons, one-third of the wage to be paid unskilled workers (12 Kz per day) should be included as a project benefit (12 Kz $\times \frac{1}{3} = 4$ Kz as the benefit). In the specific instance, the benefit is the difference between the wage paid and the shadow wage rate, but this would not always be the case.

The commission is less helpful with product prices. They point out that one of the major products—pulpwood—is not presently produced in Dondela. All paper (the expected end product) is currently imported, so any paper produced should be valued according to the foreign exchange savings. "Perhaps if one knew the proportion of the final paper price that is normally available for pulpwood. . . . But the paper plant hasn't yet been designed, let alone approved and built."

Sawtimber prices are equally difficult. No one in Dondela has sawed eucalyptus logs, nor has anyone used eucalyptus lumber. Prices given by FAO had been used in the financial analysis, and there seems little reason to make any changes for the economic analysis, even though the quantity to be produced may be enough to affect lumber prices. Ideally, lumber prices will be held without further rise.

Indirect benefits must first be identified. The most important indirect benefit, nationally, will be the effective control of the reservoir siltation, so the DFS analysts visited the agricultural economists who had stated irrigation would lead to a doubling of the crop output on the plains and asked "What value do you place on the increase in agricultural output from the irrigated area?" They were told that the World Bank, in their evaluation of the irrigation project, had estimated the value of increased crop production at $10 million dollars—120 million Kruzeros. Part of this was due to import savings, for there would no longer be a need to import certain cereals and meats; part was due to lower food prices for everyone, hence consumption would increase and so would the consumer's surplus. Finally, a good many families in the subsistence part of the economy would simply have more to eat.

The DFS analysts decided they should be conservative, but to explain everything. They started with four assumptions:

1. No benefits will accrue to the eucalpytus project from increased crop output on the plains until after 25 years.
2. These benefits will continue for only an additional 25 years.
3. Because of uncertainties, 50% of this will be credited to the eucalyptus plantings; the rest should go to the Leucaena project.
4. There are other risks involved: the dam might not hold; crops might fail; some insect attack might effectively destroy the plantation, and so on. For these reasons, the present value of this benefit, received 25 years hence, will be discounted by 15%.

Calculation of this indirect benefit was as follows: $120{,}000{,}000/2 \times 2 = 30{,}000{,}000$ Kz as the annual saving to be attributed to the project, but beginning 25 years from now as an annual payment for 25 years, which is to be discounted at a 15% rate. The net present value of this indirect benefit is

$$30{,}000{,}000 \frac{(1.08)^{25}-1}{.08(1.08)^{25}\,(1.08)^{25}} = 9{,}728{,}220$$

and the present value per hectare of the plantation's benefit of controlling siltation of the reservoir is

$$9{,}728{,}200/22{,}500 = 432.37 \text{ Kz per hectare}$$

Other indirect values would include the important power production aspect, for its life would also be prolonged because of the project's control of siltation. The analysts decided to point this out in their report, but not to give it a monetary value because neither the price nor the social value of the power generated has been determined. (The approach to its valuation would probably be similar to that taken for the agricultural benefit.)

Other indirect benefits to be described included the improved living conditions and productivity of villagers working in pleasant surroundings, the much improved outlook for young people in the villages, the gradual introduction of workers to mechanical aids, the educational effect of the project, and so on.

Indirect costs would also be noted: eucalyptus is not a good habitat for game animals (although only one-third of the area will be changed in this way); the introduction of regular work habits is not an unmixed blessing, traditional ways of life will necessarily be changed, and so forth. There would undoubtedly be some inhabitants who would have to move, and some would necessarily give up their tradition of grazing any and all animals where and when they so desired. None of these were valued.

The economic analysis, using revised labor costs, and including both the expanded life of the reservoir and 4 Kz per day of the wage paid unskilled workers as benefits, is shown in Table 18.3. Note the treatment of labor costs. The unskilled labor costs were reduced by one-third (from 12 Kz to 8 Kz per day). On average, 58.38% of all costs are for unskilled labor (Table 18.1). Rather than recalculating the entire series of costs, the cost for each year was reduced by 58.38/3, or by 19.46%.[12]

The results of the economic analysis look good. At an 8% discount rate the project shows a benefit/cost ratio of

$$\frac{194{,}653}{142{,}013} = 1.37$$

which is considerably above that of most projects. The social internal rate of return is 12.6%. Because the financial analysis had shown that the project is not very sensitive to price changes, there seemed little point in rerunning the analysis with increased prices, although both the B/C ratio and the IRR would be increased. From the national viewpoint, the project is highly beneficial, and while the numbers could be improved by including (for example) an indirect benefit from the extended availability of power, there might be some psychological and political advantage in being conservative.

[12]Since skilled and unskilled labor were kept separated in Table 18.2, a more accurate calculation could be made. In view of the other uncertainties, and because the percentage of unskilled labor each year is reasonably similar, this approximation seems justified.

TABLE 18.3
Economic Analysis Calculations for Dondela Plantation Project

Year	-1	0	1	2	3	4	5	6	7	8	9	10	11	12	13	14
Costs																
Nursery		27	29	31	34	37	40	43	46	50	54	58	63	68	73	79
First Planting		4600	1092	774	774	774	1068	934	774	1048	774	774	1070	774	774	774
Second Planting			4600	1092	774	774	774	1068	934	774	1048	774	774	1070	774	774
Third Planting				4600	1092	774	774	774	1068	934	774	1048	774	774	1070	774
Fourth Planting					4600	1092	774	774	774	1068	934	774	1048	774	774	1070
Fifth Planting						4600	1092	774	774	774	1068	934	774	1048	774	774
Sixth Planting							4600	1092	774	774	774	1068	934	774	1048	774
Seventh Planting								4600	1092	774	774	774	1068	934	774	1048
Eighth Planting									4600	1092	774	774	774	1068	934	774
Ninth Planting										4600	1092	774	774	774	1068	934
Tenth Planting											4600	1092	774	774	774	1068
Eleventh Planting												4600	1092	774	774	774
Twelfth Planting													4600	1092	774	774
Thirteenth Planting														4600	1092	774
Fourteenth Planting															4600	1092
Fifteenth Planting																4600
Total Costs	0	4627	5721	6496	7273	8049	9120	10056	10833	11884	12662	13439	14514	15292	16071	16850
Adjusted total cost		3727	4608	5232	5858	6483	7345	8099	8725	9572	10198	10824	11689	12316	12944	13571
Revenues																
First Planting							850			1570			2195			
Second Planting								850			1570			2195		
Third Planting									850			1570			2195	
Fourth Planting										850			1570			2195
Fifth Planting											850			1570		
Sixth Planting												850			1570	
Seventh Planting													850			1570
Eighth Planting														850		
Ninth Planting															850	
Tenth Planting																850
Eleventh Planting																
Twelfth Planting																
Thirteenth Planting																
Fourteenth Planting																
Fifteenth Planting																
Benefits																
Revenues		0	0	0	0	0	850	850	850	2420	2420	2420	4615	4615	4615	4615
To unskilled labor		900	1113	1264	1415	1566	1775	1957	2108	2313	2464	2615	2824	2976	3127	3279
For silt control				432	865	1297	1730	2162	2594	3027	3459	3892	4324	4756	5189	5621
Total benefits		900	1113	1697	2280	2864	4354	4969	5552	7760	8343	8927	11763	12347	12931	13515

TABLE 18.3 *(continued)*
Economic Analysis Calculations for Dondela Plantation Project

1	Year	15	16	17	18	19	20	21	22	23	24	25	26	27	28	29
2	Costs															
3	Nursery															
4	First Planting	1264	1092	774	774	774	1068	934	774	1048	774	774	1070	774	774	774
5	Second Planting	774	1264	1092	774	774	774	1068	934	774	1048	774	774	1070	774	774
6	Third Planting	774	774	1264	1092	774	774	774	1068	934	774	1048	774	774	1070	774
7	Fourth Planting	774	774	774	1264	1092	774	774	774	1068	934	774	1048	774	774	1070
8	Fifth Planting	1070	774	774	774	1264	1092	774	774	774	1068	934	774	1048	774	774
9	Sixth Planting	774	1070	774	774	774	1264	1092	774	774	774	1068	934	774	1048	774
10	Seventh Planting	774	774	1070	774	774	774	1264	1092	774	774	774	1068	934	774	1048
11	Eighth Planting	1048	774	774	1070	774	774	774	1264	1092	774	774	774	1068	934	774
12	Ninth Planting	774	1048	774	774	1070	774	774	774	1264	1092	774	774	774	1068	934
13	Tenth Planting	934	774	1048	774	774	1070	774	774	774	1264	1092	774	774	774	1068
14	Eleventh Planting	1068	934	774	1048	774	774	1070	774	774	774	1264	1092	774	774	774
15	Twelfth Planting	774	1068	934	774	1048	774	774	1070	774	774	774	1264	1092	774	774
16	Thirteenth Planting	774	774	1068	934	774	1048	774	774	1070	774	774	774	1264	1092	774
17	Fourteenth Planting	774	774	774	1068	934	774	1048	774	774	1070	774	774	774	1264	1092
18	Fifteenth Planting	1092	774	774	774	1068	934	774	1048	774	774	1070	774	774	774	1264
19																
20	Total Costs	13435	13435	13435	13435	13435	13435	13435	13435	13435	13435	13435	13435	13435	13435	13435
21	Adjusted total cost	10820	10820	10820	10820	10820	10820	10820	10820	10820	10820	10820	10820	10820	10820	10820
22																
23	Revenues															
24	First Planting	20485					978			1805			2524			
25	Second Planting		20485					978			1805			2524		
26	Third Planting			20485					978			1805			2524	
27	Fourth Planting				20485					978			1805			2524
28	Fifth Planting	2195				20485					978			1805		
29	Sixth Planting		2195				20485					978			1805	
30	Seventh Planting			2195				20485					978			1085
31	Eighth Planting	1570			2195				20485					978		
32	Ninth Planting		1570			2195				20485					978	
33	Tenth Planting			1570			2195				20485					978
34	Eleventh Planting	850			1570			2195				20485				
35	Twelfth Planting		850			1570			2195				20485			
36	Thirteenth Planting			850			1570			2195				20485		
37	Fourteenth Planting				850			1570			2195				20485	
38	Fifteenth Planting					850			1570			2195				20485
39																
40	Benefits															
41	Revenues	25100	25100	25100	25100	25100	25228	25228	25228	25463	25463	25463	25792	25792	25792	25792
42	To unskilled labor	2614	2614	2614	2614	2614	2614	2614	2614	2614	2614	2614	2614	2614	2614	2614
43	For silt control	6054	6486	6486	6486	6486	6486	6486	6486	6486	6486	6486	6486	6486	6486	6486
44	Total benefits	33768	34200	34200	34200	34200	34328	34328	34328	34563	34563	34563	34892	34892	34892	34892
45																
46																
47																
48																
49																
50																
51																

TABLE 18.3 *(continued)*
Economic Analysis Calculations for Dondela Plantation Project

1	Year	30	31	32	33	34	35	36	37	38	39	40	41	42	43	44
2	Costs															
3	Nursery															
4	First Planting	1264	1092	774	774	774	1068	934	774	1048	774	774	1070	774	774	774
5	Second Planting	774	1264	1092	774	774	774	1068	934	774	1048	774	774	1070	774	774
6	Third Planting	774	774	1264	1092	774	774	774	1068	934	774	1048	774	774	1070	774
7	Fourth Planting	774	774	774	1264	1092	774	774	774	1068	934	774	1048	774	774	1070
8	Fifth Planting	1070	774	774	774	1264	1092	774	774	774	1068	934	774	1048	774	774
9	Sixth Planting	774	1070	774	774	774	1264	1092	774	774	774	1068	934	774	1048	774
10	Seventh Planting	774	774	1070	774	774	774	1264	1092	774	774	774	1068	934	774	1048
11	Eighth Planting	1048	774	774	1070	774	774	774	1264	1092	774	774	774	1068	934	774
12	Ninth Planting	774	1048	774	774	1070	774	774	774	1264	1092	774	774	774	1068	934
13	Tenth Planting	934	774	1048	774	774	1070	774	774	774	1264	1092	774	774	774	1068
14	Eleventh Planting	1068	934	774	1048	774	774	1070	774	774	774	1264	1092	774	774	774
15	Twelfth Planting	774	1068	934	774	1048	774	774	1070	774	774	774	1264	1092	774	774
16	Thirteenth Planting	774	774	1068	934	774	1048	774	774	1070	774	774	774	1264	1092	774
17	Fourteenth Planting	774	774	774	1068	934	774	1048	774	774	1070	774	774	774	1264	1092
18	Fifteenth Planting	1092	774	774	774	1068	934	774	1048	774	774	1070	774	774	774	1264
19																
20	Total Costs	13435	13435	13435	13435	13435	13435	13435	13435	13435	13435	13435	13435	13435	13435	13435
21	Adjusted total cost	10820	10820	10820	10820	10820	10820	10820	10820	10820	10820	10820	10820	10820	10820	10820
22																
23	Revenues															
24	First Planting	23558					978			1805			2524			
25	Second Planting		23558					978			1805			2524		
26	Third Planting			23558					978			1805			2524	
27	Fourth Planting				23558					978			1805			2524
28	Fifth Planting	2524				23558					978			1805		
29	Sixth Planting		2524				23558					978			1805	
30	Seventh Planting			2524				23558					978			1085
31	Eighth Planting	1805			2524				23558					978		
32	Ninth Planting		1805			2524				23558					978	
33	Tenth Planting			1805			2524				23558					978
34	Eleventh Planting	978			1805			2524				23558				
35	Twelfth Planting		978			1805			2524				23558			
36	Thirteenth Planting			978			1805		978	2524	2524	1805		23558	2524	
37	Fourteenth Planting				978			1805							23558	
38	Fifteenth Planting					978			1805			2524				23558
39																
40	Benefits															
41	Revenues	28865	28865	28865	28865	28865	28865	28865	29843	28865	28865	30670	28865	28865	31389	28865
42	To unskilled labor	2614	2614	2614	2614	2614	2614	2614	2614	2614	2614	2614	2614	2614	2614	2614
43	For silt control	6486	6486	6486	6486	6486	6486	6486	6486	6486	6486	6486	6486	6486	6486	6486
44	Total benefits	37965	37965	37965	37965	37965	37965	37965	38943	37965	37965	39770	37965	37965	40489	37965
45																
46																
47																
48																
49																
50																
51																

TABLE 18.3 *(continued)*
Economic Analysis Calculations for Dondela Plantation Project

1	Year	45	46	47	48	49	50	51	52	53	54	55	56	57	58	59
2	Costs'															
3	Nursery															
4	First Planting	1264														
5	Second Planting	774	1264													
6	Third Planting	774	774	1264												
7	Fourth Planting	774	774	774	1264											
8	Fifth Planting	1070	774	774	774	1264										
9	Sixth Planting	774	1070	774	774	774	1264									
10	Seventh Planting	774	774	1070	774	774	774	1264								
11	Eighth Planting	1048	774	774	1070	774	774	774	1264							
12	Ninth Planting	774	1048	774	774	1070	774	774	774	1264						
13	Tenth Planting	934	774	1048	774	774	1070	774	774	774	1264					
14	Eleventh Planting	1068	934	774	1048	774	774	1070	774	774	774	1264				
15	Twelfth Planting	774	1068	934	774	1048	774	774	1070	774	774	774	1264			
16	Thirteenth Planting	774	774	1068	934	774	1048	774	774	1070	774	774	774	1264		
17	Fourteenth Planting	774	774	774	1068	934	774	1048	774	774	1070	774	774	774	1264	
18	Fifteenth Planting	1092	774	774	774	1068	934	774	1048	774	774	1070	774	774	774	1264
19																
20	Total Costs	13435	12343	11570	10796	10023	8955	8022	7248	6201	5427	4654	3584	2811	2037	1264
21	Adjusted total cost	10820	9941	9318	8695	8072	7212	6461	5838	4994	4371	3748	2887	2264	1641	1018
22																
23	Revenues															
24	First Planting	23558								5000						4000
25	Second Planting		23558								5000					3000
26	Third Planting			23558								5000				2000
27	Fourth Planting				23558								5000			500
28	Fifth Planting	2524				23558								5000		
29	Sixth Planting		2524				23558								5000	
30	Seventh Planting			2524				23558								5000
31	Eighth Planting	1805			2524				23558							4500
32	Ninth Planting		1805			2524				23558						3000
33	Tenth Planting			1805			2524				23558					2500
34	Eleventh Planting	978			1805			2524				23558				2000
35	Twelfth Planting		978			1805			2524				23558			100
36	Thirteenth Planting			978			1805			2524				23558		500
37	Fourteenth Planting				978			1805			2524				23558	
38	Fifteenth Planting					978			1805			2524				23558
39																
40	Benefits															
41	Revenues	28865	28865	28865	28865	28865	27887	27887	27887	31082	31082	31082	28558	28558	28558	51558
42	To unskilled labor	2614	2402	2251	2101	1950	1743	1561	1410	1207	1056	906	697	547	396	246
43	For silt control	6486	6486	6486	6486	6486	6486	6486	6486	6486	6486	6486	6486	6486	6486	6486
44	Total benefits	37965	37753	37602	37452	37301	36116	35934	35783	38775	38624	38474	35741	35591	35440	58290
45																
46	NPV total ben. 8%	194653	Total costs at 8%			142013	Net benefits at 8%=			52641		B/C ratio 1.37				
47	NPV total ben. 10%	132726	Total costs at 10%			111657	Net benefits at 10%=			21070		B/C ratio 1.19			IRR-12.6%	
48	NPV total ben. 11%	111574	Total costs at 11%			100386	Net benefits at 11%=			11188		B/C ratio 1.11				
49	NPV total ben. 12%	94779	Total costs at 12%			90959	Net benefits at 12%=			3820		B/C ratio 1.04				
50	NPV total ben. 14%	70321	Total costs at 14%			76180	Net benefits at 14%=			-5859		B/C ratio .92				
51	NPV total ben. 18%	42416	Total costs at 18%			56831	Net benefits at 18%=			-14416		B/C ratio .75				

A Final Comment

Two characteristics of plantation projects are particularly obvious from the example: (1) the initial establishment costs play a significant role in determining the discounted returns of a plantation project; (2) the returns—the benefits—start at a low level, and although they increase to eventually greatly exceed the costs on a yearly basis, these late returns have relatively little influence on the rate of return. Positive cash flow from plantations is long delayed. In fact, not until the first harvest cut in year 15 do the social cash flows (or the financial cash flows) become positive. Forestry does have its economic peculiarities, and they are not always favorable.

This characteristic time lag in traditional forest returns inspired early workers of the social forestry movement to depart from traditional forestry approaches. In India (Gugarat State), where communal grazing lands were often so overused that one could often look in vain for a single blade of grass in a square meter plot, initial plantings on demonstration areas (4-hectare plots) were designed to produce heavy yields of grass the *first year*. Fruit trees and fodder trees were included in the plantings, for they would also yield results quickly. Fuelwood was available from thinnings, often within 3 or 4 years, while timber for building purposes came later, but within 15 years.

Forestry can and does adapt to changing needs—and so must foresters.

QUESTIONS AND PROBLEMS

1. What are the characteristics that seem common to developing nations?
2. There is no universal agreement on the requirements for economic development, but a few basics would seem essential. Name some.
3. Why is a supply of investment capital deemed essential to economic development?
4. What role, positive or negative, is played today by population growth in the developing nations?
5. Why do you suppose that, during the 1950s and 1960s, it was generally assumed that industrialization was a necessary part of, if not really a synonym for, economic development?
6. Contrast the Western philosophy of gradual change with that espoused by Karl Marx, and comment on the seeming attraction that command economies have for some developing nations.
7. What are some of the major opportunities open to forestry as an active agent to promoting development of the poorest of the nations, and to the poorest segment of those populations, the rural inhabitants?
8. Can forest-based exports be of assistance to developing nations? How must this be qualified?
9. Developing nations are beginning now to classify forests into three major groups. Describe each of these, and the role each forest group might play in aiding development.
10. What is social forestry and what does it involve?

11. What is agroforestry?

12. What is the Taungya method?

13. Many writers have stressed the shortage of fuelwood in Africa and Asia, often claiming this to be of crisis proportions. What is the situation with respect to fuelwood? What barriers must be overcome if the crisis is to be successfully overcome? What do you believe the prospects to be for solving the fuelwood shortages of Asia and Africa?

14. How does project analysis in developing nations differ from that in the United States? Would social benefit/cost analysis in the United States differ markedly from social benefit/cost analysis in a developing nation? If so, why—and how?

BIBLIOGRAPHY

ADAMS, DARIUS M., and RICHARD W. HAYNES. 1980. "The Softwood Timber Assessment Model: Structure, Projections, and Policy Simulations." *Forest Science*, Monograph 22. 62 p.

Agricultural Law Letter. 1985. See article, Vol. 11, No. 2 (February).

AINSWORTH, J. H. 1958. *Paper and You*. Book I in series *Paper, the Fifth Wonder*. Kaukauna, Wisc. Thomas Publishing Co.

ALLEN, STEWART. 1981. "Comment: No-Rescue Wilderness—A Risky Proposition." *Journal of Forestry*, Vol. 79, No. 3 (March), p. 153–154.

ARROW, KENNETH J. 1963. *Social Choice and Individual Values*. 2nd Ed. Cowles Commission Monograph No. 15. New York: Wiley.

AYRES, ROBERT U., and ALLEN V. KNEESE. 1969. "Production, Consumption, and Externalities." *American Economic Review*, Vol. 59, No. 3 (June 1969). Also in RFF Reprint No. 76, Resources for the Future, Washington, D.C.

BAUMOL, WILLIAM J. 1965. *Economic Theory and Operations Analysis*. Englewood Cliffs, NJ: Prentice-Hall. 606 p.

BAUMOL, WILLIAM J., and WALLACE L. OATES. 1975. *The Theory of Environmental Policy*. Englewood Cliffs, N.J.: Prentice-Hall. 272 p.

BENNINGHOFF, BRUCE, and CORYELL OHLANDER. 1978. "Integrating Timber Harvest Scheduling Economics and Environment Quality." *Journal of Forestry*, Vol. 76, No. 6 (June), p. 348–351.

BERGSON, ABRAM. 1938. "A Reformulation of Certain Aspects of Welfare Economics." *Quarterly Journal of Economics*, Vol. 512 (February), p. 310–334.

BEYERHELM, CARL. 1985. "The Real Story about Deficit Sales." *Journal of Forestry*, Vol. 83, No. 4 (April), p. 203.

BIESTERFELDT, ROBERT C., and STEPHEN G. BOYCE. 1978. "Systematic Approach to Multiple-Use Management." *Journal of Forestry*, Vol. 76, No. 6 (June), p. 342–345.

BILEK, EDWARD M. 1981a. "Production Concentration in the North American Lumber Industry, 1966–1980." Paper presented at the 1981 Midwest Forest Economics Meeting, Cedar, Mich. September 22–24, p. 3.

BILEK, EDWARD M. 1981b. Unpublished manuscript titled "Economic Concentration in the Wood-Based Industries."

BILEK, EDWARD M. and PAUL V. ELLEFSON. 1981. "The Wood-Based Industry: Trends in Selected Structural and Economic Factors Through 1977." Staff Paper Series No. 22, College of Forestry and the Agricultural Experiment Station, University of Minnesota, St. Paul. 16 p., 4 appendixes.

BILEK, EDWARD M. 1983. "Joint Ventures: Evidence in the Wood-based Industries." Paper presented at the 1983 Midwest Forest Economists Meeting, Madison, Wisc. August 29–31.

BINKLEY, C. S. 1980. "Economic Analysis of the Allowable Cut Effect." *Forest Science*, Vol. 26, No. 4, p. 633–642.

BINKLEY, C. S. 1981. *Timber Supply From Private Forests: A Micro-Economic Analysis of Landowner Behavior*. New Haven, Conn.: Yale University, Bulletin 2. 97 p.

BIRCH, THOMAS W., DOUGLAS G. LEWIS, and H. FRED KAISER. 1982. "The Private Forest Land Owners of the United States." Forest Service Resource Bulletin WO-1, USDA For. Serv., Washington, D.C. 64 p.

BISHOP, RICHARD C. 1982. "Option Value: An Exposition and Extension." *Land Economics*, Vol. 58, No. 1 (February), p. 1–15.

BOUNGIORNO, JOSEPH, LANCE BRANNMAN, and TAEHO BARK. 1984. "Econometric Versus Univariate and Bivariate Time-Series Forecasts: The Case of Softwood Lumber Imports." *Forest Science*, Vol. 30, No. 1, p. 194–208.

BOWES, MICHAEL D., and JOHN KRUTILLA. 1985. "The Economics of Multiple Use Public Forest Land Management." (Draft.) Resources for the Future, Washington, D.C.

BOYD, ROY. 1983. "The Effects of FIP and Forester Assistance on Nonindustrial Private Forests." In *Nonindustrial Private Forests: A Review of Economic and Policy Studies.* Symposium Proceedings, April 19–20, Duke University, School of Forestry and Environmental Studies, Durham, N.C., p. 189–203.

BRIGGS, C. W. 1962. *Tax Treatment of Timber*. 4th Ed. Washington, D.C.: Forest Industries Committee on Timber Valuation and Taxation.

BROOKSHIRE, DAVID S., MARK A. THAYER, WILLIAM W. SCHULTZ, and RALPH C. D'ARGE. 1982. "Valuing Public Goods: A Comparison of Survey and Hedonic Approaches." *American Economic Review*, Vol. 72, No. 1 (March), p. 165–177.

BROOKSHIRE, DAVID S., LARRY S. EUBANKS, and ALAN RANDALL. 1983. "Estimating Option Prices and Existence Values for Wildlife Resources." *Land Economics*, Vol. 59, No. 1 (February), p. 1–15.

BROWN, GARDNER, JR., and ROBERT MENDELSOHN. 1984. "The Hedonic-Travel Cost Method." *The Review of Economics and Statistics*, Vol. LXVI, No. 3 (August), p. 427–433.

BUSBY, R. J. N. 1985. *A Guide to Financial Analysis of Tree Growing*. Food and Agriculture Organization of the United Nations, Rome. 104 p.

CALISH, STEVEN, ROGER D. FIGHT, and DENNIS E. TEEGUARDEN. 1978. "How Do Nontimber Values Affect Douglas-fir Rotations?" *Journal of Forestry*, Vol. 76, No. 4 (April), p. 217–221.

CARLAN, S. A., H. L. HANEY, JR., and J. E. GUNTER. 1982. "State Guides for Assessing Forest Land and Timber: 1977." USDA Forest Service, SA-TP 20, Washington, D.C. 187 p.

CARLSON, SUNE. 1952. *A Study on the Pure Theory of Production*. London: S. King & Sons, Ltd. 128 p.

CARSON, RACHEL. 1962. *Silent Spring*. Boston: Houghton Mifflin.

CHANG, SUN JOSEPH. 1983. "Rotation Age, Management Intensity, and the Economic Factors of Timber Production: Do Changes in the Stumpage Price, Interest Rate, Regeneration Cost, and Forest Taxation Matter?" *Forest Science*, Vol. 29, No. 2, p. 267–278.

CHAPPELLE, D. E., M. MANG, and R. C. MILEY. 1976. "Evaluation of Timber RAM as a Forest Management Planning Tool." *Journal of Forestry*, Vol. 74, No. 5 (May), p. 288–293.

CIRIACY-WANTRUP, S. VON. 1938, "Multiple and Optimum Use of Wildland under Different Economic Conditions." *Journal of Forestry*, Vol. 36, No. 7 (July), p. 665–674.

CIRIACY-WANTRUP, S. VON. 1952. *Resource Conservation, Economics and Policies.* Berkeley and Los Angeles: University of California Press. 395 p.

CLAWSON, MARION. 1959. "Methods of Measuring the Demand for and Value of Outdoor Recreation." Reprint No. 10, RFF Series, Resources for the Future, Washington, D.C.

CLAWSON, MARION. 1974. "Conflicts, Strategies, and Possibilities for Consensus in Forest Land Use and Management." In *Forest Policy for the Future.* RFF Working Paper LW1, Resources for the Future, Washington, D.C.

CLAWSON, MARION. 1975. *Forests for Whom and For What?* Washington, D.C.: Resources for the Future.

CLAWSON, MARION. 1976. "The National Forests." *Science,* Vol. 191, No. 4227 (Feb. 1976), p. 762–762.

CLAWSON, MARION. 1977. "Economic Timber Production Characteristics of Nonindustrial Private Forests in the United States." In Background Papers from Workshop on Policy Alternatives for Nonindustrial Private Forests, Resources for the Future, Washington, D.C., p. 1–94.

CLAWSON, MARION. 1979. "Forests in the Long Sweep of American History." *Science,* Vol. 204, No. 4398 (June). Also in Resources for the Future Reprint Series, No. 164.

CLAWSON, MARION. 1979b. *The Economics of United States Nonindustrial Private Forests.* RFF Working Paper P-14 (June 1974), Resources for the Future, Washington, D.C.

CLAWSON, MARION. 1984. "Effective Acreage for Outdoor Recreation." In *Resources,* No. 78 (Fall 1984), Resources for the Future, Washington, D.C., p. 2–7.

CLAWSON, M., and J. L. KNETCH. 1966. *Economics of Outdoor Recreation.* Washington, D.C.: Resources for the Future. 328 p.

CLAWSON, MARION, and CARLTON S. VANDOREN (eds.). 1984. *Statistics on Outdoor Recreation.* Washington, D.C.: Resources for the Future. 368 p.

CLEPHANE, T. P. 1978. "Ownership of Timber: A Critical Component in Industrial Success." *Forest Industries,* Vol. 105, No. 9 (August), p. 30–32.

CLUTTER, J. L., J. C. FORTSON, L. V. PIENAAR, G. H. BRISTER, and R. L. BAILEY. 1983. *Timber Management: A Quantitative Approach.* New York: Wiley. 333 p.

Committee of Scientists Report. 1979. "Final Report of the Committee of Scientists for Implementation of Section 6 of the National Forest Management Act of 1976, Feb. 22, 1979." Published in the *Federal Register,* Part V, May 4, 1979, p. 26628. (Taken from USDA Forest Service *ROS Users Guide.* No date, 38 p.)

CONNAUGHTON, KENT P., and RICHARD W. HAYNES. 1983 "An Evaluation of Three Simplified Approaches to Modeling the Regional Demand for National Forest Stumpage." *Forest Science,* Vol. 29, No. 1, p. 3–12.

CONNAUGHTON, KENT P., and ROGER D. FIGHT. 1984. "Applying Trade-off Analysis to National Forest Planning." *Journal of Forestry,* Vol. 86, No. 11 (November), p. 680–683.

CROWELL, JOHN B., JR. 1983. "Improved Productivity from Nonindustrial Private Forests: Dimensions of the Problem." In Jack C. Royer and Christopher D. Risbrudt (eds.), *Nonindustrial Private Forests: A Review of Economic and Policy Studies.* Symposium Proceedings, April 19–20, Duke University, Durham, N.C. 398 p.

CUBBAGE, FREDRICK W., THOMAS M. SKINNER, and CHRISTOPHER D. RISBRUDT. 1985. *An Economic Evaluation of the Georgia Rural Forestry Assistance Program.* Research Bulletin 322, College of Agriculture Experiment Stations, University of Georgia, Athens. 59 p.

DANA, SAMUEL, and SALLY FAIRFAX. 1980. *Forest and Range Policy.* 2nd Ed. New York: McGraw-Hill. 458 p.

DANE, C. W., NANCY C. MEADOR, and JOHN B. WHITE. 1977. "Goal Programming in Land-Use Planning." *Journal of Forestry,* Vol. 75, No. 6 (June), p. 325–329.

DARR, DAVID R. u.d. "United States Trade in Timber Products Within the Context of Policy and Law." Paper presented at International Forestry Congress, Quebec City, Canada. 6 p.

Dasgupta, Partha, Amartya Sen, and Stephen Marglin. 1972. *Guidelines for Project Evaluation*. Sales No. E.72.II.B.11. New York: United Nations. 383 p.

Davis, Kenneth P. 1966. *Forest Management*. 2nd Ed. New York: McGraw-Hill. 519 p.

de Steiguer, J. E. 1984. "Impact of Cost-Share Programs on Private Reforestation Investment." *Forest Science*, Vol. 30, No. 3, p. 697–704.

Dennis, Donald F. 1985. *Capital Gains Treatment of Timber Income: An Economic Assessment*. USDA Northeastern Forest Experiment Station Research Paper NE-556, Broomall, Pa. 5 p.

Domar, E. 1946. "Capital Expansion, Rate of Growth and Employment." *Econometrica*, Vol. 14, No. 2 (April), p. 137–147.

Driver, B. L. 1985. "Specifying What Is Produced by Management of Wildlife by Public Agencies." *Leisure Sciences*, Vol. 7, No. 3, p. 281–295.

Duerr, William A. 1986. "Forestry's Upheaval." *Journal of Forestry*, Vol. 84, No. 1 (January), p. 20–26.

Duerr, W. A., J. Fedkiw, and S. Guttenburg. 1956. *Financial Maturity: A Guide to Profitable Timber Growing*. USDA Tech. Bull. 1146.

Ellefson, Paul V., and Michael E. Chopp. 1978. "Systematic Analysis of the Economic Structure of the Wood-based Industry." Staff Paper No. 3. St. Paul: College of Forestry, University of Minnesota. 335 p.

Ellefson, Paul V., and Michael E. Chopp. 1979. "U.S. Industrial Ownership of Forest Land: Extent, Strategies, and Management Intensity. Staff Paper Series No. 16. St. Paul: College of Forestry and Agricultural Experiment Station, University of Minnesota. 27 p.

Emerson, Peter, Anthony T. Stout, and Deanne Kloepfer. 1985. "The Feds Can't See Their Losses in the Trees." *Wall Street Journal* (November 14, 1985), Editorial page. (See also reply letter by Steve Hanke, Joint Economic Committee, U.S. Congress, in *Wall Street Journal*, November 19, 1985.)

Enk, G. A. 1975. *A Description and Analysis of Strategic and Land-Use Decision Making by Large Corporations in the Forest Products Industry*. PhD Dissertation, Yale University, New Haven, Conn. 161 p.

Ervast, Seppo., Lauri Heikinheimo, Kallervo Kuusela, and Veikko O. Makinen. 1970. *Forestry and Forest Industry Production Alternatives in Finland, 1970–2015*. Folia Forestalia 88 Helsinki, Finland 65 p.

Fight, Roger D., and Robert M. Randall. 1980. "Visual Cost and the Cost of Growing Timber." *Journal of Forestry*, Vol. 78, No. 9 (September), p. 546–547.

Flora, Donald F. 1966. *Economic Guides for Ponderosa Pine Dwarf Mistletoe Control in Young Stands of the Pacific Northwest*. USDA Forest Service Research Paper PNW-29. Pacific Northwest Forest and Range Experiment Station, Portland, Ore. 16 p.

Food and Agriculture Organization of the United Nations. 1984. *1982 Yearbook of Forest Products 1971–1982*. FAO, Rome. 408 p.

Food and Agriculture Organization of the United Nations. 1986. *1984 Yearbook of Forest Product Statistics*. FAO, Rome. 400 p.

Forest Industries. 1981. "Lumber Produced by the Top 100 North American Producers." May, 1981.

Forest Industries. 1984a. "Leading Lumber Producers, U.S. and Canada, 1983." July, p. 16.

Forest Industries. 1984b. "Harry Merlo Speaks Out, Dialogue #4." September, p. W-3.

Forster, Robert. u.d. "Review of Allowable Cut Effect." Unpublished and undated report prepared for the Provincial Forest Service, Quebec, Canada.

Fortson, James C., and Leon A. Hargreaves. 1974. "Capital Gains Taxation and the Industrial Forests of the South." *Journal of Forestry*, Vol. 72, No. 6 (June), p. 435–448.

Francis, David R. 1987. "Canadian-U.S. 'courtship' a risky business." *Christian Science Monitor*, Jan. 15, 1987. p. 9 & 11.

FREEMAN, MYRICK. 1979. *The Benefits of Environmental Improvement: Theory and Practice.* Washington, D.C.: Resources for the Future. 272 p.

GAFFNEY, M. 1960. *Concepts of Financial Maturity of Timber and Other Assets.* Raleigh, N.C.: Dept. of Agr. Economics, State College.

GALBRAITH, JOHN KENNETH. 1967. *The New Industrial State.* Boston: Houghton Mifflin. 427 p.

GAMACINE, ADREIN E. (ed.). 1984. *Selling the Federal Forests.* Seattle: University of Washington Press. 273 p.

GITTINGER, J. PRICE. 1982. *Economic Analysis of Agricultural Projects.* 2nd Ed. Baltimore and London. The Johns Hopkins University Press, 505 p.

GRAHAM-TOMASI, THEODORE. 1984. *The Economics of Wilderness Preservation and Timber Supply: Problems in Dynamic Optimization and Uncertainty.* Ph.D. Dissertation, University of Michigan. 220 p.

GRANT, EUGENE L., W. GRANT IRESON, and RICHARD S. LEAVENWORTH. 1976. *Principles of Engineering Economy.* 6th Ed. New York: Wiley. 624 p.

GRAY, JOHN W. 1983. *Forest Revenue Systems in Developing Countries.* FAO Forestry Paper No. 43, Food and Agriculture Organization of the United Nations, Rome.

GREGERSEN, HANS, M. 1975. "Effect of Inflation on Evaluation of Forestry Investments." *Journal of Forestry,* Vol. 73, No. 9, p. 570–572.

GREGERSEN, HANS, M. 1982. "Village Forestry Development in the Republic of Korea: A Case Study." Report GCP/INT/347/SWE, FAO, Rome.

GREGERSEN, HANS. M., and ARNOLDO H. CONTRERAS. 1979. *Economic Analysis of Forestry Projects.* FAO, Forestry Paper No. 17, Rome. 193 p.

GREGORY, G. ROBINSON. 1955. "An Economic Approach to Multiple Use." *Forest Science,* Vol. 1, No. 1, p. 6–13.

GREGORY, ROBIN S. 1986. "Interpreting measures of economic loss: evidence from contingent valuation and experimental studies." *Journal of Environmental Economics and Management,* Vol 13, p. 325–337.

GRETHER, DAVID M., and CHARLES R. PLOTT. 1979. "Economic Theory of Choice and the Preference Reversal Phenomenon." *American Economic Review* (September), p. 623–638.

GUNTER, JOHN E., and HARRY L. HANEY, JR. 1978. "A Decision Tree for Compound Interest Formulas." *Southern Journal of Applied Forestry,* Vol. 2, No. 3 (August), p. 106–107.

GUNTER, JOHN E., and HARRY L. HANEY, JR. 1983. *Essentials of Forestry Investment Analysis.* Corvallis: Oregon State University Book Stores. 337 p.

GUTHRIE, J. A. 1950. *The Economics of Pulp and Paper.* Pullman: State College of Washington Press.

HALL, C. R. 1935. *The Forest Tax Problem and its Solution Summarized.* Washington, D.C.: U.S. Department of Agriculture Circular No. 358, U.S. Forest Service.

HALL, D. F., and J. D. STARR. 1985. "Developing State Forest Policy: Virginia's Pine Reforestation Program." Reprinted in *A Focused Discussion of Assistance and Incentives for Private Forest Owners,* Michigan Dept. of Natural Resources, Forest Management Division, July 1985, p. 35–54.

HALL, WILLIAM K., ARCHER MCWHORTER, and W. ALLEN SPIVEY. 1980. "Optimization Programs at the University of Michigan." 5th Ed. Ann Arbor: Graduate School of the University of Michigan.

HANEY, HARRY L. "Income Tax Treatment of Reforestation Costs." *Alabama's Treasured Forests.* Special tax supplement (undated), p. 10–12.

HARDIN, GARRETT. 1968. "The Tragedy of the Commons." *Science,* Vol. 162 December. Copyright 1968 by the AAAS, pp. 1243–1248. Used by permission.

HARROD, ROY. 1948. *Toward a Dynamic Economics, Some Recent Developments of Economic Theory and Their Application to Policy.* Lectures delivered on February 1947 at University of London. London: Macmillan, 1948. 168 p.

HARTMAN, R. 1976. "The Harvesting Decision When a Standing Forest Has Value." *Economic Inquiry,* Vol. 14, p. 52–58.

HAYNES, RICHARD. 1977. "A Derived Demand Approach to Estimating the Linkage Between Stumpage and Lumber Markets." *Forest Science,* Vol. 23, No. 2, p. 281–288.

HEINER, RONALD A. 1983. "The Origin of Predictable Behavior." *American Economic Review* (September), p. 560–595.

HEINRICHS, JAY. 1983. "The Winged Snail Darter." *Journal of Forestry,* Vol. 85, No. 4 (April), p. 212–214, 262.

HERFINDAHL, ORRIS C., and ALLEN V. KNEESE. 1974. *Economic Theory of Natural Resources.* Columbus, Ohio: Charles E. Merrill Publishing Co. 405 p.

HICKS, J. R. 1939. "The Foundations of Welfare Economics." *The Economic Journal,* Vol. 49, No. 196 (December), p. 696–712.

HICKS, J. R. 1943. "The Four Consumer Surpluses." *Review of Economic Studies,* Vol. XI, Vol. 1 (Winter), p. 31–41.

HICKS, J. R. 1946. *Value and Capital.* Oxford: Clarenden Press. 340 p.

HOBLEY, MARY. 1985. "Common Property Does Not Cause Deforestation." *Journal of Forestry,* Vol. 83, No. 11 (November), p. 663–664.

HOWE, CHARLES W. 1979. *Natural Resource Economics.* New York: Wiley. 350 p.

HYDE, WILLIAM F. 1980. *Timber Supply, Land Allocation, and Economic Efficiency.* Baltimore: Johns Hopkins University Press. MD. 224 p.

International Bank for Reconstruction and Development. 1984. *World Tables.* Washington, D.C.: The World Bank.

JACKSON, DAVID H. 1980. *The Microeconomics of the Timber Industry.* Boulder, Colo. Westview Press. 136 p.

JOHNSON, K. NORMAN, and H. LYNN SCHEURAMAN. 1977. "Techniques for Prescribing Optimal Timber Harvest and Investment Under Different Objectives: Discussion and Synthesis." *Forest Science* Monograph No. 18. 31 p.

JOHNSON, K. NORMAN, B. JONES, and B. M. KENT. 1980. *User's Guide and Operations Manual.* Washington, D.C.: U.S. Forest Service.

Journal of Forestry. 1986a. "Issues to Watch in 1986." Unsigned editorial under "National Update." *Journal of Forestry,* Vol. 84, No. 1 (January), p. 5.

Journal of Forestry. 1986b. "Canadian Softwood Imports: How Great a Problem." Unsigned editorial under "Legislative Update." *Journal of Forestry,* Vol. 84, No. 6 (June), p. 4.

JUST, RICHARD E., DARRELL L. HUETH, and ANDREW SCHMITZ. 1982. *Applied Welfare Economics and Public Policy.* Englewood Cliffs, N.J.: Prentice-Hall. 491 p.

KALDOR, NICHOLAS. 1939. "Welfare Propositions of Economics and Interpersonal Comparisons of Utility." *The Economic Journal,* Vol. 49, No. 195 (September), p. 549–552.

KAPP, L. WILLIAM. 1950. *The Social Costs of Private Enterprise.* Cambridge: Harvard University Press. 287 p.

KELLISON, ROBERT C. 1985. "Weak Wood Flaws" in "Letters" section, *Journal of Forestry,* Vol. 83, No. 11 (November), p. 662.

KEMPER, ROBERT E., and LAWRENCE S. DAVIS. 1976. "Costs of Environmental Constraints on Timber Harvesting and Regeneration." *Journal of Forestry,* Vol. 74, No. 11 (November), p. 754–751.

KENT, BRIAN M. 1980. "Linear Programming in Land-Management Planning on Natural Forests." *Journal of Forestry,* Vol. 78, No. 8 (August), p. 469–471.

KLEMPERER, DAVID W. 1981. "Rejoinder to Schallau and Wirth." *Journal of Forestry,* Vol. 79, No. 9 (September), p. 618.

KLEMPERER, DAVID W. 1983. "Ambiguities and Pitfalls in Forest Productivity Taxation." *Journal of Forestry*, Vol. 81, No. 1 (January), p. 16–19.

KRUTILLA, JOHN V. 1967. "Conservation Reconsidered." *American Economic Review* (September), p. 777–786.

KRUTILLA, JOHN V. 1980. "Reflections of an Applied Welfare Economist." Presidential Address at the Annual Meeting of the Association of Environmental and Resource Economics, Denver, Colo. (September).

KRUTILLA, JOHN V., and A. FISHER. 1975. *The Economics of Natural Environments*. Baltimore: Johns Hopkins University Press.

LANGE, WILLIAM J. 1981. "The 1980 Hardwood Timber Assessment Model." Pacific Northwest Forest and Range Experiment Station, Portland, Ore.

LEOPOLD, A. S. 1978. "Wildlife and Forest Practice." p. 108–120 In H. P. Brokaw (ed.), *Wildlife and America*. Washington, D.C.: Council on Environmental Quality. 592 p.

LEUSCHNER, WILLIAM A. 1984. *Introduction to Forest Resources Management*. New York: Wiley. 298 p.

LE MASTER, DENNIS C. 1976. "The Resources Planning Act as Amended by the National Forest Management Act." *Journal of Forestry*, Vol. 74, No. 12 (December), p. 798, 836–840.

LITTLE, IAN, M.D., and JAMES A. MIRRLEES. 1969. *Manual of Industrial Project Analysis, Vol. II, Social Cost Benefit Analysis*. Development Centre of the Organization for Economic Cooperation and Development (OECD), Paris. 280 p.

LITTLE, IAN, M.D., and JAMES A. MIRRLEES. 1974. *Project Appraisal and Planning for Developing Countries*. New York: Basic Books.

LUPPOLD, WILLIAM G. 1982. "An Econometric Model of the Hardwood Lumber Market." Northeastern Forest Experiment Station Research Paper NE-512. 15 p.

LYONS, JAMES R. 1985. "Reflections on the Image of the Forest Service." *Journal of Forestry*, Vol. 83, No. 8 (August), p. 461–462.

MAJOR, MIKE. 1985. "What's the Big Deal About Corporate Mergers?" *Journal of Forestry*, Vol. 83, No. 7 (July), p. 406–410.

MANSFIELD, EDWIN. 1983. *Principles of Microeconomics*. 4th Ed. New York: Norton. 456 p.

MARQUIS, RALPH W. 1952. *Forest Yield Taxes*. Circular No. 899. Washington, D.C.: U.S. Dept. of Agriculture.

MARSHALL, ALFRED. 1947. *Principles of Economics*. 8th Ed. London: Macmillan.

MCAVOY, LEO A., and DANIEL L. DUSTIN. 1981. "The Right to Risk in Wilderness." *Journal of Forestry*, Vol. 83, No. 3 (March), p. 150–152.

MCCARL, BRUCE A., and RICHARD W. HAYNES. 1985. "Exchange Rates Influence Softwood Lumber Trade." *Journal of Forestry*, Vol. 83, No. 6 (April), p. 368–370.

MCCRACKEN, PAUL. 1985. *The Wall Street Journal*, March 19, 1985, p. 32.

MCGEE, GEORGE T., HARRY L. HANEY, JR., and WILLIAM C. SIEGEL. 1984. "State Income Tax Implications for Nonindustrial Private Forestry in the United States." *Forest Products Journal*, Vol. 34, No. 3 (March), p. 27–36.

MCGUIRE, JOHN. 1976. "National Forest Policy and the 94th Congress." *Journal of Forestry*, Vol. 74, No. 12 (December), p. 800–805.

MEAD, J. W. 1966. *Competition and Oligopsony in the Douglas Fir Industry*. Berkeley: University of California Press.

MEADOWS, JOHN, BRUCE BARE, KEN WARE, and CLARK ROW (eds.). 1975. *Systems Analysis and Forest Resource Management: Proceedings of a Workshop at the University of Georgia*. Washington, D.C.: Systems Analysis Working Group, Society of American Foresters. 457 p.

MERRETT, A. J., and ALLEN SYKES. 1963. *The Finance and Analysis of Capital Projects*. London: Longman. (Cited in Gittinger, p. 339).

MERRETT, A. J., and ALLEN SYKES. 1985. "The Takeover of St. Regis." *Journal of Forestry*, Vol. 83, No. 7 (July), p. 409.

Michigan Department of Natural Resources, Forest Management Division. 1985. *A Focused Discussion of Assistance and Incentives for Private Forest Owners*. Proceedings of a special meeting of the Michigan Forest Product Industry Development Council, July.

MILLS, THOMAS J., and DARIA CAIN. 1976. "1974 Forestry Incentive Program: Indicators of Cost Effectiveness." *Journal of Forestry*, Vol. 74, No. 10 (October), p. 678, 683.

MISHAN, E. J. 1971. *Cost-Benefit Analysis*. London: George Allen and Unwin. 364 p.

Multiple Use-Sustained Yield Act. 1960. United States Code 16.

MURRAY, ALAN, and JEFFREY H. BIRNBAUM. 1986. "Senate Panel's Tax Overhaul Effort Is Betraying Reform Impetus by Preserving Many Tax Breaks." *The Wall Street Journal*, April 7, p. 42.

MUSGRAVE, RICHARD, and PEGGY MUSGRAVE. 1984. *Public Finance in Theory and Practice*. 4th Ed. New York: McGraw-Hill. 824 p.

NAVON, DANIEL I. 1967. *Timber RAM . . . A Long-Range Planning Method for Commercial Timber Lands Under Multiple-Use Management*. USDA For. Service Research Paper PSW-70.

NEFF, PAUL E. 1973. "Calculation of Allowable Harvest for the National Forests." *Journal of Forestry*, Vol. 72, No. 2 (February), p. 86–89.

New York Times. 1985. "An Exodus From the Timberlands." Kathleen Sharp, Business Editor of the *Port Angeles Daily News*, Sunday, June 16, 1985.

O'LAUGHLIN, JAY. 1981. "Antitrust in the Era of Diversification: The Wood-Based Industry as a Case Example." Paper prepared for presentation at 51st Annual Conference of the Southern Economic Association, New Orleans, November 5, 1981. 151 p.

O'LAUGHLIN, JAY, and PAUL V. ELLEFSON. 1979. "Diversified Firms Entering the Wood-Based Industry by Merger." Staff Paper Series No. 7, College of Forestry and Agricultural Experiment Station, University of Minnesota, St. Paul. 121 p.

O'LAUGHLIN, JAY, and PAUL V. ELLEFSON. 1982a. *New Diversified Entrants Among U. S. Wood-Based Companies: A Study of Economic Structure and Corporate Strategy*. Ag. Exp. Station Bulletin 541, Forestry Series 37, University of Minnesota, p. 63.

O'LAUGHLIN, JAY, and PAUL V. ELLEFSON. 1982b. "Strategies for Corporate Timberland Ownership and Management." *Journal of Forestry*, Vol. 80, No. 12 (December), p. 784–788.

OLSEN, SCOTT C., HARRY L. HANEY, JR., and WILLIAM C. SIEGEL. 1981. "State Death Tax Implications for Private Nonindustrial Forestry." *Forest Products Journal*, Vol. 31, No. 7 (July), p. 28–37.

PARETO, VILFREDO. 1896. *Cours d'Economie Politique*. Reprinted in Volume 1, (Economics) of *Complete Works of Vilfredo Pareto*, Giovanni Busino. Geneve: Druz, 1964.

PEARSE, P. H. 1965. "Distortions in the Market for Forest Land." *Forestry Chronicle*, Vol. 41 (December 1965), p. 406–418.

PELINCK, E., P. K. MANANDHAR, and R. H. GECOLEA. 1984. "Forestry Extension: Community Development in Nepal." *Unasylva*, Vol. 36/1, No. 143, p. 2–12.

PERRY, T. 1948. *Modern Plywood*. 2nd Ed. New York: Pitman.

PICKSTOCK, MICHAEL. 1983. "The Fuelwood Crisis in Africa: An FAO Round Table Discussion." *Unasylva*, Vol. 35, No. 141, p. 22.

PINCHOT, GIFFORD. 1947. *Breaking New Ground*. New York: Harcourt Brace.

POTTIER, DIRK. 1984. "Running Cattle Under Trees: An Experiment in Agroforestry." *Unasylva*, Vol. 36, No. 143. p. 23–27.

POULSEN, GUNNAR. 1982. "The Non-Wood Product of African Forests." *Unasylva*, Vol. 34, No. 137, p. 15–21.

PUBLIC LAND LAW REVIEW COMMISSION. 1969. *Study of Public Land Policy*. Vol. 2. Prepared for Public Land Law Review Commission, April 1969. Milwaukee: George Banzhaf and Co.

RADCLIFFE, SAMUEL J. 1978. "Barriers to International Trade in Forest Products: A U.S.

Perspective." Draft, unpublished working paper, Resources for the Future, Washington, D.C. 55 p.

RANDALL, ALAN, and JOHN R. STOLL. 1980. "Consumer's Surplus in Commodity Space." *American Economic Review*, Vol. 71, No. 3 (June), p. 449–457.

RAUP, PHILLIP M. 1978. "Reflections on the Public Interest in Private Nonindustrial Forestlands." Paper No. 1690, Miscellaneous Journal Series, Agricultural Experiment Station, University of Minnesota, St. Paul. Reprinted by Resources for the Future, in background papers from workshop in Sedjo, R. A. and D. M. Ostermeier, *Policy Alternatives for Nonindustrial Private Forests*, p. 110–126.

REID, RICHARD G. 1985. "USFS & Co." *Journal of Forestry*, Vol. 83, No. 1 (January), p. 25–27.

RICARDO, DAVID. 1817. *The Principles of Political Economy and Taxation*. 3rd Ed. London: J. M. Dent, Everyman's Library Edition.

RITTERS, KURT, J. DOUGLAS BRODIE, and DAVID W. HANN. 1982. "Dynamic Programming for Optimization of Timber Production and Grazing in Ponderosa Pine." *Forest Science*, Vol. 28, No. 3, p. 517–526.

ROCKEL, MARK L., and JOSEPH BUONGIORNO. 1982. "Derived Demand for Wood and Other Inputs in Residential Construction: A Cost Function Approach." *Forest Science*, Vol. 28, No. 2, p. 207–219.

ROSTOW, W. W. 1960. *The Stages of Economic Growth: A Non-Communist Manifesto*. Cambridge: Cambridge University Press. 178 p.

ROW, CLARK, H. FRED KAISER, and JOHN SESSIONS. 1981. "Discount Rate for Long-Term Forest Service Investments." *Journal of Forestry*, Vol. 79, No. 6 (June), p. 367–369, 376.

ROYER, JACK P., and FRANK J. CONVERY (eds.). 1980. *Nonindustrial Private Forests: Data and Information Needs*. Conference Proceedings, April 17–18, Duke University, Durham, N.C. 130 p.

ROYER, JACK P., and CHRISTOPHER D. RISBRUDT (eds.). 1983. *Nonindustrial Private Forests*. Symposium Proceedings, School of Forestry and Environmental Studies, Duke University, Durham, N.C. 398 p.

S.A.F. Study Group on Forest Taxation. 1980. "Forest Taxation." *Journal of Forestry*, Vol. 78, No. 7 (July), Special Insert of 7 pgs., following p. 408.

SAMUELSON, PAUL A. 1951. *Economics: An Introductory Analysis*. 2nd Ed. New York: McGraw-Hill. 762 p.

SAMUELSON, PAUL A. 1956. "Social Indifference Curves." *Quarterly Journal of Economics*, LXX, No. 1 (February), p. 1–22.

SAMUELSON, PAUL A. 1976. "Economics of Forestry in an Evolving Society". *Journal of Public Inquiry*, Vol. 14, No. 4 (December), pp. 466–492.

SCHALLAU, C. H., and M. E. WIRTH. 1980. "Reinvestment Rate and the Analysis of Forestry Enterprises." *Journal of Forestry*, Vol. 18, No. 12 (December), p. 740–742.

SCHMID, R. 1976. "The Jiri Multipurpose Development Project (JMDP), a Hard Experience in Mountain Ecology and a Very Important Learning-Process in Nepalese/Swiss Technical Cooperation." In *Mountain Environment and Development*, Collection of papers published on the occasion of the 20th Anniversary of the Swiss Association for Technical Assistance in Nepal (SATA). Kathmandu, Nepal: Tribhuvan University. 214 p.

SCHMITHUSEN, FRANZ. 1983. "The Weak Link in Forestry Development." *Unasylva*, Vol. 35, No. 142, p. 36–43.

SCHULER, ALBERT T., H. H. WEBSTER, and J. C. MEADOWS. 1977. "Goal Programming in Forest Management." *Journal of Forestry*, Vol. 75, No. 6 (June), p. 320–324.

SCHUMPETER, JOSEPH A. 1954. *History of Economic Analysis*. (Edited by Elizabeth B. Schumpeter.) New York: Oxford University Press. 1260 p.

SCHWEITZER, D. L., R. W. SASSAMAN, and C. H. SCHALLAU. 1972. "Allowable Cut Effect: Some Physical and Economic Implications." *Journal of Forestry*, Vol. 70, No. 7 (July), p. 415–418.

SCITOVSKY, J. 1951. *Welfare and Competition.* Chicago: Richard D. Irwin.

SEDJO, ROGER A. (ed.) 1983. *Governmental Interventions, Social Needs, and the Management of U.S. Forests.* Baltimore, Md.: Johns Hopkins University Press. 300 p.

SEDJO, ROGER A. 1985. "The Role of Forestry Trade in Economic Development." Paper prepared for the 1985 World Forestry Congress. Washington, D.C.: Resources for the Future. 55 p.

SEDJO, ROGER A., and MARION CLAWSON. 1983. "Global Forests." Washington D.C.: Working paper, Resources for the Future. 57 pp.

SEDJO, ROGER A., and KENNETH S. LYON. 1983. "Long-Term Forest Resources Trade, Global Timber Supply, and Intertemporal Comparative Advantage." *American Journal of Agricultural Economics*, Vol. 65, No. 5 (December), p. 1012–1016.

SEDJO, ROGER A., and D. M. OSTERMEIER. 1978. *Policy Alternatives for Nonindustrial Private Forests.* Society of American Foresters, Washington, D.C. 64 p.

SENFT, JOHN F., B. ALAN BENDTSEN, and WILLIAM L. GALLIGAN. 1985. "Weak Wood: Fast-Grown Trees Make Problem Lumber." *Journal of Forestry*, Vol. 83, No. 8 (August), p. 476–484.

SHOEMAKER, PAUL J. H. 1982. "The Expected Utility Model: Its Variants, Purposes, Evidence and Limitations." *Journal of Economic Literature*, Vol. XX (June), p. 529–563.

SIMON, HERBERT A. 1979. "Rational Decision Making in Business Organizations." *American Economic Review*, Vol. 69, No. 4 (September), p. 493–513.

SIZEMORE, WILLIAM R. 1977. "A 1977 Update of the 1972 President's Advisory Panel on Timber and the Environment Findings." Discussion Papers from Workshop on Policy Alternatives for Nonindustrial Private Forests, Resources for the Future, Washington, D.C.

SKOK, R. A., and H. M. GREGERSEN. 1975. "Motivating Private Forestry: An Overview." *Journal of Forestry*, Vol. 73, No. 4 (April), p. 202–205.

SPEARS, JOHN. 1982. "Preserving Watershed Environments." *Unasylva*, Vol. 34, No. 137, p. 10–14.

STANKEY, GEORGE H. 1979. "Use Rationing in Two Southern California Wildernesses." *Journal of Forestry*, Vol. 81, No. 6 (June), p. 347–349.

STEINLIN, HANS JURG. 1982. "Monitoring the World's Tropical Forests." *Unasylva*, Vol. 34, No. 137, p. 2–8.

STEVENSON, L. E. 1940. *The Background and Economics of American Paper Making.* New York: Harper and Row.

Subcommittee on Benefits and Costs. 1950. *Proposed Practices for Economic Analysis of River Projects.* Washington, D.C.: Report to the Interagency River Basin Committee.

SUNLEY, EMIL M., JR. 1976. "Capital Gains Treatment of Timber: Present Law and Proposed Changes." *Journal of Forestry*, Vol. 74, No. 2 (February), p. 75–78.

TARNAPOL, PAULA J. 1986. "Canada Charged with Subsidizing Softwood." *Journal of Forestry*, Vol. 84, No. 12 (December), p. 4.

TEEGUARDEN, DENNIS E. 1973. "The Allowable Cut Effect: A Comment." *Journal of Forestry*, Vol. 71, No. 4 (April), p. 224–227.

THUROW, LESTER. 1980. *The Zero Sum Society: Distribution and the Possibilities for Economic Change.* New York: Basic Books. 230 p.

Unsigned. 1986. "Issues to Watch in 1986." *Journal of Forestry*, Vol. 84, No. 1 (January), p. 5.

USDA Forest Service. 1981. *An Assessment of the Forest and Rangeland Situation in the United States.* Forest Resource Report No. 22. 352 p.

USDA Forest Service. 1982. *A Guide to Federal Income Tax for Timber Owners.* Agriculture Handbook No. 596. 74 p.

USDA Forest Service. 1983. *An Analysis of the Timber Situation in the United States, 1950–2030.* Forest Resource Report No. 23. 499 p.

USDA Forest Service. 1984. *America's Renewable Resource: A Supplement to the 1979 Assessment of the Forest and Rangeland Situation in the United States.* Forest Service 386 84 p.

USDA Forest Service. 1984. *Report of the Forest Service, Fiscal Year 1983.* 167 p.

USDA Forest Service. 1985. *U. S. Timber Production, Trade, Consumption, and Price Statistics, 1950–84.* Misc. Publ. No. 1450.

USDA Forest Service. 1985. *Report of the Forest Service, Fiscal Year 1984.* 152 p.

USDA Forest Service. 1987. *U. S. Timber Production, Trade, Consumption, and Price Statistics 1950–1985.* Misc. Pub. No. 1453.

U.S. Dept. of Commerce, Census Bureau. 1984. *Statistical Abstract of the United States, 1985.* (105th Ed.). U.S. Dept. of Commerce, Washington, D.C.

U.S. Dept. of Commerce, Census Bureau. 1986. *1982 Census of Manufacturers, Concentration Ratios in Manufacturing.* MC82-S-7 Subject Series. Issued April 1986.

U.S. Dept. of Interior. 1977. *Final Report of the Task Force on the Availability of Federally-Owned Lands.* 43 p. (Cited in USDA Forest Resource Report No. 22, USDA.)

U.S. Water Resources Council. 1979. "Procedures for Evaluation of National Economic Development Benefits and Costs in Water Resource Planning." Federal Register, Vol. 44, No. 242 (Dec. 14), p. 72892–73076.

U.S. Water Resources Council. 1983. *Economic and Environmental Principles and Guidelines for Water and Related Land Resources Implementation Studies.* Washington, D.C.: U.S. Govt. Printing Office. 137 p.

Vajda, S. 1962. *Readings in Mathematical Programming.* New York: Wiley. 130 p.

Vardaman, James M. 1978. "Attitude of Foresters toward the Small, Private Landowner: A Consulting Forester's Opinion." *Journal of Forestry,* Vol. 76, No. 6 (June), p. 368–369, 387.

Wagar, Alan J. 1964. *The Carrying Capacity of Wild Lands for Forest Recreation. Forest Science,* Monograph No. 7.

Wagar, Alan J. 1981. "Comment on 'The Right to Risk in Wilderness.'" *Journal of Forestry,* Vol 79, No. 3 (March), p. 151–152.

Wagner, Frederic H. 1978. "Livestock Grazing and the Livestock Industry." Chapter 9, p. 121–149 in H. P. Brokaw (ed.), *Wildlife and America: Contributions to an Understanding of American Wildlife and its Conservation.* Council on Environmental Quality. Cited in Forest Resource Report No. 22.

Wall Street Journal. 1983. February 23, p. 50.

Wall Street Journal. 1984. "Challenges to Prime Rate as Base for Loans Stirs Fears Among Banks." (Byline, Daniel Hertzberg). Copyright 1984, April 3, p. 35.

Wallace, Patrick. 1987. "Southern Pine Dealt Setback by Costlier Canadian Lumber." *Wall Street Journal,* January 27, 1987, p. 29.

Wallace, T. Dudley, and J. Lew Silver. 1983. "Public Cost-sharing and Production in Nonindustrial Private Forests: The Case of FIP in Coastal Georgia." In *Nonindustrial Private Forests: A Review of Economic and Policy Studies.* Symposium Proceedings, April 19–20, Duke University School of Forestry and Environmental Studies, Durham, N.C., p. 165–180.

Wehnstock, Joseph A. 1984. "Rattan: A Complement to Swidden Agriculture." *Unasylva,* Vol. 36, No. 143., p. 16–22.

Weisbrod, Burton A. 1964. "Collective-Consumption Services of Individual-Consumption Goods." *Quarterly Journal of Economics,* Vol. LXXVIII (August), p. 471–477.

Western Timber Association. 1985. "Remarks of R. Max Peterson, Chief, Forest Service, at the Land Management Planning Directors' Meeting, Missoula, Mont. October 29–November 2, 1984. Mimeo #11688, February 8, 1985.

Wicksteed, P. H. 1933. *The Common Sense of Political Economy.* (Edited by Lionel Robbins.) Vol. 1. London: George Routledge.

WILLIG, ROBERT D. 1976. "Consumer's Surplus Without Apology." *American Economic Review*, Vol. 66, No. 4 (September), p. 589–597.

WISEMAN, A. CLARK, and ROGER A. SEDJO. 1985. "Have Controls Reduced Log Exports in the Pacific Northwest?" *Journal of Forestry*, Vol. 83, No. 11 (November), p. 680–682.

YOHO, JAMES G. 1985. "Continuing Investments in Forestry: Private Investment Strategies." In *Investments in Forestry*. (Edited by Roger Sedjo.) Westview Press, Inc., 5500 Central Avenue, Boulder, Colo, p. 151–166. Used by permission.

YOTOPOULOS, PAN A., and JEFFREY B. NUGENT. 1976. *Economics of Development*. New York: Harper and Row. 478 p.

ZIVNUSKA, JOHN A. 1952. *Business Cycles, Building Cycles, and Commercial Forestry*. New York Institute of Public Administration. 254 p.

ZIVNUSKA, JOHN A. 1978. "Two Studies of Private Nonindustrial Forestlands—A Critical Review." *Journal of Forestry*, Vol. 76, No. 8 (August), p. 467–469.

INDEX